Water Science and Technology Library

Volume 129

The aim of the *Water Science and Technology Library* is to provide a forum for dissemination of the state-of-the-art of topics of current interest in the area of water science and technology. This is accomplished through publication of reference books and monographs, authored or edited. Occasionally also proceedings volumes are accepted for publication in the series. *Water Science and Technology Library* encompasses a wide range of topics dealing with science as well as socio-economic aspects of water, environment, and ecology. Both the water quantity and quality issues are relevant and are embraced by *Water Science and Technology Library*. The emphasis may be on either the scientific content, or techniques of solution, or both. There is increasing emphasis these days on processes and *Water Science and Technology Library* is committed to promoting this emphasis by publishing books emphasizing scientific discussions of physical, chemical, and/or biological aspects of water resources. Likewise, current or emerging solution techniques receive high priority. Interdisciplinary coverage is encouraged. Case studies contributing to our knowledge of water science and technology are also embraced by the series. Innovative ideas and novel techniques are of particular interest.

Comments or suggestions for future volumes are welcomed.

Vijay P. Singh, Department of Biological and Agricultural Engineering & Zachry Department of Civil and Environment Engineering, Texas A&M University, USA
Email: vsingh@tamu.edu

All contributions to an edited volume should undergo standard peer review to ensure high scientific quality, while monographs should also be reviewed by at least two experts in the field.

Manuscripts that have undergone successful review should then be prepared according to the Publisher's guidelines manuscripts: https://www.springer.com/gp/authors-editors/book-authors-editors/book-manuscript-guidelines

Akhilesh Kumar Yadav · Kanchan Yadav ·
Vijay P. Singh
Editors

Integrated Management of Water Resources in India: A Computational Approach

Optimizing for Sustainability and Planning

Springer

Editors
Akhilesh Kumar Yadav
Department of Mining Engineering
Indian Institute of Technology (Banaras Hindu University)
Varanasi, India

Environmental Science and Engineering Department
Indian Institute of Technology Bombay
Mumbai, India

Present Address:
Department of Environmental Engineering and Management
Chaoyang University of Technology
Taichung, Taiwan

Kanchan Yadav
Progressive Foundation
Lucknow, Uttar Pradesh, India

Vijay P. Singh
Department of Biological and Agricultural Engineering and Zachry Department of Civil and Environmental Engineering
Texas A&M University
College Station, TX, USA

ISSN 0921-092X ISSN 1872-4663 (electronic)
Water Science and Technology Library
ISBN 978-3-031-62078-2 ISBN 978-3-031-62079-9 (eBook)
https://doi.org/10.1007/978-3-031-62079-9

This Springer imprint is published by the registered company Springer Nature Switzerland AG
The registered company address is: Gewerbestrasse 11, 6330 Cham, Switzerland

Foreword

Water security is a critical challenge facing India, a nation with a vast and diverse population. Traditional siloed approaches to water management are no longer sufficient. This book, *Integrated Management of Water Resources in India: A Computational Approach*, offers a compelling and timely solution. By advocating for a computational approach to *Integrated Water Resources Management (IWRM)*, the authors break new ground. Leveraging the power of computational tools, they propose a comprehensive framework that considers the complex interplay between water quantity, quality, and the social and ecological systems it sustains.

This book is not merely a theoretical exploration. It presents a practical roadmap for implementing IWRM across India's diverse landscapes. The emphasis on stakeholder engagement ensures that the solutions developed are not only technically sound, but also meet the needs of the communities most impacted by water scarcity. *Integrated Management of Water Resources in India: A Computational Approach* is a valuable resource for water resource managers, policymakers, researchers, and anyone with a stake in India's water future. By embracing a holistic and data-driven approach, this book offers a path toward a more sustainable and equitable future for all. I congratulate the editors *Akhilesh Kumar Yadav, Kanchan Yadav, and Vijay P. Singh* for bringing out this thought-provoking work. I am sure this book will be useful to graduate students, faculty, and policymakers on environmental impact and public health.

Sabita Madhavi Singh, [Ph.D., IIT(BHU)]
Joint Director
National River Conservation Directorate
Department of Water Resources
River Development and Ganga Rejuvenation
Ministry of Jal Shakti
(Government of India)
New Delhi, India
sm.singh@gov.in

Preface

The motivation of this volume, entitled *Integrated Management of Water Resources in India: A Computational Approach*, is mainly to present techniques to protect water resources in India. Water is the lifeblood of our nation, sustaining not only our ecosystems, but also our very existence. However, India faces a growing challenge in managing its water resources. With a vast and diverse population, coupled with the effects of climate change, ensuring water security for present and future generations requires a holistic approach. This book, *Integrated Management of Water Resources in India: A Computational Approach*, proposes a paradigm shift in how we address this critical issue. It advocates for a framework that integrates various aspects of water management, moving beyond traditional siloed approaches. This framework takes advantage of the power of computational tools to analyze complex data, model water systems, and optimize decision making. This book is an important contribution for water professionals, policymakers, researchers, and students. It presents a comprehensive overview of the challenges and opportunities in Indian water management. By exploring the potential of computational approaches, the book equips readers with the knowledge and tools needed to develop and implement integrated water resource management strategies. Throughout the book, we emphasize the importance of stakeholder engagement. Sustainable water management requires the active participation of communities, government agencies, and private entities. By fostering collaboration and utilizing the power of computational tools, we can move toward a future where water resources are managed effectively and equitably for the benefit of all.

This book presents an overview of water resources in India. The book content has been categorized into four parts. Part I gives an overview of water resources fundamental and assessment with five chapters. Part II, consisting of eleven chapters, deals with water quality assessment, monitoring, and management, while Part III consists of six chapters under the title water resources management strategies and technologies; Part IV is devoted to the advanced optimization, and efficiency in water management of six chapters.

First, we thank the Almighty for His blessings throughout this project achievement, which envisioned us successfully exploring this challenging research domain with a technique perspective. We are also sincerely thankful to Springer Nature for allowing us to publish this monogram on a contemporary research domain of modeling and simulations.

Special thanks to all who contributed to making this volume a source of knowledge and reporting the latest findings in their areas. Against the title of the proposed chapter, the total number of abstracts received was 63, of which 48 were selected for a full chapter contribution. In this book, the project accommodated only **28** chapters according to the scope of the book, which were of better quality. The authors put a lot of effort during all phases of book production, starting from writing, revising based on the comments and Springer evaluation reports, and finally checking the proofs of the chapters. The project follows a review process in which the identities of the authors and reviewers are not disclosed to avoid bias decisions. We also thank and acknowledge the Springer Nature publication team for their quick responses and for providing a proper guideline on time. Finally, we acknowledge *Jaydeep Kumar (BCAS, New Delhi, India)* and *Advocate Dinesh Chandra Gupta (High Court, Allahabad, India)* for their continuous help and support during the execution of the project.

The editors would be happy to receive any comments to further improve future editions. Comments, feedback, and suggestions for further improvement or proposals for the new chapters for next editions are welcome and should be sent directly to the volume editors.

Taichung, Taiwan — Akhilesh Kumar Yadav, Ph.D., MIE
Lucknow, India — Kanchan Yadav, Ph.D.
College Station, USA — Vijay P. Singh, Ph.D., D.Sc.

Contents

About the Editors

Akhilesh Kumar Yadav Ph.D., MIE, earned a Bachelor of Technology (B.Tech.) in Electronics and Communication Engineering from Chaudhary Charan Singh University in Meerut, India. He then obtained a Master of Technology (M.Tech.) in Environmental Engineering from Madan Mohan Malaviya Engineering College in Gorakhpur, India (affiliated with Dr. A. P. J. Abdul Kalam Technical University, Lucknow). His doctoral degree, with a research focus on air pollution, was awarded by the Indian Institute of Technology (BHU), Varanasi, India.

Dr. Yadav has served at various institutions, including the Indian Institute of Technology (Banaras Hindu University), Varanasi, India; the Indian Institute of Technology Bombay, Mumbai, India; the Bhabha Atomic Research Centre in Mumbai, India (as a research collaborator); and Chaoyang University of Technology in Taichung, Taiwan. He has authored/co-authored several research and review articles, book chapters, and edited/authored books and holds a patent.

Dr. Yadav received the Young Engineer Award from the Institution of Engineers (India) and the Young Scientist Award from VDGOOD Technology Factory, Kolkata. His research interests include air, water, and soil pollution; climate change; vulnerability; human health risk assessments; and GIS applications in environmental pollution and management. He is an active member of several reputed international professional bodies, including the IEI (Kolkata), ECI (New Delhi), AEACI (Mumbai), MSI (New Delhi), and IASTA (Mumbai).

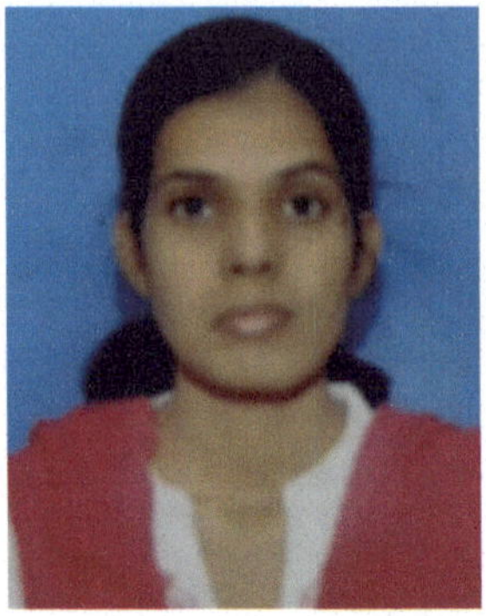

Kanchan Yadav Ph.D., possesses expertise in research writing, data analysis, study design, and methodology. She currently holds the position of the program coordinator at Progressive Foundation, where she is responsible for overseeing and executing programs in the fields of Public Health and Education. Prior to this, she served as a research associate specializing in the utilization of microwave technology for the management of hospital waste and the purification of water while retaining precious minerals. Furthermore, she has served as the research director for a project in collaboration with the Government of India, focusing on the adherence of tuberculosis medications. She has authored numerous research pieces that have been published in peer-reviewed journals and international conference proceedings.

Vijay P. Singh Ph.D., D.Sc., D.Eng. (Hon.), Ph.D. (Hon.), P.E., P.H., Hon.D.WRE, Dist.M. ASCE, Dist.Hon. M. IWRA, Dist.F. AGGS, Hon. Member AWRA, NAE, holds Caroline and William N. Lehrer distinguished chair in Water Engineering and is the distinguished professor and the regents professor, Department of Biological and Agricultural Engineering and Zachry Department of Civil and Environmental Engineering at Texas A&M University, USA. He has been recognized for four decades of leadership in research, teaching, and service to the hydrologic and water resources engineering profession. His contribution to the state of the art has been significant in many different specialty areas, including hydrologic science and engineering, hydraulic engineering, water resources engineering, environmental engineering, irrigation science, soil and water conservation engineering, entropy-based modeling, copula-based modeling, and mathematical modeling. His extensive publications in these areas include 32 textbooks, 1425 refereed journal articles, 115 book chapters, 330 conference proceedings papers, and 72 technical reports.

For his seminal contributions, he has been honored with more than 105 national/international awards from professional organizations. As a sample, he is a recipient of the Arid Land Hydraulic Engineering Award, Ven Te Chow Award, Torrens Award, Norma Medal, Lifetime Achievement Award, and OPAL Award, all given by ASCE. He was awarded the Ray K. Linsley Award for outstanding contributions to surface water hydrology and the Founders Award of AIH. He has been awarded three honorary doctorates. He is a fellow of ASCE, AWRA, IE, ISAE, IWRS, IASWC, and IAH; and a member of AGU, IAHR, IAHS, and WASER. He is a member/fellow of 12 engineering/science academies. He is licensed as the professional engineer (PE), the professional hydrologist (PH), and the honorary diplomate, AAWRE.

Part I
Water Resources Fundamental and Assessment

Chapter 1
Exploring Climatic Dynamics in Madhya Pradesh, India Utilizing Long-Term Gridded Data (1951–2021): An Integrated Statistical and GIS Modules

Amit Kumar, Tapas Ray, and T. Mohanasundari

Abstract Climate change is a global issue resulting from numerous human activities that alter hydro-climatic factors such as temperature, precipitation, river flow, and severe weather occurrences. India has seen a notable transformation in its natural resources and agricultural industries. This chapter comprehensively examines the climatic fluctuations in Madhya Pradesh, India, from 1951 to 2021 using IMD daily gridded rainfall and temperature data. Research utilizes non-parametric statistical techniques to evaluate trends, detect abrupt change points, and investigate regional patterns in seasonal precipitation and temperature. The Mann–Kendall test (MK) and Sen's slope estimator (SS) are applied to identify the trends and magnitude and understand temporal changes in meteorological variables. Furthermore, Pettitt's test detects sudden changes in long-term data, uncovering sudden patterns alterations. The findings reveal significant fluctuations in seasonal precipitation, with a discernible upward pattern in the premonsoon season (Z = 0.29) with a rate of 0.02 mm annually and downward patterns in the monsoon (Z = -0.93), winter seasons (Z = -0.73) and post-monsoon (Z = -0.76). These results have implications for the administration of water resources and the planning of agriculture. However, the investigation uncovers a consistent increase in T_{mean}, T_{max}, and T_{min} across all seasons, with notable differences in different areas. The change point analysis identifies the year when significant changes in rainfall and temperature patterns occurred. Inverse distance weighting methods allow for examining spatial changes, revealing the geographical distribution of certain climatic variables. This chapter improves the comprehension of climatic dynamics in Madhya Pradesh, providing a constructive understanding for policy makers, researchers and practitioners engaged in climate resilience and sustainable development planning.

A. Kumar · T. Ray · T. Mohanasundari (✉)
School of Humanities and Social Sciences, Indian Institute of Technology Indore, Simrol, Indore, Madhya Pradesh 453552, India
e-mail: mohana@iiti.ac.in

A. K. Yadav et al. (eds.), *Integrated Management of Water Resources in India: A Computational Approach*, Water Science and Technology Library 129,
https://doi.org/10.1007/978-3-031-62079-9_1

Keywords Climate change · Rainfall · Temperature · Trend analysis · Change point · Spatiotemporal analysis

1.1 Introduction

The rise in atmospheric CO_2 and the resulting climate warming have recently posed a significant global concern, causing shifts in the usual climate patterns (Daniel 2022; IPCC 2014; Nadeau et al. 2022). These changes in the Earth's climate are evident through various indicators, including changes in global rainfall patterns, the frequency of extreme temperatures (Kumar et al. 2023b; Tabari 2020), and variations in ground and surface water (Shirin et al. 2023). Recent studies, such as the study conducted by Daramola (2022), have shown a rise in temperature and a reduction in moisture levels in arid regions, especially in the last several decades. Global temperatures have increased by 0.032 °C yearly during the previous 40 years, but precipitation has decreased by 0.074 mm monthly (Daramola 2022). The primary climatic factors significantly impact Earth's ecosystems, agriculture, associated industries, food security, and energy stability in many locations (Malhi et al. 2021). Climate change is causing changes in precipitation patterns, resulting in unexpected and catastrophic weather events, including floods and droughts (Kumar and Mohanasundari 2024; Stein et al. 2020). India, which is highly dependent on agriculture, has negatively impacted the pH of agricultural areas due to variations in rainfall (Thakur et al. 2019) and degradation of land (Shirin et al. 2019; Yadav et al. 2014a, b). India's economic stability is intricately linked to agricultural production, leaving millions susceptible to fluctuations in rainfall patterns (Birthal and Hazrana 2019). Various research has examined rainfall patterns in India at local and provincial stages (Kumar et al. 2010, 2023a; Machiwal et al. 2017). Recent studies show notable shifts in India's precipitation patterns, characterized by a general reduction in total precipitation and an increase in severe rainfall occurrences in certain areas (Katzenberger et al. 2021). A study of the influence of environmental change on precipitation patterns in certain areas, such as Gujarat and Uttarakhand, has shown reductions in both rainy days and average precipitation levels (Shah et al. 2023; Banerjee et al. 2020). Temperature patterns in India have been analyzed together with rainfall patterns. Research has shown rises in maximum, mean, and minimum temperatures over time (Mall et al. 2021; Ray et al. 2019; Thakur et al. 2019). Madhya Pradesh, located in central India, has shown declining patterns in both rainfall and temperature, as reported by Gupta and Mishra (2019), Sam et al. (2020), and Yadav et al. (2019). Duhan and Pandey (2013) documented a steady decrease in precipitation in Madhya Pradesh from 1901 to 2002. Fluctuating temperatures and rainfall patterns impact the hydrological system of the area, influencing precipitation trends (Madhukar et al. 2021).

Comprehending the spatiotemporal flexibility of temperature and rainfall is critical to understanding the consequences of regional climate change. This chapter analyzed the evolving temperature and precipitation patterns in the study area using

IMD long-term data from 1951 to 2021. One of the key objectives of this chapter is to detect specific transition points in rainfall and temperature within the region. By pinpointing these transition points, the research aims to highlight critical variations in the climatic conditions of this study region. The findings of this chapter will provide valuable insights into how the region is responding to the impacts of environmental change. Assessing spatio-temporal changes in temperature and rainfall in this region will enable researchers and policy makers better to understand the region's approach of the region to environmental change. This knowledge will facilitate an evaluation of the potential consequences of the shifting patterns of rainfall and temperature in vital industries like agriculture and water supplies. Additionally, this helps create effective measures to mitigate climate change in the area.

1.2 Methodology

1.2.1 Study Area

The study area comprises Madhya Pradesh, which is 308,252 km^2 and accounts for 9.38% of the total area. Located between 21°03′ to 26°52′ N and 78°02′ to 82°48 E (Fig. 1.1), this region has a sub-tropical climate characterised by distinct monsoon, winter, and summer seasons. Madhya Pradesh can be divided into four biogeographical zones: low-lying regions around Gwalior, Satpura ranges, Vindhyan mountains, and the Malwa plateau. The annual average monsoon precipitation is 1194 mm; however, it varies in different places. Changing rainfall patterns make accessing water difficult, affecting agricultural and ecological stability (Kumar et al. 2024). Significant rivers like Narmada, Ken, and Son rely on rainfall, which is essential for water supplies and agriculture. The Narmada River is the boundary between the Satpura and Vindhyan mountain ranges. The Satpura range is characterized by thick forests and heights that reach 1350 m, with an average maximum temperature of 34.6 °C. Temperature and rainfall interact significantly to impact the microclimate of Madhya Pradesh.

1.2.2 Data Acquisition

The gridded data for daily rainfall and temperature with a spatial resolution of 0.25° × 0.25° and 1° × 1° respectively, were obtained from the Indian Meteorological Department (IMD) (https://www.imdpune.gov.in/). The data covers a time of 71 years, from 1951 to 2021. The current research used daily maximum and lowest temperatures from 26 grid points and rainfall from 439 grid points to cover the study area. This method has been widespread in numerous hydroclimatological investigations to assess rainfall, temperature patterns, and trends (Kumar et al. 2023a). The data from

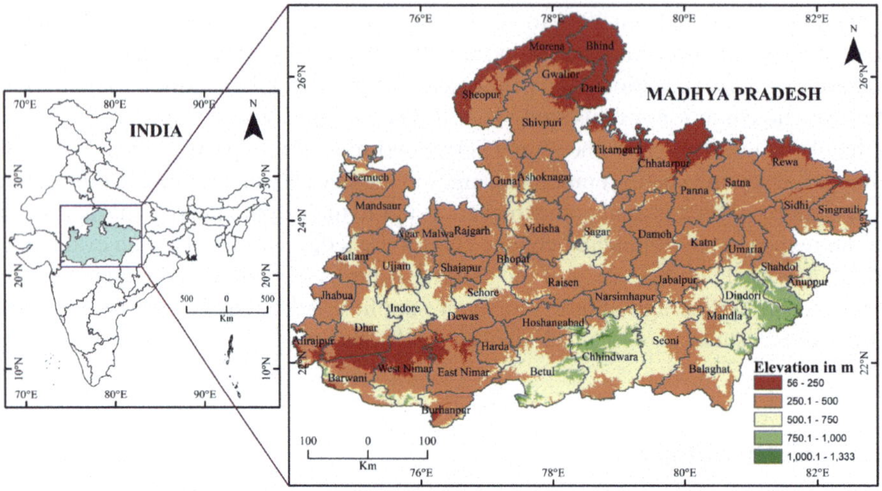

Fig. 1.1 Study area map

each year are divided into four distinct seasons: The four seasons in this region are as follows: Monsoon (J-J-A-S), premonsoon (M-A-M), Post-monsoon (O-N), and winter (D-J-F).

1.2.3 Methods

Examining data patterns necessitates parametric and nonparametric methodologies (Fig. 1.2), providing a thorough and resilient study (Shree and Kumar 2018). The study analyzed the trends and magnitude of rainfall and temperature using the Mann–Kendall (MK) test and the Sen slope (SS) estimator. The researchers used the Pettitt test to detect the abrupt discontinuities of temperature and precipitation data by examining the fluctuations from 1951 to 2021. The IDW interpolation method was employed to illustrate the temporal and spatial distribution of temperature and precipitation patterns in Madhya Pradesh. Descriptive statistical analysis was performed using MS Excel, trend and change point analysis was performed using MATLAB R2023a, and GIS analysis was performed using ArcGIS 10.8.

1.2.4 MK Test

It is a statistical technique that identifies trends or monotonic patterns in time series data (Kumar et al. 2023b; Singla et al. 2022). Equation 1.1 was used to calculate MK statistics.

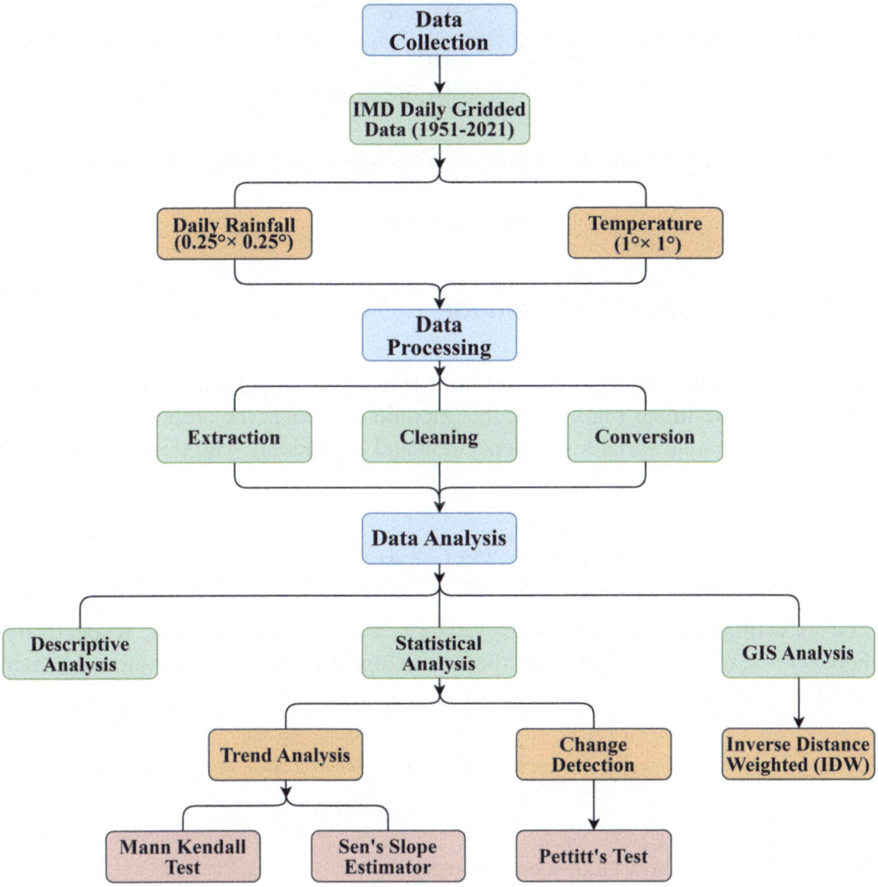

Fig. 1.2 Methodology flowchart

$$S = \sum_{j=1}^{n-1} \sum_{i=j+1}^{n} sig(x_i - x_j) \quad (1.1)$$

The normal distribution S is denoted by the mean, n represents the number of observations (≥ 10), and x_j signifies the jth observation. The sign function sig () is defined as follows: sig (α) = 1 if $\alpha > 0$; sig (α) = 0 otherwise; and sig (α) = − 1 if $\alpha \leq 0$. Equation 1.2 is used to calculate the variance of *S*.

$$var(S) = \frac{n(n-1)(2n+5) - \sum_{i=1}^{K} p_i(p_i - 1)(2p_i + 5)}{18} \quad (1.2)$$

The number of connected groups with similar values is n, and the number of data points in the ith linked group is t_i. Equation 1.3 shows the MK statistics accurately.

$$Z_s = \begin{bmatrix} \frac{S-1}{\sqrt{var(S)}}, ifS > o \\ 0, ifS = 0 \\ \frac{S+1}{\sqrt{var(S)}}, ifS < 0 \end{bmatrix} \quad (1.3)$$

Positive Z_S values indicate a rising trend, while negative Z_S numbers indicate a falling trend.

1.2.5 Sen's Slope (SS) Estimator

It identified the magnitude of the trend slope in a dataset, including N data pairs (Sen, 1968). Like basic linear regression, this method computes the slope of the median between two separate variables (independent and dependent). The slope is calculated using Eq. 1.4.

$$Q_i = \frac{t_j - t_k}{j - k} \text{ For } i = 1, 2, 3 \ldots, \text{N} \quad (1.4)$$

Given a time series data set with values t_j and t_k at time j and k, respectively, whereas $j > k$, the steps to calculate the Sen's slope estimator involve and calculated using Eq. 1.5.

$$Q_{med} = \begin{bmatrix} Q_{\frac{N+1}{2}}, ifNisodd \\ \frac{Q_{\frac{N}{2}} + Q_{\frac{N+2}{2}}}{2}, ifNiseven \end{bmatrix} \quad (1.5)$$

where, if $Q_{med} < 0$, shows a descending trend.

$Q_{med} > 0$, shows an ascending trend.

Pettitt test

It identifies abrupt changes or breakpoints in a time series dataset (Pettitt 1979), indicating a significant shift in the data values (Dhorde and Zarenistanak 2013; Kang and Yusof 2012). The stages associated with performing the Pettitt test are as follows:

The first stage includes calculating the U_K statistic using Eq. 1.6.

$$U_k = 2\sum_{i=0}^{n} p_i - k(n+1) \quad (1.6)$$

The variable P_i represents the position of the *ith* observation in the time series data when the values $x_1, x_2, x_3, x_4 \ldots\ldots x_{n-1}, x_n$ are ordered in the increasing sequence. The statistical change point test should, therefore, be defined as follows (Eq. 1.7):

$$K_n = max|U_k| \tag{1.7}$$

When U_k in a data set experiences the highest value of K, the transition point will occur. Equation 1.8 was used to get the crucial value.

$$K_\alpha = \left[-1n\alpha\left(n^3 + n^2\right)/6\right]^{1/2} \tag{1.8}$$

where α is the degree of significance and n is the number of observations that define the critical value.

Coefficient of Variation (CV)

The present chapter used the CV to assess the extent to which each data point deviated from the mean regarding temperature and precipitation variability. An increased CV value signifies more significant variability (Kumar et al. 2023a, 2023b). The CV was computed using Eq. 1.9.

$$CV = \frac{\sigma}{\mu} \times 100 \tag{1.9}$$

where μ = mean; σ = standard deviation.

A CV is categorised as high (> 30%), moderate (20–30%), and low (<20%) (Bharath et al. 2023; Sarkar et al. 2021).

Inverse Distance Weighted (IDW)

In order to demonstrate the spatio-temporal pattern of temperature and rainfall in Madhya Pradesh, we computed the mean temperature for each grid. Subsequently, the collected data were used to generate a cartographic representation that illustrates the spatial arrangement of temperature and precipitation distribution patterns. The spatiotemporal maps were generated using the Interpolation method (IDW). The method assumes that close data points are comparable regardless of distance (Kumar et al. 2023a, b, c).

1.3 Results and Discussion

Analyzing climatic variability, detecting trends and assessing regional distribution is essential to gain deep insights into the implications of environmental change and improving infrastructure and resource planning. Meteorological data are an essential source of knowledge to understand these alterations. Detecting trends in long-term data without imagining component distributions is a major benefit of nonparametric tests in climate research. Weather factors such as temperature, precipitation, wind speed, and humidity can be measured in various sizes with these tests. Examining the spatial distribution is essential to understanding local alterations in climatic variables and their historical changes (Ampofo et al., 2023).

1.3.1 Temporal Trend Analysis

Rainfall

Table 1.1 summarizes the study area's seasonal temperature and rainfall patterns of the study area for 71 years (1951–2021). The CV for monsoon, pre-monsoon, post-monsoon, and winter season rainfall are 82.40, 17.39, 84.44, and 79.25, respectively (Table 1.1a). It changes a lot during the winter, post-monsoon, and premonsoon seasons, but not as much during the monsoon season. This conclusion supports other studies conducted by Shree and Kumar (2018), Chandniha et al. (2017), Kumar et al. (2023a), and Warwade et al. (2018), highlighting the region's susceptibility to droughts and floods, as noted by Pandey and Ramasastri (2001). The chapter used the MK and SS tests to demonstrate a slight rise in premonsoon precipitation and a decline in monsoon, postmonsoon, and winter (Fig. 1.3a). The monsoon exhibits a statistically insignificant annual reduction of −0.977 mm, with a Z-score of −0.933, raising concerns about water shortages during this crucial season. On the other hand, a little increase ($Z = 0.288$) in pre-monsoon rainfall indicates some advantages for agricultural and industrial water resources. The decreasing trends in winter and postmonsoon rainfall trends of −0.067 mm and −0.116 mm per year, respectively, provide difficulties in recharging groundwater and maintaining river and reservoir levels. The results are consistent with previous research conducted by Chandniha et al. (2017), Kumar et al. (2023a), Warwade et al. (2018) and Shree and Kumar (2018), highlighting the susceptibility of the area to floods and droughts as highlighted by Pandey and Ramasastri (2001).

Table 1.1: Statistical results of seasonal rainfall, T_{mean}, T_{max}, T_{min} in Madhya Pradesh (5% significance level).

Mean temperature (T_{mean})

The premonsoon season of T_{mean} exhibits moderate variability ($CV = 2.29$), with a non-significant positive trend ($Z = 1.052$, annual magnitude $= 0.005$ °C). The monsoon season shows lower variability ($CV = 1.73$) with a similar trend ($Z = 1.092$, annual magnitude $= 0.003$ °C) (Table 1.1b). On the contrary, the post-monsoon season shows a significant increasing trend ($Z = 3.594$, annual magnitude $= 0.018$ °C) and higher variability ($CV = 3.47$), while winter also shows considerable variability ($CV = 3.23$). Despite a weaker result of the MK test ($Z = 0.149$), the SS estimator supports a positive trend (0.050 °C annually). All seasons show warming trends (Fig. 1.3b), aiding in adaptation strategies to climate change.

Maximum Temperature (T_{max})

The CV shows moderate to high variability in T_{max} across all seasons: Premonsoon ($CV = 2.37$), Monsoon ($CV = 3.50$), Postmonsoon ($CV = 3.38$), and Winter ($CV = 3.77$) (Table 1.1c). The MK and SS analysis shows increasing trends in T_{max} throughout all seasons. Pre-monsoon has a notable trend ($Z = 1.241$, change: 0.009 °C/yr), while Monsoon, Post-Monsoon, and Winter have less pronounced trends. Although there is some variation, all seasons in Madhya Pradesh consistently

Table 1.1 Statistical results of seasonal rainfall, T_{mean}, T_{max}, T_{min} in Madhya Pradesh (5% significance level)

Season	CV	Z-Stat	Q-Stat	P-value	Trend
Rainfall (a)					
Pre-monsoon	82.40	0.288	0.018	0.773	Increasing
Monsoon	17.39	−0.933	−0.977	0.351	Decreasing
Post-monsoon	84.44	−0.764	−0.116	0.445	Decreasing
Winter	79.25	−0.735	−0.067	0.462	Decreasing
T_{mean} *(b)*					
Pre-monsoon	2.29	1.052	0.005	0.293	Increasing
Monsoon	1.73	1.092	0.003	0.275	Increasing
Post-monsoon	3.47	3.594	0.018	3.26e−04	Increasing
Winter	3.23	0.149	0.050	0.882	Increasing
T_{max} *(c)*					
Pre-monsoon	2.37	1.241	0.009	0.215	Increasing
Monsoon	3.50	1.449	0.011	0.148	Increasing
Post-monsoon	3.38	0.834	0.007	0.404	Increasing
Winter	3.77	0.074	0	0.941	Increasing
T_{min} *(d)*					
Pre-monsoon	17.49	2.422	0.026	0.015	Increasing
Monsoon	5.88	0.551	0.002	0.582	Increasing
Post-monsoon	20.38	3.891	0.033	9.96e-05	Increasing
Winter	37.20	2.750	0.018	0.006	Increasing

increase T_{max} (Fig. 1.3c), indicating a long-term warming trend with significant consequences for industries susceptible to climate change.

Minimum Temperature (T_{min})

The CV shows different levels of fluctuation of T_{min} throughout the seasons (Table 1.1d). The premonsoon has a coefficient of variation of 17.49, while the monsoon has a coefficient of variation of 5.88. The postmonsoon season has a CV of 20.38, whereas the winter season has a CV of 37.20, showing significant fluctuation. The MK test and SS analysis validate increasing minimum temperature trends throughout all seasons. The premonsoon is seeing a notable rise with a Z value of 2.422 and a temperature change of 0.026 °C each year. The monsoon at Z = 0.551 shows a diminishing yet ascending pattern with a temperature change of 0.002° C. The postmonsoon and winter seasons show notable upward tendencies, with the postmonsoon having a Z value of 3.891 and a temperature change of 0.033 °C and winter having a Z value of 2.750 and a temperature change of 0.018 °C. The data highlights the increasing temperature trend in Madhya Pradesh (Fig. 1.3d), which affects many industries that face climate change problems.

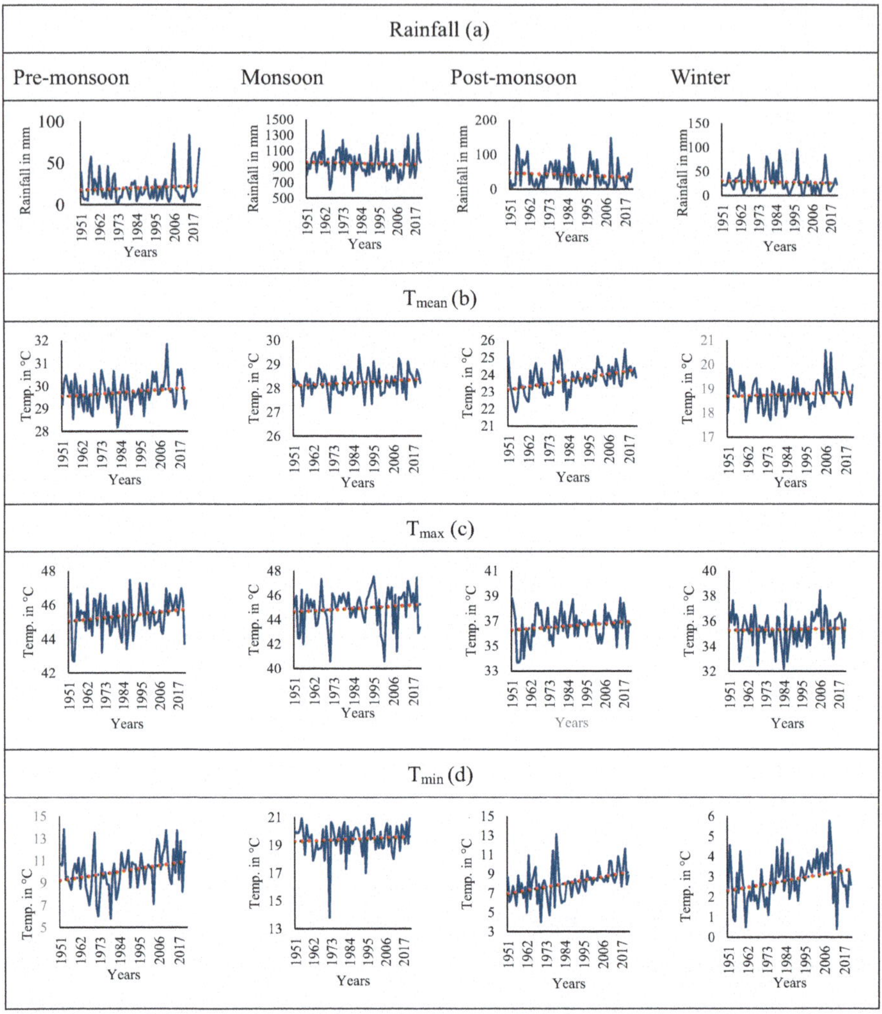

Fig. 1.3 Trends in seasonal rainfall, T_{mean}, T_{max}, T_{min} of Madhya Pradesh

Our results are consistent with previous research in the Central West and Central India. Kundu et al. (2017) noted a decrease in monsoon, annual, winter, and post-monsoon precipitation from 1901 to 2011, although seeing an opposite increase in premonsoon precipitation. Kumar et al. (2010) noticed a yearly and monsoonal rainfall reduction for 135 years (1871–2005) but found an increase in premonsoonal precipitation. Devi et al. (2020) also documented a decline in rainfall patterns in Madhya Pradesh. Duhan and Pandey (2013), Shukla and Khare (2013), and Shukla et al. (2017) observed notable increases in annual average temperatures, including T_{mean}, T_{max}, and T_{min}. Devi et al. (2020) noted increasing trends in T_{mean}, T_{max}, and T_{min} in Central India between 1971 and 2015. Kundu et al. (2017) observed a steady rise in both T_{max}, and T_{min} over 105 years (1901–2005), with the most notable increases occurring in the winter and post-monsoon periods. Singla et al. (2022) observed a steady rise in both T_{max}, and T_{min} in Northwestern India since 1970. The results align with previous studies by Dubey et al. (2022), Kumar et al. (2023b), Pal and Al-Tabbaa (2010), and Radhakrishnan et al. (2017), which reported increasing temperatures in different parts of India. Consistent patterns highlight the importance of comprehending and dealing with variations in precipitation and temperature for successful climate adaptation and mitigation initiatives.

1.3.2 Change Point Analysis

Identifying sudden changes in climate historical data is crucial to revealing abrupt changes in trends. Identification of change points uses several techniques, and this chapter used Pettitt's test to find sudden breaking points within the temperature and rainfall data in the study area. The Pettitt test detected abrupt changes in seasonal temperature and rainfall trends (Fig. 1.4).

Rainfall

The outcomes of Pettitt's test for rainfall are graphically represented in Fig. 1.4a. The results of the Pettitt test show 1955, 1998, 1987, and 1986 for all seasons of rainfall, according to the data. These transition points indicate the years when significant changes in rainfall patterns occurred, signalling modifications in the corresponding seasons.

Mean Temperature (T_{mean})

The test successfully identified abrupt changes in the seasonal T_{mean} trends (Fig. 1.4b). The analysis revealed the breaking points for 1998, 2007, 1987, and 1959 for all seasons, respectively. These identified change points signify pivotal moments when substantial changes occurred in the average temperature patterns, indicating noteworthy alterations in the respective seasons.

Maximum Temperature (T_{max})

Based on the findings, 2008, 1990, 1963, and 1966 were identified as change points for all seasons (Fig. 1.4c). These points denote significant shifts in the T_{max} patterns, indicating alterations within the corresponding seasons.

Minimum Temperature (T_{min})

The analysis revealed change points in 2003, 1960, 2008, and 1995 for all seasons, respectively (Fig. 1.4d). These change points indicate the years when significant changes occurred in the T_{min} patterns, signifying notable alterations in the corresponding seasons.

Research results underscore significant alterations in rainfall and temperature during specific time intervals. Similar investigations have been conducted in different

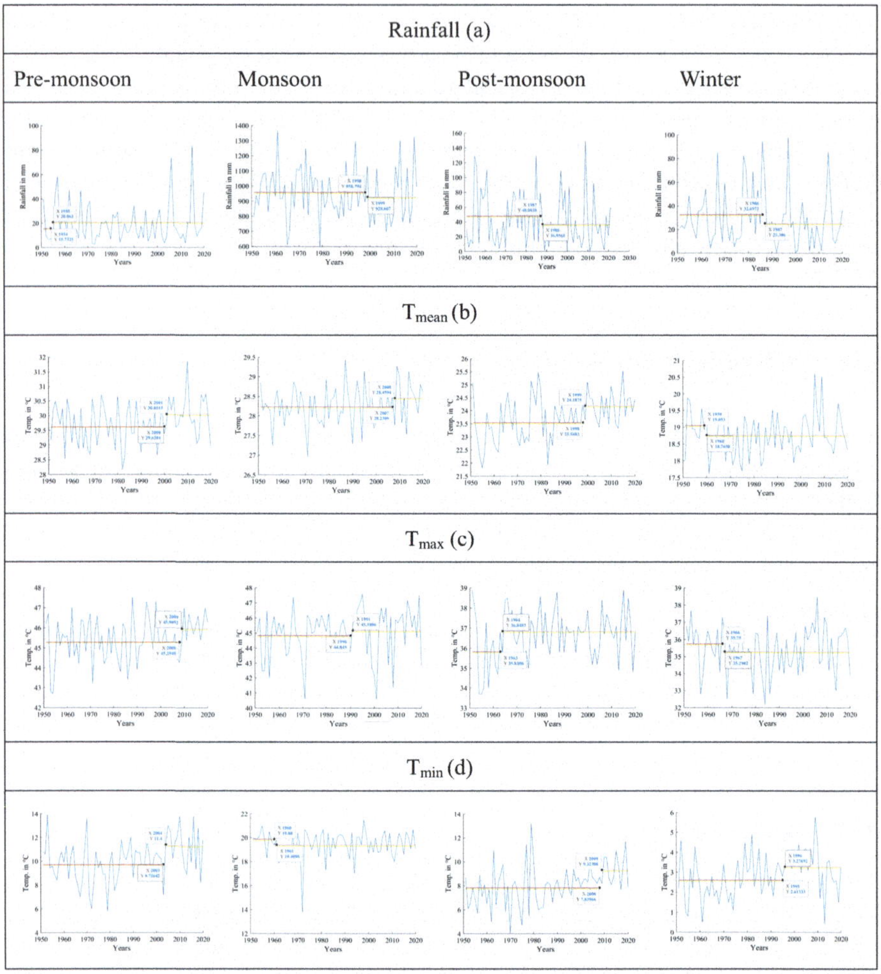

Fig. 1.4 Change point results of seasonal rainfall, T_{mean}, T_{max}, T_{min}

geographical areas to pinpoint notable transition points in climate variables. For instance, Zarenistanak et al. (2014) found a significant transition in Iran's climate in 1973. Kumar et al. (2023a, b, c) identified abrupt changes in annual rainfall trends at three stations in India: 2004, 1998, and 1991. These findings highlight the dynamic nature of climatic conditions and their implications for ecosystems and human activities. In their study, Shukla et al. (2017) found 1963 to be the critical moment when the T_{mean} data, covering 1901–2005, significantly changed the study area. Chandole and Joshi (2023) observed a favourable pattern in the T_{min}, with a consistent change occurring in two specified districts of Gujarat after 1986. Zarenistanak et al. (2014) identified significant changes in temperature patterns throughout the 1980s and 1990s. These investigations in several areas enhance our understanding of the intricacies of climate change with its temporal expressions.

1.4 Spatial Change Analysis

This chapter used interpolation methods to evaluate changes in rainfall and temperature using the spatial change analysis approach. Pettitt's test used the yearly rainfall, T_{mean}, T_{max}, and T_{min} data to compare seasonal changes, using the base year as a reference. The elongated time-series data were divided into 2 separate parts: 1951–1998 and 1999–2021 for rainfall, 1951–2004 and 2005–2021 for T_{mean}, 1951–2020 and 2011–2011 for T_{max}, and 1951–1999 and 2021 for T_{min}, offering valuable insights into spatial and temporal trends. Figures 1.5 and 1.6 provide IDW maps showing the seasonal distribution of rainfall and temperature in Madhya Pradesh, allowing for comparison. These maps provide valuable information on spatial and temporal precipitation and temperature fluctuations in various regions within the state.

Rainfall

This chapter analyzes seasonal precipitation patterns in Madhya Pradesh, India, explicitly emphasizing changes after 1998. Premonsoon rainfall considerably rises in central and northern regions, but monsoon rainfall falls after 1998, particularly in places with above 1000 mm of rainfall. During the winter and post-monsoon seasons,

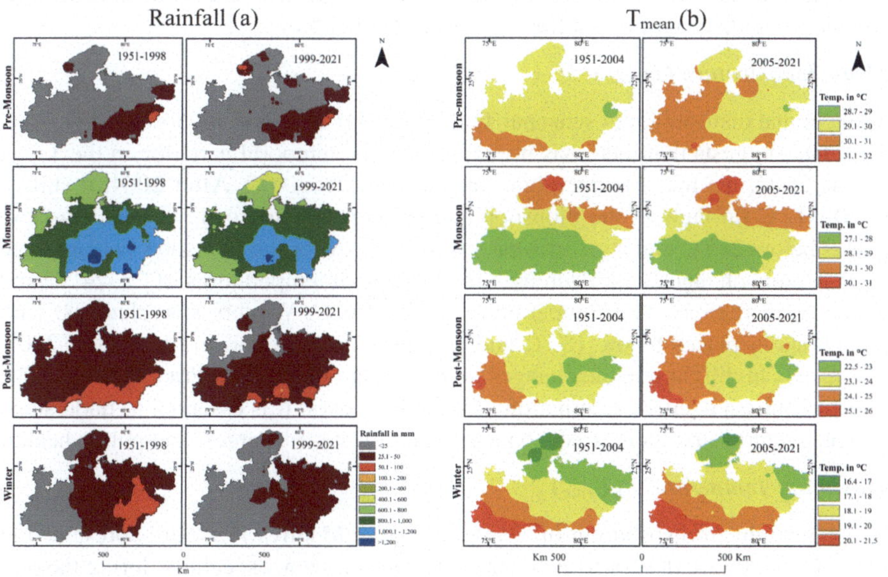

Fig. 1.5 Spatiotemporal distribution maps of seasonal rainfall and T_{mean}

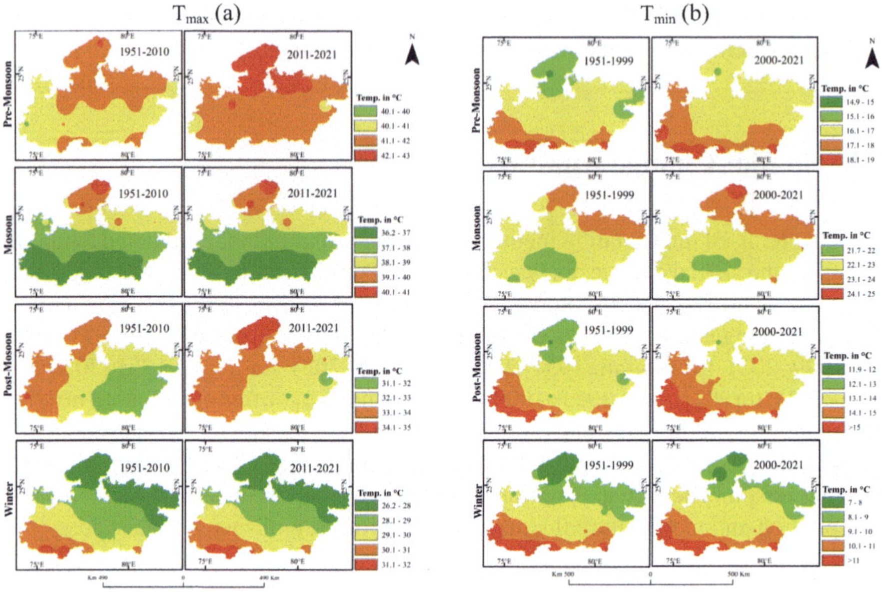

Fig. 1.6 Spatiotemporal distribution maps of seasonal T_{max} and T_{min}

there has been a decrease in rainfall patterns. Areas receiving less rainfall (<25 mm) have increased, while those receiving more than 50 mm have decreased since 1998. The research uses statistical tests to analyze rainfall patterns, focusing on the spatial and temporal distribution in different seasons (Fig. 1.5a).

Mean Temperature (T_{mean})

The regional distribution of seasonal T_{mean} in Madhya Pradesh is shown in Fig. 1.5b. This data suggests that there has been a substantial temporal variation in the T_{mean}. Before 2004, the mean temperature ranged from 29–30 °C. After 2004, it shifted to 30–31 °C throughout the monsoon and premonsoon seasons, with decreases in places below 28 °C and rises in areas over 28 °C. After the monsoon season, there were notable changes with decreases in places with temperatures of 23–24 °C and increases in regions with temperatures of 24–26 °C. Winters saw changes, with reductions in places below 18 °C and increases in areas above 19 °C after 2004. The results highlight notable changes in the climate of Madhya Pradesh after 2004, which require a deeper examination of the root causes. It is essential to understand climate patterns and their effects on agriculture, water resources, and public health.

Maximum Temperature (T_{max})

The geographical distribution of seasonal T_{max} in Madhya Pradesh is presented in Fig. 1.6a. After 2010, significant changes have been seen, especially during the pre- and post-monsoon periods. Before the monsoon season, the T_{max} notably rises in regions ranging from 41–43 °C. After the monsoon season, areas with temperatures over 34 °C increase, while regions below 32 °C disappear entirely. Minor alterations in the distribution of T_{max} in winter. The results highlight the fluctuating nature of T_{max} during different seasons in Madhya Pradesh, offering valuable information on the effects of climate change. Additional research is warranted into the root causes of these alterations.

Minimum Temperature (T_{min})

Figure 1.6b illustrates the spatiotemporal distribution of seasonal T_{min} in Madhya Pradesh. Analysis conducted after 1999 shows significant temperature fluctuations over different seasons. Before the monsoon season, there is a decrease in places with temperatures below 16 °C and an increase in areas with temperatures over 17 °C. Monsoon: the appearance of regions with temperatures above 24 °C in the northern zones. After the monsoon season, temperatures have decreased in places below 13 °C and increased in areas over 14 °C. Winter temperatures have increased somewhat in locations where the temperature is above 9 °C, while they have decreased in regions where the temperature is below 9 °C since 1999. The shifts highlight the changing temperature patterns, indicating possible climate changes. Additional research is required to comprehend the root causes and their effects on the ecology and residents of the area.

The study shows altered precipitation patterns in Madhya Pradesh: lower monsoon, post-monsoon, winter rainfall, and higher premonsoon rainfall. Yadav et al. (2014a, b) observed a decrease in the occurrence of intense rainfalls in recent times. Duhan and Pandey (2013) analysed 102 years of temperature data (1901–2002) and identified several geographical temperature patterns. Nabi et al. (2023) found that temperatures decrease with increasing height and increase with decreasing elevation. Sengupta and Thangavel (2023) examined data from 1990 to 2015, establishing a connection between temperature changes and drought intensity in Maharashtra, highlighting the influence of height.

1.5 Conclusion

This chapter analyses the extended precipitation and temperature patterns in the study area, using long-term gridded data from 1951 to 2021. Seasonal precipitation patterns exhibit substantial fluctuations, particularly before and after the monsoon and throughout winter, with the monsoon season showing less variability. The temperature trend analysis shows a uniform increase in T_{mean}, T_{max}, and T_{min} in all seasons, suggesting a widespread warming trend throughout the research period. Pettitt's technique detects significant transition years in rainfall and temperature trends. Transition points for rainfall in the monsoon, post-monsoon, pre-monsoon, and Winter seasons occurred in 1998, 1987, 1955, and 1986, respectively. The temperature change points for the same seasons are designated as 1998, 2007, 1987, and 1959, respectively. Analysis of spatial and temporal data shows changes in rainfall distribution after 1998, significantly increasing premonsoon rainfall in central and northern areas.

There is a noticeable increase in seasonal mean temperature in the state, particularly after 2004, with temperatures ranging between 29 and 30 °C before that year.

The results highlight significant changes in rainfall and temperature trends in Madhya Pradesh, suggesting a changing climate. The observed patterns potentially harm agriculture, water resources, and ecosystems. This chapter provides useful information for creating well-informed adaptation and mitigation plans to tackle the problems presented by changing climatic conditions.

Acknowledgements The authors thank the School of Humanities and Social Sciences, Indian Institute of Technology Indore. They believe that their results will help to understand the changes in rainfall and temperature changes and solve water issues for a sustainable future.

Data Availability Statement All relevant data are available from an online repository or repositories. (https://www.imdpune.gov.in/).

Conflict of Interest The authors declare that there are no conflicts of interest.

References

Ampofo S, Annor T, Aryee JNA, Atiah WA, Amekudzi LK (2023) Long-term spatio-temporal variability and change in rainfall over Ghana (1960–2015). Sci Afr 19:e01588. https://doi.org/10.1016/j.sciaf.2023.e01588

Banerjee A, Chen R, Meadows ME, Singh RB, Mal S, Sengupta D (2020) An analysis of long-term rainfall trends and variability in the Uttarakhand Himalaya using google earth engine. Rem Sens 12(4):709. https://doi.org/10.3390/rs12040709

Bharath A, Maddamsetty R, Manjunatha M, Reshma TV (2023) Spatiotemporal rainfall variability and trend analysis of Shimsha river basin, India. Environ Sci Pollut Res 30(49):107084–107103

Birthal PS, Hazrana J (2019) Crop diversification and resilience of agriculture to climatic shocks: Evidence from India. Agric Syst 173:345–354. https://doi.org/10.1016/j.agsy.2019.03.005

Chandniha SK, Meshram SG, Adamowski JF, Meshram C (2017) Trend analysis of precipitation in Jharkhand State, India: Investigating precipitation variability in Jharkhand State. Theoret Appl Climatol 130(1–2):261–274. https://doi.org/10.1007/s00704-016-1875-x

Chandole V, Joshi GS (2023) Case study of rainfall and temperature assessment through trend and homogeneity analyses in Vadodara and Chhotaudepur district of Gujarat State, India. Environ Monit Assess 195(5):1–25. https://doi.org/10.1007/s10661-023-11089-w

Daniel TKH (2022) Climate Change policies of the four largest global emitters of greenhouse gases: their similarities, differences and way forward. J Energy Res Rev 10(2):19–35. https://doi.org/10.9734/jenrr/2022/v10i230251

Daramola MT (2022) Recent changes in global dryland temperature and precipitation. Int J Climatol 42:1267–1282. https://doi.org/10.1002/joc.7301

Devi RM, Patasaraiya MK, Sinha B, Bisaria J, Dimri AP (2020) Analysing precipitation and temperature trends of Kanha and Satpura Tiger Reserve, Central India. Theoret Appl Climatol 140(3–4):1435–1450. https://doi.org/10.1007/s00704-020-03134-2

Dhorde AG, Zarenistanak M (2013) Three-way approach to test data homogeneity: an analysis of temperature and precipitation series over southwestern Islamic Republic of Iran. J Ind Geophys Union 17(3):233–242

Dubey SK, Ranjan RK, Misra AK, Wanjari N, Vishwakarma S (2022) Variability of precipitation extremes and drought intensity over the Sikkim State, India, during 1950–2018. Theoret Appl Climatol 148(1–2):1–14. https://doi.org/10.1007/s00704-022-03931-x

Duhan D, Pandey A (2013) Statistical analysis of long term spatial and temporal trends of precipitation during 1901–2002 at Madhya Pradesh, India. Atmos Res 122:136–149. https://doi.org/10.1016/j.atmosres.2012.10.010

Gupta R, Mishra A (2019) Climate change induced impact and uncertainty of rice yield of agro-ecological zones of India. Agric Syst 173(January):1–11. https://doi.org/10.1016/j.agsy.2019.01.009

IPCC (2014) Intergovernmental panel on climate change

Kang HM, Yusof F (2012) Homogeneity tests on daily rainfall series in Peninsular Malaysia. Int J Contemp Math Sci 7(1):9–22

Katzenberger A, Schewe J, Pongratz J, Levermann A (2021) Robust increase of Indian monsoon rainfall and its variability under future warming in CMIP6 models. Earth Syst Dyn 12(2):367–386. https://doi.org/10.5194/esd-12-367-2021

Kumar A, Kumar S, Rautela KS, Kumari A, Shekhar S, Thangavel M (2023a) Exploring temperature dynamics in Madhya Pradesh: a spatial-temporal analysis. Environ Monit Assess 195(11):1313. https://doi.org/10.1007/s10661-023-11884-5

Kumar A, Kumar S, Rautela KS, Shekhar S, Ray T, Thangavel M (2023b) Assessing seasonal variation and trends in rainfall patterns of Madhya Pradesh, Central India. J Water Clim Change 14(10):3692–3712. https://doi.org/10.2166/wcc.2023.280

Kumar A, Mohanasundari T (2024) Assessing district-level climate vulnerability in Madhya Pradesh, Central India: an integrated environmental and socio-economic approach. In: Theoretical and applied climatology. https://doi.org/10.1007/s00704-023-04814-5

Kumar U, Singh DK, Panday SC, Bisht JK, Kant L (2023) Spatio-temporal trend and change detection of rainfall for Kosi River basin, Uttarakhand using long-term (115 years) gridded data. Arab J Geosci 16(3). https://doi.org/10.1007/s12517-023-11244-0

Kumar V, Jain SK, Singh Y (2010) Analysis of long-term rainfall trends in India. Hydrol Sci J 55(4):484–496. https://doi.org/10.1080/02626667.2010.481373

Kumar A, Kumar A, Mann K, Mohanasundari T (2024) Analysing district-level climate vulnerability pattern in Madhya Pradesh, India, based on agricultural and socio-economic indicators. Environ Monit Assess 196(6):528. https://doi.org/10.1007/s10661-024-12646-7

Kundu S, Khare D, Mondal A (2017) Interrelationship of rainfall, temperature and reference evapotranspiration trends and their net response to the climate change in Central India. Theoret Appl Climatol 130(3–4):879–900. https://doi.org/10.1007/s00704-016-1924-5

Machiwal D, Dayal D, Kumar S (2017) Long-term rainfall trends and change points in hot and cold arid regions of India. Hydrol Sci J 62(7):1050–1066. https://doi.org/10.1080/02626667.2017.1303705

Madhukar A, Dashora K, Kumar V (2021) Climate trends in temperature and water variables during wheat growing season and impact on yield. Environ Process 8:1047–1072

Malhi GS, Kaur M, Kaushik P (2021) Impact of climate change on agriculture and its mitigation strategies: a review. Sustainability 13:1318

Mall RK, Chaturvedi M, Singh N, Bhatla R, Singh RS, Gupta A, Niyogi D (2021) Evidence of asymmetric change in diurnal temperature range in recent decades over different agro-climatic zones of India. Int J Climatol 41:2597–2610. https://doi.org/10.1002/joc.6978

Nabi S, Ahmad M, Abdul A, Dar Q (2023) Spatio-temporal variability of temperature and precipitation in a Himalayan watershed. Nat Hazards 115(2):1743–1760. https://doi.org/10.1007/s11069-022-05616-0

Nadeau KC, Agache I, Jutel M, Annesi Maesano I, Akdis M, Sampath V, D'Amato G, Cecchi L, Traidl-Hoffmann C, Akdis CA (2022) Climate change: a call to action for the United Nations. Allerg Eur J Allerg Clin Immunol 77(4):1087–1090. https://doi.org/10.1111/all.15079

Pal I, Al-Tabbaa A (2010) Long-term changes and variability of monthly extreme temperatures in India. Theoret Appl Climatol 100(1):45–56. https://doi.org/10.1007/s00704-009-0167-0

Pandey RP, Ramasastri KS (2001) Relationship between the common climatic parameters and average drought frequency. Hydrol Process 15(6):1019–1032. https://doi.org/10.1002/hyp.187

Pettitt AN (1979) A non-parametric approach to the change-point problem. J R Stat Soc Ser C (Appl Stat) 28(2):126–135

Radhakrishnan K, Sivaraman I, Jena SK, Sarkar S, Adhikari S (2017) A climate trend analysis of temperature and rainfall in India. Clim Change Environ Sustain 5(2):146. https://doi.org/10.5958/2320-642x.2017.00014.x

Ray LK, Goel NK, Arora M (2019) Trend analysis and change point detection of temperature over parts of India. Theoret Appl Climatol 138:153–167

Sam AS, Padmaja SS, Kächele H, Kumar R, Müller K (2020) Climate change, drought and rural communities: understanding people's perceptions and adaptations in rural eastern India. Int J Dis Risk Reduct 44:101436. https://doi.org/10.1016/j.ijdrr.2019.101436

Sarkar A, Saha S, Sarkar D, Mondal P (2021) Variability and trend analysis of the rainfall of the past 119 (1901–2019) years using statistical techniques: a case study of Uttar Dinajpur, India. J Clim Change 7(2):49–61. https://doi.org/10.3233/JCC210011

Sen PK (1968) Estimates of the regression coefficient based on Kendall's Tau. J Am Stat Assoc 63(324):1379–1389

Sengupta A, Thangavel M (2023) Analysis of the effects of climate change on cotton production in Maharashtra State of India using statistical model and GIS mapping. J Sustain Agric 38(1):1–12. https://doi.org/10.20961/carakatani.v38i1.64377

Shah NH, Shukla BP, Chahal J (2023) Yearly variation of rainfall rate and drop size distribution of RD-80 disdrometer for Ahmedabad and Thiruvananthapuram. In: Research sqare, pp 1–17. https://doi.org/10.21203/rs.3.rs-3095080/v1

Shree S, Kumar M (2018) Analysis of seasonal and annual rainfall trends for Ranchi district, Jharkhand, India. Environ Earth Sci 77(19):1–9. https://doi.org/10.1007/s12665-018-7884-6

Shirin S, Jamal A, Emmanouil C, Singh VP, Yadav AK (2023) Assessment and characterization of waste material used as backfilling in an abandoned mine. Int J Coal Prep Util 43(8):1402–1410

Shirin S, Jamal A, Ranjan P, Yadav AK (2019) Study on assessment of slope stability and mixed disposal of overburden in voids of Singrauli Coalfield. Environ Qual Manage 28(3):131–139

Shukla R, Khare D (2013) Historical trend investigation of temperature variation in Indira Sagar canal command area in Madhya Pradesh (1901–2005). Int J Adv Inf Sci Technol (IJAIST) ISSN 2(7):5–13. https://doi.org/10.15693/ijaist/2013.v2i7.5-13

Shukla R, Khare D, Tiwari P, Mishra P, Gupta S (2017) Analysis of long term temperature trend for Madhya Pradesh, India (1901–2005). Curr World Environ 12(1):68–79. https://doi.org/10.12944/CWE.12.1.09

Singla C, Aggarwal R, Kaur S, Sharma R (2022) Analysis of meteorological parameter changes using Mann-Kendall statistical tests in Indian Punjab. MAUSAM 74(1):207–213. https://doi.org/10.54302/mausam.v74i1.1440

Stein L, Pianosi F, Woods R (2020) Event-based classification for global study of river flood generating processes. Hydrol Process 34(7):1514–1529. https://doi.org/10.1002/hyp.13678

Tabari H (2020) Climate change impact on flood and extreme precipitation increases with water availability. Sci Rep 10(1):1–10. https://doi.org/10.1038/s41598-020-70816-2

Thakur A, Mishra PK, Nema AK, Sahoo SP (2019) Spatio-temporal trends and shift analysis of temperature for Wainganga sub-basin, India. Int J Environ Stud, pp 1–16. https://doi.org/10.1080/00207233.2019.1686884

Warwade P, Tiwari S, Ranjan S, Chandniha SK, Adamowski J (2018) Spatio-temporal variation of rainfall over Bihar State, India. J Water Land Dev 36(1):183–187. https://doi.org/10.2478/jwld-2018-0018

Yadav AK, Sahoo SK, Mahapatra S, Kumar AV, Pandey G, Lenka P, Tripathi RM (2014a) Concentrations of uranium in drinking water and cumulative, age-dependent radiation doses in four districts of Uttar Pradesh, India. Toxicol Environ Chem 96(2):192–200

Yadav R, Tripathi SK, Pranuthi G, Dubey SK (2014b) Trend analysis by Mann-Kendall test for precipitation and temperature for thirteen districts of Uttarakhand. J Agrometeorol 16(2):164–171

Yadav S, Bhattacharya P, Srivastava K (2019) Analysing long term seasonal and annual trends for precipitation and temperature in Central India. Mausam 3(July):523–532

Zarenistanak M, Dhorde AG, Kripalani RH (2014) Trend analysis and change point detection of annual and seasonal precipitation and temperature series over southwest Iran. J Earth Syst Sci 123(2):281–295. https://doi.org/10.1007/s12040-013-0395-7

Chapter 2
Mathematical Models and Computational Approaches for Assessing Water Quality in Rural Area of India

Deepak Kumar, Pooja Khurana, and Devender Kumar

Abstract Water, one of the valuable resources of nature, nurtures and takes care of almost all forms of life, without which life is impossible. Once known to be abundant in quantity, potable water is now scarce for a variety of reasons, growing population, climate change, increasing levels of pollution, and urbanization being some of them. Irrespective of the great importance and irreplaceable nature that it holds, humans have taken it for granted, making it now a scarce resource that should be used judiciously. Water pollution is one of the major causes involved in the wastage of huge amounts of water, leading to devastating effects not only on the lives of humans, but also possess threat to the lives of other coexisting living beings on the planet. Water pollution not only limits the amount of potable water, but also spreads water-borne diseases and causes various other health problems which can cause temporary or permanent impairment. To achieve the utmost simplicity and accuracy in understanding problems related to the quality of water and to assess water quality, several results mathematical models and computational approaches are designed and implemented which are a means of finding the best social, economic, and efficient solutions to water problems. This chapter will highlight various statistical, mathematical, and computational models that are potent enough to tackle the above-mentioned issues related to water and assess the quality of water with special emphasis on the most current topics.

D. Kumar (✉) · P. Khurana
Department of Applied Sciences, School of Engineering and Technology, Manav Rachna International Institute of Research and Studies, Faridabad 121004, India
e-mail: deepakman12@gmail.com

P. Khurana
e-mail: pooja.set@mriu.edu.in

D. Kumar
Information Technology Department, University of Technology and Applied Sciences-Shinas, 324 Muscat, Oman
e-mail: devendra.kumar@shct.edu.om

A. K. Yadav et al. (eds.), *Integrated Management of Water Resources in India: A Computational Approach*, Water Science and Technology Library 129,
https://doi.org/10.1007/978-3-031-62079-9_2

Keyword Mathematical models · Water resource management · Water quality · Analytical analysis

2.1 Introduction

2.1.1 Background

Water resources, essential for the maintenance of life and ecosystems, face increasing challenges due to population growth, climate change, and evolving land use patterns. In this era of complexity, computational models have become indispensable tools for comprehensively assessing and managing water resources. This chapter introduces the fundamental concepts, significance, and evolving landscape of computational models in the domain of water resource assessment.

2.1.2 Significance of Water Resources Assessment

The evaluation of water resources is at the forefront of sustainable development efforts, covering the quantification, evaluation and prediction of water availability, quality, and usage patterns. The importance of accurate assessments cannot be overstated, as they underpin informed decision making for water allocation, infrastructure development, and environmental protection.

As global challenges, including changing precipitation patterns and increasing demand for water-intensive activities, intensify, there is a growing need for sophisticated tools that can integrate diverse data sources, simulate complex hydrological processes, and provide actionable insights. Computational models, grounded in applied mathematics, hydrology, and computer science, have emerged as powerful instruments to meet these demands.

2.1.3 The Evolving Landscape of Computational Models

The landscape of computational models for the assessment of water resources has evolved significantly in the last few decades. Traditional methods, while valuable, often fail to address the intricacies of contemporary challenges. Computational models bridge this gap by offering a dynamic and flexible framework capable of incorporating diverse data types, considering the impacts of climate change, and facilitating scenario-based analysis.

Evolution encompasses advances in data assimilation techniques, machine learning applications, and the integration of remote sensing data. These developments allow for more accurate and real-time assessments, fostering a paradigm shift in how water resources are understood, managed, and preserved (Lamon et al. 2019).

2.1.4 Objectives

This chapter serves as a comprehensive introduction to the field of computational models for assessing water resources. The primary objectives are as follows:

Contextualization of Computational Models: Provide a clear understanding of the contextual relevance of computational models in the broader landscape of water resource assessment, highlighting their role in addressing contemporary challenges.

Significance in Decision-Making: Emphasize the critical role of computational models in supporting decision-making processes related to water allocation, infrastructure planning, and environmental conservation.

Overview of Evolving Technologies: Provide insights into the evolving technologies and methodologies within the realm of computational modeling for the assessment of water resources, illustrating how these innovations contribute to improved accuracy and applicability.

2.1.5 Structure of the Chapter

This chapter is structured to facilitate a gradual exploration of computational models for the assessment of water resources. It unfolds in the following sections:

Introduction and Application of Applied Science and Management of Water Resources Assessment: Establishes the foundational concepts and importance of water resource assessment in the broader context of sustainable development.

Role of Computational Models: Explore the unique contributions of computational models to addressing the complexities of water resource assessment, comparing them to traditional methodologies.

Mathematical Models: Provides information on the models that enable the simulation of various processes related to pollutant dispersion, transformation, and removal.

This chapter sets the stage for the subsequent sections, providing readers with a robust foundation for exploring the intricate world of computational models in the assessment and management of water resources.

2.2 Application of Applied Science and Management in the Assessment of Water Resources

Assessment of water resources involves the application of applied science and effective management strategies to ensure sustainable use, conservation and protection of this critical natural asset. Here are key applications of applied science and management in the assessment of water resources.

2.2.1 *Hydrological Modeling*

Hydrological modeling is a fundamental component of water resource assessment, providing a systematic and quantitative approach to understanding the complex processes that govern the movement and distribution of water within a watershed. This modeling technique integrates principles from hydrology, meteorology, and environmental science to simulate the behavior of various components of the hydrological cycle, such as precipitation, runoff, evaporation, and groundwater flow (Cherif et al. 2023). The primary objective of hydrological modeling is to represent the intricate interactions between these components, allowing scientists, engineers, and water resource managers to make informed decisions about water availability, flood forecasting, and sustainable water resource management. Hydrological models come in various forms, ranging from simple empirical models to sophisticated, physically-based models. These models use mathematical equations and algorithms to represent the processes that influence the movement and distribution in a landscape. Geographic Information System (GIS) technologies are often integrated with hydrological models to incorporate spatial data, enhancing the accuracy and applicability of simulations (Abbaspour et al. 2007, 2015; Moges et al. 2021).

2.2.2 *Integrated Water Resource Management*

Integrated water resource management (IWRM) is a holistic and sustainable approach to water resources that recognizes the interconnectedness of social, economic, and environmental factors within a given watershed or region. At its core, IWRM integrates various disciplines such as hydrology, ecology, economics, and sociology to address the complexity of water-related challenges. Applied science contributes to understanding the dynamics of water availability, quality, and use, while management practices guide the formulation and implementation of comprehensive policies. IWRM emphasizes the importance of considering the needs of various stakeholders, balancing competing water uses, and preserving ecosystems. Through the coordinated management of surface water, groundwater, and related resources, an IWRM aims to ensure the equitable and sustainable distribution of water for agriculture, industry, domestic use, and environmental conservation. By fostering collaboration among government agencies, communities, and the private sector, IWRM serves as a strategic framework for navigating the challenges posed by growing populations, urbanization, and climate change, ultimately promoting responsible and efficient use of water resources for present and future generations (Meran et al. 2021, Kulwant et al. 2024).

2.2.3 Water Quality Assessment

Water quality assessment is a critical process that integrates applied science methodologies to monitor and analyze various parameters that determine the condition of water. These parameters include the chemical composition and microbial content of water sources (Uddin et al. 2021). Applied science provides the tools and techniques to collect accurate data through laboratory testing and field measurements. The results of these assessments form the basis for management actions grounded in scientific evidence. Based on these assessments, management strategies are developed to prevent pollution, improve water quality, and ultimately protect public health. This collaborative approach ensures that decisions regarding water resource management are not only based on empirical data, but are also aligned with the overarching goal of maintaining the safety and integrity of water for both human and environmental well-being.

2.2.4 Remote Sensing and GIS Technologies

Remote sensing and Geographic Information System (GIS) technologies play a crucial role in water resource management. Applied science employs remote sensing techniques to collect spatial data that provide insight into land cover, vegetation patterns, and alterations in water bodies. These technologies provide a comprehensive and real-time view of the landscape. Management, in turn, capitalizes on these advances for various purposes, including mapping, monitoring, and decision making in water resource planning and conservation efforts (Dutta and Goel 2017). GIS tools enable the integration of diverse datasets, facilitating a more holistic understanding of the geographical dynamics that influence water availability and quality. Using the capabilities of remote sensing and GIS technologies, water resource managers can make informed decisions, implement effective conservation strategies, and respond proactively to changes in the environment, ensuring the sustainable use and preservation of water resources (McKinney and Cai 2002).

2.2.5 *Public Awareness and Education*

Public awareness and education are vital components of effective water resource management. Applied science provides evidence-based information on water conservation, pollution prevention, and sustainable water use. Management practices focus on the development and implementation of educational initiatives to communicate this knowledge to the public. Through media campaigns, community events, and educational programs, efforts are directed to raise awareness of the value of water resources and promote environmentally conscious behaviors. Informed and engaged communities are more likely to adopt water-friendly practices, participate in conservation efforts, and advocate for policies that prioritize sustainable water resource management. Continuing public awareness and education initiatives are crucial to building a collective consciousness that values and preserves this precious resource (EL-Nwsany et al. 2019).

The collaborative application of applied science and management practices is essential to address the complex and dynamic nature of water pollution. This integrated approach facilitates the development of evidence-based policies, effective pollution control measures, and sustainable water resource management strategies that protect water quality and protect ecosystems (Shirin and Yadav 2021).

2.3 Computational Models in Controlling and Managing Water Pollution

2.3.1 *Continuous Models*

Continuous models refer to mathematical structures that describe phenomena or processes that change continuously over time or space. These models are often expressed using differential equations that capture the rates of change of variables concerning one or more independent variables, such as time or spatial coordinates.

Continuous models in the assessment of water resources are mathematical representations that describe the continuous variation of water-related variables over time and space. These models use partial differential equations to simulate physical processes that govern the movement, distribution, and quality of water.

The key equations that represent continuous models for the quality assessment of water are described in the following.

2.3.1.1 Groundwater Flow Models

They are mathematical representations used to simulate and analyze groundwater movement within subsurface aquifers. These models play a crucial role in the assessment of water resources, providing insights into groundwater behavior, aquifer characteristics, and the impact of various factors on groundwater flow (Thangarajan 2007).

Key concepts associated with groundwater flow models include the following:

a. ***Basic Groundwater Flow Equation***

Dupuit-Forchheimer assumption: This simplifies the three-dimensional groundwater flow equations into a two-dimensional form, neglecting the vertical flow components. It is commonly applied in unconfined aquifers to facilitate simpler modeling (Okuyade et al. 2022).

$$\frac{\partial^2 h}{\partial x^2} + \frac{\partial^2 h}{\partial y^2} = 0 \tag{2.1}$$

where, x and y are Horizontal and Vertical spatial coordinates respectively. h indicates the hydraulic head, representing the potential energy of the groundwater.

b. ***Confined aquifer flow models***

Confined aquifer flow equation: It describes the movement of groundwater within a confined aquifer, considering factors such as transmissivity, well contributions, and specific storage (Wang 2020). It is expressed as follows:

$$\nabla.(\mathrm{T}\nabla \mathrm{h}) + \mathrm{W} = S_s \frac{\partial h}{\partial t} \tag{2.2}$$

This equation describes the flow of groundwater in confined aquifers. It considers the transmissivity (T) of the aquifer, the contributions of the well (W), and specific storage (S_s).

The equation indicates how the hydraulic head (h) changes over time and space.

The Equation (for confined aquifers): is used to model drawdown (s) in a confined aquifer around a pumping well. It depends on parameters such as pumping rate (Q) and aquifer transmissivity (T).

$$s(r,t) = \frac{Q}{4\pi T} \int_0^\infty \frac{e^{-\frac{s(r,t)}{4Tt}}}{r} dr \tag{2.3}$$

c. ***Unconfined Aquifer Flow Models***

Modflow: It is a widely used numerical model to simulate groundwater flow. It considers both confined and unconfined aquifers, allowing the simulation of complex hydrogeological conditions (Khadri and Pande 2016; Leake 1997; Panagopoulos 2012).

d. ***Dupuit-Forchheimer assumption (modified for unconfined aquifers)***:

In unconfined aquifers, the Dupuit-Forchheimer assumption is modified to account for the effect of the specific yield (S), which represents the volume of water released from storage per unit volume of aquifer per unit decrease in the hydraulic head (Zerihun 2018).

$$\frac{\partial^2 h}{\partial x^2} + \frac{\partial^2 h}{\partial y^2} = \frac{S}{T}\frac{\partial h}{\partial t} \tag{2.4}$$

2.3.1.2 Surface Water Flow Models

They are mathematical representations used to simulate and analyze the movement of water across the Earth's surface. These models are essential tools in hydrology and water resource management, providing information on river and stream behavior, flood prediction, and watershed dynamics (Badjana et al. 2023). This has been represented by the Saint–Venant equations (Mujumdar 2001).

$$\frac{\partial A}{\partial t} + \frac{\partial Q}{\partial x} = P - E \tag{2.5}$$

$$\frac{\partial Q}{\partial t} + \frac{\partial}{\partial x}\left(\frac{Q^2}{A}\right) + gA\frac{\partial h}{\partial x} = 0 \tag{2.6}$$

These partial differential equations describe the flow in open channels. The first equation represents the continuity equation, which expresses the conservation of mass, where A is the cross-sectional area, Q is the flow rate, P is precipitation, and E is evaporation. The second equation represents the momentum equation, incorporating terms for changes in flow velocity, water surface elevation (h), and gravitational effects.

2.3.1.3 Water Quality Models

They are mathematical representations designed to simulate and predict the distribution and dynamics of pollutants in water bodies. These models help to understand how various factors affect water quality, enabling researchers and water resource managers to make informed decisions about pollution control and environmental management. The most common type of water quality model is explained by the diffusion–advection equation, which describes the transport of pollutants in a flowing water system. Consider advection (transport with flow), diffusion (dispersion due to random molecular motion), and removal processes that affect pollutant concentrations over time and space (Loucks et al. 2017; Atshan et al. 2020, Patel et al. 2024). In one dimension, the equation is given by:

$$\frac{\partial C}{\partial t} + u\frac{\partial C}{\partial x} = D\frac{\partial^2 C}{dx^2} - R \tag{2.7}$$

where,

C is the pollutant concentration,
t is time,
u is the flow velocity,
D is the dispersion coefficient,
R represents any removal processes (e.g., decay, settling).

2.3.2 *Distributed Models*

Distributed models refer to mathematical representations that consider spatial variations in a study area. These models are used to describe physical and environmental processes that vary between different locations within a region. In the context of the assessment of water resources, distributed models play a crucial role in capturing the heterogeneity of factors such as precipitation, land use, soil properties, and topography.

2.3.2.1 Distributed Hydrological Models

These models simulate the movement of water throughout the landscape, considering the spatial distribution of rainfall, land use, and soil properties (Dehotin and Braud 2008).

Surface runoff has been expressed as

$$Q = \iint \frac{dS}{dT} dA \tag{2.8}$$

This equation represents the calculation of surface runoff (Q) on a distributed area (A). It involves the integration of the local slope $\left(\frac{dS}{dT}\right)$ with respect to both space and time.

2.3.2.2 Rainfall-Runoff Distribution Models

These kinematic wave models simulate the overland flow of water, considering the slope and flow resistance across the watershed (Reddy et al. 2007). The model has been represented as follows:

$$\frac{\partial h}{\partial t} + \frac{\partial Q}{\partial x} = 0 \tag{2.9}$$

This kinematic wave equation describes the movement of water in the overland flow. Consider the change in flow depth (h) with respect to time and space.

2.3.2.3 Distributed Hydrodynamic Models

These models simulate the flow of water in rivers or channels, considering the spatial distribution of water depth, velocity, and bed elevation (Guo et al. 2023). The 2-D shallow-water equation is illustrated as

$$\frac{\partial h}{\partial t} + \frac{\partial hu}{\partial x} + \frac{\partial hv}{\partial y} = S_f - S_b \tag{2.10}$$

The 2D shallow-water equations describe the evolution of water depth (h), velocity components (u) and (v), and bed elevation (S_b). . Source and friction terms (S_f) account for external inputs and energy dissipation.

2.3.2.4 Distributed Water Quality Models

These models account for the spatial distribution of pollutant concentrations, considering advection, diffusion, and removal processes in flowing water (Gomes et al. 2023; Yadav et al. 2014). This model has been described in the form of an advection–diffusion equation that is expressed as

$$\frac{\partial C}{\partial t} + \nabla.(uC) = \nabla.(D\nabla C) - R \tag{2.11}$$

This advection–diffusion equation describes the transport of a pollutant concentration (C) in a flowing-water system (Fig. 2.1). Incorporates advection, diffusion, and removal processes.

2.3.3 *Empirical Models*

Empirical models are statistical models that describe relationships between variables based on observed data. These models are derived from empirical evidence rather than theoretical considerations or physical principles.

2.3.3.1 Empirical Stream-Flow Prediction Model

It is a statistical model used to predict the flow of a stream based on observed data. Streamflow prediction models are valuable tools in hydrology to forecast water

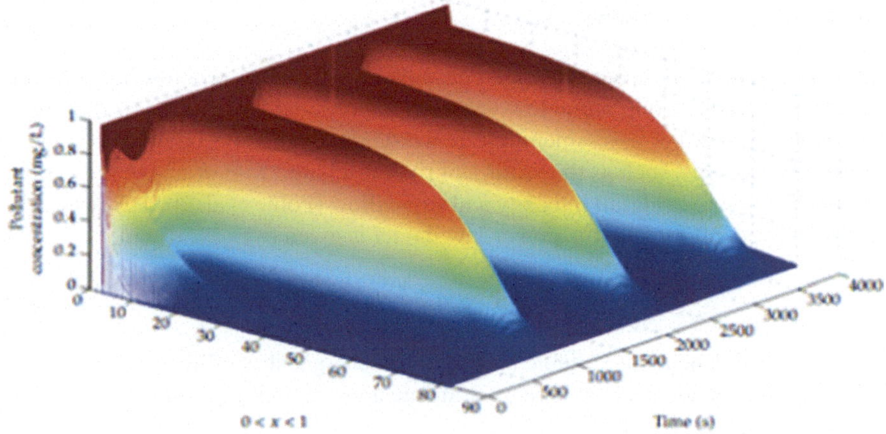

Fig. 2.1 The pollutant concentration over time (Gomes et al. 2023)

discharge in rivers and streams (Cho and Kim 2022). One common form of an empirical streamflow prediction model is a linear regression model (Irving et al. 2018). The general equation for a linear regression model is as follows:

$$Q = a + b.P \tag{2.12}$$

where, Q is the stream flow or discharge,

P is the precipitation,

a is the intercept coefficient,

b is the precipitation coefficient.

This simple linear regression model assumes a linear relationship between precipitation and streamflow.

2.3.3.2 Exponential Growth Model

This is applied to simulate the filling and stabilization of water storage systems such as reservoirs or aquifers. Carrying capacity (L) would represent the maximum sustainable water storage capacity and the growth rate (k) would determine how quickly the system fills up.

The logistic growth curve would exhibit an initial phase of rapid filling, followed by a slowdown as the storage approaches its maximum capacity. The midpoint (t_0) represents the time at which the storage is halfway to its maximum, indicating the inflection point of the curve.

$$Y = \frac{L}{1 + e^{-k(t-t_0)}} \tag{2.13}$$

2.3.3.3 Power-Law Model

It is a mathematical representation that describes a non-linear relationship between two variables, where one variable's value is proportional to a power of the other. In the context of water assessment, the power-law model could be used to capture certain relationships, such as the scaling of river discharge with drainage area or the relationship between rainfall intensity and streamflow (Pohle et al. 2021; Shirin et al. 2021). The general form of the power-law model is given by:

$$Y = a.X^b \tag{2.14}$$

where, Y is the dependent variable (e.g., river discharge, streamflow),
X is the independent variable (e.g., drainage area, rainfall intensity),
a is a scaling coefficient,
b is the exponent that determines the power-law relationship.

2.3.3.4 Time Series Models

Such as autoregressive Integrated Moving Average (ARIMA) models, are effective in capturing temporal patterns in pollutant concentrations over time. By examining historical data, these models provide predictions for future pollutant levels, facilitating proactive management strategies (Hasnain et al. 2022).

2.3.3.5 Multiple Regression Models

Incorporate several predictor variables to explain variations in pollutant concentrations (Li et al. 2019). This approach considers the influence of multiple environmental factors ($X_1, X_2, \ldots, X_n$) in predicting pollutant levels (C):

$$C = a + b_1X_1 + b_2X_2 + \cdots + b_nX_n + \varepsilon \tag{2.15}$$

Here, a is the intercept, b_i are coefficients, and ε accounts for the error term.

Machine learning techniques, including random forests or support vector machines, are increasingly being applied in water pollution modeling. These models can capture complex relationships among variables without relying on predefined equations, making them suitable for handling intricate and non-linear patterns in pollution data (Ghobadi and Kang 2023).

2.4 Mathematical Models in Controlling and Managing Water Pollution

Mathematical models are mathematical representations of real-world systems designed to capture their essential features and relationships. In the field of water pollution, these models enable simulation of various processes related to dispersion, transformation, and removal of pollutants. From simple mass balance equations to complex numerical simulations, mathematical models provide a means to analyze and optimize strategies to control and managing water pollution.

Mathematical models serve as indispensable tools in the field of environmental engineering, providing a quantitative framework for understanding, predicting, and mitigating complex systems. In the context of water pollution control and management, these models play a crucial role in assessing the sources, transport, and fate of pollutants in aquatic systems. Several models include the following:

2.4.1 Pollutant Transport Models

They are crucial tools to understanding and mitigating the movement of contaminants in aquatic systems. A prominent model used for this purpose is the Advection–Diffusion model, a partial differential equation that describes the transport of pollutants in flowing water (Kachiashvili et al. 2007). The equation is given by:

$$\frac{\partial C}{\partial t} + U\frac{\partial C}{\partial x} = D\frac{\partial^2 C}{dx^2} \tag{2.16}$$

where, C is the concentration of the pollutant

t is time, x is the spatial coordinate along the flow direction.

U is the water velocity, and D is the dispersion coefficient.

This model incorporates the processes of advection (transport by bulk water movement) and diffusion (dispersion due to turbulence), offering a comprehensive understanding of how pollutants travel downstream. Such models help identify critical areas of pollutant accumulation, determine the impact of discharges, and guide the formulation of effective pollution control and management strategies. By quantifying the temporal and spatial distribution of pollutants, these models contribute to informed decision making for the sustainable protection of water quality and ecosystems.

2.4.2 Water Quality Models

They are indispensable tools for assessing and managing the health of aquatic systems and offer a comprehensive understanding of the factors that influence water quality. The Water Quality Index (WQI) model is a notable example, providing a single numerical value that reflects the overall quality of water based on various physical, chemical, and biological parameters (Chidiac et al. 2023). The WQI is often calculated using the formula:

$$WQI = \sum_{i=1}^{n} w_i \times I_i \tag{2.17}$$

where, n is the number of parameters

w_i is the weight assigned to each parameter

I_i is the sub-index for each parameter.

Water quality models that incorporate the WQI facilitate the synthesis of complex information into a readily understandable metric, aiding in the communication of water quality status to policymakers and the public. Furthermore, these models offer a systematic approach to pollution monitoring, helping to identify trends, assess compliance with water quality standards, and guide the formulation of targeted pollution control measures. By integrating multiple parameters, water quality models provide a holistic perspective on the state of aquatic environments, enabling informed decision-making for the protection and sustainable management of water resources (Adelagun et al. 2021).

2.4.3 Eutrophication Models

They are crucial tools for understanding and mitigating the impacts of nutrient enrichment on water bodies, a phenomenon known as eutrophication. One common approach involves nutrient loading models, where the nutrient concentration dynamics is represented over time (Yang et al. 2008). An illustrative equation for nutrient loading can be expressed as

$$N_t = N_0 + L - R \tag{2.18}$$

where, N_t is the nutrient concentration at a given time,

N_0 is the initial nutrient concentration,

L represents the loading of nutrient from external sources (such as agricultural runoff or point source discharges) and

R represents nutrient removal processes (such as sedimentation or uptake by aquatic organisms).

Eutrophication models help simulate nutrient cycles, predict the potential for algae blooms, and assess the risk of oxygen depletion in water bodies. By considering factors such as nutrient inputs, hydrodynamics, and ecological interactions, these models contribute to the development of strategies to prevent and control eutrophication. These models play a critical role in designing management practices that aim to reduce nutrient loads, restore water quality, and promote overall ecological health of aquatic ecosystems.

2.4.4 Biological Models

They are essential tools for assessing and managing water quality, particularly in the context of wastewater treatment and ecological health. One prominent example is the biochemical oxygen demand (BOD) model, which quantifies the rate at which microorganisms break down organic matter in water (Vigiak et al. 2019; Shirin et al. 2023). The BOD model can be represented by the following equation:

$$\frac{dL}{dt} = k \times (L_0 - L) \tag{2.19}$$

where, $\frac{dL}{dt}$ is the rate of change of BOD concentration over time,

k is the BOD reaction rate constant,

L_0 is the initial BOD concentration, and

L is the BOD concentration at a given time.

This model provides insight into the oxygen demand of organic pollutants, aiding in the design and optimization of wastewater treatment processes. Biological models play a crucial role in understanding the dynamics of microbial processes, the nutrient cycle, and the general health of aquatic ecosystems. By simulating the interactions between microorganisms and organic pollutants, these models contribute to the design and operational strategies of effective wastewater treatment plants, ensuring compliance with water quality standards and sustainable management of water resources (Song et al. 2014; Shirin and Yadav 2014). Investigates the consequences of drinking water contamination that can be harmful to human health and the health of other animals, as well as the effects of fluoride contamination on the natural environment (Kumar et al. 2024).

2.4.5 Source Identification Models

They are pivotal tools for quantifying the contributions of different sources of pollution to a water body, allowing for targeted pollution control measures. A common method involves the use of mass balance equations (Thuy et al. 2016; Hua et al. 2019). An illustrative equation for source identification can be expressed as follows.

$$C_{total} = \sum_{i=1}^{n} (C_i \times F_i) \tag{2.20}$$

where, C_{total} is the observed concentration of a pollutant in the water body,

C_i is the concentration of the pollutant from a specific source,

F_i is the fraction of the total flow contributed by the ith source, and

n is the number of pollution sources.

This equation enables the estimation of pollutant contributions from various sources on observed concentrations and flow proportions. Source identification models help environmental managers and policymakers prioritize efforts to control specific sources of pollution. Using mathematical relationships to trace pollutants back to their origins, these models contribute to informed decision making for effective pollution source management, pollution prevention, and the preservation of water quality in receiving environments (Shirin et al. 2019).

2.5 Conclusion and Future Suggestions

In conclusion, the discussion of computational models for water resources underscores the pivotal role of applied science and management strategies in understanding, predicting and sustainably managing water systems. From hydrological modeling to groundwater assessments, the integration of advanced computational models allows a comprehensive understanding of water dynamics. These models, coupled with Geographic Information System (GIS) technologies and remote sensing, technologies and remote sensing, provide valuable information on spatial patterns, water quality, and the impact of climate change on water resources. Management practices, guided by these computational models, play a critical role in translating scientific knowledge into effective policies. Integrated water resource management (IWRM) emerges as a holistic approach, considering the interconnected nature of water uses and users, fostering sustainability, and balancing diverse needs. In addition, water quality assessments, informed by computational models, guide management actions to prevent pollution and protect public health.

As we navigate the challenges of population growth, climate change, and increasing water demand, synergy between applied science and management becomes imperative. The collaborative efforts outlined in this chapter emphasize the need for evidence-based decision-making, adaptive strategies, and a proactive approach to ensure the resilience and sustainability of water resources. Using the power of computational models, we pave the way for informed, efficient and equitable water resource management, crucial to meeting the needs of current and future generations.

Acknowledgements The authors extend their heartfelt appreciation to our esteemed institute and all individuals who have played a role, directly or indirectly, in contributing to this study.

Chapter 3
Hydrological Modeling and Simulation for Water Resource Assessment

Nirmalya Kumar Nath, Pritam Das, Leena Rani Mishra, Abhinav Kumar, Sachin L. Suryawanshi, and Vinay Kumar Gautam

Abstract Water resources are vital to sustainable development, and understanding their availability, distribution, and management is crucial for efficient utilization. Hydrological modeling plays a significant role in assessing water resources by simulating the movement of water through various hydrological components, such as rainfall, evaporation, infiltration, and runoff. The basic principles and components of a hydrological model include data collection, model calibration, and validation. Various types of hydrological models, such as empirical, conceptual, and physically-based models, are presented, highlighting their strengths and limitations in different applications. The chapter explores the process of simulating hydrological processes using these models. It covers topics such as rainfall-runoff modeling, evapotranspiration estimation, and streamflow routing. The challenges and uncertainties associated with hydrological modeling are also discussed, emphasizing the importance of data quality, model parameterization, and model selection. Additionally, the chapter discusses the role of GIS in data management, visualization, and spatial analysis and how remote sensing data can provide valuable input for hydrological models. The final section of the chapter examines the application of hydrological modeling in water resource assessment. It discusses the use of models to estimate water availability, design water infrastructure, and assess the impacts of climate change on water resources. Case studies from different regions are presented to demonstrate the practical applications of hydrological modeling in real-world scenarios.

Keywords Water resources · Hydrological modeling · Simulation · Data management · Climate change impacts

N. K. Nath (✉) · P. Das · L. R. Mishra · A. Kumar · S. L. Suryawanshi
Maharana Pratap University of Agriculture and Technology, Udaipur, Rajasthan 313001, India
e-mail: nirmalyakumarnath11@gmail.com

V. K. Gautam
College of Post Graduate Studies in Agricultural Sciences, CAU (Imphal), Umiam, Meghalaya 793103, India

A. K. Yadav et al. (eds.), *Integrated Management of Water Resources in India: A Computational Approach*, Water Science and Technology Library 129,
https://doi.org/10.1007/978-3-031-62079-9_3

3.1 Introduction

Water resources play a central role in the promotion of sustainable development, exerting influence on the economic well-being of nations (Callejas Moncaleano et al. 2021), the ecological health of ecosystems, and the general welfare of communities. Given the current trajectory of global population growth and the consequential alterations in climatic patterns, there is an ever-increasing imperative to acquire a comprehensive understanding of the accessibility, distribution, and governance of water (Sarkar and Bharat 2021; Yadav et al. 2014). Disparities in the availability of water, shaped by factors such as precipitation, evaporation, and anthropogenic activities, result in inequalities in the attainment of clean and potable water. The mounting strain on water resources arising from population expansion, compounded by the ramifications of climate change, presents formidable obstacles, including increased demand for water and possible alterations in precipitation patterns (Lal et al. 2021; Rehana and Monish 2021; Shirin and Yadav 2021).

Effective practices in resource management, conservation efforts, and advancements in technologies for water treatment and purification assume crucial significance in grappling with these challenges (Pani et al. 2021). In addition, international collaboration on the utilization of transboundary water resources is indispensable in the resolution of conflicts and the assurance of sustainable utilization. Ultimately, an all-encompassing and multifaceted approach encompassing sustainable governance, conservation, research, and global cooperation emerges as an imperative in guaranteeing water security and the well-being of both the present and future generations (Sharma et al. 2021).

Water, in its various manifestations, serves as a fundamental cornerstone of sustainable development, providing support for agricultural productivity, industrial operations, and ecological equilibrium (Lakshmi et al. 2024). As societies strive for advancement, responsible governance of water resources becomes essential to attain sustainable development objectives. However, the universal importance of water is accompanied by unequal distribution and numerous obstacles, including water scarcity exacerbated by climate change and population growth. Addressing these intricacies requires a profound comprehension of water availability, distribution, and administration.

Hydrological modeling assumes a central role in this endeavor by simulating the movement of water through its multiple components, such as precipitation, evaporation, infiltration, and runoff (Pani et al. 2021). These models, distinguished by their ability to capture essential hydrological processes, offer powerful instruments for comprehensively evaluating water resources (Dutta and Sarma 2021). They face challenges and uncertainties, including accurately representing natural processes and the influence of climate change (Rehana et al. 2020), and are closely integrated with Geographic Information Systems (GIS) and remote sensing technologies. Various categories of hydrological models, ranging from rainfall-runoff models to physically based models, cater to specific applications, facilitating flood risk assessment, water supply planning, and ecosystem management (Jaiswal et al. 2020; Shekar et al.

2023). Through case studies from diverse regions, hydrological modeling demonstrates its practical importance in tackling real-world challenges in water resource management (Chinnasamy et al. 2018). Hydrological modeling is indispensable for comprehending, managing, and mitigate the obstacles associated with water resource management, providing invaluable tools to formulate effective strategies to achieve sustainable development (Galkate et al. 2022).

3.2 Fundamentals of Hydrological Modeling

Understanding the fundamentals of hydrological modeling is paramount for unscrambling the complexities of water movement and distribution. This section delves into the core principles that underpin hydrological modeling, elucidating the crucial components that contribute to accurate assessments of water resources.

3.2.1 Basic Principles

Hydrological Components Overview: Hydrological modeling centers around the complex interplay of various constituents, which includes precipitation, vaporization, permeation, and streamflow (Pareta 2023). Each of these constituents assumes a distinctive role in moulding the motion and accessibility of water within a designated catchment area. Precipitation represents the volume of condensation that descends upon a specific location during a designated timeframe (Sokol et al. 2021), while vaporization signifies the process through which liquid water is transformed into a gaseous state and released into the atmosphere (Singh et al. 2020). Permeation entails the progression of water from the surface of the earth into the soil, while streamflow denotes the progression of water across the terrestrial surface and into streams and rivers (Ansari et al. 2020; Kumari et al. 2021). These constituents are indispensable to understand the motion and allocation of water within a catchment area.

Simulating Water Movement: In its essence, hydrological modeling encompasses the imitation of water flow through various constituents such as precipitation, evaporation, infiltration, and discharge. These models serve as digital laboratories, giving scientists a means to observe and scrutinize the dynamic processes that govern the conduct of water in diverse environmental circumstances (Wei et al. 2022). Through simulation of water flow, scientists can gain valuable insights into the intricate interplay between different hydrological components and their influence on the overall water system. This analytical approach facilitates the examination of how water behaves under different circumstances, contributing to a more profound understanding of the intricate dynamics within watersheds and other hydrological systems (Arulbalaji and Maya 2019; Jain and Singh 2020).

3.2.2 Components of a Hydrological Model

Data Collection: Effective hydrological modeling commences with the procurement of precise and all-encompassing data. This section delves into the techniques and methodologies used in the aggregation of indispensable data, encompassing meteorological observations, terrain characteristics, and land use information. Meteorological observations, which encompass precipitation, temperature, humidity, and wind patterns, serve as the basis for comprehending the climatic conditions that propel hydrological processes (Thakur et al. 2017). Terrain characteristics, such as topography, soil properties, and land cover, provide vital information on the physical attributes of the landscape and the intricate interplay between water and the environment. Furthermore, land use information, which encompasses human activities and infrastructure, exerts influence over water hydrological cycle and the availability (Alawi and Özkul 2023; Naha et al. 2021).

Model Calibration: Calibration constitutes a crucial step in the refinement of hydrological models to guarantee their precision and dependability. This subsection undertakes a comprehensive examination of the methodologies and practices that are employed in the adjustment of model parameters to correspond to observed data, thereby augmenting the model's prognostic capabilities. Through the act of calibration, the model parameters are meticulously adjusted to mimic real-world hydrological processes, including streamflow, groundwater levels, and water storage dynamics. This iterative process encompasses the comparison of simulated outputs with observed data, the identification of disparities, and the modification of the model structure and parameters to minimize discrepancies. By attaining a close correspondence between model simulations and observed hydrological behavior, calibrated models are able to provide more accurate predictions regarding water quantity, quality, and flow patterns (Chouksey and Gupta 2014; Tigabu et al. 2023).

Model Validation: The validation process examines the performance of the model in relation to independent datasets, thus confirming its robustness and generalizability. This particular component delves into the significance of validation in establishing the trustworthiness of hydrological models and ensuring their suitability beyond the calibration domain. Validation of the model is indispensable to evaluate its capability to capture hydrological processes under various conditions and over different time periods. Through comparison of model outputs with independent observations, such as streamflow measurements, data on water quality, or the behavior of hydrological phenomena during extreme events, the validation process offers insight into the predictive accuracy of the model and its capacity to depict real-world occurrences (Dubey et al. 2023; Refsgaard and Storm 1990).

3.3 Types of Hydrological Models

Hydrological modeling encompasses diverse methodologies, each offering unique approaches to simulate the intricate processes of water movement within a watershed. This section provides a detailed exploration of various hydrological models, characterized by distinct advantages and limitations, which encapsulate the complexity of watershed dynamics.

3.3.1 Empirical Models

Empirical hydrological models leverage statistical correlations derived from observed data, making them well-suited for scenarios where a thorough understanding of the underlying physical mechanisms is limited. Empirical models rely on historical data to establish empirical associations between input variables, such as precipitation, land usage, and soil characteristics, and the resulting hydrological response, such as streamflow or runoff. By capturing these statistical patterns, empirical models offer valuable insights into hydrological behavior without necessitating a profound understanding of the underlying physical processes. This simplicity makes them particularly valuable for situations where obtaining comprehensive data on intricate hydrological interactions may pose challenges.

Advantages: Empirical models excel in situations where there is limited knowledge of the underlying physical processes. They rely on statistical relationships derived from observed data, which makes them well-suited for various hydrological contexts. Their simplicity and versatility allow their application in situations where data availability is limited.

Limitations: However, empirical models may lack the ability to capture the intricate interactions and physical mechanisms that drive hydrological processes. This limitation can restrict their accuracy and applicability in environments with distinct characteristics or under dynamic conditions.

3.3.2 Conceptual Models

Conceptual hydrological models aim to depict the processes occurring in a watershed by utilizing simplified conceptualizations of the underlying physical mechanisms. Striking a delicate balance between simplicity and accuracy, these models prove to be valuable in situations where detailed data is limited. The objective of conceptual models is to capture the fundamental hydrological processes within a watershed by representing key elements such as precipitation, evapotranspiration, infiltration, and runoff using simplified conceptual frameworks. Although they do not explicitly mimic the intricate physical interactions taking place in the natural environment, conceptual models offer a practical compromise by providing a structured representation of hydrological processes that balances simplicity with a reasonable level of accuracy. This approach has distinct advantages in scenarios where comprehensive data on specific physical properties or processes are scarce, as it allows for the development of predictive models that can inform water resource management, flood forecasting, and environmental impact assessments.

Advantages: Conceptual models provide a balanced representation of hydrological processes, offering simplicity along with a reasonable level of accuracy. They are particularly valuable in situations where detailed data is limited as they allow for the simulation of essential hydrological phenomena within a watershed. Their adaptability to various geographical and environmental settings enhances their applicability in scenarios where comprehensive data may be scarce.

Limitations: Nevertheless, conceptual models may encounter difficulties in capturing the intricate complexities of hydrological interactions due to their simplified representations. This limitation can potentially impact their precision in certain applications.

3.3.3 Physically-Based Models

Physically based hydrological models aim to replicate watershed processes by integrating a thorough representation of the underlying physical principles. These models are particularly suitable for extensively studied environments, where comprehensive data on terrain, soil properties, and land cover are available. In doing so, these models can navigate the intricate nature of hydrological interactions. This subsection examines the complexities and benefits of physically-based modeling, highlighting its ability to capture precise hydrological phenomena. Physically-based models make use of extensive knowledge of the physical processes that govern hydrological behavior, including principles of fluid mechanics, heat transfer, and mass balance equations. By incorporating detailed representations of terrain characteristics, soil properties, vegetation, and land use, these models strive to replicate the intricate interplay of hydrological processes within a watershed. This approach is particularly advantageous in extensively studied environments where comprehensive data on various terrain attributes and hydrological parameters are accessible, enabling the development of robust models that can accurately capture detailed hydrological phenomena. Physically based models play a crucial role in simulating complex interactions such as surface runoff, infiltration, groundwater flow, and streamflow dynamics, thereby supporting applications in flood forecasting, water resources management, and environmental impact assessments.

Advantages: Physically based models incorporate detailed representations of the underlying physical principles, allowing them to effectively simulate complex hydrological phenomena in well-studied environments with comprehensive data. They have the ability to capture intricate hydrological processes with a high level of precision, supporting applications in flood forecasting, water resource management, and environmental impact assessments.

Limitations: However, the reliance of physically-based models on comprehensive data and computational complexity can pose challenges in environments with limited data availability or constrained computational resources.

3.4 Simulating Hydrological Processes

Hydrological modeling extends its impact by delving into the intricate processes that govern the movement and distribution of water within a watershed. This section explores specific aspects of hydrological processes and provides insight into the techniques and methodologies used in the simulation of critical phenomena.

3.4.1 Rainfall-Runoff Modeling

Rainfall-runoff modeling occupies a central position within hydrological simulations, capturing the process of transforming precipitation into runoff. Including a wide range, these approaches range from simple statistical methods to more intricate models such as the Soil Conservation Service Curve Number method and the Rational Method. This article covers considerations for model selection, data requirements, and calibration and validation processes.

Techniques and Methodologies

Simple Statistical Methods: These methods often serve as the fundamental basis for rainfall-runoff modeling, utilizing statistical relationships to estimate runoff based on rainfall characteristics and historical data. Their appeal lies in their simplicity and ease of application, making them particularly valuable in situations where detailed data and computational resources are limited.

Soil Conservation Service Curve Number Method: Developed by the United States Department of Agriculture, this method considers various factors such as soil type, land use, and hydrological soil group to estimate direct runoff from rainfall events. Its extensive usage in hydrological modeling stems from its ability to incorporate soil and land cover characteristics into the process of runningoff estimation.

Rational Method: The rational method presents a simplified approach to estimate maximum runoff rates from urban areas by considering rainfall intensity, drainage areas, and runoff coefficients. This method offers a practical means of evaluating stormwater runoff in urban hydrology and hydraulic design.

Considerations for Model Selection: The selection of an appropriate rainfall-runoff model involves an evaluation of the specific characteristics of the study area, the available data, and the objectives of the modeling. Factors such as computational complexity, data requirements, and the desired level of detail in representing hydrological processes play a decisive role in model selection.

Data Requirements and Calibration/Validation Processes: Successful rainfall-runoff modeling necessitates robust data on precipitation, land use, soil properties, and hydrological characteristics of the area under study. Additionally, the calibration and validation processes are vital steps in refining and verifying the performance of the model against observed data, ensuring its accuracy and reliability in simulating runoff generation.

3.4.2 Evapotranspiration Estimation

Evapotranspiration, the combined process of evaporation of water from surfaces and transpiration of vegetation, plays a fundamental role in the equilibrium of water. This section delves into the methodologies and considerations involved in estimating evapotranspiration. From simpler empirical approaches like the Penman–Monteith equation to remote sensing techniques based on satellite observation, the discourse encompasses a wide range of strategies. Considerations encompass factors that influence the rates of evapotranspiration, such as land cover, climate, and soil characteristics, as well as the challenges associated with achieving accurate estimations.

Methodologies and Considerations

Penman–Monteith Equation: The Penman–Monteith equation, a widely employed method for estimating evapotranspiration, amalgamates meteorological data such as temperature, humidity, wind speed, and solar radiation to calculate the potential evapotranspiration. It offers a comprehensive approach that considers the physical processes inherent in evapotranspiration and is applicable to various environmental conditions.

Satellite-Based Remote Sensing Techniques: Remote sensing technologies, including satellite data, provide valuable insights into the spatial and temporal fluctuations of evapotranspiration. These techniques take advantage of satellite observations of land surface temperature, vegetation indices, and solar radiation to estimate evapotranspiration at different scales, thus augmenting the comprehension of regional dynamics in water equilibrium.

Factors Influencing Rates of Evapotranspiration: Land cover, climate patterns, soil characteristics, and vegetation attributes exert a significant influence on evapotranspiration rates. Comprehending the interaction among these factors is of utmost importance to achieve accurate estimations and effective management of water resources.

Challenges Associated with Accurate Estimation: Accurate estimation of evapotranspiration presents challenges due to the spatial and temporal variability of the factors that influence this process. Furthermore, the incorporation of data from various sources and scales, addressing uncertainties in satellite measurements, and accounting for changes in land use and vegetation dynamics of vegetation introduce complexities in the estimation of evapotranspiration.

3.4.3 Streamflow Routing

The process of simulating the movement of water through river channels, known as streamflow routing, is of significant importance within hydrological modeling. It plays a vital role in understanding the dynamics of water travel within a river system, which is crucial for various applications, including flood forecasting, water resource management, and environmental impact assessments. The streamflow routing process encompasses a wide array of techniques, ranging from the relatively straightforward Muskingum-Cunge method to more intricate distributed hydrological models. These techniques consider factors such as the spatial and temporal scales of the modeling, the unique characteristics of the river channel, and the potential repercussions for downstream water availability. By taking these factors into account, hydrologists and water resource managers can acquire valuable insights into the movement of water within river systems, empowering them to make well-informed decisions regarding water-related issues and challenges.

3.5 Challenges and Uncertainties in Hydrological Modeling

Hydrological modeling, while a powerful tool, encounters challenges and uncertainties that necessitate careful consideration for accurate and reliable results. This section explores the intricacies associated with data quality, model parameterization, model selection, and the broader task of addressing uncertainties inherent in hydrological modeling.

3.5.1 Importance of Data Quality

The importance of data quality in hydrological modeling cannot be overstated, as it establishes the basis for the precision and dependability of model outcomes. Erroneous or unfinished data can considerably undermine the dependability of hydrological models, thus accentuating the need for rigorous measures to ensure data precision and completeness. Addressing these challenges involves implementing robust protocols for data collection, employing validation techniques to detect and rectify errors, and incorporating methods to quantify and consider uncertainties in the data. By highlighting the importance of data quality and advocating for comprehensive practices in data management, this subsection underscores the fundamental role of high-quality data in generating precise and reliable hydrological models, thus increasing their usefulness in various applications related to water resource management and environmental assessment.

3.5.2 Model Parameterization

Model parameterization is an extremely important element of hydrological modeling, wherein the determination of parameter values within the model is carried out to accurately mirror real-world conditions. This section delves into the intricacies associated with model parameterization, shedding light on the complexities that arise due to limited field measurements and the ever-changing nature of environmental processes. By emphasizing the impact of uncertainties, the discussion explores strategies for achieving optimal parameter estimation, calibration, and validation, reflecting the ongoing challenge of striking a balance between the intricacy and simplicity of the model. The inherent uncertainties, which arise from the limited availability of field measurements and the dynamic behavior of environmental processes, highlight the need for robust parameter estimation techniques that take into account the variability and complexity of hydrological systems. Effective calibration and validation procedures are of utmost importance in ensuring that the model accurately portrays the observed behavior of the hydrological system, thus enhancing its reliability and predictive capability.

3.5.3 Model Selection

The selection of an appropriate hydrological model is a crucial decision that has a profound impact on the accuracy and significance of the results of the modeling process. By taking into account factors such as the timeframe in which the study is conducted, the availability of data, and the specific hydrological processes involved, this discussion emphasizes the importance of having a comprehensive understanding of the area under study in order to effectively inform the decision-making process for model selection. The spatial and temporal scales, along with the presence of pertinent data, play significant roles in determining the suitability of a particular model for a given application. Therefore, it is imperative to have a thorough understanding of the complexities associated with the study area and its hydrological processes in order to make well-informed decisions regarding model selection.

3.5.4 Addressing Uncertainties in Modeling

The consideration of uncertainties in hydrological modeling is a critically important aspect due to the inherently dynamic and complex nature of environmental processes. The analysis of sensitivity plays a vital role in evaluating the influence of input parameters on the outputs of the model, thereby identifying the parameters that have a significant impact on the behavior of the model and those that have a less influential effect. The propagation of uncertainty involves the assessment of how uncertainties in input data and model parameters propagate through the modeling process, ultimately affecting the reliability of the outcomes. In addition, the incorporation of probabilistic approaches enables the quantification of uncertainties and the generation of probabilistic forecasts, providing decision-makers with valuable information for assessing and managing risks. Transparent reporting and communication of uncertainties are of utmost importance in order to ensure that the limitations and potential sources of error in the modeling process are effectively conveyed to stakeholders and decision-makers, thus enhancing the credibility and usefulness of the results obtained from the model.

3.6 Integration of GIS and Remote Sensing in Hydrological Modeling

Integration of Geographic Information Systems (GIS) and Remote Sensing Technologies Enhances the capabilities of hydrological modeling by providing spatially explicit information and real-time data inputs. This section explores the symbiotic relationship between hydrological modeling, GIS, and remote sensing.

3.6.1 Role of GIS in Data Management and Visualization

GIS (Geographic Information System) plays an important role in hydrological modeling by facilitating efficient management, integration, and visualization of data. Through the use of GIS, hydrological modelers can effectively integrate diverse geospatial datasets, including digital elevation models, soil types, and land cover, in order to construct a comprehensive representation of the study area. This integrated approach allows for a more precise depiction of the spatial distribution of hydrological processes and characteristics, which is essential in the development of reliable hydrological models. In addition, GIS allows for the visual exploration of intricate spatial data, thereby aiding in the identification of spatial patterns and relationships that are crucial to comprehending the interactions between various environmental factors and hydrological processes.

3.6.2 Spatial Analysis in Hydrological Modeling

Spatial analysis plays a crucial role in hydrological modeling within the context of GIS, as it allows for the examination of spatial connections and patterns within a given watershed. Using GIS-based spatial analysis techniques, hydrologists can accurately define watershed boundaries, which in turn enables a comprehensive understanding of the drainage area and the identification of areas that contribute to specific water bodies. This level of precision is essential to gain insight into the spatial distribution of hydrological processes and to effectively model the movement of water within the watershed. Moreover, spatial analysis aids in the identification of flow paths, which, in turn, allows for the simulation of surface runoff and the evaluation of potential flood risk zones. Furthermore, the process of characterizing changes in land cover through spatial analysis offers valuable insights into the impact of modifications in land use on hydrological processes, which supports the assessment of alterations in water flow patterns and sediment transportation.

3.6.3 Utilizing Remote Sensing Data for Model Inputs

Remote sensing technologies play a vital role in providing valuable data inputs for hydrological models, offering high-resolution real-time information on the conditions of the land surface. Using remote sensing data, hydrologists can access a vast amount of information regarding precipitation patterns, thus enabling the estimation of spatially distributed precipitation inputs for hydrological models. Furthermore, remote sensing data provide valuable insights into the moisture content, facilitating the characterization of spatial variability in soil moisture levels, which is crucial to accurately representing the processes of infiltration and runoff processes. Furthermore, the application of remote sensing data to analyze the dynamics of land cover supports the evaluation of changes in land use and their impact on hydrological processes, thus contributing to a more comprehensive and dynamic approach to hydrological modeling.

3.7 Application of Hydrological Modeling in Water Resource Assessment

The ultimate objective of hydrological modeling is to inform water resource assessment, guiding sustainable water management practices. This section explores the diverse applications of hydrological modeling in assessing water resources and addressing critical challenges related to water availability, infrastructure design, and the impacts of climate change.

3.7.1 Estimating Water Availability

Hydrological modeling plays a fundamental role in the calculation of water availability within a specific watershed. Through the use of hydrological models, researchers and water resource managers can estimate the quantity and temporal distribution of water resources within a watershed, taking into account various factors such as precipitation inputs, evapotranspiration, soil moisture dynamics, and interactions between groundwater and surface water. These models allow for the evaluation of water availability under different scenarios, including varying climatic conditions and land use patterns, thereby providing valuable insights for the sustainable management of water resources. Furthermore, the estimation of water availability through hydrological modeling facilitates the identification of areas that may experience water stress, thereby helping in the development of strategies to alleviate water scarcity and ensure adequate water supply for agriculture, industry, and domestic use.

3.7.2 Designing Water Infrastructure

The design of water infrastructure, which includes dams, reservoirs, and irrigation systems, is highly dependent on the use of accurate hydrological models. This section delves into how hydrological models contribute to the design process through the simulation of water flow, evaluation of potential risks, and optimization of infrastructure layouts. Using hydrological modeling, engineers and planners can simulate multiple scenarios, thus gaining an understanding of the behavior of water systems, including the impact of infrastructure on river flow, sediment transport, and flood routing. Furthermore, these models play a pivotal role in the evaluation of potential risks associated with infrastructure design, such as flooding, sedimentation, and ecological impacts, providing valuable insights for decision-making processes aimed at improving the safety and sustainability of water infrastructure projects.

3.7.3 Assessing Climate Change Impacts on Water Resources

Climate change introduces a layer of complexity to the dynamics of water resources, which requires a proactive approach to comprehending and mitigating its effects. Hydrological models play a crucial role in simulating alterations in the patterns of precipitation, temperature, and extreme weather events, thus providing valuable insights into potential shifts in the accessibility of water and the susceptibility of water systems to changing climatic circumstances. By incorporating climate change scenarios into hydrological models, researchers and decision-makers can enhance their understanding of potential impacts on key hydrological variables, such as streamflow, groundwater recharge, and water quality (Shaikh et al. 2022). This information is essential for formulating adaptation strategies to address the challenges posed by climate change, including modified water availability, alterations in flood and drought, and potential consequences for ecosystems and sectors relying on water.

Applications of hydrological modeling in the assessment of water resources underscore its significance as a decision-support tool for informed and sustainable water management practices (Dutta and Sarma 2021). By addressing challenges, integrating advanced technologies, and exploring diverse applications, hydrological modeling emerges as a key asset in ensuring the resilience and adaptability of water systems in a changing world.

3.8 Conclusion

Throughout this chapter, we have embarked on an expedition through the intricate realm of hydrological modeling, acquiring valuable insights into its fundamental principles, diverse methodologies, and pivotal role in evaluating water resources. We have elucidated the fundamental principles of hydrological modeling, delving into the various categories of models, and simulating essential processes such as rainfall runoff and streamflow routing, each contributing to our understanding of water resources.

In this exploration, we have also grappled with the challenges and uncertainties inherent in hydrological modeling, acknowledging the critical significance of data quality, model complexity, and the imperative to address uncertainties. Furthermore, we have witnessed how the integration of GIS and remote sensing revolutionizes hydrological modeling, augmenting our capacity to manage and visualize data, ultimately culminating in more precise simulations. As we have discovered the multifaceted nature of hydrological modeling in evaluating water resources, encompassing the estimation of water availability and the anticipation of climate change impacts, it has become evident that this instrument is indispensable in our endeavors to sustainably manage water. As we progress, the knowledge acquired from this expedition will serve as an illuminating beacon, empowering us to make well-informed decisions and adapt to the challenges of the future.

Acknowledgements The authors thank our prestigious institute and other people who are involved in this study for their direct and indirect contributions.

References

Alawi SA, Özkul S (2023) Evaluation of land use/land cover datasets in hydrological modelling using the SWAT model. H2Open J 6(1):63–74. https://doi.org/10.2166/H2OJ.2023.062

Ansari MI, Thakural LN, Anbu Kumar S (2020) Streamflow modelling for a peninsular basin in India. Lecture Notes Civil Eng 58:645–659. https://doi.org/10.1007/978-981-15-2545-2_53/COVER

Arulbalaji P, Maya K (2019) Effects of land use dynamics on hydrological response of watershed: a case study of Chittar watershed, Vamanapuram River Basin, Thiruvananthapuram District, Kerala, India. Water Conserv Sci Eng 4(1):33–41. https://doi.org/10.1007/S41101-019-00066-5/METRICS

Callejas Moncaleano DC, Pande S, Rietveld L (2021) Water use efficiency: a review of contextual and behavioral factors. Front Water 3:685650. https://doi.org/10.3389/FRWA.2021.685650/BIBTEX

Chinnasamy P, Muthuwatta L, Eriyagama N, Pavelic P, Lagudu S (2018) Modeling the potential for floodwater recharge to offset groundwater depletion: a case study from the Ramganga basin, India. Sustain Water Resour Manage 4(2):331–344. https://doi.org/10.1007/S40899-017-0168-6/TABLES/4

Chouksey A, Gupta A (2014) Hydrological modelling of upper and middle Narmada river basin, India using geospatial tools. In: Hydraulics, water resources, coastal and environmental engineering. https://www.academia.edu/11780483/Hydrological_Modelling_of_Upper_and_Middle_Narmada_River_Basin_India_using_Geospatial_Tools

Dubey SK, Kim J, Her Y, Sharma D, Jeong H (2023) Hydroclimatic impact assessment using the SWAT model in India—State of the art review. Sustainability 15(22):15779. https://doi.org/10.3390/SU152215779

Dutta P, Sarma AK (2021) Hydrological modeling as a tool for water resources management of the data-scarce Brahmaputra basin. J Water Clim Change 12(1):152–165. https://doi.org/10.2166/WCC.2020.186

Galkate RV, Yadav S, Pandey RP, Negm AM, Yadava RN (2022) An overview: water resource management aspects in India, pp 29–55. https://doi.org/10.1007/978-3-030-95687-5_3

Jain CK, Singh S (2020) Impact of climate change on the hydrological dynamics of River Ganga, India. J Water Clim Change 11(1):274–290. https://doi.org/10.2166/WCC.2018.029

Jaiswal RK, Ali S, Bharti B (2020) Comparative evaluation of conceptual and physical rainfall–runoff models. Appl Water Sci 10(1):1–14. https://doi.org/10.1007/S13201-019-1122-6/TABLES/7

Kumari N, Srivastava A, Sahoo B, Raghuwanshi NS, Bretreger D (2021) Identification of suitable hydrological models for streamflow assessment in the Kangsabati River Basin, India, by using different model selection scores. Nat Resour Res 30(6):4187–4205. https://doi.org/10.1007/S11053-021-09919-0/METRICS

Lakshmi R, Thomas J, Joseph S (2024) Impacts of recent rainfall changes on agricultural productivity and water resources within the Southern Western Ghats of Kerala, India. Environ Monit Assess 196(2):1–18. https://doi.org/10.1007/S10661-023-12270-X/FIGURES/7

Lal P, Shekhar A, Kumar A (2021) Quantifying temperature and precipitation change caused by land cover change: a case study of India using the WRF model. Front Environ Sci 9:766328. https://doi.org/10.3389/FENVS.2021.766328/BIBTEX

Naha S, Rico-Ramirez MA, Rosolem R (2021) Quantifying the impacts of land cover change on hydrological responses in the Mahanadi river basin in India. Hydrol Earth Syst Sci 25(12):6339–6357. https://doi.org/10.5194/HESS-25-6339-2021

Pani A, Ghatak I, Mishra P (2021) Understanding the water conservation and management in India: an integrated study. Sustain Water Resour Manage 7(5):1–16. https://doi.org/10.1007/S40899-021-00556-2/FIGURES/6

Pareta K (2023) Hydrological modelling of largest braided river of India using MIKE Hydro River software with rainfall runoff, hydrodynamic and snowmelt modules. J Water Clim Change 14(4):1314–1338. https://doi.org/10.2166/WCC.2023.484

Refsgaard JC, Storm B (1990) Construction, calibration and validation of hydrological models, pp 41–54. https://doi.org/10.1007/978-94-009-0257-2_3

Rehana S, Rajulapati CR, Ghosh S, Karmakar S, Mujumdar P (2020) Uncertainty quantification in water resource systems modeling: case studies from India. Water 12(6):1793. https://doi.org/10.3390/W12061793

Rehana S, Monish NT (2021) Impact of potential and actual evapotranspiration on drought phenomena over water and energy-limited regions. Theoret Appl Climatol 144(1–2):215–238. https://doi.org/10.1007/S00704-021-03521-3/TABLES/6

Sarkar SK, Bharat GK (2021) Achieving sustainable development goals in water and sanitation sectors in India. J Water Sanit Hyg Dev 11(5):693–705. https://doi.org/10.2166/WASHDEV.2021.002

Shaikh MM, Lodha P, Lalwani P, Mehta D (2022) Climatic projections of Western India using global and regional climate models. Water Pract Technol 17(9):1818–1825. https://doi.org/10.2166/WPT.2022.090

Sharma SK, Seetharaman A, Maddulety K (2021) Framework for sustainable urban water management in context of governance, infrastructure, technology and economics. Water Resour Manage 35(12):3903–3913. https://doi.org/10.1007/S11269-021-02916-1/TABLES/1

Shekar PR, Mathew A, Gopi VP (2023) Rainfall-runoff modelling using SWAT and eight artificial intelligence models in the Murredu Watershed, India. Environ Monit Assess 195(9):1–28. https://doi.org/10.1007/S10661-023-11649-0/TABLES/3

Shirin S, Yadav AK (2021) Assessment and suitability for the dumping of flyash in an abandoned mine. Environ Challenges 4:100111. https://doi.org/10.1016/J.ENVC.2021.100111

Singh R, Mishra V, Narasimhan B, Ghosh S, Sharma A, Dutta S, Mujumdar PP (2020) Hydrological modeling in India. Proc Indian Natl Sci Acad 86(1):479–494. https://doi.org/10.16943/PTINSA/2020/49802

Sokol Z, Szturc J, Orellana-Alvear J, Popová J, Jurczyk A, Célleri R (2021) The role of weather radar in rainfall estimation and its application in meteorological and hydrological modelling—a review. Remote Sens 13(3):351. https://doi.org/10.3390/RS13030351

Thakur PK, Nikam BR, Garg V, Aggarwal SP, Chouksey A, Dhote PR, Ghosh S (2017) Hydrological parameters estimation using remote sensing and GIS for Indian Region: a review. Proc Natl Acad Sci India Sect A Phys Sci 87(4):641–659. https://doi.org/10.1007/S40010-017-0440-Z

Tigabu TB, Wagner PD, Narasimhan B, Fohrer N (2023) Pitfalls in hydrologic model calibration in a data scarce environment with a strong seasonality: experience from the Adyar catchment, India. Environ Earth Sci 82(15):1–17. https://doi.org/10.1007/S12665-023-11047-2/FIGURES/10

Wei N, Xie J, Lu K, He S, Gao Y, Yang F (2022) Dynamic simulation of ecological flow based on the variable interval analysis method. Sustainability 14(13):7988. https://doi.org/10.3390/su14137988

Yadav AK, Sahoo SK, Mahapatra S, Kumar AV, Pandey G, Lenka P, Tripathi RM (2014) Concentrations of uranium in drinking water and cumulative, age-dependent radiation doses in four districts of Uttar Pradesh, India. Toxicol Environ Chem 96(2):192–200. https://doi.org/10.1080/02772248.2014.934247

Chapter 4
Challenges and Opportunities for Water Resources in India: An Exploratory Story of Resilience

Manoj Kumar Mahto

Abstract Embarking on a profound exploration of India's intricate water landscape, the chapter titled "Water Resources in India: Challenges and Opportunities" delves into a multifaceted narrative, unravelling the complexities that intricately define our nation's vital aquatic lifelines. From the conspicuous disparities in the distribution of water wealth across diverse regions to the disconcerting Specter of depleting groundwater levels, the challenges unfold like chapters in a gripping novel. The canvas darkens as we confront the insidious influence of pollution emanating from industrial, agricultural, and urban sources, coupled with the ominous shadow of climate change, intensifying the stress on an already overburdened system. However, amid these challenges, hopeful rays pierce through the gloom. Technological advancements, particularly the promise of computational approaches, emerge as beacons illuminating the path toward more efficient water management. Yet, beneath the surface, the true heartbeat of narrative pulses in the resilience of communities and the power of shared responsibility. Through vivid stories of grassroots initiatives, the chapter weaves a compelling narrative of empowerment and collective consciousness, painting a poignant picture of the transformative role community involvement plays in shaping the destiny of water resources. As the narrative unfolds, it leaves behind not merely a documentation of challenges, but a resounding call to action—an impassioned invitation to integrate cutting-edge technology, recalibrate policies, and nurture a cultural ethos that cherishes every droplet of water. In this symphony of challenges and opportunities, the chapter beckons us to participate in the harmonious orchestration of India's water management, where every stroke of innovation and each community-driven initiative contribute to the evolving masterpiece of sustainability.

Keywords Water resources · Sustainable management · Community engagement · Technological innovations · International collaboration

M. K. Mahto (✉)
Vignan Institute of Technology and Science, Deshmukhi(V), Pochampally(M), Yadadri-Bhuvanagiri District, Telangana 508284, India
e-mail: manojkr.bit@gmail.com

A. K. Yadav et al. (eds.), *Integrated Management of Water Resources in India: A Computational Approach*, Water Science and Technology Library 129,
https://doi.org/10.1007/978-3-031-62079-9_4

4.1 Introduction

Water, a fundamental resource crucial for maintaining life and fostering socioeconomic development, is of immense significance in the context of India. With diverse climatic regions and a growing population, the country faces a myriad of challenges in managing its water resources efficiently (Kulwant et al. 2024). Rapid urbanization, industrialization, and agricultural practices have led to an increase in water demand, placing tremendous stress on existing water bodies and groundwater reservoirs (Singh and Mishra 2019; Kumar et al. 2024). This chapter aims to provide a comprehensive overview of the challenges and opportunities associated with water resources in India, addressing critical issues such as water scarcity, pollution, and the impact of climate change.

India's water landscape is characterized by a complex interplay of surface water bodies and groundwater reservoirs, each with its unique set of challenges. The intricate balance between water demand and availability requires careful consideration to ensure equitable distribution and environmental conservation. The need for a holistic understanding of these challenges is crucial for the formulation of effective policies and sustainable water management practices (Singh and Mishra 2019; Shirin et al. 2023). Water resources play an essential role in supporting agriculture, industry and domestic needs, making their sustainable utilization essential for India's socio-economic development. This chapter delves into the complexities of water management in India, exploring the challenges that hinder efficient resource utilization while highlighting opportunities for innovative solutions. By examining existing policy frameworks, governance structures, and technological interventions, our aim to provide insights that contributes to the ongoing discourse on sustainability of water in India.

4.1.1 Background

The intricate dynamics of water resources in India is shaped by a multitude of factors, ranging from geographic diversity to anthropocentric influences. With a rapidly growing population, urbanization, and industrialization, the demand for water has increased, placing unprecedented stress on existing surface and groundwater reservoirs. Understanding the historical and contemporary context of water resource management is crucial for navigating the complex challenges facing the country.

Diverse climatic regions, each presenting unique challenges and opportunities (Tiwari et al. 2020), mark India's water landscape. From the perennial rivers of the Himalayas to the arid regions of the west, the distribution and availability of water resources vary significantly throughout the country. The historical context of water management practices, including traditional harvest and distribution, has influenced the current state of water resources in India.

4.1.2 Importance of Water Resources in India

The water resources of India play a pivotal role in sustaining life, supporting agriculture, and driving socioeconomic development. With a population exceeding a billion, the demand for water has intensified, necessitating a delicate balance between utilization and conservation. Agricultural practices, a significant contributor to India's economy, rely heavily on water for irrigation, making the availability of water crucial for food security (Tiwari et al. 2020).

The industrial sector, another major component of India's economic landscape, requires substantial water resources for various processes. Therefore, efficient water management is essential to ensure the sustainability of industries and the overall economic growth of the nation. Additionally, water resources support domestic needs, sanitation, and hygiene, directly affecting public health and well-being.

The environmental importance of water resources cannot be overstated. Ecosystems, biodiversity, and overall health of the environment intricately related to the availability and quality of water. Rivers, lakes, and wetlands not only provide habitats for diverse species, but also contribute to ecological balance.

Understanding the importance of water resources in India is fundamental for devising effective policies and sustainable management practices. This chapter aims to explore the multifaceted significance of water resources in India, highlighting the need for holistic approaches that consider ecological, social and economic dimensions.

4.2 Overview of Water Resources in India

The water resources of India exhibit a complex interplay of surface water and groundwater systems, distributed in various climatic regions. Surface water, which encompasses rivers, lakes, and reservoirs, constitutes a critical component of the country's water landscape. The Himalayan rivers, such as the Ganges and Brahmaputra, contribute significantly to surface water resources, while peninsular rivers, such as Godavari and Krishna, play a vital role in water availability for the southern regions.

Groundwater, sourced from aquifers, is a key supplement to surface water and is widely used for agricultural irrigation and domestic purposes. However, unregulated extraction and declining groundwater levels pose significant challenges to the sustainability of this resource (Yadav et al. 2014).

The distribution of water resources in India is uneven, with some regions facing abundance of water, while others grapple with scarcity. The arid and semi-arid regions in the northwest and central parts of the country are particularly prone to water stress, which requires careful management and conservation efforts. Additionally, the variability in rainfall patterns and the contribute of climate change impacts further to the complexity of India's water resources.

4.2.1 Surface Water

Surface water constitutes a vital component of India's water resources, which includes rivers, lakes, reservoirs, and other water bodies. The Himalayan rivers, including the Ganges, Brahmaputra, and Indus, play an important role in contributing to the surface water resources of the country. These rivers not only serve as a primary source of drinking water but also support agricultural irrigation and industrial processes, and contribute to the ecological balance of the regions they traverse.

Peninsular rivers, such as Godavari, Krishna, and Cauvery, contribute to surface water availability in the southern parts of India (Tiwari et al. 2009). These rivers are crucial to maintaining agriculture and meeting the water needs of both rural and urban populations. Additionally, India's vast network of lakes and reservoirs further enhances its surface water storage capacity, supporting water supply, hydroelectric power generation, and flood control.

Despite the abundance of surface water resources, challenges such as pollution, overextraction, and inter-state water disputes pose threats to sustainable management. Understanding the distribution, utilization, and challenges associated with surface water is essential for formulating effective water policies and management strategies in India.

4.2.2 Groundwater

Groundwater, sourced from aquifers, is a crucial component of India's water resources, supporting agriculture, industry, and domestic needs. It serves as a key supplement to surface water and plays a vital role in sustaining water availability during periods of low rainfall or surface water scarcity.

India's dependence on groundwater for irrigation is substantial, with a significant portion of agricultural land relying on this resource for crop cultivation. However, unregulated extraction and declining groundwater levels pose significant challenges to the sustainability of this vital water source (Tiwari et al. 2009). Overextraction has led to issues such as land subsidence and the depletion of aquifer reserves.

Understanding the distribution, recharge mechanisms, and challenges associated with groundwater is essential for sustainable water resource management in India. This chapter delves into the complexities of groundwater dynamics, exploring the impact of human activities, climate change, and the need for effective conservation measures to ensure long-term availability of this critical resource. Figure 4.1 illustrates the predicament of groundwater management in India.

Fig. 4.1 Groundwater challenges in India

4.3 Challenges in Water Management

Water management in India faces a spectrum of challenges that require a comprehensive strategy for sustainable use and conservation. Foremost among these challenges is water scarcity, accentuated by rising demand, erratic rainfall patterns, and overextraction, particularly in arid and semi-arid regions. The contamination of water bodies poses a significant threat, with industrial discharges, agricultural runoff, and inadequate sanitation infrastructure contributing to pollution in rivers and lakes (Kumar et al. 2021). Climate change exacerbates water scarcity problems, altering precipitation patterns and intensifying extremes such as floods and droughts. Water-related conflicts that arise from competing demands among regions and sectors pose governance challenges. Aging infrastructure in urban and rural areas contributes to water inefficiencies and losses. Unregulated groundwater extraction, which exceeds recharge rates, further threatens long-term sustainability. Addressing these challenges is crucial for formulating effective water management policies considering both current and future needs in India.

4.3.1 Water Scarcity

Water scarcity stands as a pressing challenge in India's water management landscape, reflecting a delicate imbalance between increasing demand and diminishing water resources, particularly in arid and semi-arid regions. Erratic rainfall patterns, aggravated by impacts of climate change, contribute to the exacerbation of water scarcity issues, with periods of drought becoming more frequent and severe (Tiwari et al. 2020). This scarcity poses a significant threat to agricultural practices, industrial

Fig. 4.2 Water scarcity

processes, and the domestic water supply. Competition for water resources among various sectors intensifies, leading to challenges in equitable distribution and allocation. Effective management of water scarcity requires a multifaceted approach that considers climate-resilient strategies, efficient water use practices, and the development of alternative water sources. Figure 4.2 illustrates the global problem of water scarcity in the environment.

4.3.2 Pollution of Water Bodies

Water bodies in India face significant pollution challenges, which pose threats to human health and ecosystem integrity. Industrial discharges, agricultural runoff containing pesticides and fertilizers, and inadequate sanitation infrastructure contribute to the contamination of rivers and lakes. The Ganges, one of India's major rivers, is particularly susceptible to pollution from various sources, impacting its water quality lakes (Kumar et al. 2021). This pollution not only affects the availability of clean drinking water, but also disrupts aquatic ecosystems, leading to a decline of biodiversity. Management of water pollution requires comprehensive measures, including the enforcement of stringent regulations, the adoption of sustainable agricultural practices, and the development of effective wastewater treatment infrastructure to safeguard the health of water bodies and the well-being of communities relying on them. Figure 4.3 shows the pollution of water bodies in various environments.

4.3.3 Impacts of Climate Change

India faces pronounced impacts of climate change on its water resources, which introduce challenges that require adaptive strategies for sustainable water management. Altered precipitation patterns, rising temperatures, and changing hydrological cycles contribute to the increased frequency and intensity of extreme weather events,

Fig. 4.3 Pollution of water bodies

including floods and droughts. These changes disrupt traditional water management practices and exacerbate water scarcity issues, particularly in vulnerable regions. The Himalayan glaciers, a crucial source of river water, are also experiencing accelerated melting, which affects the long-term availability of water resources (Saleh et al. 2022). Understanding and mitigating the impacts of climate change on water resources require interdisciplinary approaches, including climate-resilient infrastructure, sustainable water use practices, and integrated water resource management strategies.

4.3.4 Water-Related Conflicts

India grapples with intricate water-related conflicts arising from competing demands across diverse sectors, geographical regions, and states, necessitating adept governance and diplomatic solutions Rajeevan et al. 2021). Interstate disputes over transboundary rivers, such as the Cauvery and Mahanadi, underscore the challenges in achieving equitable water allocation (Saleh et al. 2022). Additionally, intra-state conflicts arise from disparities in water availability and usage patterns among rural and urban areas. The complexities of water-related conflicts demand integrated water resource management strategies, legal frameworks, and collaborative mechanisms to ensure fair distribution and mitigate potential socioeconomic and environmental repercussions.

4.3.5 *Aging Infrastructure*

India's water management sector is grappling with the challenge of aging and inadequate infrastructure, particularly in urban and rural areas, necessitating urgent attention and strategic interventions (Kumar and Samadder 2022). Urban centers often face water losses, inefficiencies, and service disruptions due to outdated distribution networks and treatment plants. In rural areas, insufficient infrastructure hampers access to safe and reliable drinking water sources, affecting the health and well-being of lakes from the local community (Kumar et al. 2021). Upgrading and maintaining water infrastructure is imperative for enhancing resilience to climate change, ensure water security, and meeting the growing demands of a burgeoning population. Comprehensive investment, technological innovation, and effective governance are essential components of a strategy to address the challenges posed by the aging water infrastructure in India.

4.3.6 *Unregulated Extraction*

India's water management sector grapples with the challenge of aging and inadequate infrastructure, particularly in urban and rural areas, necessitating urgent attention and strategic interventions (Kumar and Samadder 2022). Urban centers often face water losses, inefficiencies, and service disruptions due to outdated distribution networks and treatment plants. In rural areas, insufficient infrastructure hinders access to safe and reliable drinking water sources, affecting the health and well-being of local communities. Upgrading and maintaining water infrastructure is imperative for enhancing resilience to climate change, ensure water security, and meeting the growing demands of a burgeoning population. Comprehensive investment, technological innovation, and effective governance are essential components of a strategy to address the challenges posed by the aging water infrastructure in India.

4.4 Policy and Governance

Effective management of water resources in India requires robust policy frameworks and governance structures to address the complex challenges associated with water scarcity, pollution, and unregulated extraction (Bandyopadhyay et al. 2021). Policy interventions are essential to promote sustainable water use, enforce regulations, and facilitate integrated water resource management. Additionally, governance mechanisms play a pivotal role in fostering collaboration between different stakeholders, ensuring equitable distribution, and resolving water-related conflicts, both at the interstate and intrastate levels.

In recent years, there has been a growing recognition of the need for adaptive governance structures that can respond to the dynamic challenges posed by climate change and population growth. Such governance models should integrate scientific insights, community participation, and technological innovations to ensure sustainable water management practices. This chapter explores the existing policy landscape and governance structures in India's water sector, highlighting strengths, weaknesses, and potential avenues for improvement.

4.4.1 National Water Policy

The India National Water Policy serves as a cornerstone in shaping the country's approach to water resource management, providing a comprehensive framework to address various challenges. Enacted by the Ministry of Water Resources, the policy emphasizes sustainable and equitable use of water resources, promoting water conservation, and ensures integrated water planning at the basin level. It underscores the importance of stakeholder participation, efficient water use, and the incorporation of technological advances in water management practices.

The policy acknowledges the importance of climate change and aims to improve resilience through adaptive strategies, emphasizing the need for robust infrastructure, efficient water governance, and the preservation of ecological integrity. This chapter delves into the key tenets of the National Water Policy, assessing its effectiveness in addressing contemporary water challenges and outlining potential avenues for policy enhancement.

4.4.2 Legal Framework

India's legal framework for water resources plays a pivotal role in guiding the sustainable management, allocation, and utilization of this critical resource. Encompassing a range of statutes, acts, and regulations, the legal framework addresses issues such as water rights, pollution control, and interstate water disputes. Key legislation includes the Water (Prevention and Control of Pollution) Act of 1974, the River Boards Act of 1956, and the Inter-State Water Disputes Act of 1956.

This chapter explores the intricacies of India's legal framework for water resources, analyzing its effectiveness in addressing contemporary challenges. It evaluates the strengths and limitations of existing laws, examines their enforcement mechanisms, and considers opportunities for legal reforms to improve water governance in the country.

4.4.3 Institutional Mechanisms

India's water governance is highly dependent on institutional mechanisms that facilitate coordinated planning, implementation, and management of water resources (Raju et al. 2019). This includes institutions at the central, state, and local levels, each contributing to various aspects of water management. Key organizations such as the Central Water Commission (CWC) and the Central Ground Water Board (CGWB) play crucial roles in the evaluation and development of water resources.

The chapter examines the effectiveness of existing institutional mechanisms, examining their roles, responsibilities, and interactions in addressing water challenges. It also explores potential enhancements and innovations in institutional frameworks to adapt to evolving water management needs.

4.5 Opportunities for Sustainable Water Use

Among the challenges faced by India's water resources, there exist significant opportunities for sustainable water use, encompassing innovative technologies, community-based initiatives, and policy reforms (Saleh et al. 2022). Integrated water resource management, coupled with the adoption of efficient irrigation practices and rainwater harvesting, presents avenues to improve water use efficiency in agriculture. The integration of smart technologies, such as sensor-based irrigation systems and remote sensing, can contribute to precision water management. In addition, community-led conservation efforts, awareness campaigns, and decentralized water governance models offer promising approaches to address local water challenges.

4.5.1 Water Conservation Practices

Water conservation practices play a central role in mitigating water scarcity and promoting sustainable water use in India. Adopting efficient irrigation techniques, such as drip irrigation and sprinklers, helps optimize water usage in agriculture. Rainwater harvesting, at the individual and community levels, presents an effective means of increasing water supply (Tiwari et al. 2009). Furthermore, the implementation of water saving technologies in industries and the promotion of water-efficient practices in households contribute to overall conservation efforts.

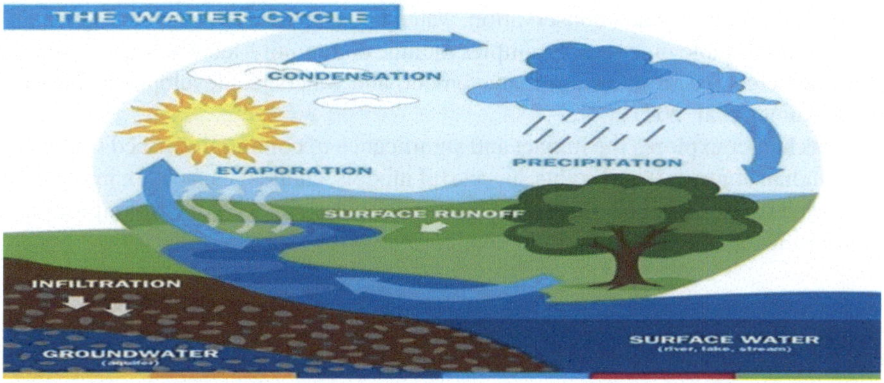

Fig. 4.4 Water resource management

4.5.2 *Integrated Water Resource Management*

Integrated Water Resource Management (IWRM) emerges as a holistic approach to address the complexities of water management in India, encompassing social, economic, and environmental dimensions. IWRM involves coordinated planning, development, and sustainable use of water resources, considering the entire hydrological cycle. This approach emphasizes stakeholder participation, intersectoral coordination, and adaptive strategies to cope with changing climatic conditions lakes (Kumar et al. 2021). Figure 4.4 provides insight into the management practices implemented for water resources across the Earth.

4.5.3 *Harnessing Rainwater and Alternative Sources*

In the face of water scarcity challenges, harnessing rainwater and exploring alternative water sources emerge as critical strategies to increase water supply in India. Rainwater harvesting, a traditional practice revived and modernized, allows the capture and storage of rainwater for various purposes, reducing dependence on conventional water sources. Additionally, exploring alternative sources such as treated wastewater and brackish water for no potable uses presents opportunities to alleviate pressure on freshwater resources lakes (Kumar et al. 2021).

4.5.4 *Community-Based Initiatives*

Community-based initiatives play a central role in sustainable water management, promoting local participation, awareness, and ownership. These initiatives involve local communities in water conservation, watershed management, and the implementation of sustainable practices. Examples include community-led water user associations, participatory groundwater management, and watershed development projects lakes (Kumar et al. 2021).

This chapter explores the impact and significance of community-based initiatives in the Indian context. It highlights successful models, challenges faced in implementation, and the potential for scaling up these initiatives to empower communities and build resilience in the face of water challenges.

4.5.5 Efficient Use of Agricultural Water

Efficient use of agricultural water is paramount for sustainable farming practices, food security, and water conservation in India. Adopting precision irrigation techniques, such as drip and sprinkler systems, minimizes water wastage and optimizes water delivery to crops. In addition, promoting water-saving crop varieties, implementing rainwater harvesting in agriculture, and utilizing real-time data for irrigation scheduling contribute to enhancing water use efficiency.

4.6 Technological Innovations

Technological innovations are at the forefront of addressing water challenges in India, providing advanced tools and solutions for efficient monitoring, conservation, and sustainable use. Remote sensing technologies, including satellite imagery and ground-based sensors, enable precise monitoring of water resources, facilitating accurate assessments of availability and distribution. These innovations provide water resource managers with real-time data, helping them make effective decisions for optimal water allocation and use.

In agriculture, the advent of smart irrigation systems equipped with sensor technologies has transformed traditional farming practices. These systems enable farmers to apply water precisely where and when it is needed, minimizing wastage and improving overall water use efficiency (Sharma et al. 2022). In addition, technologies such as precision agriculture, which integrates data analytics, GPS, and automation, offer a holistic approach to resource management, optimizing crop yields while conserving water resources.

Water treatment technologies also play a central role in ensuring the availability of clean and safe water. Advanced processes such as membrane filtration, reverse osmosis, and advanced oxidation contribute to the purification of water for various purposes, including drinking, industrial use, and agriculture. Integration of these technologies helps address challenges in water quality, promoting public health and environmental sustainability.

Fig. 4.5 Smart irrigation systems

4.6.1 Water Purification Technologies

Ensuring access to clean and safe water is imperative for public health and sustainable development in India. Advances in water purification technologies contribute significantly to addressing water quality challenges. Membrane-based technologies, including reverse osmosis and ultrafiltration, play a crucial role in removing contaminants and the provision of potable water (Gupta et al. 2020; Tripathi et al. 2019). Furthermore, advanced oxidation processes, such as photocatalysis, offer effective means of eliminating pollutants and pathogens from water sources.

4.6.2 Smart Irrigation Systems

Smart irrigation systems represent a transformative approach to water management in agriculture, leveraging technology to optimize water usage, enhance crop yields, and promote sustainability (Mishra et al. 2021). These systems integrate sensor technologies, real-time data analytics, and automation to precisely deliver water to crops based on their actual needs, reducing water waste and improving overall efficiency. Remote sensing tools, such as satellite imagery and drones, contribute to monitoring soil moisture levels and crop health, enabling informed irrigation decisions. Figure 4.5 depicts the implementation of smart irrigation systems.

4.6.3 Remote Sensing in Water Management

Remote sensing technologies have become key tools in the realm of water management, offering invaluable insight into the dynamic and complex interactions within hydrological systems. Satellite imagery, aerial photography, and unmanned aerial vehicles (UAVs) serve as instrumental platforms for the capture of high-resolution spatial data, facilitating a comprehensive understanding of water resources, land use patterns, and environmental changes (Thenkabail et al. 2017).

In the Indian context, these technologies have been instrumental in monitoring and managing water resources in diverse landscapes. Satellite-based observations enable the estimation of key hydrological parameters, such as transportation, providing critical information for water balance assessments and agricultural water management. The ability to monitor changes in land cover and land use dynamics through remote sensing aids in assessing the impact of human activities on water availability and quality.

Additionally, remote sensing plays a crucial role in water quality assessment by detecting and mapping parameters such as turbidity, chlorophyll concentration, and pollutants. This capability is particularly significant in addressing challenges related to pollution and ensuring the sustainability of water sources for human consumption and ecosystem health (Sharma and Gupta 2017).

Although remote sensing technologies offer unparalleled advantages in water management, challenges persist. These include the need for specialized training, data accessibility, and integration with ground-based observations. The chapter explores these challenges, discusses ongoing research initiatives, and outlines the potential to use remote sensing in shaping evidence-based policies for sustainable water resource management in India. By bridging the gap between spatial data acquisition and informed decision-making, remote sensing continues to be a transformative force in improving the efficiency and resilience of water management practices.

4.6.4 Block-Chain for Water Governance

Blockchain technology, known for its decentralized and secure nature, has emerged as a disruptive force in reshaping traditional water governance systems, offering novel solutions to complex challenges. In the context of water resource management in India, where issues of equitable distribution, transparency and accountability are paramount, the block chain presents an array of applications that can revolutionize how water-related transactions and decisions are recorded, monitored, and executed.

One of the key advantages of the block chain in water governance lies in its ability to provide a transparent and immutable ledger. This ensures secure recording of water transactions, including water allocation, rights, and usage. Using the tamper-proof nature of the block chain, the risk of fraudulent activities and unauthorized alterations in water records was significantly reduced, fostering a more trustworthy and accountable water governance framework (Bhatia and Wagle 2019). This transparency is particularly crucial in the Indian context, where disputes over water rights and usage are not common.

Smart contracts, another facet of block chain technology, introduce automation and enforceability into water-related agreements. These self-executing contracts,

Fig. 4.6 Block-chain for water

embedded in the block chain, enable the automatic execution of predefined conditions, streamlining administrative processes, and minimizing the potential for disputes (Hassan et al. 2021). In the Indian water governance landscape, where administrative inefficiencies and manual processes have historically been challenges, the implementation of smart contracts through the block chain can lead to more efficient and equitable water management practices.

Several pilot projects and initiatives around the world, and possibly in India, have explored the application of block chain in water governance. These projects include the use of block chain to verify water rights, monitoring water quality, and ensure fair and transparent water distribution. This chapter examines these case studies, drawing insight into the successes, challenges faced, and lessons learned, thus providing a comprehensive overview of the current state and future potential of the block chain in the Indian water governance context. Figure 4.6 visualizes the application of blockchain technology for water management.

However, challenges persist, including regulatory uncertainties, technical complexities, and the need for widespread adoption. Regulatory frameworks must evolve to accommodate block-chain applications in water governance, ensuring legal validity and compliance. Moreover, technical considerations, such as scalability and interoperability, must be addressed for the block chain to seamlessly integrated into existing water management systems.

4.7 Future Prospects

Future Prospects of Water Resources Management in India

As we navigate the complex challenges of water resource management, the future holds both opportunities and imperatives for sustainable practices. Figure 4.7 presents an outlook on the future prospects of water resources management in India, indicating potential strategies, innovations, and developments in the field.

The following considerations outline potential pathways for shaping the future of water management in India.

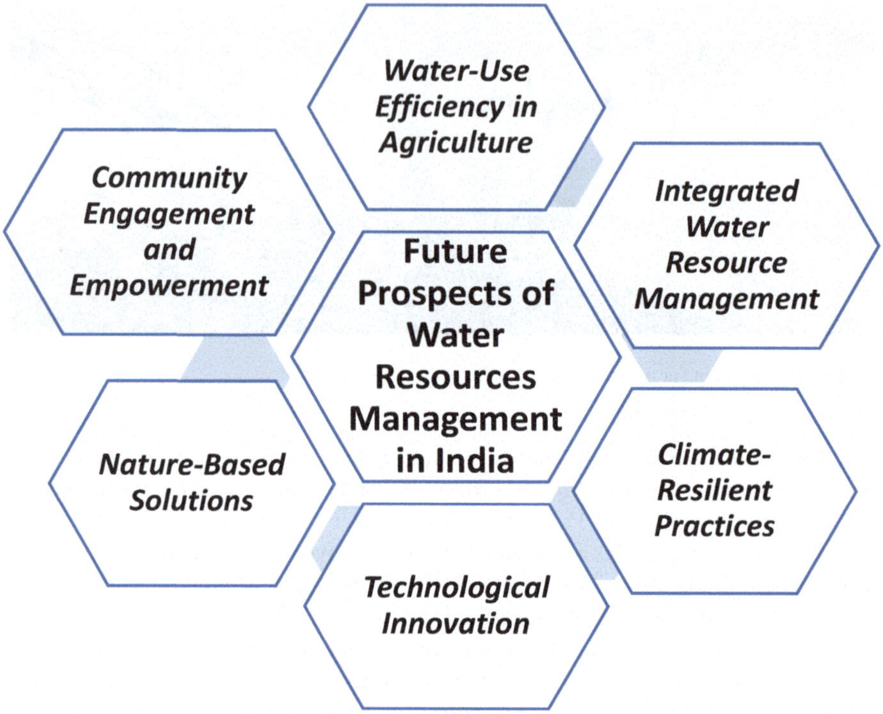

Fig. 4.7 Future prospects of water resources management in India

Integrated Water Resource Management (IWRM): Putting emphasis on the need for a holistic approach, future strategies should prioritize integrated water resource management. This involves the coordination of various aspects, including surface water and groundwater management, watershed development, and the integration of social, economic and environmental considerations.

Climate-Resilient Practices: Given the increasing impacts of climate change on water availability and quality, future prospects hinge on the adoption of climate-resilient practices. This includes implementing adaptive measures, developing early warning systems, and fostering community resilience to cope with changing precipitation patterns and extreme weather events.

Technological Innovation: The integration of advanced technologies, such as remote sensing, artificial intelligence, and the Internet of Things (IoT), presents a promising avenue for efficient water management. These technologies can improve monitoring, data-driven decision-making, and optimizing water use across various sectors.

Nature-Based Solutions: Incorporating nature-based solutions, such as afforestation, wetland restoration, and sustainable land management, can play a crucial role in water conservation. These strategies contribute to ecosystem health, biodiversity conservation, and sustainable replenishment of water resources.

Community Engagement and Empowerment: Active participation of local communities remains crucial to the success of water management initiatives. Future prospects should prioritize community engagement, fostering awareness, education, and participatory decision making to ensure sustainable water practices at the grassroots level.

Water Use Efficiency in Agriculture: Agriculture, a major consumer of water, requires innovative approaches to improve water use efficiency. Future prospects include promoting precision agriculture, adopting drought resistant crops, and incentivizing sustainable agricultural practices that optimize water use.

By embracing these prospects and committing to transformative actions, India can pave the way for a sustainable and secure future for its water resources. It requires a concerted effort from policymakers, communities, researchers, and industries to navigate the complexities and uncertainties surrounding water management in the years to come.

4.7.1 Role of Technology in the Future

Technology plays an important role in shaping the future of water management in India, offering innovative solutions to address challenges related to the scarcity, quality, and efficient use of water resources. Figure 4.8 highlights the pivotal role of technology in shaping the future landscape. The following outlines the key roles technology is expected to play in the future.

Smart Water Infrastructure: Implementing smart technologies, including sensors, meters, and IoT devices, will improve the efficiency of water distribution systems. Smart meters provide real-time data on water consumption, enabling consumers and utilities to make informed decisions, detect leaks promptly, and optimize water use.

Remote Sensing and GIS: Advanced remote sensing technologies, coupled with Geographic Information System (GIS) applications, provide comprehensive information on water resources. These tools facilitate the monitoring of surface water, groundwater and changes in land use, supporting informed decision-making for sustainable water management.

Precision Agriculture: Precision agriculture technologies, such as drones, sensors, and data analysis, will revolutionize water use in farming. These tools allow farmers to precisely manage irrigation, monitor soil moisture levels, and optimize crop water requirements, resulting in increased agricultural productivity and reduced water waste.

Artificial Intelligence (AI) for Water Prediction: AI algorithms will play a crucial role in predicting and managing water-related events. Machine learning models can analyze historical data to predict rainfall patterns, river flows, and potential flooding, aiding in early warning systems and adaptive water management strategies.

Block-Chain for Transparent Water Transactions: Blockchain technology is expected to bring transparency and accountability to water transactions. By creating

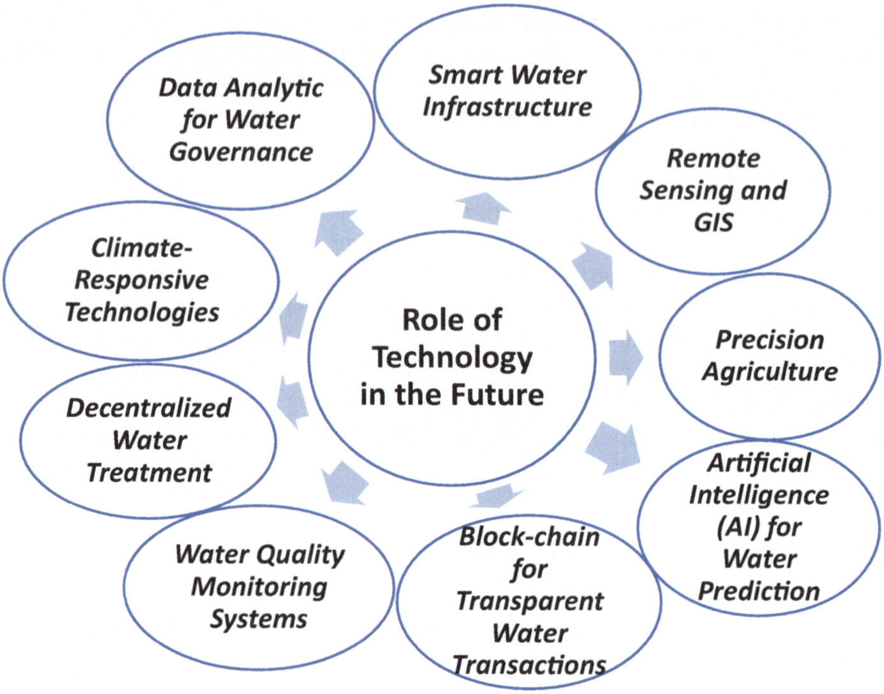

Fig. 4.8 Role of technology in the future

secure, decentralized ledgers, the block chain can streamline water trading, verify water rights, and automate contracts, ensuring fair and transparent water governance.

Water Quality Monitoring Systems: Continuous monitoring of water quality using advanced sensors and real-time monitoring systems will become widespread. These technologies help to detect contaminants early, ensure the safety of drinking water and support timely interventions to prevent water pollution.

Decentralized Water Treatment: Compact, decentralized water treatment systems that utilize advanced filtration, membrane technologies, and solar-powered solutions will become more prevalent. These systems provide access to clean water in remote areas, reducing reliance on centralized infrastructure and minimizing water losses during distribution.

Climate-Responsive Technologies: Technologies designed to address the impacts of climate change on water resources will gain prominence. Climate-resilient water infrastructure, weather forecasting models, and adaptive management strategies will be essential to mitigate the effects of changing climate patterns on water availability.

Data Analytic for Water Governance: Advanced data analysis will be instrumental in processing large amounts of water-related data. This includes analyzing usage patterns, predicting demand, and optimizing water distribution networks, contributing to more efficient and data-driven water governance.

The integration of these technologies into water management has the potential to transform the sector, making it more resilient, efficient, and sustainable. Continued research, investment, and collaboration between government, industry, and research institutions will be crucial to realize the full benefits of technology in securing India's water future.

4.7.2 International Collaboration for Water Solutions

International Collaboration for Water Solutions: Navigating Global Challenges

In addressing the complex and interconnected challenges related to water resources, international collaboration emerges as a vital avenue for sharing knowledge, expertise, and resources. Collaborative efforts can contribute to the development and implementation of innovative solutions that ensure sustainable water management on a global scale. Figure 4.9 demonstrates the importance of international collaboration in navigating and resolving global challenges related to water resources, emphasizing the need for coordinated efforts between nations.

Fig. 4.9 International collaboration for water solutions: navigating global challenges

The following highlights the key aspects and benefits of international collaboration for water solutions.

Data Sharing and Research Collaboration: Collaborative initiatives facilitate the exchange of water-related data, research findings, and best practices across borders. International partnerships improve the collective understanding of water dynamics, contributing to improved modeling, forecasting, and evidence-based decision making.

Transboundary Water Management: Many rivers and aquifers span multiple countries, necessitating cross-border cooperation for effective management. Collaborative frameworks enable the negotiation of equitable water sharing agreements, joint monitoring, and coordinated responses to transboundary water challenges, fostering regional stability and cooperation.

Climate Change Adaptation: The impacts of climate change on water resources are a global concern. Collaborative efforts allow nations to pool resources for climate adaptation strategies, such as developing resilient infrastructure, implementing early warning systems, and sharing technologies for sustainable water use in the face of changing climatic conditions.

Technology Transfer and Capacity Building: International partnerships facilitate the transfer of water-related technologies and expertise between countries. Collaborative projects can focus on capacity building, training programs, and knowledge exchange, empowering nations to address water challenges using the latest innovations and best practices.

Humanitarian Assistance in Water Crises: Collaborative efforts are crucial to responding to water-related humanitarian crises. International organizations and alliances play a key role in mobilizing resources, providing emergency relief, and coordinating efforts to ensure access to clean water after natural disasters, conflicts, or public health emergencies.

Global Governance and Policy Harmonization: Harmonizing water governance policies on a global scale is essential for effective resource management. International collaboration supports the development of common standards, frameworks, and protocols, fostering a shared commitment to sustainable water use and management practices.

Funding and Investment: Collaboration can unlock financial resources for water projects through international funding mechanisms and partnerships. Multilateral development banks, international organizations, and donor countries can pool financial resources to support infrastructure development, research initiatives, and community water projects in regions facing water challenges.

Public Awareness and Education: Collaborative efforts improve global awareness of water issues and promote education on sustainable water practices. International partnerships can leverage communication channels to share success stories, raise public awareness, and inspire collective action toward responsible water stewardship.

Inclusive Decision-Making: International collaboration fosters inclusive decision-making processes that involve diverse stakeholders, including governments, NGOs, academia, and local communities. Inclusive approaches ensure that the perspectives and needs of all stakeholders are considered, leading to more equitable and sustainable water solutions.

Bridging Knowledge Gaps: Water challenges often differ between regions, and collaboration helps bridge knowledge gaps. By bringing together experts from diverse geographical and cultural contexts, collaborative initiatives can foster a comprehensive understanding of water issues and promote context-specific solutions.

In conclusion, international collaboration serves as a cornerstone to address the global water crisis. By fostering partnerships, sharing resources, and collectively tackling water challenges, nations can work toward ensuring equitable access to water, protecting ecosystems, and building a resilient future for generations to come.

4.8 Conclusion

In conclusion of this comprehensive exploration of "Water Resources in India: Challenges and Opportunities," the intricate tapestry of India's water landscape unfolds, revealing a mosaic of challenges and prospects that demand nuanced and collective responses. As we navigate the labyrinth of water scarcity, pollution, and the ever-growing impact of climate change, the imperative for transformation change becomes increasingly apparent. The chapter highlights the indispensable role of communities in water management, from the wisdom embedded in traditional practices to the dynamism of modern community-led initiatives. Technological innovations, including smart meters, precision agriculture, and block chain applications, present a glimpse into a future where data-driven efficiency coexists with sustainable practices. Amidst this, robust policy frameworks stand out as guiding stars, directing the nation toward inclusive, efficient, and sustainable water practices. The international collaboration thread, which recognizes water challenges as a global concern, weaves seamlessly, highlighting the importance of shared solutions and mutual support. As this chapter ends, it is echoed not just as a conclusion, but also as a resonant call to action. It calls for a collective ethos, inspiring innovative thinking, steadfast commitment to water stewardship, and an unwavering commitment to building a sustainable legacy for future generations. This chapter serves as a pivotal moment in India's ongoing water narrative, each initiative contributing to an evolving story of resilience, adaptation, and sustainability, ensuring that the nation's water resources remain a cherished heritage for generations to come.

References

Bandyopadhyay S, Mondal A, Mujumdar PP (2021) Regional strategies for solving inter-state water conflicts in India. Environ Sci Policy 115:61–70. https://doi.org/10.1016/j.envsci.2020.11.003

Bhatia R, Wagle S (2019) Rejuvenation of traditional water harvesting systems in Rajasthan: a case study of Johads in Alwar District. In: Water conservation and harvesting in rural and urban areas. Springer, pp 129–139

Gupta A, Chattopadhyaya MC, Tripathi SN (2020) A review on advancements in the field of water purification technology. Environ Technol Innov 19:100924. https://doi.org/10.1016/j.eti.2020.100924

Hassan Z, Irshad A, Javaid N, Alam MM (2021) A review on the role of block-chain technology in the water sector. J Hydrol 590:125462. https://doi.org/10.1016/j.jhydrol.2020.125462

Kulwant M, Patel D, Shirin S, Sharma SN, Yadav AK (2024) Future frameworks for fluoride and algorithms for environmental system. Advanced treatment technologies for fluoride removal in water: water purification, pp 343–364. https://doi.org/10.1007/978-3-031-38845-3_19

Kumar A, Samadder SR (2022) Assessing the drinking water quality and its impact on human health in rural areas of the Ganges Delta. Chemosphere 287(Pt 2):132102. https://doi.org/10.1016/j.chemosphere.2021.132102

Kumar P, Kumar R, Malaviya P (2021) Water quality assessment and apportionment of pollution sources in the Ganges River, India. Environ Sci Pollut Res 28(2):1752–1770. https://doi.org/10.1007/s11356-020-11149-7

Kumar A, Chava R, Gupta S, Shirin S, Jamal A, Yadav AK (2024) Environmental and health effects of fluoride contamination and treatment of wastewater using various technologies. In: Advanced treatment technologies for fluoride removal in water: water purification. Springer Nature Switzerland, Cham, pp 323–341. https://doi.org/10.1007/978-3-031-38845-3_18

Mishra A, Hopp SB, Wang J (2021) Smart irrigation technologies: a critical review of research needs and approaches. J Clean Prod 319:128536. https://doi.org/10.1016/j.jclepro.2021.128536

Rajeevan M, Srivastava AK, RaviKumar K (2021) Climate change scenarios for the Indian subcontinent: a review. Geosci Front 12(5):101381. https://doi.org/10.1016/j.gsf.2020.101381

Raju KV, Soundararajan B, Shah T (2019) Water management in India: policy reforms, institutional structure, and emerging challenges. Econ Polit Wkly 54(44):39–46. Retrieved from https://www.epw.in/engage/article/water-management-india

Saleh N, Gupta VK, Qamar MTU (2022) Water pollution in developing countries and its management. Curr Pollut Rep 8(1):36–46. https://doi.org/10.1007/s40726-021-00181-0

Sharma R, Gupta M (2017) Application of remote sensing in hydrology and water management: a review. J Hydrol: Reg Stud 9:174–193. https://doi.org/10.1016/j.ejrh.2016.07.003

Sharma R, Shukla A, Chakraborty D (2022) Precision agriculture for sustainable water management: a review. Environ Res Lett 17(1):013001. https://doi.org/10.1088/1748-9326/ac5df4

Shirin S, Jamal A, Emmanouil C, Singh VP, Yadav AK (2023) Assessment and characterization of waste material used as backfilling in an abandoned mine. Int J Coal Prep Util 43(8):1402–1410. https://doi.org/10.1080/19392699.2022.2118259

Singh A, Mishra A (2019) Water resources in India: current status and future challenges. J Environ Manage 248:109227. https://doi.org/10.1016/j.jenvman.2019.109227

Thenkabail PS, Lyon JG, Huete A (2017) Hyperspectral remote sensing of vegetation. CRC Press

Tiwari VM, Wahr J, Swenson S (2009) Terrestrial water storage contributions to sea level rise and variability. J Geophys Res: Solid Earth 114(B9):B09406. https://doi.org/10.1029/2008JB006048

Tiwari VM, Wahr J, Swenson S (2020) Dwindling groundwater resources in northern India, from satellite gravity observations. Geophys Res Lett 47(19):e2020GL088445. https://doi.org/10.1029/2020GL088445

Tripathi SN, Gurjar BR, Chauhan A (2019) Advanced oxidation processes for water treatment and reuse: emerging opportunities, challenges, and future directions. Environ Rev 27(3):295–317. https://doi.org/10.1139/er-2018-0094

Yadav AK, Sahoo SK, Mahapatra S, Kumar AV, Pandey G, Lenka P, Tripathi RM (2014) Concentrations of uranium in drinking water and cumulative, age-dependent radiation doses in four districts of Uttar Pradesh, India. Toxicol Environ Chem 96(2):192–200. https://doi.org/10.1080/02772248.2014.934247

Chapter 5
Challenges and Opportunities in the Management of India's Water Resources

Rashmi Raghav, Kamal Kant Tiwari, and Pankaj Kandwal

Abstract The issues of climate change have a direct influence on agriculture and ecology. Also, the foundation of water is an important resource for life support and a variety of socioeconomic activities. When considering India, a nation with a rapidly growing population and a varied topography, water resources are crucial in determining the course of the country's growth. In this abstract an overview of the potential and problems related to India's management of its water resources is given.

Challenges:

- Water shortage is a problem in many parts of India due to the rapid expansion of the population, urbanization, and industry that increases the water demand. The problem is further complicated by the unequal distribution of water resources, which results in differences in the availability of water in various states.
- Impact of Climate Change: The disruption of conventional precipitation patterns due to climate change has resulted in unpredictable rainfall, prolonged droughts, and more violent floods. This increased the demand for water supplies.
- Groundwater Depletion: Aquifers, especially in agricultural areas, are being depleted due to excessive groundwater extraction for home and agricultural use. Uncontrolled extraction methods increase the long-term risk to a sustainable water supply by lowering water tables.
- The quality in rivers and groundwater sources has been affected by pollution from industrial discharges, agricultural runoff, and inadequate sewage treatment. Infrastructure for water conservation and treatment must be invested because contaminated water poses health dangers.

Opportunities:

- Management: There is a chance to maximize water usage and reduce scarcity by putting into practice effective water conservation strategies such as rainwater collection, watershed management, and effective irrigation techniques.

R. Raghav (✉) · K. K. Tiwari · P. Kandwal
Department of Chemistry, National Institute of Technology Uttarakhand, Srinagar, Garhwal 246174, India
e-mail: raghavrashmi13@gmail.com

A. K. Yadav et al. (eds.), *Integrated Management of Water Resources in India: A Computational Approach*, Water Science and Technology Library 129,
https://doi.org/10.1007/978-3-031-62079-9_5

- Technology adoption: Using cutting-edge technologies may improve water resource management, monitor consumption, and support data-driven decision making. Examples of these technologies include sensor-based irrigation, smart water grids, and remote sensing.
- Reforming policies: All-encompassing water policies that address issues of price, governance, and water distribution can support sustainable water management. Long-term water security can be attained via fostering equitable distribution and community involvement.
- Renewable Energy Integration: Investigating the mutual benefits of renewable and water sources, such as solar-powered desalination and hydropower technologies can offer long-term solutions to problems related to energy and water.

In essence, the management of water resources in India requires a multifaceted strategy that makes use of both the benefits and the existing problems. Sustainable growth in the nation and water security depends on innovative technology, innovative policy, and well-thought-out planning.

Keyword Water resources · Groundwater depletion

5.1 Introduction

Water is a precious and indispensable resource that sustains life and supports various sectors of the economy. In the context of India, a country known for its diverse geographical and climatic conditions, the management of water resources is of paramount importance. India faces a unique set of challenges and opportunities related to its water resources, and understanding these dynamics is crucial to ensuring sustainable development and the well-being of its population (Shiklomanov et al. 2003; Yadav et al. 2014).

Geographically, India is endowed with an extensive network of rivers, lakes, and groundwater reservoirs, making it one of the most water-rich nations. However, the distribution of water resources across regions and seasons is uneven, leading to challenges such as water scarcity and overexploitation in certain areas. The intricate interplay of climate change, population growth, urbanization, and agricultural practices further complicates the landscape of water management in the country (Shah et al. 2014; Kulwant et al. 2024; Patel et al. 2024).

This introduction will explore the key aspects of water resources in India, shedding light on the challenges faced and the opportunities that lie ahead. From traditional water harvesting practices to modern technological interventions, India's approach to water resource management is evolving, reflecting the dynamic nature of the country's socio-economic landscape. By examining these factors, we can gain insight into the complexities of water governance in India and the efforts undertaken to achieve a sustainable and equitable water supply (Bharati et al. 2014; Shirin et al. 2014, 2019; Yadav et al. 2014).

5.2 Surface and Subsurface Water Assets in Indian River Basins

India's water resources are largely characterized by surface water bodies, including rivers, lakes, and reservoirs, as a well as significant reliance on groundwater. The distribution and availability of these resources vary across different river basins in the country, influencing water management strategies and the overall water security scenario (Kumar et al. 2015). Here, we will briefly discuss the surface and groundwater resources in some of the major river basins in India.

1. Ganges–Brahmaputra-Meghna (GBM) Basin:
   - *Surface water:* The Ganges, Meghna, and Brahmaputra rivers form one of the most extensive river basins worldwide. The Ganges, in particular, is a major source of surface water for eastern and northern India, both.
   - *Groundwater:* The basin also has substantial groundwater resources, with the northern plains highly dependent on aquifers for agriculture and domestic use (Mishra et al. 2003).
2. Indus Basin:
   - *Surface water:* The Indus River and its tributaries provide vital surface water resources to northwest India, particularly in the states of Punjab, Jammu and Kashmir, Himachal Pradesh.
   - *Groundwater:* Groundwater is crucial for supplementing irrigation needs in this basin, especially in arid and semi-arid regions (Postel et al. 2003).
3. Brahmaputra Basin:
   - *Surface water:* The Brahmaputra, which originates in Tibet and flows through northeastern India, contributes significantly to surface water resources in the region, supporting agriculture and ecosystems (Rabindranath et al. 2004; Rodell et al. 2009).
   - *Groundwater:* Groundwater is also an essential component, especially in the lower Brahmaputra basin.
4. Yamuna Basin:
   - *Surface water:* The Yamuna, a major tributary of the Ganges, is an important surface water source for the northern plains, including Delhi and parts of Uttar Pradesh and Haryana.
   - *Groundwater:* Like many other basins, the Yamuna basin relies on groundwater for various purposes, including agriculture and urban water supply (Rabindranath et al. 2004; Rodell et al. 2009).
5. Godavari Basin:

- *Surface water:* Godavari, the second longest river in India, is a significant source of surface water for the states of Maharashtra, Telangana, Andhra Pradesh, and Chhattisgarh.
- *Groundwater:* Groundwater resources are extensively used in the Godavari basin for irrigation and drinking water supply.

6. Krishna Basin:

   - *Surface water:* The Krishna River is a vital surface water source for the states of Maharashtra, Karnataka, Telangana, and Andhra Pradesh.
   - *Groundwater:* Groundwater is utilized to supplement surface water for agricultural and domestic purposes (Shah et al. 2009).

7. Cauvery Basin:

   - *Surface water:* The Cauvery River is a key surface water resource for Karnataka, Tamil Nadu, and parts of Kerala and Puducherry.
   - *Groundwater:* Groundwater plays a crucial role in maintaining agriculture and meeting domestic water needs in the Cauvery basin (Shah et al. 2009).

5.3 Water Requirements of India

India's water demands are intricate and varied, influenced by factors such as population growth, agricultural practices, industrial development, and climate fluctuations.

5.3.1 Agriculture

India's need for agricultural water has always been a big, complicated problem. A significant percentage of the workforce is employed in agriculture, which is a vital component of the Indian economy. However, water shortages, poor water resource management, and the rising demand for water resources pose problems for the industry, nevertheless (Mukherji et al. 2005; Kumar et al. 2005).

Several factors contribute to the high agricultural water demand in India are as follows.

1. Traditional irrigation methods: In India, a large number of farmers continue to use outdated and ineffective irrigation techniques such as flood irrigation, which can use an excessive amount of water.
2. Crop Patterns: The crops chosen also have an impact on water demand. Crops that require a lot of water, such as sugarcane and rice, are a major cause of excessive water use. This problem can be addressed by supporting effective irrigation methods or the growth of less water-intensive crops.

3. Population Growth: The increasing demand for food due to population growth puts pressure on agriculture to produce more, which also increases the water requirement for irrigation (Grafton et al. 2018).
4. Climate Change: Changes in temperature and precipitation patterns caused by climate change can have an influence agriculture and water supply. For instance, modifications to monsoon patterns may have an effect on the amount and timing of rainfall, which may change the amount of water needed for crops (Kumar et al. 2015).
5. Water Management Practices: The total water demand is influenced by poor water management techniques, such as waste and ineffective delivery networks.

5.3.2 Industrial Sector

Water consumption in India's industrial sector is substantial and a number of factors have an impact on the water demand of this sector. Industries need water for operations including cleaning, cooling, and manufacturing. India's industrial water demand is plagued by issues such as resource rivalry, water shortages, and the requirement for sustainable water management techniques (Boyd et al. 2015).

5.3.3 Urban and Rural Domestic Consumption

In India, the amount of water required for residential use varies greatly between urban and rural locations. These differences are caused by a number of variables, including population density, infrastructural growth, and accessibility to sanitation facilities and water supplies (Gray et al. 2010).

Urban Domestic Water Demand

1. Population Density: India's urban regions often have higher population densities than rural ones, which raises the need for household water. The density of residential and commercial buildings in cities and towns results in a greater need for water (Chapman et al. 1996).
2. Infrastructure Development: Compared to rural regions, urban areas usually have more advanced water supply and distribution networks. However, current water infrastructure may be strained by rapid urbanization and population expansion (Eaton et al. 2005).
3. Lifestyle and Standard of Living: Urban homes often have better living standards and water-intensive fixtures, such as dishwashers, showers, and washing machines. This leads to an increase in the amount of water used per person (Sawyer et al. 2002).

4. Demand from the business and Industrial Sectors: In addition to meeting residential demands, urban areas also facilitate business and industrial operations, increasing the total demand for water.
5. Leakage and Wastage: In some metropolitan locations, aging water infrastructure can lead to leakage and water waste, which reduces the overall effectiveness of water delivery systems (Metcalf and Eddy et al. 2013).

Rural Domestic Water Demand

1. Population distribution: Rural regions, which are made up of scattered communities, make up a sizable section of India's terrain. Rural families may have less individual water demands, but the total demand of a large number of scattered households is substantial.
2. Limited Infrastructure: Access to reliable and secure water delivery systems may be restricted in rural locations. The inability to satisfy household water demands may result from the dependence on conventional water sources such as ponds and wells.
3. Awareness and Education: To sustainably manage water demand in rural regions, programs focused on cleanliness, water conservation, and effective water usage are essential (United States Environmental Protection Agency (USEPA) 2021).

5.4 Parameters of Water Quality

The level of water quality in the country can be used to estimate the degree of water contamination. Countless contaminants, such as trace metals, pesticides, inorganic ions, and other micropollutants, are found in water bodies. Many characteristics of water quality serve as the foundation for the water quality criterion. Water quality guidelines sometimes set at least allowable concentrations, while other times they define a maximum limit for a substance's concentration in a certain medium beyond which it is dangerous. These standards are often created for raw water and its relevant applications in irrigation, drinking, household use, aquatic ecosystems, and habitats for biological species (Jal Shakti Ministry of Agriculture, Government of India). The Central Pollution Control Board (CPCB) has divided water resources into five classes (AE) based on how they are intended to be used to establish water-quality objectives (Table 5.1) (Hester et al. 2008). These five classifications have been used along the Yamuna and Ganga to establish goals and carry out surveys to compare their quality with that which is necessary for the greatest possible use. With the assessment in hand, it is possible to identify any shortcomings and develop appropriate plans to mitigate pollution from the main sources. A range of factors in water quality, including pH, DO, BOD, TC, and others, are shown in Table 5.1 over a duration of 5–7 days. Total coliform (TC) and fecal coliform (FC) are indicators of the presence in water, whereas biochemical oxygen demand (BOD) and dissolved oxygen (DO) are used to quantify organic pollution (Hester et al. 2008; India Water Portal, United Nations (2019). World Water Development Report 2019,2020 United Nations 2020).

Table 5.1 Water-quality indicators

Use of water	pH	DO (mg/L)	TC (MPN/100 mL)	BOD
Parameter				
Drinking water source without conventional treatment but after disinfection	6.5–8.5	≥ 6	≤ 50	≤ 2
Outdoor bathing	6.5–8.5	≥ 5	≤ 500	≤ 3
Drinking water source with conventional treatment and disinfection	6.5–8.5	≥ 4	≤ 5000	≤ 3
Propagation of wildlife and fisheries	6.5–8.5	≥ 4	Free ammonia ¼ ≤ 1.2 mg/L	
Irrigation and industrial purposes	6.0–8.5	Electrical conductivity ¼ ≤ 2250 mʊ/cm Boron content ¼ ≤ 2.0 mg/L Sodium absorption ratio ¼ ≤ 26		

DO—dissolved Oxygen
TC—total coliform bacteria
BOD—biological oxygen demand

5.5 Factors Contributing to the Degradation of Water Quality

Factors that contribute to the degradation of water quality can be diverse and are often influenced by natural processes and human activities. Here are some key factors:

1. Industrial discharges: Industrial activities release pollutants into water bodies through discharges of chemicals, heavy metals, and other harmful substances, contributing to water pollution (Hudson et al. 2008; Allan et al. 2007).
2. Agricultural runoff: The use of fertilizers, pesticides, and herbicides in agriculture can result in runoff that carries these chemicals into water sources, affecting water quality and posing risks to aquatic ecosystems.
3. Urban runoff: Urbanization leads to increased impervious surfaces such as roads and buildings, preventing water from infiltrating the soil. This results in increased surface runoff, which transports pollutants such as oil, heavy metals, and debris into water bodies.
4. Wastewater Disposal: Improper disposal of untreated or partially treated wastewater from households, industries, and wastewater treatment plants can introduce pathogens, nutrients, and pollutants into water sources (Dodds et al. 2010).
5. Deforestation: The removal of natural vegetation, such as forests, reduces the ability of the land to filter and retain water. This can lead to increased runoff, soil erosion, and sedimentation in rivers and lakes, affecting water quality.

6. Mining Activities: Mining operations can introduce sediments, heavy metals, and chemicals into nearby water bodies, causing contamination and altering the chemistry of the water.
7. Oil spills: Accidental or deliberate release of oil into water bodies can have severe consequences for water quality, impacting aquatic ecosystems, and causing long-term damage.
8. Climate Change: Altered precipitation patterns, temperature changes, and extreme weather events associated with climate change can affect water quality. For example, increased temperatures can influence the growth of harmful bacteria in water.
9. Groundwater Contamination: Improper disposal of hazardous waste, leaking underground storage tanks, and certain agricultural practices can lead to contamination of groundwater, affecting both surface water and drinking water supplies.
10. Eutrophication: Excessive nutrient runoff, primarily phosphorus and nitrogen from agricultural activities and wastewater, can cause eutrophication in water bodies. This leads to algae blooms, oxygen depletion, and harm to aquatic life.
11. Illegal Dumping: The improper disposal of household and industrial waste directly into water bodies contributes to pollution and degradation.
12. Invasive Species: The introduction of non-native species to water ecosystems can disrupt natural balances and contribute to deterioration of water quality.
13. Lack of environmental regulation: Weak or inadequate environmental regulations and enforcement can allow irresponsible practices that contribute to water pollution to go unchecked (Davies et al. 2003).

Numerous activities responsible for water-quality deterioration are as follows in Fig. 5.1.

5.6 Responsibilities of Various Agencies

Table 5.2 lists the several agencies that were established to evaluate the quality of the water along with their corresponding duties.

5.7 Challenges and Opportunities

Pollution can be absorbed by water up to a point where it does not affect its quality. However, as the natural regeneration capacity of water has been surpassed by the pollution loads caused by various human activities, we must recognize the problems facing the current system and develop comprehensive, human-centered solutions for water quality and sustainability (Jal Jeevan Mission, Ministry of Jal Shakti. (Official website).

Fig. 5.1 Numerous activities accountable for water-quality deterioration

5.7.1 Water Stress and Scarcity

In India, water shortage and stress pose serious problems for both the agricultural and non-agricultural sectors. The nation's water shortage and stress are caused by a number of issues, including (Famiglietti et al. 2014):

1. Unequal Distribution: Different parts of India have varying levels of access to water resources. While some areas have an abundance of water resources, others don't. Parts of Maharashtra, Rajasthan and Gujarat are among the states most susceptible to water shortages (Wada et al. 2012).
2. Population Growth: Water supplies are being strained by India's rapidly expanding population. Water stress is exacerbated by increasing urbanization and industry, which also increase water consumption.
3. Agricultural Water Use: India's agriculture uses a lot of water, and conventional irrigation techniques are not always effective. Water resources are under stress due to the high water demand for agriculture caused by water-intensive crops and antiquated irrigation techniques (Gleick et al. 2014).
4. Climate Change: The water supply may be impacted by changes in rainfall patterns and weather patterns caused by climate change. This includes higher temperatures, more frequent extreme weather events, and changed monsoon patterns (Ducrotoy et al. 2001).
5. Groundwater Depletion: Aquifers are depleted in several regions of India as a result of excessive groundwater extraction for home and agricultural uses. Water stress over an extended period of time can be caused by an overreliance on

Table 5.2 Multiple agencies' responsibilities in the water-quality assessment

Concerned Ministry	Name of agency	Responsibility
Ministry of Water Resources (MoWR)	Central Water Commission (CWC)	Handles all surface water-related issues; controls how surface water is used for business, agriculture, and drinking water; and resolves disputes over interstate water allocation
Ministry of Water Resources (MoWR)	Central Groundwater Board (CGWB)	Develops strategies for recharge augmentation, provides scientific inputs for groundwater resource management, research, periodic evaluation, and regulation
Ministry of Rural Development (MoRD)	Rajiv Gandhi National Drinking Water Mission (RGNDWM) (under the Department of Drinking Water Supply)	Establishes standards, creates regulations, and gives the state's financial support and technical help for initiatives related to rural water supply and sanitation
Ministry of Urban Development (MoUD)		Formulates regulations and offers direction to the urban water supply and sanitation industry
Ministry of Health and Family Welfare	Central Bureau of Health Intelligence (CBHI)	Gathers, aggregates, evaluates, and distributes data about health issues brought on by waterborne illnesses
Ministry of Agriculture (MoA)		Participates in the conception, formulation, monitoring, and evaluation of projects aimed at improving watersheds
Ministry of Environment and Forests (MoEF)	Central Pollution Control Board (CPCB)	Promotes pollution management techniques and conducts studies regarding the state of pollution, puts national environmental laws into effect and establishes criteria
Ministry of Environment and Forests (MoEF)	State Pollution Control Board (SPCB)	Demands data from all industries regarding their adherence to the CPCB's acts and has the authority to penalize noncompliant businesses
Ministry of Environment and Forests (MoEF)	National Rivers Conservation Directorate (NRCD)	Puts ideas into action to enhance India's river quality

groundwater without adequate recharging systems (Central Ground Water Board (CGWB). (Official Website); Konikow et al. 2005; Scanlon et al. 2012).

6. Urbanization and Industrialization: As a result of rapid industrial development and urbanization, there is a greater need for water in these regions. Water supply systems are frequently under stress due to growing urban populations and industrial activity (Central Groundwater Board (CGWB). (Official Website)).
7. Pollution: The amount of potable water is further reduced by pollution from industrial discharges, untreated sewage, and agricultural runoff. Water stress is exacerbated by contaminated water sources, which can pose health hazards.

5.7.2 Water Pollution

Water resource regulations must be continuously evaluated to meet the challenge of water pollution, which has become a worldwide concern. Around 14,000 people worldwide die from the impacts of water pollution every day, demonstrating the impact of this problem on health and mortality rates (Srikanth et al. 2009; Sharma et al. 2014; Kathpalia et al. 2020). Water contamination is an issue in both industrialized and developing countries (Kumar et al. 2005). Water quality is influenced by a number of variables, including precipitation, climate, soil type, vegetation, geology, flow conditions, ground water, and human activity. The main danger to the quality of the water comes from point sources, such as towns and businesses. In addition, mining, urbanization, and agriculture have a big impact on the quality of water. Furthermore, fertilizers, sediments, and hazardous pollutants are introduced via nonpoint source pollution (Chatterjee et al. 2014).

The classification of various sources of pollution is shown in Fig. 5.2.

The sources of water contaminant and their impact on human health are shown in Table 5.3.

5.7.3 Depletion of Groundwater

The natural rate of recharge is exceeded by the pace of groundwater extraction, resulting in a decrease in the groundwater table. Groundwater depletion is a major global problem. There are major environmental, social, and economic issues associated with this occurrence. In many areas, groundwater is a key supply of water for industrial, agricultural, and drinking needs; therefore, its appropriate management is critical (Konikow et al. 2005; Scanlon et al. 2012).

The following are important causes of groundwater depletion:

1. Over-Extraction: Without adequate control and management, excessive groundwater extraction for industrial, drinking and agricultural purposes can result in depletion.

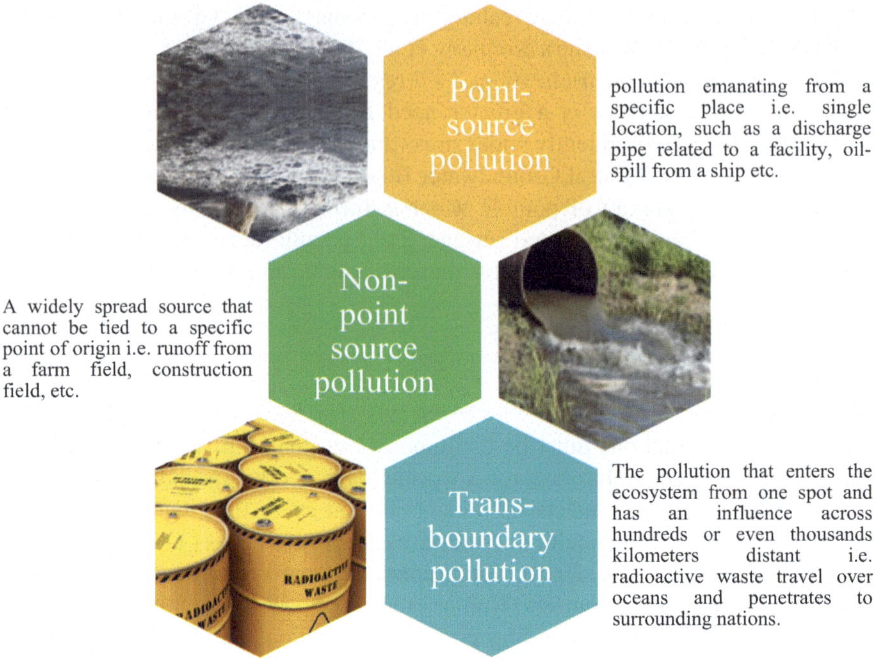

Fig. 5.2 Classification of various sources of pollution

2. Agricultural Practices: Especially in areas with limited access to surface water, agriculture is a significant user of groundwater. Overextraction of groundwater is a result of uncontrolled and ineffective irrigation techniques (Konikow et al. 2005; Scanlon et al. 2012).
3. Urbanization: Cities that depend heavily on groundwater without sufficient recharge systems may experience depletion as a result of the increasing demand for water that comes with rapid urbanization (Konikow et al. 2005; Scanlon et al. 2012).
4. Lack of Recharge: Rainfall and surface water penetration are two natural processes that help recharge groundwater. Urbanization, land use changes, and deforestation may reduce a land's capacity to absorb water and replenish aquifers.
5. Climate Change: The distribution and availability of groundwater may be affected by variations in temperature and precipitation patterns caused by climate change.
6. Industrial Demand: When extraction rates are excessive and are not balanced with recharge, industries that primarily rely on groundwater for their operations may be responsible for its depletion.

Table 5.3 Water contaminant sources and their impact on human health

Element	Source/cause	Health impact
Fluoride	Due to contamination of groundwater with fluorine rich rocks	High concentrations can result in anemia, skin conditions, digestive problems, skeletal fluorosis (severe bone deterioration), dental fluorosis (tooth enamel discoloration), stiffness in the back, and difficulties moving naturally
Nitrate	Application of manure and fertilizer rich in nitrogen, nitrogen compounds released into the air by vehicles and industries, and septic and sewage discharge	Causes of Blue Baby illness and raises cancer risk
Arsenic	Because to the extensive groundwater excavation, geogenic contamination	Acute symptoms include diarrhea, esophageal and stomach discomfort, and vomiting Chronic consequences include pigmentation changes, skin, lung, kidney, and urinary bladder cancer, as well as skin sores that can progress to arsenicosis
Iron	Because of the groundwater's absorption of iron from soil (in heavy rainfall areas)	Damage to blood tissues, skin problems, tooth troubles, and digestive ailments
Heavy metals	Owing to effluents from industry	Cause harm to the kidneys, central nervous system, and other metabolic disturbances
Persistent organic pollutants (POPs)	From both industrial sources and agricultural fertilizers	Growth retardation, hormonal disruption, and hypertension
Salinity	Excessive groundwater extraction and saltwater seepage	Water with an unpleasant flavor that interferes with osmotic flow and fluid movement
Pesticides	From both industrial sources and agricultural fertilizers	Reduced immunity, development of tumors, harm to the reproductive system, and endocrine system
Coliform bacteria such as Escherichia coli	Because the water is contaminated by sewage	Cause cholera, gastroenteritis, typhoid, and jaundice, among other gastrointestinal illnesses induce gastrointestinal inflammation, nausea, vomiting, and fever

5.8 Impact on Water Resource

Numerous factors related to water resources are affected by climate change, including distribution, availability, quality, and total demand for water. These effects differ between geographical areas and are frequently linked to additional environmental and socioeconomic factors (IPCC (Intergovernmental Panel on Climate Change) 2014; Bierkens et al. 2015). The following are some significant ways in which water demand is impacted by climate change.

1. Modified precipitation patterns: Water availability can be impacted by modifications to precipitation patterns, such as increased unpredictability and modified rainfall timing. Extended dry spells or heavy rains can cause shortages or flooding, affecting the availability and use of water.
2. Increasing Temperatures: As a result of global warming, there may be a rise in evaporation rates and a corresponding rise in the amount of water needed for and agricultural, home, and industrial purposes. Temperature increases may also have an impact on the operation of water distribution and treatment systems.
3. Changes in Snowmelt Timing: In locations depending on snowmelt for water supply, changes in the timing and amount of snowmelt owing to warmer temperatures could impact the availability of water during particular seasons, influencing agricultural and hydroelectric water demand (Wada et al. 2020).
4. Sea level rise: Coastal regions are susceptible to rising sea levels, which can cause freshwater supplies to become contaminated with saltwater. The quality of water supplies may be affected by this incursion, requiring the use of different sources or treatment techniques (Van Dijk et al. 2013).
5. Extreme weather events: Storms, hurricanes, and cyclones are examples of more frequent and powerful extreme weather events that can interfere with water distribution and supply networks, increasing the need for emergency response and recovery work.
6. Impact on Agriculture: Crop yields, growth seasons, and patterns are all impacted by climate change. Crop water requirements can be changed by variations in temperature and precipitation patterns, which can impact the demand for agricultural water.
7. Changes in Ecosystems: The accessibility and quality of water in rivers and lakes can be impacted by ecosystems due to climate change. This in turn may have an impact on the need for water for a variety of uses, such as urban supplies, industry, and agriculture.
8. Melting Glaciers: The long-term supply of water can be affected by the melting of glaciers brought on by global warming in areas where glaciers are a major source of freshwater. Variations in river flows fed by glaciers can impact the amount of water needed for hydropower and agriculture (Huntington et al. 2006).

5.9 Water Conservation Efforts

To solve challenges connected to water scarcity and depletion and maintain the sustainable use of water resources, conservation activities are essential. To save water, communities, businesses, governments, and people may contribute. Here are a few typical initiatives to conserve water:

1. Public awareness campaigns: These are attempts to increase knowledge about the value of conserving water, the effects of water shortages, and easy ways people may save water. Events in the community, media campaigns, and educational initiatives in schools may be useful resources for raising awareness.
2. Efficient irrigation practices: Agriculture is a major water user. By encouraging and implementing water-efficient irrigation techniques such as drip irrigation and sprinkler systems, water use in agriculture may be greatly reduced.
3. Water-Efficient Technologies: Promoting the use of water-efficient washing machines, low-flow toilets, and other water-saving equipment in homes and businesses, as well as industrial procedures that reduce water usage.
4. Rainwater Harvesting: Installing systems to collect and store rainwater for use in landscaping, irrigation, and nonpotable household requirements at both the individual and communal levels.
5. Leak Detection and Repair: Water waste can be avoided by routinely inspecting water supply systems for leaks and quickly fixing any that are found. This is true for both private homes and the systems that deliver water to cities.
6. Greywater Recycling: The need for fresh water can be decreased by treating and recycling graywater, or wastewater from domestic activities other than toilet flushing, for nonpotable uses like watering plants or cleaning toilets.
7. Xeriscaping: sing native plants that withstand drought to create a landscape that requires less water and less irrigation. Using mulch and other water-saving landscaping techniques is another aspect of xeriscaping.
8. Water Pricing and Metering: Putting in place water pricing systems that encourage conservation and accurately represent the cost of water. Water meters are installed to monitor water consumption and promote optimal use for both homes and businesses.
9. Wastewater Treatment and Reuse: Reusing treated wastewater for irrigation, cooling systems, and industrial activities, as well as for nonpotable uses, helps lower the demand for fresh water.
10. Policy and Regulations: Governments have the authority to enact and enforce laws and policies related to water conservation. Limitations in water consumption limitations, appliance water efficiency, and financial incentives for companies and people to use water saving technology are a few examples of this.
11. Community Engagement: Including nearby communities in programs to save water, promoting teamwork, and cultivating a feeling of accountability for water resources.
12. Educational Programs: Including lessons on water conservation in the school curriculum and hosting seminars to instruct on sustainable water use techniques.
13. Incentive Programs: These entice people and companies to participate in water conservation measures by providing cash incentives or refunds for the adoption of water-efficient technology and practices.

14. Corporate Responsibility: Industry can assist in water conservation through the adoption of sustainable water management techniques, the reduction of water consumption during production processes, and the acquisition of water-efficient technology.

Benefits of wastewater recycling and reuse and tips for the optimal use of water is depicted in Figs. 5.3 and 5.4.

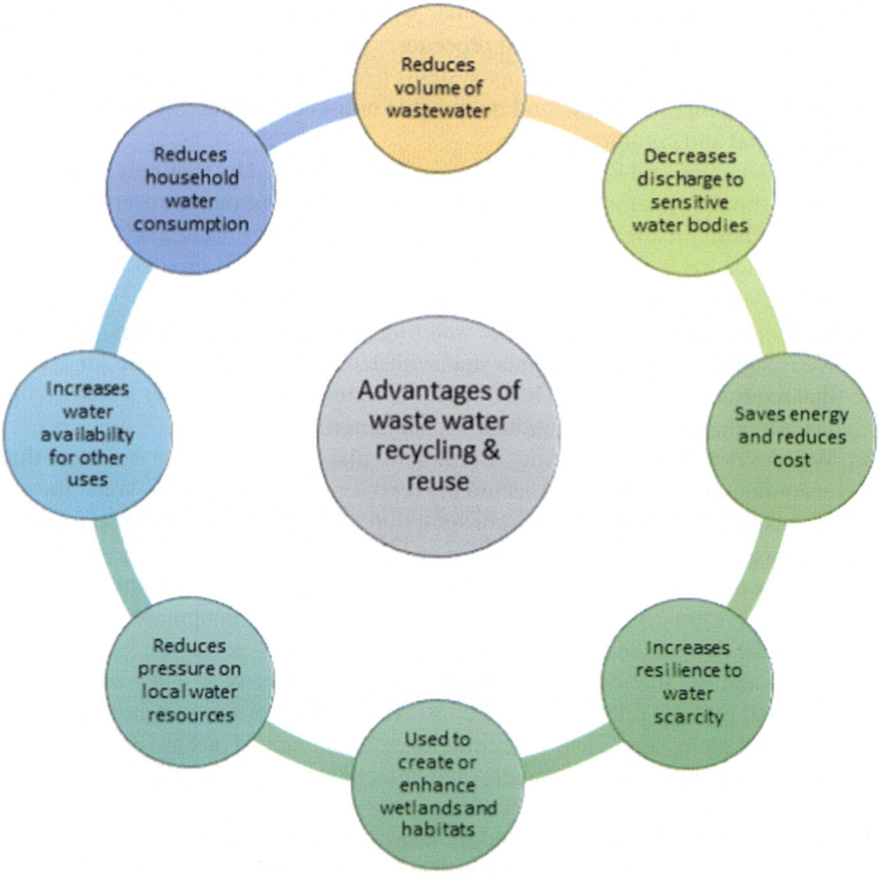

Fig. 5.3 Benefits of wastewater recycling and reusing

Fig. 5.4 Tips for the optimal utilization of water

5.10 Government Initiatives

To alleviate water demand, encourage sustainable water management, and alleviate water-related issues, the Indian government has implemented a number of projects and programs. These programs include a wide range of industries, such as environmental preservation, rural communities, urban development, and agriculture. Important government programs related to India's water requirements include:

1. National Water Mission (NWM): The National Water Mission (NWM) was established as a component of the National Action Plan on Climate Change. Its objectives are to ensure a more effective use of water in various sectors, preserve water, and reduce waste.
2. Pradhan Mantri Krishi Sinchai Yojana (PMKSY): Increasing irrigation coverage and enhancing agricultural water management techniques, this program aims to increase farm output and water usage. The Per Drop More Crop (PDMC) component and the Accelerated Irrigation Benefit Program (AIBP) are examples of components.

3. Jal Jeevan Mission (JJM): JJM prioritizes the use of nearby water sources, water conservation, and community participation in water management. It was established with the aim of providing clean and drinkable water to all rural families by 2024.
4. Atal Bhujal Yojana (ABHY): The Atal Bhujal Yojana (ABHY) aims to manage groundwater resources sustainably, encourage community involvement, and promote effective groundwater use in regions that have been classified as important and overexploited.
5. National Rural Drinking Water Programme (NRDWP): This program aims to supply clean drinking water to rural regions, focusing on building community capacity, establishing infrastructure, and water quality monitoring.
6. Smart Cities Mission: To guarantee sustainable urban water management, some Indian cities are implementing programs pertaining to stormwater management, sewage treatment, and water supply.
7. Swachh Bharat Mission (SBM): Although the Swachh Bharat Mission is primarily concerned with sanitation, it also addresses water-related concerns by encouraging behaviours that improve water quality via cleanliness and sanitation.
8. National River Conservation Plan (NRCP): The National River Conservation Plan (NRCP) seeks to enhance the quality of water from India's major rivers by implementing watershed management, sewage treatment, and pollution control strategies.
9. Peri-Urban, Rural and Urban Water Supply Schemes: The government funds several initiatives, such as piped water delivery projects and water quality monitoring, to improve the infrastructure for water distribution in both rural and urban regions.
10. Mahatma Gandhi National Rural Employment Guarantee Act (MGNREGA): MGNREGA has provisions for water management and conservation efforts to improve rural livelihoods, however, it is not only focused on water.

5.11 Conclusion

In summary, India's water resources situation is at a crucial crossroads, characterized by a careful balancing act between possibilities and problems. The nation faces challenges that include pollution, interstate conflicts, groundwater depletion, and water shortages, all of which require decisive action and immediate response. Nevertheless, there are many chances for progress among these difficulties. India can have a more resilient and equitable water future by adopting sustainable water management methods, using technology developments, implementing effective legislation into place, and encouraging community participation. International involvement and cooperation may improve the overall efficacy of water resource management and promote regional stability, particularly when it comes to shared river basins.

Additionally, the necessity for climate-resilient measures cannot be emphasized, considering the unpredictable influence of climate change on water distribution.

In the end, managing the intricate nature of India's water resources requires an all-encompassing and comprehensive strategy that incorporates responsible governance, policy changes, community involvement and scientific innovation. India can strive toward sustainable water security, guaranteeing a higher standard of living for its citizens, and protecting its ecosystems for future generations by taking advantage of the possibilities present themselves and putting smart solutions.

Acknowledgements The authors thank NIT Uttarakhand for supporting this.

References

Allan JD, Castillo MM (2007) Stream ecology: structure and function of running waters. Springer

Bharati L, Ramanathan AL (2014) A review of challenges and opportunities in estimating water use in India. Water Resour Manage 28(4):999–1013

Bierkens MF, Bell VA, Burek P, Chaney N, Condon LE, David CH, de Roo A (2015) Modeling the role of groundwater in the larger land surface system: a white paper of the NSF-supported community workshop modeling the water cycle. Link

Boyd CE (2015) Water quality: an introduction. Springer

Central Ground Water Board (CGWB). (Official Website). Link

Central Water Commission (CWC). (2018). Water and related statistics. Government of India

Chapman D (1996) Water quality assessments: a guide to the use of biota, sediments and water in environmental monitoring. CRC Press

Chatterjee K (2014) Water resources of India

Davies JM, Mazumder A (2003) Health and environmental policy issues in Canada: the role of watershed management in sustaining clean drinking water quality at surface sources. J Environ Manage 68(3):273–286

Dodds WK, Whiles MR (2010) Freshwater ecology: concepts and environmental applications. Academic Press

Ducrotoy JP (ed) (2001) Water pollution control: a guide to the use of water quality management principles. E & FN Spon

Eaton AD, Clesceri LS, Rice EW, Greenberg AE (eds) (2005) Standard methods for the examination of water and wastewater. American Public Health Association (APHA)

Famiglietti JS (2014) The global groundwater crisis. Nat Clim Chang 4(11):945–948

Gleick PH (2014) Water, drought, climate change, and conflict in Syria. Weather, Clim Soc 6(3):331–340

Grafton RQ, Williams J, Perry CJ, Molle F, Ringler C, Steduto P, Udall B (2018) The paradox of irrigation efficiency. Science 361(6404):748–750

Gray NF (2010) Water technology: an introduction for environmental scientists and engineers. Elsevier

Hester RE, Harrison RM (eds) (2008) Water quality: guidelines, standards and health. Royal Society of Chemistry

Hudson LA, Baker DB, Capel PD, Weller DE (2008) Trends in pesticides in streams of the United States, 1992–2003. J Am Water Resour Assoc 44(6):1308–1329

Huntington TG (2006) Evidence for intensification of the global water cycle: review and synthesis. J Hydrol 319(1–4):83–95

Metcalf & Eddy, Inc., Tchobanoglous G, Burton FL, Stensel HD (2013) Wastewater engineering: treatment and resource recovery. McGraw-Hill

India Infrastructure Report (2011) Water: policy and performance for sustainable development. Oxford University Press

India Water Portal (Website). Link

IPCC (Intergovernmental Panel on Climate Change) (2014) Climate change 2014: impacts, adaptation, and vulnerability. In: Working Group II contribution to the fifth assessment report of the intergovernmental panel on climate change

Jal Jeevan Mission, Ministry of Jal Shakti. (Official Website). Link

Kathpalia G, Kapoor R (2002) Water policy and action plan for India 2020: an alternative. Water Policy Action Plan 2020:14e7

Konikow LF, Kendy E (2005) Groundwater depletion: a global problem. Hydrogeol J 13(1):317–320

Kulwant M, Patel D, Shirin S, Sharma SN, Yadav AK (2024) Future frameworks for fluoride and algorithms for environmental system. In: Advanced treatment technologies for fluoride removal in water: water purification, pp 343–364

Kumar R, Singh R, Sharma K (2005) Water resources of India. Curr Sci 89:794e811

Kumar MD, Singh OP (2005) Indicators for water resources development in India: an overview. Water Policy 7(1):1–28

Kumar D, Singh RD (2015) Impact assessment of climate change on water resources in the Upper Ganga River Basin, India. Stoch Env Res Risk Assess 29(5):1337–1355

Ministry of Environment, Forest and Climate Change, Government of India (Official Website). Link

Ministry of Jal Shakti, Government of India (2019) Composite water management index: a tool for water management and analysis

Ministry of Jal Shakti, Government of India (Official Website). Link

Mukherji A, Shah T (2005) Groundwater socio-ecology and governance: a review of institutions and policies in selected countries. Hydrogeol J 13(1):328–345

Patel D, Kulwant M, Shirin S, Varshney R, Pandey G, Yadav AK (2024) Fluoride removal from aqueous solution using iron-based materials: preparation, characterization, and applications. In: Advanced treatment technologies for fluoride removal in water: water purification. Springer Nature Switzerland, Cham, pp 71–92

Postel SL, Richter BD (2003) Rivers for life: managing water for people and nature. Island Press

Ravindranath NH, Rao S (2004) Climate change impacts and vulnerability assessment in India. Clim Change 66(1–2):77–92

Rodell M, Velicogna I, Famiglietti JS (2009) Satellite-based estimates of groundwater depletion in India. Nature 460(7258):999–1002

Sawyer CN, McCarty PL, Parkin GF (2002) Chemistry for environmental engineering and science. McGraw-Hill

Scanlon BR, Faunt CC, Longuevergne L, Reedy RC, Alley WM, McGuire VL, McMahon PB (2012) Groundwater depletion and sustainability of irrigation in the US High Plains and Central Valley. Proc Natl Acad Sci 109(24):9320–9325

Shah T (2014) Groundwater governance in India: the case of a 40-year crisis in Gujarat. Water Policy 16(4):605–622

Sharma RK, Gulati S, Puri A (2014) Green chemistry solutions to water pollution. In: Ahuja S (ed) Water reclamation and sustainability. Elsevier Publications, pp 57–75

Shiklomanov IA, Rodda JC (2003) World water resources at the beginning of the 21st century. Cambridge University Press

Shirin S, Yadav AK (2014) Physico chemical analysis of municipal wastewater discharge in Ganga River, Haridwar District of Uttarakhand, India. Curr World Environ 9(2):536

Shirin S, Jamal A, Ranjan P, Yadav AK (2019) Study on assessment of slope stability and mixed disposal of overburden in voids of Singrauli Coalfield. Environ Qual Manage 28(3):131–139

United Nations (2019) World water development report 2019: leaving no one behind. Link

United Nations (2020) World water development report 2020: water and climate change. Link

United States Environmental Protection Agency (USEPA) (2021) Water quality standards handbook

Van Dijk AI, Beck HE, Crosbie RS, de Jeu RA, Liu YY, Podger GM, Viney NR (2013) The Millennium Drought in southeast Australia (2001–2009): natural and human causes and implications for water resources, ecosystems, economy, and society. Water Resour Res 49(2):1040–1057

Wada Y, Wisser D, Bierkens MF (2020) Global modeling of withdrawal, allocation and consumptive use of surface water and groundwater resources. Earth Syst Dyn 5(1):15–40

Yadav AK, Sahoo SK, Mahapatra S, Kumar AV, Pandey G, Lenka P, Tripathi RM (2014) Concentrations of uranium in drinking water and cumulative, age-dependent radiation doses in four districts of Uttar Pradesh, India. Toxicol Environ Chem 96(2):192–200

Part II
Water Quality Assessment, Monitoring and Management

Chapter 6
Assessment of Water Quality in Terms of the Water Quality Index

Jaiprakash Nayak, Reena Singh, and Rajiv Ganguly

Abstract Water is a necessary resource for life to exist. This study includes, the water quality of the Ganga River in Kanpur is evaluated in terms of WQI. The Water Quality Index (WQI) is a useful and distinctive ranking that helps choose the most effective treatment strategy to meet the current difficulties by giving a comprehensive picture of the condition of its water in a single term. The parameters like temperature, pH, electrical conductivity, turbidity, TS, TSS, TDS, acidity, alkalinity, magnesium, calcium, total hardness, chlorides, dissolved oxygen, BOD, chromium, and sulphate used for analysis work and also compared with the Indian Standard (IS 10500:2012). The need for clean water for human use and an ecologically balanced environment makes water quality assessment important. It is necessary to manage and safeguard freshwater resources because of the ongoing contamination of water bodies caused by both point and non-point causes. The modeling process involves a thorough and intricate analysis of the surface water body, involving a vast amount of quality parameter data that must be gathered over an extended time. To access the river water quality, WQI has been used extensively. The Ganga River in Kanpur was subjected to a WQI evaluation, and then the river resource's suitability for human consumption was discussed. The proposed chapter will present an extensive examination of the quality of the Ganga River, keeping in mind the development of a new, well-recognized "WQI" in a simplified presentation that may be applied broadly and offer an accurate picture of the water's condition, and may offer a useful method for regulating and evaluating the quality of the water.

Keywords Surface water · Water quality assessment · Water quality index. Weighted arithmetic · CCME

J. Nayak (✉) · R. Singh
Department of Civil Engineering, National Institute of Technology, Patna, Bihar 800005, India
e-mail: jaiprakashn.ph21.ce@nitp.ac.in

R. Ganguly
Department of Civil Engineering, Harcourt Butler Technical University, Kanpur 208002, India

A. K. Yadav et al. (eds.), *Integrated Management of Water Resources in India: A Computational Approach*, Water Science and Technology Library 129,
https://doi.org/10.1007/978-3-031-62079-9_6

6.1 Introduction

Water is a characteristic asset on which all living creatures depend. Ancient civilizations were frequently founded alongside the banks of lakes and rivers because water is necessary for survival (Vasistha and Ganguly 2020a, b). One of nature's most valuable resources, water is crucial to all living things (Rana and Ganguly 2020). The expansion in populaces, higher expectations for everyday comforts, urbanizations, and so forth, have placed a ton of weight on the regular water body. Though water is an essential part of the ecological system, human activity and natural processes have long been affecting both sources of water quality to deteriorate (Uddin et al. 2021; Patel et al. 2024). Rivers offer significant water resources for household, industrial, and irrigation uses (Wu et al. 2021). Due to a number of circumstances, including anthropogenic activity, overdevelopment, and natural processes, water quality degradation is a severe and widespread problem in the world (Todd et al. 2012; Carpenter et al. 1998; Wang et al. 2008;). To safeguard and regulate water quality, it is crucial to gather pertinent data regarding the water's characteristics and its geographical and temporal variability (Romero et al. 2016; Zhang et al. 2019). Surface water quality has, however, significantly declined in recent decades due to its increased susceptibility to contaminants originating from diverse sources, including municipal sewage, home waste disposal, and irrigation drainage water (Asha et al. 2020; Rahman et al. 2021). Recently, as developing nations have worked to increase the supply of water and sanitation, they have encountered substantial challenges in maintaining water quality (Carvalho et al. 2011; Yadav et al. 2014; Debels et al. 2005; Kannel et al. 2007; Ortega et al. 2016).

The portability of various sources is impacted by water contamination, a serious environmental problem. Waterbody contamination has increased as a result of an increase in development activity (Vasistha and Ganguly 2022). The Water Quality Index (WQI) is a practical and helpful tool for evaluating the quality of water, and it can be applied to ascertain the standard of river water. One important mathematical tool that is commonly used to aggregate multiple water quality data items into a single number is the Water Quality Index (Kumar and Singh 2021). The comprehensive tool known as the WQI has been used in the assessment of the Ganga River's water quality in Kanpur, Uttar Pradesh. Using the WQI, multiple measurements of water quality are combined to create a single numerical expression that represents the total water quality. Assessing and synthesizing the vast datasets needed for water quality management can be challenging, which must be gathered and analyzed (Uddin et al. 2021; Kumar et al. 2024). The index for water quality model is one of the various methods created to assess data on water quality. A single value, or water quality index, reflecting the state of water quality can be obtained by analyzing large datasets with varying temporal and spatial water quality thanks to aggregation functions, which form the basis of WQI models. Water-related management and supply businesses find them appealing as they simplify complex data on water quality into a single, easily comprehensible value metric that is also simple to implement.

A WQI typically has four steps or components, first, the pertinent metrics for water quality are identified. Secondly, water quality parameters are analyzed, and all parametric values are transformed into a single-valued, non-dimensional sub-index. In step three, each water quality parameter's weighting factor is determined. In step four, an aggregation function integrates the sub-indices and weighting factors for all parameters to create an overall water quality index value (Shukla et al. 2017).

This work aims to achieve the ensuing objectives:

1. To determine the physical–chemical and some heavy metal characteristics of water samples taken from specific locations
2. To estimate the WQI of identified sites.

6.2 Materials and Methods

6.2.1 Study Area

Kanpur district is situated in Uttar Pradesh, India, between 80,021″ East longitude and 260 28″ North latitude (Arya and Gupta 2013). According to the census conducted in 2023, Kanpur's population is currently estimated to be at 32.86 lakh, and by 2025, it might reach 56 lakhs. Eight sampling locations were chosen in Kanpur along the bank of banks of the Ganga River (Fig. 6.1).

6.2.2 Collection of Samples

Samples were collected from 8 different sites such as S1, S2, S3, S4, S5, S6, S7, and S8 during November 2022 (Fig. 6.1). The length of the river segment taken into consideration for this study is approximately 10 km and the samples were collected two times in the month during the second week and fourth week of November 2022 from the upper, Middle, and Down surfaces at a total distance of 1.5 m by using grab sampling of the eight selected monitoring locations covering the study stretch as shown in Fig. 6.1, sample mapping coordinate as shown in Table 6.1.

The rationale behind selecting these study locations for sampling is that human activities along riverbanks include bathing, doing laundry in launderettes, and disposing of sewage. The sampling stations are situated in the middle of the city and receive domestic and commercial garbage, resulting in highly contaminated areas completely covered in dirt and mud. Seesamau Nala is near to the Kanpur Jail. Every small drain that carries sewage over the jail property empties into the Ganga. There are additional drains in addition to the sewer pipes. This neighborhood's drains are regularly clogged, which allows sewage to overflow and eventually reach the Ganga. Below this point, domestic wastewater is dumped into the Ganga

Fig. 6.1 Locations for water sampling

Table 6.1 Details of locations

S.No.	Sites	Location	Latitude	Longitude
1	S1	Ganga Barrage	26.50313	80.318813
2	S2	Atal Ghat	26.500443	80.320008
3	S3	Rani Ghat	26.496029	80.324662
4	S4	Bhairav Ghat	26.494685	80.327009
5	S5	Seesasamau Nala	26.491729	80.332735
6	S6	Parmath Ghat	26.486709	80.345538
7	S7	Sarsaiya Ghat	26.479928	80.35953
8	S8	Dabka Ghat	26.44479	80.394389

in significant volumes because of the area's high population density (Naseema et al. 2013). At the sampling locations (Fig. 6.1), boats were available to take samples.

6.2.3 Sample Analysis

The used sample vial had been immersed in 10% HNO_3 for an overnight period before being washed. To prevent the assembly, unfiltered samples were acidified by adding 2 ml of concentrated HNO_3 per liter of the sample to prevent the accumulation of heavy metals. For protection against evaporation, the bottles were firmly sealed and kept at 4 °C. (Ahmad and Chaurasia 2019).

The Physical, chemical, and biological characteristics, as well as how they interact with one another, have a substantial effect on the water quality in bodies of water. (Vasistha and Ganguly 2020a, b). A total of 48 water samples (25 cm at the upper surface, 55 cm at the middle surface, and 70 cm at the lower surfacing) were gathered from eight sampling locations within the research region throughout November 2022, and detailed information about the sampling sites was provided in Table 6.1 (Wu et al. 2021). All 19 parameters analyzed by standard techniques were according to the prescribed guidelines (APHA et al. 2012; Parween et al. 2022).

6.2.4 Water Quality Index

This study uses two techniques to compute the WQI: the Weight Arithmetic Water Quality Index (WAWQI), which is derived from BIS 10500, and the CCME respectively. By integrating all variables into a common value and streamlining the classification of factors, the WQI by BIS 10500 provides a scoring system that facilitates water quality analysis (Jitha and Vijaya 2022). Based on WAWQI, Eq. 6.1 provides a mathematical expression for the WQI approach.

$$WQI = \frac{\sum W_n Q_n}{\sum W_n} \tag{6.1}$$

where W_n is the unit weight (W_n), and Q_n is the sub-index. The quantified information from WQI is linked to water quality Brown et al. 1970.

Three factors are used to calculate the CCME WQI: the scope (F1), frequency (F2), and amplitude (F3), as indicated by the following Eqs. (6.2)–(6.8) (Miyittah et al. 2020).

$$\boldsymbol{CCMEWQI} = 100 - \left(\frac{\sqrt{\boldsymbol{F}1^2 + \boldsymbol{F}2^2 + \boldsymbol{F}3^2}}{1.732} \right) \tag{6.2}$$

The resulting WQI value is normalized to fall between zero (worst quality) and 100 (highest quality) using the denominator 1.732. The vector length that is produced can go up to 173.2. That is:

$$\sqrt{100^2 + 100^2 + 100^2} = \mathbf{173.2} \tag{6.3}$$

100 is the maximum vector length, hence division by 1.732 gets this number.

Over the appropriate time frame, F1 represents an extent or scope of non-compliance with standards for water quality.

$$F1 = \frac{Numberoffailedvariables}{Totalnumberofvariables} \times 100 \quad (6.4)$$

F2 is the frequency, which expresses the percentage of each test that does not comply (pass) with the relevant requirements.

$$F2 = \frac{NumberoffailedTests}{Totalnumberoftests} \times 100 \quad (6.5)$$

The amount that the failed tests did not meet the appropriate requirements is represented by the amplitude, or F3.

$$F3 = \frac{nse}{(0.01nse + 0.01)} \quad (6.6)$$

where nse denotes Normalised Sum of Excursions, is the total amount that each failed test deviates from compliance

$$nse = \frac{\sum_{i=1}^{n} excrusion_i}{totalnumberoftests} \quad (6.7)$$

Excursion is determined by Eq. (6.7)

$$excrusion_i = \frac{Failedtestvalue_i}{Objective_i} - 1 \quad (6.8)$$

Table 6.2 shows the results of classifying water quality using the BIS and CCME standards, respectively.

6.3 Results and Discussion

6.3.1 Water Quality Assessment

The average observed values and weighted average values for several physical, chemical, and heavy metals characteristics are listed in Table 6.3. The value of Temp varies from 20 to 24 °C at all the selected locations and periods and a similar study was found by (Shirin and Yadav 2014). The highest temperature was 24 °C at station S2 and the Lowest Temperature was 20 °C at S8 as shown in Fig. 6.1. The value of pH was found within the permissible limit for all the selected locations and a similar

Table 6.2 WQI value

BIS WQI value, developed by Brown et al. (1970)	
WQI rating	Status of water quality
0–25	Excellent
26–50	Good
51–75	Poor
76–100	Very poor
More than 100	Unfit for consumption
Rating description of CCME WQI scores (CCME); Water quality index (WQI)	
95–100	Excellent
80–94	Good
65–79	Fair
45–64	Marginal
0–44	Poor

study was found in Nirmala et al. (2022). Conductivity is a useful and quick way to calculate the total number of dissolved ions and is correlated with the total amount of solids. The value of EC was determined to be within the allowable range for each of the selected locations and a similar result was found within the permissible limit (Arya and Gupta 2013).

At various tropical elevations, the energy connection is influenced by water clarity, which is a significant element. It essentially depends on light reflection from the surface and is influenced by the way light is absorbed by water as well as by suspended and dissolved matter (Stepane 1959). The mean quantified turbidity value exceeds the permissible limit (Arya and Gupta 2013). When analysing industrial wastewater, total dissolved solids are especially helpful. It represents the amount of inorganic pollution in the aqueous system (Usha et al. 2008). The mean value of TDS varied from 192 (S3) to 646 (S8). The values of TDS were found within permissible limits except for the S8 location due to effluent entry via Jajmau Nala. The same for TSS and TS was found in this study. Alkalinity is instrumental in determining the water quality of rivers (Rai 2019). A variation of the alkalinity varied in the permissible limit for S4 and S5 but except S4 and S5, all the locations did not meet the permissible limit of Alkalinity. Additionally, because untreated waste was being injected into the river near the industrial area, the overall alkalinity was higher. The elevated levels of alkalinity at S1, S2, S3, S6, S7, and S8 could perhaps be attributed to a rise in free carbon dioxide in the Ganges River (Namrata 2010; Praveen et al. 2013). The value of Total hardness was found within permissible limits for all the selected locations of the Ganga River. To support aquatic life, dissolved oxygen is needed in the water. The most basic amount of DO require to sustain aquatic life is 4 mg/l (Singh and Kumar 2021). To ensure the survival of aquatic life, a higher value of minimum DO must be found in all of these sites, and a similar study was found (Singh and Kumar 2021). The amount of oxygen used by the microbes to break down this waste

Table 6.3 Average and weighted average values of various physicochemical and heavy metals parameters of the study area

S.No.	Parameter	Stretch	S1	S2	S3	S4	S5	S6	S7	S8	Permissible limit
1	T		22	24	22	23	23	22	23	20	20
2	pH	U/S	8	8	8	8	8	8	8	8	8.5
		M/S	8	7	8	8	8	8	8	8	
		D/S	8	7	8	8	8	8	8	8	
		Weighted average values	8	7	8	8	8	8	8	8	
3	EC	U/S	36	35	35	34	32	34	49	100	300
		M/S	36	35	32	32	34	32	50	101	
		D/S	41	36	33	35	34	32	52	110	
		Weighted average values	40	40	30	30	30	03	50	100	
4	Turbidity	U/S	77	88	174	111	91	150	90	99	5
		M/S	83	113	109	84	105	106	90	101	
		D/S	109	110	90	90	85	117	83	173	
		Weighted average values	94	107	111	91	93	118	87	134	
5	TDS	U/S	203	203	199	197	194	197	306	619	500
		M/S	207	204	187	187	182	197	302	629	
		D/S	239	208	193	201	201	196	308	669	
		Weighted average values	221	206	192	195	193	197	305	646	
6	TSS	U/S	22	25	23	23	22	31	47	76	500
		M/S	22	26	22	24	23	33	47	98	
		D/S	23	26	23	24	22	100	50	82	
		Weighted average values	23	26	22	24	22	64	48	87	
7	TS	U/S	225	228	222	219	216	229	353	695	500
		M/S	230	230	209	211	205	230	349	727	
		D/S	262	234	216	225	223	297	357	751	
		Weighted average values	244	231	214	219	215	261	353	733	
8	Alkalinity	U/S	201	205	225	199	200	210	250	359	200

(continued)

Table 6.3 (continued)

S.No.	Parameter	Stretch	S1	S2	S3	S4	S5	S6	S7	S8	Permissible limit
		M/S	192	205	200	198	188	216	243	350	
		D/S	210	211	228	191	201	214	265	352	
		Weighted average values	202	208	217	195	196	214	254	352	
9	Acidity	U/S	24	21	19	26	24	26	35	140	200
		M/S	27	20	21	27	22	27	107	135	
		D/S	24	19	25	24	23	26	55	157	
		Weighted average values	25	20	23	25	23	27	71	146	
10	Total Hardness	U/S	180	157	155	150	155	158	204	246	300
		M/S	166	163	150	158	155	157	194	236	
		D/S	176	164	162	146	152	154	202	263	
		Weighted average values	173	163	156	151	154	156	199	251	
11	Calcium	U/S	51	45	53	43	45	54	52	70	75
		M/S	51	47	48	48	55	51	50	86	
		D/S	51	51	55	48	53	48	52	100	
		Weighted average values	51	48	52	47	52	50	51	90	
12	Magnicium	U/S	31	27	25	26	27	25	37	42	30
		M/S	27	28	25	26	24	26	35	36	
		D/S	30	27	26	24	24	25	36	39	
		Weighted average values	29	27	25	25	24	25	36	39	
13	Chlorides	U/S	30	20	17	18	20	19	26	84	250
		M/S	26	23	19	12	17	19	25	65	
		D/S	44	21	18	18	19	24	28	92	
		Weighted average values	35	22	18	16	18	21	27	81	
14	D.O	U/S	7.73	7.60	7.85	7.60	7.88	7.85	8.38	8.30	5
		M/S	7.90	7.25	8.23	8.13	7.70	8.18	8.10	7.55	
		D/S	7.68	7.60	8.03	7.98	7.90	8.15	7.85	8.00	

(continued)

Table 6.3 (continued)

S.No.	Parameter	Stretch	S1	S2	S3	S4	S5	S6	S7	S8	Permissible limit
		Weighted average values	8	7	8	8	8	8	8	8	
15	B.O.D. 5@ 20	U/S	11	8	10	7	6	8	10	10	10
		M/S	11	9	9	12	10	7	7	12	
		D/S	10	7	8	12	7	9	7	13	
		Weighted average values	10	8	8	11	8	8	7	12	
16	COD	U/S	102	114	122	124	182	150	157	202	200
		M/S	115	122	133	129	161	164	145	197	
		D/S	112	119	129	113	167	129	121	173	
		Weighted average values	112	119	129	121	167	145	136	186	
17	Chromium	U/S	2	2	3	3	3	3	3	4	0.5
		M/S	2	2	3	3	3	3	3	3	
		D/S	2	2	3	3	3	3	3	4	
		Weighted average values	2	2	3	3	3	3	3	4	
18	Sulphate	U/S	444	627	652	393	438	387	497	669	200
		M/S	415	448	395	401	457	416	451	659	
		D/S	436	442	401	408	439	405	516	655	
		Weighted average values	430	475	441	403	446	406	489	659	
19	Iron	U/S	503	595	645	803	809	733	1844	1849	300
		M/S	637	628	611	1153	1086	781	1465	2122	
		D/S	607	657	598	770	803	1075	1312	1581	
		Weighted average values	600	636	611	916	907	911	1457	1824	

is indicated by the BOD. The BOD levels in all the samples were higher than what was allowed, primarily because drinking water should have extremely minimal BOD levels. The values obtained in this study were found very high (Shukla et al. 2021). The concentration of COD in the water provides both quantitative and qualitative data regarding the amount of pollution present (Arya and Gupta 2013; Shirin and

Table 6.4 Water quality index using BIS 10500 methods for selected locations and quality of water

Site	WQI value	Water quality status
S1	**107**	Unfit for consumption
S2	**109**	Unfit for consumption
S3	**112**	Unfit for consumption
S4	**106**	Unfit for consumption
S5	**103**	Unfit for consumption
S6	**108**	Unfit for consumption
S7	**92**	Very Poor
S8	**129**	Unfit for consumption

Yadav 2014). The obtained value of COD was very high in the current study which shows a high pollutant level present and a similar result was studied (Rai 2019).

The value of chlorides was found within permissible limits for the selected locations. The discharge of local effluents or the dumping of domestic sewage at various locations can be the cause of the value of chloride in the river Ganges waters, which may lead to a slight increase in chloride levels (Mishra 2010). It is a significant inorganic anion found in both water and wastewater. During the period of heavy rainfall, the levels of chloride were consistently higher (Rai 2019).

The value of Chromium and Iron exceeded the permissible limit for all the selected locations.

6.3.2 Water Quality Index

WQI was determined with the help of weighted Average values depending on relevant parameters of the selected locations as shown in Fig. 6.1

The Water Quality Index for selected locations and the Water Quality Index (using WAWQI and the CCME approach) as shown in Tables 6.4 and 6.5 respectively.

Table 6.5 Water quality index using CCME methods for selected locations and quality of water

Site	F1	F2	F3	CCME WQI value	Water quality status
S1	42	42	18	64	Marginal
S2	33	33	78	47	Marginal
S3	33	33	79	47	Marginal
S4	33	33	17	71	Fair
S5	25	25	18	77	Fair
S6	33	33	20	70	Fair
S7	42	42	21	64	Marginal
S8	67	67	34	42	Poor

This study also computed the Weighted Arithmetic WQI, which is displayed in Table 6.4. WQI of the Ganga River was seen to be growing and above the 100-point scale, indicating the unsuitability of Gangesh water for oral consumption at every sample site except S7 (92). Similar results were also noted by Gangwar et al. (2012) in the Ramganga River (Bareilly), Dutta and Sarma (2018) in the Kolong River (Nagaon, Assam), and Jindal and Sharma (2011) in the Sutlej River (Ludhiyana).

The Ganga River's water quality was marginal for locations S1, S2, S3, and S7, fair for S4, S5, and S6, and poor for S8, according to the CCME WQI evaluation. The values are shown in Table 6.5. In this situation, the marginal score as determined by the CCME (2001) shows the departure of relevant parameters from desired values (Figs. 6.2 and 6.3).

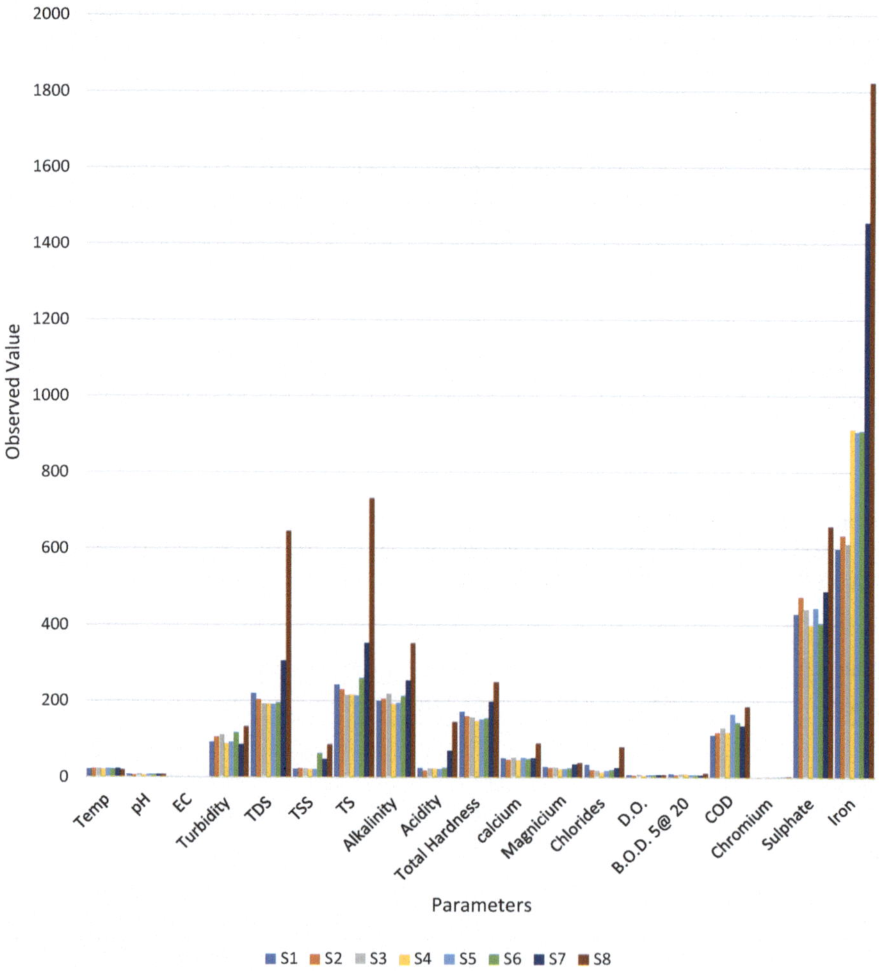

Fig. 6.2 Values of selected locations

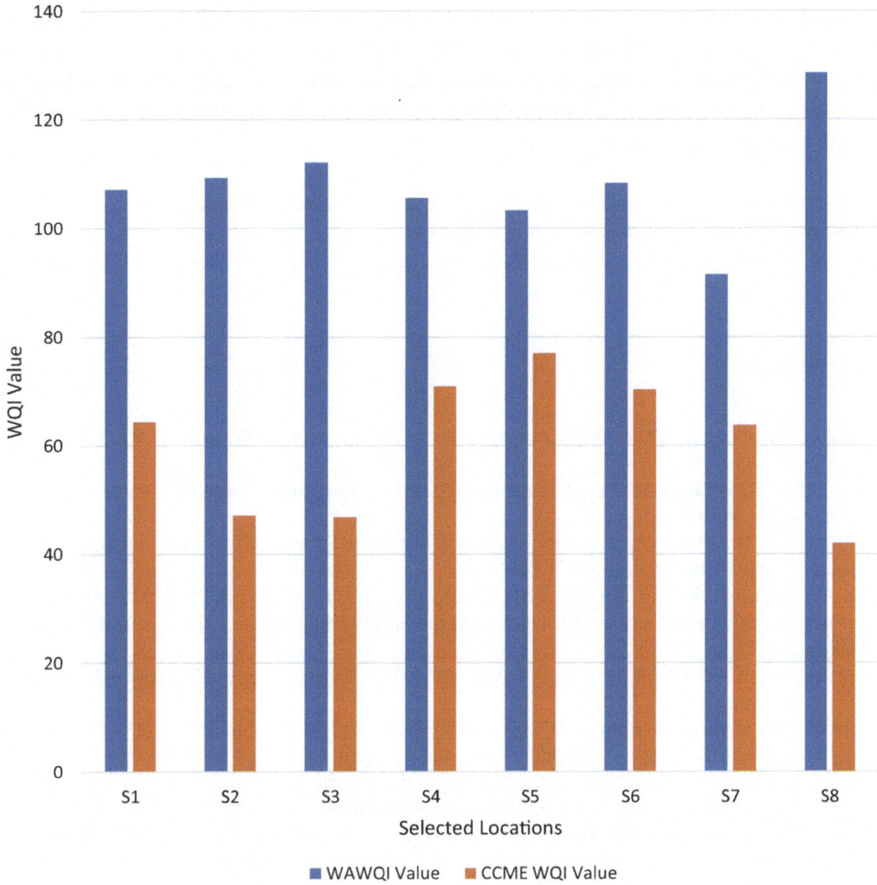

Fig. 6.3 Calculated WAWQI and CCME WQI score at various locations

6.4 Conclusions

Fish and plants that grow underwater can survive in Ganga water safely since, according to our analysis, the DO concentration is greater than 4 mg/l. Due to the tannery, sewage drain, and residential area drains discharging into the Ganga River, all of the chosen locations had the highest levels of contaminants, as determined by the weighted arithmetic mean approach, and CCME approach in which S8 location was the most pollutant locations because A drain at the S8 location routinely transports hazardous substances from almost 100 tanneries near the Ganga and dumps them into the river's main channel without any screening, thus exacerbating the already severe water pollution. It was found that the BOD level was high at S8 locations. Even if various properties of the Gangesh River water are within acceptable limits, it is determined that the water is not suitable for direct usage. While it is not recommended for intake, the water is still suitable for irrigation as well as bathing. The appropriate layout of wastewater treatment facilities must be done by the government for sewage water to be eliminated following treatment. It is necessary to periodically monitor the quality of the river water.

Acknowledgements The authors thank our prestigious institute and other people who are involved in this study for their direct and indirect contributions.

References

Ahmad I, Chaurasia S (2019) Water quality index of Ganga River at Kanpur (U.P.). Thematics J Geogr 8(11):66–77

APHA, AWWA, WEF (2012) Standard methods for the examination of water and wastewater, 22nd edn. APHA, Washington

Arya S, Gupta R (2013) Water quality evaluation of Ganga River from up to downstream area at Kanpur City. J Chem Chem Sci 3(2):54–63

Asha AS, Saifullah ASM, Uddin MG, Sheikh MS, Uddin MJ, Diganta MTM (2020) Assessment of trace metal in macroalgae and sediment of the Sundarban mangrove estuary. Appl Water Sci 10(1):1–13

Brown RM, McClelland NI, Deininger RA, Tozer RG (1970) A water quality index-do we dare. Water Sewage Work 117(10)

Carpenter SR, Caraco NF, Correll DL, Howarth RW, Sharpley AN, Smith VH (1998) Nonpoint pollution of surface waters with phosphorus and nitrogen. Ecol Appl 8(3):559–568

Carvalho L, Cortes R, Bordalo AA (2011) Evaluation of the ecological status of an impaired watershed by using a multi-index approach. Environ Monit Assess 174:493–508

CCME (2001) Canadian water quality guidelines for the protection of aquatic life: CCME water quality index 1.0. Technical report. In: Canadian environmental quality guidelines, 1999. Canadian Council of Ministers of the Environment, Winnipeg

Debels P, Figueroa R, Urrutia R, Barra R, Niell X (2005) Evaluation of water quality in the Chillán River (Central Chile) using physicochemical parameters and a modified water quality index. Environ Monit Assess 110:301–322

Dutta B, Sarma B (2018) Assessment of the Water Quality Index of the Kolong River of Nagaon District of Assam, India. IJERA 8(6):29–38

Gangwar RK, Khare P, Singh J, Singh AP (2012) Assessment of physicochemical properties of water: River Ramganga at Bareilly, UP. J Chem Pharm Res 4(9):4231–4234

Indian standards for drinking water Speciation. Bureau of Indian Standard, New Delhi (BIS 10500: 2012).

Jindal R, Sharma C (2011) Studies on the water quality of the Sutlej River around Ludhiana concerning physicochemical parameters. Environ Monit Assess 174(1–4):417–425

Jitha P, Vijaya MS (2022) River water quality prediction and index classification using machine learning

Kannel PR, Lee S, Lee YS, Kanel SR, Khan SP (2007) Application of water quality indices and dissolved oxygen as indicators for river water classification and urban impact assessment. Environ Monit Assess 132:93–110

Kumar S, Singh R (2021) Qualitative assessment and corrosiveness of the Ganga water: A comparative assessment. Materials Today: Proc 45:5695–5701

Kumar A, Chava R, Gupta S, Shirin S, Jamal A, Yadav AK (2024) Environmental and health effects of fluoride contamination and treatment of wastewater using various technologies. In:

Advanced treatment technologies for fluoride removal in water: water purification. Springer, Cham, pp 323–341. https://doi.org/10.1007/978-3-031-38845-3_18
Mishra A (2010) Assessment of water quality using principal component analysis: a case study of the river Ganges. J Water Chem Technol 32:227–234
Miyittah MK, Tulashie SK, Tsyawo FW, Sarfo JK, Darko AA (2020) Assessment of surface water quality status of the Aby Lagoon System in the Western Region of Ghana. Heliyon 6(7)
Namrata S (2010) Physicochemical properties of polluted water of river Ganga at Varanasi. Int J Energy Environ 1(5):823–832
Naseema K, Masihur R, Husain KA (2013) Study of seasonal variation in the water quality among different ghats of river Ganga, Kanpur, India. J Environ Res Dev 8(1):1
Nirmala K, Kumar PS, Ambujam NK, Srinivasalu S (2022) Assessment of physico- chemical parameters of surface waters of a tropical brackish water lake in South Asia. Environ Res 214:113958
Ortega DJ, Pérez DA, Américo JH, De Carvalho SL, Segovia JA (2016) Development of an index of resilience for surface water in watersheds. J Urban Environ Eng 10(1):72–82
Parween S, Siddique NA, Diganta MTM, Olbert AI, Uddin MG (2022) Assessment of urban river water quality using a modified NSF water quality index model at Siliguri City, West Bengal, India. Environ Sustain Indicators 16:100202
Patel D, Kulwant M, Shirin S, Varshney R, Pandey G, Yadav AK (2024) Fluoride removal from aqueous solution using iron-based materials: preparation, characterization, and applications. In: Advanced treatment technologies for fluoride removal in water: water purification. Springer, Cham, pp 71–92. https://doi.org/10.1007/978-3-031-38845-3_4
Praveen A, Kumar R, Pratima KR (2013) Physico-chemical properties of the water of river Ganga at Kanpur. Int J Comput Eng Res 3:134–137
Rahman A, Jahanara I, Jolly YN (2021) Assessment of physicochemical properties of water and their seasonal variation in an urban river in Bangladesh. Water Sci Eng 14(2):139–148
Rai A (2019) Physiochemical analysis of Ganga River at Kanpur in Uttar Pradesh, India. Int J Adv Sci Res 4(4):21–26
Rana R, Ganguly R (2020) Water quality indices: challenges and applications—An overview. Arab J Geosci 13(22):1190
Rice EW, Baird RB, Eaton AD, Clesceri LS (2012) Standard methods for the examination of water and wastewater
Romero E, Le Gendre R, Garnier J, Billen G, Fisson C, Silvestre M, Riou P (2016) Long-term water quality in the lower Seine: lessons learned over 4 decades of monitoring. Environ Sci Policy 58:141–154
Shukla AK, Ojha CSP, Garg RD (2017) Surface water quality assessment of Ganga River Basin, India using index mapping. In: 2017 IEEE international geoscience and remote sensing symposium (IGARSS). IEEE, pp 5609–5612
Shukla AK, Ojha CSP, Shukla S, Garg RD (2021) Water quality challenges in Ganga River Basin, India. The Ganga River Basin: Hydrometeorol Approach 1–19
Singh R, Kumar S (2021) Assessment and suitability analysis of water quality of River Ganga in Patna, Bihar. In: The F-EIR conference on environment concerns and its remediation. Springer, Cham, pp 251–258
Shirin S, Yadav AK (2014) Physico-chemical analysis of municipal wastewater discharge in Ganga River, Haridwar district of Uttarakhand, India. Curr World Environ 9(2):536–543
Stepane KM (1959) Limnological study of the reservoir Sedlice near Zeliv. IX. Transmission and Transparency of water. Sci Pap Inst Chem Tedinol 363–430
Todd AS, Manning AH, Verplanck PL, Crouch C, McKnight DM, Dunham R (2012) Climate-change-driven deterioration of water quality in a mineralized watershed. Environ Sci Technol 46(17):9324–9332
Uddin MG, Nash S, Olbert AI (2021) A review of water quality index models and their use for assessing surface water quality. Ecol Ind 122:107218

Usha NM, Jayaram KC, Lakhmi Kantha H (2008) Proceedings of Tall 2007: the 12th world lake conference, pp 1737–1741
Vasistha P, Ganguly R (2020a) Assessment of spatio-temporal variations in lake water body using the indexing method. Environ Sci Pollut Res 27(33):41856–41875
Vasistha P, Ganguly R (2020b) Water quality assessment of natural lakes and its importance: an overview. Mater Today: Proc 32:544–552
Vasistha P, Ganguly R (2022) Water quality assessment in two lakes of Panchkula, Haryana, using GIS: a case study on seasonal and depth-wise variations. Environ Sci Pollut Res 29(28):43212–43236
Wang J, Da L, Song K, Li BL (2008) Temporal variations of surface water quality in urban, suburban, and rural areas during rapid urbanization in Shanghai, China. Environ Pollut 152(2):387–393
Wu Z, Lai X, Li K (2021) Water quality assessment of rivers in Lake Chaohu Basin (China) using the water quality index. Ecol Ind 121:107021
Yadav AK, Sahoo SK, Mahapatra S, Kumar AV, Pandey G, Lenka P, Tripathi RM (2014) Concentrations of uranium in drinking water and cumulative, age-dependent radiation doses in four districts of Uttar Pradesh, India. Toxicol Environ Chem 96(2):192–200
Zhang J, Li S, Dong R, Jiang C, Ni M (2019) Influences of land use metrics at multi-spatial scales on seasonal water quality: a case study of river systems in the Three Gorges Reservoir Area, China. J Clean Prod 206:76–85

Chapter 7
Challenges and Opportunities for Water Quality Monitoring and Management in India

Mridu Kulwant and Akhilesh Kumar Yadav

Abstract In India, there are many problems associated with monitoring and managing water quality, for example, pollution from industrial and agricultural sources, a lack of wastewater treatment infrastructure, and a limited supply of safe drinking water. Population expansion, urbanization, and furthermore, the effects of climate change all compound these challenges. However, there is a need to improve the monitoring and management. Advances in technology, present monitoring data, and warning systems can be provided by remote sensing and statistical analysis of data. Collaboration among government agencies, stakeholders in the corporate sector, academia, and civil society can also lead to more successful policy. Integrating scientific knowledge, policy interventions, community involvement, and sustainable behaviors is critical for protecting water quality in India. By solving these difficulties and capitalizing on emerging opportunities, India can strive for an era that ensures all citizens have access to clean safe water.

Keywords Wastewater · Monitoring · Agricultural sources · Quality · Management

M. Kulwant (✉)
Department of Environmental Studies, The Maharaja Sayajirao University of Baroda, Vadodara 390002, India
e-mail: mridu.k-envphd@msubaroda.ac.in

A. K. Yadav (✉)
Department of Mining Engineering, Indian Institute of Technology (Banaras Hindu University), Varanasi 221005, India
e-mail: yadavbasti@gmail.com

Department of Environmental Engineering and Management, Chaoyang University of Technology, Taichung 413310, Taiwan

Environmental Science and Engineering Department, Indian Institute of Technology Bombay, Mumbai 400076, India

A. K. Yadav et al. (eds.), *Integrated Management of Water Resources in India: A Computational Approach*, Water Science and Technology Library 129,
https://doi.org/10.1007/978-3-031-62079-9_7

7.1 Introduction

Water scarcity and pollution present substantial problems to the monitoring and management of India's water quality, compounded by climate change. These factors can affect rainfall patterns, glacier melting, floods, and droughts. Establishing water quality monitoring systems, implementing pollution control measures, encouraging water conservation habits, and investing in wastewater treatment facilities are critical to solving these concerns. Collaboration among stakeholders, such as government agencies, water management authorities, research institutes, and local populations, is critical to the sustainable management of water resources in India. Integrating modern technologies and data-driven approaches into water quality monitoring can offer decision makers real-time information (Jadeja et al. 2022). Public understanding of water quality and conservation might encourage a proactive attitude. Despite the limitations of improving water quality screening and Maintenance, there are opportunities for collaboration and technical breakthroughs. The monitoring and management are key components in ensuring public health, environmental sustainability, and economic growth. These characteristics are essential in India, which is dealing with major water issues such as pollution and scarcity. This introduction will look at the challenges India faces in monitoring and maintaining water quality, as well as the opportunities available to overcome these issues (Seelen et al. 2019; Shirin and Yadav 2014). India's water quality issue is complex, with natural and anthropogenic elements contributing to the problems. Industrial pollution, agricultural runoff, urbanization, and inadequate wastewater treatment facilities contribute to poor water quality throughout the country. Climate change exacerbates these difficulties, resulting in changes in water supply and quality. Despite these limitations, India has the opportunity to improve its monitoring and management (Negm et al. 2022). Technological developments, governmental reforms, community participation, and international collaborations provide opportunities to solve water quality challenges more effectively. Understanding and resolving these difficulties, as well as harnessing existing opportunities, will help India achieve a sustainable and healthy future of water for its population (Ajith et al. 2022; Yadav et al. 2014).

There are several elements that cause the poor quality of water in India, including pollution, scarcity, and inadequate infrastructure. Industrial pollution, agricultural runoff, household trash, and mining contaminate water. Industrial waste, including heavy metals and chemicals, is released, while agricultural runoff, which includes pesticides, fertilizers, and animal manure, can contaminate bodies of water. Domestic garbage, particularly in cities, is not properly handled and disposed of (Dickin and Gabrielsson 2023; He et al. 2021; Kumar et al. 2024). Mining activities emit pollutants that degrade water quality. Several arid and semi-arid parts of India are facing water shortages, leading to overexploitation of resources. Pollution, scarcity, inadequate infrastructure, and population expansion all have a consequence on the condition of water in India. Industrial pollution, agricultural runoff, household trash, and mining all contaminate the water (Ay et al. 2018; Shirin and Yadav 2014). Industrial waste including heavy metals and chemicals is released, while agricultural runoff, which includes pesticides, fertilizers, and animal manure, can contaminate bodies of water. Domestic garbage, particularly in cities, is not properly handled and disposed of. Mining activities emit pollutants, which degrades water quality. Despite the availability of water, India faces water scarcity, mainly in arid and semi-arid regions, causing overexploitation (Alengebawy et al. 2021; Pandeya et al. 2021). Unequal distribution of water resources between regions and nations exacerbates water scarcity issues and can lead to disputes over consumption. There are several elements that cause poor of Indian water quality, including pollution, scarcity, and inadequate infrastructure. Industrial pollution, agricultural runoff, household trash, and mining contaminate water (Okereafor et al. 2020). Industrial waste including heavy metals and chemicals is released, while agricultural runoff, which includes pesticides, fertilizers, and animal manure, can contaminate bodies of water. Domestic garbage, particularly in cities, is not properly handled and disposed of. The unequal distribution of water resources between regions and nations exacerbates water scarcity issues and can lead to disputes over consumption (Al-Hazmi et al. 2023). This chapter explores how India can navigate the challenges and opportunities of water resource management.

7.2 Challenges in Water Quality Monitoring and Management

Water quality needs to be observed and managed to provide clean drinking water to public consumption. It is still important to address several challenges in this field. The challenges include:

1. Data Collection: To effectively manage water quality, it is essential to collect a large amount of data. However, it is time consuming and labour-intensive to collect data manually in the field.
2. In water quality datasets, missing data and heterogeneity are common, making it difficult to analyze and draw accurate conclusions.
3. Limited Resources, Rural public health offices, and smaller water providers may not have the resources necessary to meet water quality monitoring regulations.
4. The water level, temperature, and conductivity are the only three parameters used for real-time sensor monitoring, despite the increasing use of the technology. Therefore, a complete assessment of the quality of water and identification of potential contaminants is restricted.

7.2.1 Lack of Comprehensive Monitoring Infrastructure

The lack of extensive monitoring infrastructure is a serious obstacle to India's water quality monitoring and management efforts. A common problem is the lack of monitoring stations or inadequate coverage that results in data gaps for many water bodies, for example rivers, lakes, and groundwater sources, located in rural and distant areas. The lack of extensive monitoring of water quality measures, also pH, turbidity, dissolved oxygen, and fecal coliform, is a major concern (Kumar et al. 2022). This lack of monitoring also has an impact on the frequency and timeliness of data reporting, making it difficult to take corrective action on time. The availability of advanced monitoring systems and qualified personnel, particularly in rural regions, influences the precision and reliability of the data obtained, thus reducing the efficacy of water quality control initiatives (Aneesh 2021). Additionally, there is a lack of coordination and integration between various monitoring agencies and stakeholders, resulting in duplication of efforts, inconsistencies in data interpretation, and difficulties in sharing data among agencies for comprehensive water quality management. Budgetary constraints and resource limits contribute to a lack of extensive water quality monitoring infrastructure (Abdeen et al. 2021). Adequate financing is required for monitoring stations, equipment purchases, people training, and data management systems. Despite existing policies, implementation and enforcement are ineffective. Improving the monitoring infrastructure requires a stronger institutional framework and more capacity at all levels. Addressing this issue requires a holistic approach that includes technical investment, capacity building, data sharing, regulatory reforms, and collaboration among academic institutions, government agencies, nongovernmental organisations (NGOs), and the private sector. This allows more accurate and rapid monitoring of water quality indicators (Hamel et al. 2024; Pinder et al. 2019).

7.2.2 *Pollution and Its Impact on Water Quality*

Pollution from industrial, agricultural, and domestic sources poses considerable hurdles to India's water quality monitoring and management efforts. Industrial pollution, including heavy metals, chemicals, organic compounds, and thermal effluents, contaminates and degrades water quality (Nelliyat 2016). This pollution reduces oxygen levels, increases toxicity, causes contaminants to bioaccumulate in aquatic creatures, and disrupts aquatic ecosystems. Agricultural pollution, household trash, and mining operations all contribute to water contamination. Pesticides, fertilizers, herbicides, and animal manure are released into water bodies by agricultural runoff, resulting in nutrient pollution, eutrophication, and contamination with agrochemical residues (Ingrao et al. 2023). This causes algae blooms, oxygen depletion, loss of biodiversity, and poor water quality. Inappropriate sewage treatment and improper disposal of household waste introduce pathogens, nutrients, organic matter, and pollutants into water bodies, resulting in waterborne infections, fecal contamination, and health hazards. Pollutants are released during mining activities through drainage, leaching, and runoff, affecting aquatic habitats and ecosystems. Urban pollution, such as sewage, industrial effluents, solid waste, chemicals, and plastics, pollutes aquatic bodies through stormwater runoff and direct discharge. This causes pollution hotspots, increased nutrient loads, microbial contamination, and impairment of water purity (Khan et al. 2022; Patel et al. 2023). Polluted water sources endanger human health and cause waterborne diseases such as cholera, typhoid, diarrhea, and skin infections, specifically in locations due to the lack of availability to clean water and sanitary supplies. Environmental degradation, which includes loss of biodiversity, habitat damage, and a drop in water quality indices, is also a major concern. Pollution must be addressed through tougher legislation, improved pollution control methods, sustainable behavior, better sewage treatment infrastructure, public awareness programs, and investments in clean technologies (Babuji et al. 2023).

7.3 Current Approaches and Initiatives

Water quality challenges are being addressed through various approaches and initiatives around the world (Suresh 2012). In Fig. 7.1, multiple strategies are listed to represent initiatives and innovation in the field of interest.

7.3.1 Government Policies and Regulations

The Water Prevention and Control Pollution Act of (1974) and the Environment Protection Act, 1986 is essential government policies in India that oversee water quality. The Water Act (1974) authorizes federal and state pollution control agencies to establish standards, regulate industrial and sewage emissions, and impose fines for non-compliance. The Environment Act (1986) establishes a comprehensive framework for environmental protection, covering water quality management, pollution prevention, environmental impact assessments, and resource conservation. Both statutes increase regulatory procedures, promote sustainable development methods, and address environmental concerns, such as water pollution (UK 2021). The 2012 National Water Policy establishes concepts for sustainable water management, such as considerations of water quality, integrated resource management, conservation, and equitable distribution. It directs national and state water policies, promotes water quality monitoring, funds pollution control measures, and encourages public participation (UNEP 2012). The Water Cess Act of 1977 levies a cess on industries that discharge pollutants into bodies of water, producing cash for pollution control, monitoring, and conservation projects. The National River Conservation Plan (NRCP) aims to improve water quality in key rivers by implementing pollution controls, sewage treatment plants, industrial effluent laws, riverside development, and public awareness initiatives (Rajaram and Das 2008). Focuses on specific sources of pollution, restores contaminated rivers, encourages sustainable behaviours, and improves public access to clean water bodies. State-level policies and initiatives address regional water quality concerns, while also bolstering national efforts, facilitating decentralized governance, and encouraging collaboration between state and

Fig. 7.1 Various approaches and initiatives are being undertaken globally to address the challenges associated with water quality (Scanlon et al. 2023)

central authorities. India is committed to international agreements and conventions on water quality and environmental protection, such as Sustainable Development Goals (SDGs), the Paris Agreement (PA), and the Convention on Biological Diversity (CBD) (Zhang et al. 2023). These policies are consistent with global interests, encourage international cooperation, and facilitate information transfer. The Indian government's policies provide a framework for monitoring water quality, pollution control, sustainable water management, and stakeholder participation, all of which contribute to long-term environmental sustainability (Leal Filho et al. 2022).

7.3.2 Use of Technology in Monitoring and Managing Water Quality

India is increasingly using technology to monitor and manage water quality. Remote Sensing and Geographic Information System are performed to locate pollution sources, evaluate changes in land use, and determine water quality indicators. GIS technology merges spatial and water quality data, allowing spatial analysis, mapping of pollution hotspots, trend visualization, and informed water management decisions (Singh et al. 2023; Srivastava and Sharma 2021). Online monitoring systems are installed in water bodies, industrial effluent outlets, and wastewater treatment plants to continuously monitor water quality parameters, allowing for immediate detection of pollution events, timely interventions, data-driven decision making, and adherence to water quality standards. Advanced sensors and probes are used for in-situ measurements of water quality parameters, resulting in accurate and trustworthy results. These technologies include optical, electrochemical, spectrophotometer, fluorometer, and conductivity meters (Lakshmikantha et al. 2021).

Water quality management and monitoring are crucial for the safety and cleanliness of our water resources, particularly rivers (Chapman and Sullivan 2022). Data on these parameters can be collected efficiently using manual field sampling and real-time sensor monitoring. A machine learning algorithm with Artificial Neural Networks is then used to analyze the data to predict future water quality issues and anticipate potential problems. Water quality monitoring, transmission, and management can also be enhanced by using information and communication technology and sensor technology (Shalu and Singh 2023). These technologies are used by practitioners to monitor water health mentioned in Fig. 7.2.

Water quality modeling, which employs mathematical models and simulation tools, helps predict water quality dynamics, assessing pollutant transport, simulate scenarios, and evaluating pollution management methods. The WASP, QUAL2K and MIKE models are useful for understanding complicated water body interactions and optimizing management tactics (Ejigu 2021). Crowd-sourced data collection, citizen science programs, public reporting of water quality issues, and community awareness raising are all achieved through the use of mobile apps and platforms. These platforms centralize water quality data, make stakeholder sharing easier, and promote collaborative decision making (Hognogi et al. 2023). AI algorithms and machine learning approaches examine vast datasets to find patterns, forecast trends, optimize monitoring procedures, and provide early warning systems for pollution problems. AI-powered tools improve data interpretation, automate tasks, and facilitate adaptive management strategies. Smart water management systems combine IoT devices, sensors, data analytics, and cloud computing to track, control, and optimize water use, quality, and distribution in real time. They enable smart metering, leak detection, water conservation, demand forecasting, and effective resource distribution in both urban and rural regions (Amador-Castro et al. 2024; Božić 2023). The adoption of technology provides benefits such as increased data accuracy, decision support,

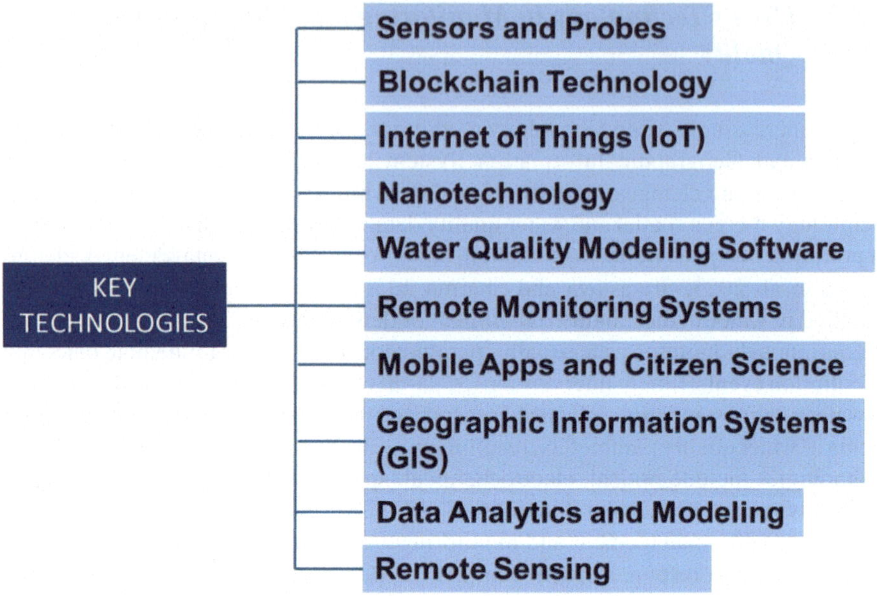

Fig. 7.2 Several key technologies are used for monitoring and managing water quality (Schullehner et al. 2024)

early warning capabilities, cost savings, and sustainable water resource management. However, obstacles such as technological constraints, data protection concerns, capacity-building requirements, and interoperability issues require ongoing attention and investment (Al-Emran and Griffy-Brown 2023).

7.4 Opportunities for Improvement

Monitoring and managing water quality in India requires sustainable water management methods. These include measures for water conservation and efficiency, such as rainwater collection, water reuse, and efficient irrigation techniques that reduce freshwater demand, reduce pollution, and improve water quality (Mohanty 2023).

7.4.1 Promoting Sustainable Water Management Practices

Another technique is integrated water resource management (IWRM), which comes into consideration interconnectivity of water resources, land use, ecosystems, with socioeconomic aspects to promote sustainable use, allocation, and conservation of water resources, ultimately leading to improved water quality. Investing in natural infrastructure such as wetlands and green areas promotes natural filtration, flood mitigation, groundwater recharge, and biodiversity conservation, all of which improve water quality and ecosystem health (Meran et al. 2021). Implementing pollution mitigation techniques at the source, such as cleaner production practices, proper waste management, and reduction of agricultural runoff through best management practices, all help to lower pollutant loads entering aquatic bodies. Engaging local communities and stakeholders in decision-making and water quality monitoring activities promotes ownership and accountability for sustainable water management practices (Khanam et al. 2023). Promoting regenerative agriculture practices, soil conservation measures, organic farming techniques and agroecological approaches increases soil health, decreases nutrient runoff, reduces pesticide pollution, and improves water quality in agricultural areas. Investing in innovative wastewater treatment technology and decentralized systems helps minimize pollution and safeguard aquatic bodies. Building climate resilience through water design for cities, ecologically sound infrastructure planning, and managing floodplains, including adaptive water management measures, assists communities in dealing with climate change impacts, preserving water quality, and ensuring water security. Strengthening policy frameworks, regulatory processes, and institutional capacity at all levels promotes sustainable water management, encourages innovation, and drives continual progress in water quality monitoring and management (Khangura et al. 2023). Promoting sustainable water management requires a comprehensive approach that considers environmental, social, economic, and governance factors, encourages innovation, and prioritizes continuous sustainability and resilience in water-related assets and ecosystems (Hasan et al. 2023).

7.4.2 Innovation in Water Treatment Technologies

Innovative water treatment technologies are critical to improving water quality monitoring and management in India. Advanced Oxidation Processes (AOP) Apply reactive radicals composed of hydroxyl to break apart organic contaminants and pathogens, including toxins of water. Examples such as ozone-based oxidation, UV/H_2O_2 therapy, and the Fenton reaction. Membrane filtration technologies such as reverse osmosis, nanofiltration, and ultrafiltration remove dissolved solids, pathogens, bacteria, and pollutants from water, resulting in clean water for drinking and industrial use (Van der Bruggen 2021). Biological treatment technologies, such as activated sludge processes, biofilters, and bioremediation procedures, rely on microorganisms to digest water contaminants and nutrients. These environmentally friendly and cost-effective solutions are appropriate for treating wastewater from a variety of sources (Saeed et al. 2022). Adsorption and absorption processes, as well as activated carbon adsorption and ion-exchange resins, remove pollutants and organic compounds from water, improving its quality and treatment efficiency. Electrochemical treatment procedures use an electric current to clean water by precipitating contaminants, sterilizing pathogens, and decomposing organic molecules

(Wang et al. 2022). Examples include electrocoagulation, electrooxidation, and electroflocculation. These energy-efficient approaches can be scaled up and applied to a wide range of water treatment applications. Nanotechnology-based solutions, such as nanofiltration membranes, nanoparticles, and nanocomposites, provide superior filtration, adsorption, and catalytic capabilities, hence increasing water quality, efficiency, and energy consumption. Smart water treatment systems combine IoT devices, automation, and data analytics to monitor and optimize treatment operations, increasing operational efficiency and lowering costs (Barrera-Díaz et al. 2018). Green water treatment technologies, including phytoremediation, solar powered systems, environmentally friendly coagulants, and natural filtering processes, prioritize sustainability, energy efficiency, and low environmental effects. These technologies provide solutions to India's water quality concerns, such as pollution management, wastewater treatment, drinking water supply and environmental conservation. Continuous research and improvement are essential for providing clean and safe water to all (Ghernaout et al. 2023).

7.5 Case Studies

The case studies highlight various water quality monitoring and management systems used throughout India, emphasizing the importance of integrated strategies, stakeholder collaboration, technology innovation, and community participation in protecting water resources (Perveen and Amar Ul 2023) case studies are follows:

7.5.1 Ganga Action Plan

The GAP (National Mission for Clean Ganga), which was initiated (1985), aims to revitalize the Ganga River and its tributaries by reducing pollution, promoting sustainable practices, and improving water quality. The plan asks for continuous monitoring and management of water quality, with a network of stations collecting data on physicochemical parameters, biological indicators, and pollution levels. Innovative technology such as real-time sensors, satellite-based monitoring systems, and pollution tracking tools are used to assess trends, prioritize pollution hotspots, and guide pollution control strategies (Kumar 2019).

7.5.2 Yamuna River Revitalization Project

The Delhi government initiated the Yamuna River Revitalization Project, which aims to restore river water quality and ecosystem health by establishing sewage treatment plants, intercepting untreated sewage drains, implementing pollution control measures, and encouraging public participation in conservation efforts. Water quality monitoring stations along the Yamuna River collect real-time data, allowing officials to assess pollution levels, identify contamination sources, and perform targeted interventions to improve water quality (Joshi et al. 2022).

7.5.3 Community-Based Water Quality Monitoring

Community-based water quality monitoring in Kerala is working in collaboration with local people, nongovernmental organizations (NGOs) and government agencies to conduct community-based water quality monitoring programs (Antony and Remya 2022). Members are educated in water quality testing methods and given portable testing kits to monitor water sources for pH, turbidity, bacterial contamination, and chemical contaminants. This grassroots strategy builds local capacity, creates community ownership, and encourages action to safeguard water resources (Flores-Díaz et al. 2018).

7.5.4 Bangalore Water Supply and Sewerage Board Initiatives

BWSSB (Bangalore Water Supply and sewerage board initiatives) Initiatives, Karnataka has adopted several measures to improve water quality, monitoring, and management in the city. These include the construction of online monitoring systems in reservoirs, lakes, and treatment facilities, as well as mobile applications and citizen portals that encourage the public to report water quality issues and promote conservation behaviors (Karnataka 2024; Manasi and Umamani 2013).

7.5.5 Technological Innovations for Rural Water Quality Monitoring

Technological innovations for rural water quality monitoring in Uttar Pradesh use cutting-edge technologies such as solar powered devices, low-cost sensors, and IoT-based data collection systems to monitor water quality in wells, hand pumps, and communal water sources. These devices provide real-time data, aid in contamination identification, and allow prompt measures to ensure safe drinking water for rural communities (Zulkifli et al. 2022).

7.6 Future Prospects

India's water quality monitoring and management faces difficulties and opportunities that must be handled. Future considerations include the integration of advanced technologies such as remote sensing, IoT devices, AI, ML (machine learning), and big data analytics for real-time monitoring, as well as innovative sensor technologies, smart water treatment systems, nanomaterials, and sustainable pollution control, wastewater treatment and water reuse solutions (Sharma et al. 2017). Data-driven approaches, such as big data analytics, cloud computing, and digital platforms, can be used to centrally manage data, analyze trends, and anticipate water quality factors. Data-driven decision-making processes and adaptive management strategies can be used to continuously improve water quality management procedures. Additionally, citizen science projects, crowd-sourced data collection programs, mobile apps, and participatory monitoring methodologies can be broadened to include local communities, nongovernmental organisations, educational institutions, and volunteers in monitoring and management efforts (Kulwant Mridu et al. 2023; Sarker 2021). The future of water quality monitoring and management in India will require a comprehensive strategy that includes technical innovation, regulatory reforms, community participation, industrial partnership, climate resilience, and capacity building. This includes improving regulatory frameworks, incorporating water quality concerns into national and state water policies, and encouraging green technologies, circular economy concepts, sustainable practices, and corporate social responsibility efforts (Behmel et al. 2016). Climate resilience measures, disaster risk management frameworks, and natural solutions will be incorporated into water quality monitoring and management strategies. Capacity-building programs, training workshops, technical skill development, and education campaigns will be funded to encourage learning, dissemination of innovation, and continuous learning in water quality management. This comprehensive strategy will aid in addressing difficulties, seizing opportunities, and achieving sustainable water resource management for current and future generations (Shammin et al. 2022).

7.7 Conclusion

India presents enormous problems in monitoring and managing water quality, such as pollution, water scarcity, and poor infrastructure, including the consequences of warming temperatures. However, there is potential to improvement with innovation. By properly tackling these difficulties and capitalizing on these opportunities, India may make significant progress in providing clean and safe water to its population, safeguarding ecosystems, promoting public health, and supporting sustainable development. Key recommendations and initiatives include implementing an integrated approach that incorporates technical breakthroughs, regulatory reforms, community engagement, and capacity building. Advanced technology such as remote sensing, IoT devices, artificial intelligence, and big data analytics can dramatically improve water quality control operations (Patel et al. 2022). Effective governance requires policy reforms, strengthened regulatory frameworks, enforcement of pollution control measures, and the incorporation of water quality considerations into national and state policies. Citizen science programs, participatory monitoring, and awareness campaigns can promote community ownership, accountability, and collaborative action. Industry collaboration, research alliances, technology transfer, and innovation centers promote innovation and collaboration in the creation of green technologies, long-term solutions, and cost-effective water treatment techniques. Climate resilience strategies, natural solutions, and disaster risk management are incorporated into water quality management plans to mitigate the impacts of climate change and improve water security. Investing in capacity building, training programs, skill development and information-sharing platforms for water quality professionals, policymakers, and the general public will help India overcome obstacles and set the stage for more sustainable and resilient water conditions.

Acknowledgements The authors express their gratitude to the reviewers and the editor for dedicating their valuable time to enhancing this work.

References

Abdeen FN, Fernando T, Kulatunga U, Hettige S, Ranasinghe KDA (2021) Challenges in multi-agency collaboration in disaster management: A Sri Lankan perspective. Int J Disaster Risk Reduct 62:102399. https://doi.org/10.1016/j.ijdrr.2021.102399

Ajith V, Mohan ASR, Vinodini Ramesh M (2022) Empowering communities in addressing drinking water challenges using a systematic, participatory and adaptive approach and sustainable PPP model. Technol Forecast Soc Change 185:121970. https://doi.org/10.1016/j.techfore.2022.121970

Al-Emran M, Griffy-Brown C (2023) The role of technology adoption in sustainable development: overview, opportunities, challenges, and future research agendas. Technol Soc 73:102240. https://doi.org/10.1016/j.techsoc.2023.102240

Al-Hazmi HE, Mohammadi A, Hejna A, Majtacz J, Esmaeili A, Habibzadeh S, Saeb MR, Badawi M, Lima EC, Mąkinia J (2023) Wastewater reuse in agriculture: prospects and challenges. Environ Res 236:116711. https://doi.org/10.1016/j.envres.2023.116711

Alengebawy A, Abdelkhalek ST, Qureshi SR, Wang MQ (2021) Heavy metals and pesticides toxicity in agricultural soil and plants: ecological risks and human health implications. Toxics 9(3). https://doi.org/10.3390/toxics9030042

Amador-Castro F, González-López ME, Lopez-Gonzalez G, Garcia-Gonzalez A, Díaz-Torres O, Carbajal-Espinosa O, Gradilla-Hernández MS (2024) Internet of Things and citizen science as alternative water quality monitoring approaches and the importance of effective water quality communication. J Environ Manage 352:119959. https://doi.org/10.1016/j.jenvman.2023.119959

Aneesh MR (2021) Quality of drinking water and sanitation in India. Ind J Human Dev 15(1):138–152. https://doi.org/10.1177/09737030211003658

Antony E, Remya RS (2022) Water quality analysis of different water sources in Kerala, India. 9:6–11. https://doi.org/10.14445/23942568/IJAES-V9I4P102

Ay U, Ajibade F, Ajibade T (2018) Water pollution resulting from mining activity: an overview.

Babuji P, Thirumalaisamy S, Duraisamy K, Periyasamy G (2023) Human health risks due to exposure to water pollution: A review. Water 15(14):2532. https://www.mdpi.com/2073-4441/15/14/2532

Barrera-Díaz CE, Balderas-Hernández P, Bilyeu B (2018) Chapter 3—Electrocoagulation: fundamentals and prospectives. In: Martínez-Huitle CA, Rodrigo MA, Scialdone O (eds) Electrochemical water and wastewater treatment. Butterworth-Heinemann, pp 61–76. https://doi.org/10.1016/B978-0-12-813160-2.00003-1

Behmel S, Damour M, Ludwig R, Rodriguez MJ (2016) Water quality monitoring strategies—review and future perspectives. Sci Total Environ 571:1312–1329. https://doi.org/10.1016/j.scitotenv.2016.06.235

Božić V (2023) AI and predictive analytics. https://doi.org/10.13140/RG.2.2.23798.47682

Chapman DV, Sullivan T (2022) The role of water quality monitoring in the sustainable use of ambient waters. One Earth 5(2):132–137. https://doi.org/10.1016/j.oneear.2022.01.008

Dickin S, Gabrielsson S (2023) Inequalities in water, sanitation and hygiene: Challenges and opportunities for measurement and monitoring. Water Secur 20:100143. https://doi.org/10.1016/j.wasec.2023.100143

Ejigu MT (2021) Overview of water quality modeling. Cogent Eng 8(1):1891711. https://doi.org/10.1080/23311916.2021.1891711

Flores-Díaz AC, Quevedo Chacón A, Páez Bistrain R, Ramírez MI, Larrazábal A (2018) Community-based monitoring in response to local concerns: creating usable knowledge for water management in rural land. Water 10(5):542. https://www.mdpi.com/2073-4441/10/5/542

Ghernaout D, Irki S, Elboughdiri N, Ghernaout B (2023). Solar-driven water treatment: new technologies, challenges, and futures. 13:110–152. https://doi.org/10.4236/gsc.2023.132007

Hamel P, Ding N, Cherqui F, Zhu Q, Walcker N, Bertrand-Krajewski J-L, Champrasert P, Fletcher TD, McCarthy DT, Navratil O, Shi B (2024) Low-cost monitoring systems for urban water management: lessons from the field. Water Res X 22:100212. https://doi.org/10.1016/j.wroa.2024.100212

Hasan N, Pushpalatha R, Manivasagam VS, Arlikatti S, Cibin R (2023) Global sustainable water management: a systematic qualitative review. Water Resour Manage 37(13):5255–5272. https://doi.org/10.1007/s11269-023-03604-y

He C, Liu Z, Wu J, Pan X, Fang Z, Li J, Bryan BA (2021) Future global urban water scarcity and potential solutions. Nat Commun 12(1):4667. https://doi.org/10.1038/s41467-021-25026-3

Hognogi G-G, Meltzer M, Alexandrescu F, Ştefănescu L (2023) The role of citizen science mobile apps in facilitating a contemporary digital agora. Hum Soc Sci Commun 10(1):863. https://doi.org/10.1057/s41599-023-02358-7

Ingrao C, Strippoli R, Lagioia G, Huisingh D (2023) Water scarcity in agriculture: an overview of causes, impacts and approaches for reducing the risks. Heliyon 9(8):e18507. https://doi.org/10.1016/j.heliyon.2023.e18507

Jadeja NB, Banerji T, Kapley A, Kumar R (2022) Water pollution in India—Current scenario. Water Secur 16:100119. https://doi.org/10.1016/j.wasec.2022.100119

Joshi P, Chauhan A, Dua P, Malik S, Liou Y-A (2022) Physicochemical and biological analysis of river Yamuna at Palla station from 2009 to 2019. Sci Rep 12(1):2870. https://doi.org/10.1038/s41598-022-06900-6

Karnataka GO (2024) BWSSB. https://bwssb.karnataka.gov.in/info-1/About%20BWSSB/en

Khan WA, Ali S, Shah S (2022) Water pollution: sources and its impact on human health, control and managing. J Int Cooper Dev 5:69. https://doi.org/10.36941/jicd-2022-0005

Khanam Z, Sultana FM, Mushtaq F (2023) Environmental pollution control measures and strategies: an overview of recent developments. In: Mushtaq F, Farooq M, Mukherjee AB, Ghosh Nee Lala M (eds) Geospatial analytics for environmental pollution modeling: analysis, control and management. Springer, pp 385–414. https://doi.org/10.1007/978-3-031-45300-7_15

Khangura R, Ferris D, Wagg C, Bowyer J (2023) Regenerative agriculture—A literature review on the practices and mechanisms used to improve soil health. Sustainability 15(3):2338. https://www.mdpi.com/2071-1050/15/3/2338

Kulwant M, Patel D, Shirin S, Sharma SN, Kumar YA (2023) Future frameworks for fluoride and algorithms for environmental system. In: Yadav AK, Shirin S, Singh VP (eds) Advanced treatment technologies for fluoride removal in water: water purification. Springer, pp 343–364. https://doi.org/10.1007/978-3-031-38845-3_19

Kumar A, Chava R, Gupta S, Shirin S, Jamal A., Yadav AK (2024) Environmental and health effects of fluoride contamination and treatment of wastewater using various technologies. In: Advanced treatment technologies for fluoride removal in water: water purification. Springer, Cham, pp 323–341. https://doi.org/10.1007/978-3-031-38845-3_18

Kumar D, Ram S, Srivastav AL (2022) Chapter 11—Urban water quality and wastewater discharges in mega and metro cities of India—Challenges, impacts, and management: an overview. In: Srivastav AL, Madhav S, Bhardwaj AK, Valsami-Jones E (eds) Curr Directions Water Scarcity Res 6:223–244. https://doi.org/10.1016/B978-0-323-91838-1.00006-3 (Elsevier)

Kumar S (2019) National mission for clean ganga, pp 177–188

Lakshmikantha V, Hiriyannagowda A, Manjunath A, Patted A, Basavaiah J, Anthony AA (2021) IoT based smart water quality monitoring system. Glob Trans Proc 2(2):181–186. https://doi.org/10.1016/j.gltp.2021.08.062

Leal Filho W, Wall T, Barbir J, Alverio GN, Dinis MAP, Ramirez J (2022) Relevance of international partnerships in the implementation of the UN Sustainable Development Goals. Nat Commun 13(1):613. https://doi.org/10.1038/s41467-022-28230-x

Manasi S, Umamani KS (2013) Water conservation in urban areas: a case study of rain water harvesting initiative in Bangalore City. In: Nautiyal S, Rao KV, Kaechele H, Raju KV, Schaldach R (eds) Knowledge systems of societies for adaptation and mitigation of impacts of climate change. Springer, Berlin, Heidelberg, pp 303–328. https://doi.org/10.1007/978-3-642-36143-2_19

Meran G, Siehlow M, von Hirschhausen C (2021) Integrated water resource management: principles and applications. In: Meran G, Siehlow M, von Hirschhausen C (eds) The economics of water: rules and institutions. Springer, pp 23–121. https://doi.org/10.1007/978-3-030-48485-9_3

Mohanty S (2023) Sustainable water management in India

Negm AM, Omran E-SE, Yadav S, Yadava RN (2022) Update, conclusions and recommendations for "water quality, assessment and management in India". In: Yadav S, Negm AM, Yadava RN (eds) Water quality, assessment and management in India. Springer, pp 393–412. https://doi.org/10.1007/978-3-030-95687-5_19

Nelliyat P (2016) Water pollution: extent, impact, and abatement. In: Narain V, Narayanamoorthy A (eds) Indian water policy at the crossroads: resources, technology and reforms. Springer, pp 131–151. https://doi.org/10.1007/978-3-319-25184-4_8

Okereafor U, Makhatha M, Mekuto L, Uche-Okereafor N, Sebola T, Mavumengwana V (2020) Toxic metal implications on agricultural soils, plants, animals, aquatic life and human health. Int J Environ Res Public Health 17(7). https://doi.org/10.3390/ijerph17072204

Pandeya B, Buytaert W, Potter C (2021) Designing citizen science for water and ecosystem services management in data-poor regions: challenges and opportunities. Curr Res Environ Sustain 3:100059. https://doi.org/10.1016/j.crsust.2021.100059

Patel D, Kulwant M, Shirin S, Kumar A, Ansari M, Yadav A (2022) Artificial intelligence for air quality and control systems, p 20. https://doi.org/10.1201/9781003203445

Patel D, Kulwant M, Shirin S, Varshney R, Pandey G, Yadav AK (2023) Fluoride removal from aqueous solution using iron-based materials: preparation, characterization, and applications. In: Yadav AK, Shirin S, Singh VP (eds) Advanced treatment technologies for fluoride removal in water: water purification. Springer, pp 71–92. https://doi.org/10.1007/978-3-031-38845-3_4

Perveen S, Amar ULH (2023) Drinking water quality monitoring, assessment and management in Pakistan: a review. Heliyon 9(3):e13872. https://doi.org/10.1016/j.heliyon.2023.e13872

Pinder RW, Klopp JM, Kleiman G, Hagler GSW, Awe Y, Terry S (2019) Opportunities and challenges for filling the air quality data gap in low- and middle-income countries. Atmos Environ (1994) 215. https://doi.org/10.1016/j.atmosenv.2019.06.032

Rajaram T, Das A (2008) Water pollution by industrial effluents in India: discharge scenarios and case for participatory ecosystem specific local regulation. Futures 40(1):56–69. https://doi.org/10.1016/j.futures.2007.06.002

Saeed MU, Hussain N, Sumrin A, Shahbaz A, Noor S, Bilal M, Aleya L, Iqbal HMN (2022) Microbial bioremediation strategies with wastewater treatment potentialities—A review. Sci Total Environ 818:151754. https://doi.org/10.1016/j.scitotenv.2021.151754

Sarker IH (2021) Data science and analytics: an overview from data-driven smart computing, decision-making and applications perspective. SN Comput Sci 2(5):377. https://doi.org/10.1007/s42979-021-00765-8

Scanlon BR, Fakhreddine S, Rateb A, de Graaf I, Famiglietti J, Gleeson T, Grafton RQ, Jobbagy E, Kebede S, Kolusu SR, Konikow LF, Long D, Mekonnen M, Schmied, HM, Mukherjee A, MacDonald A, Reedy RC, Shamsudduha M, Simmons CT, Sun A, Taylor RG, Villholth KG, Vörösmarty CJ, Zheng C (2023) Global water resources and the role of groundwater in a resilient water future. Nat Rev Earth Environ 4(2):87–101. https://doi.org/10.1038/s43017-022-00378-6

Schullehner J, Cserbik D, Gago-Ferrero P, Lundqvist J, Nuckols JR (2024) Integrating different tools and technologies to advance drinking water quality exposure assessments. J Expos Sci Environ Epidemiol 34(1):108–114. https://doi.org/10.1038/s41370-023-00588-0

Seelen LMS, Flaim G, Jennings E, De Senerpont Domis LN (2019) Saving water for the future: public awareness of water usage and water quality. J Environ Manage 242:246–257. https://doi.org/10.1016/j.jenvman.2019.04.047

Shalu, Singh G (2023) Environmental monitoring with machine learning. EPRA Int J Multidisc Res (IJMR) 208–212. https://doi.org/10.36713/epra13330

Shammin MR, Haque AKE, Faisal IM (2022) A framework for climate resilient community-based adaptation. In: Haque AKE, Mukhopadhyay P, Nepal M, Shammin MR (eds) Climate change and community resilience: Insights from South Asia. Springer, pp 11–30. https://doi.org/10.1007/978-981-16-0680-9_2

Sharma RK, Yadav M, Gupta R (2017) Chapter five—Water quality and sustainability in India: challenges and opportunities. In Ahuja S (ed) Chem Water. Elsevier, pp 183–205. https://doi.org/10.1016/B978-0-12-809330-6.00005-2

Shirin S, Yadav AK (2014) Physico-chemical analysis of municipal wastewater discharge in Ganga river, Haridwar district of Uttarakhand, India. Curr World Environ 9(2):536–543. https://doi.org/10.12944/CWE.9.2.39

Singh D, Dohare S, Yadav G, Pandey H, Singh V (2023) Chapter 6—polluted water bodies remediation by using GIS and remote sensing approach: a deeper insight. In: Chatterjee U, Pradhan B, Kumar S, Saha S, Zakwan M, Fath BD, Fiscus D (eds) Water, land, and forest susceptibility and sustainability, vol 1. Elsevier, pp 155–169. https://doi.org/10.1016/B978-0-323-91880-0.00021-0

Srivastava A, Sharma VP (2021) Water quality monitoring and management: importance, applications, and analysis, pp 421–440. https://doi.org/10.1002/9781119725237.ch16

Suresh S (2012) Global challenges need global solutions. Nature 490(7420):337–338. https://doi.org/10.1038/490337a

UK LA (2021) Environment Act 2021. https://www.legislation.gov.uk/ukpga/2021/30/contents/enacted (legislation.gov.uk)

UNEP. (2012). National Water Policy 2012. https://leap.unep.org/en/countries/in/national-legislation/national-water-policy-2012 (UNEP)

Van der Bruggen B (2021) Sustainable implementation of innovative technologies for water purification. Nat Rev Chem 5(4):217–218. https://doi.org/10.1038/s41570-021-00264-7

Wang X, Cheng H, Ye G, Fan J, Yao F, Wang Y, Jiao Y, Zhu W, Huang H, Ye D (2022) Key factors and primary modification methods of activated carbon and their application in adsorption of carbon-based gases: a review. Chemosphere 287:131995. https://doi.org/10.1016/j.chemosphere.2021.131995

Yadav AK, Sahoo SK, Mahapatra S, Kumar AV, Pandey G, Lenka P, Tripathi RM (2014) Concentrations of uranium in drinking water and cumulative, age-dependent radiation doses in four

districts of Uttar Pradesh, India. Toxicol Environ Chem 96(2):192–200. https://doi.org/10.1080/02772248.2014.934247
Zhang H, Cao X, Huo S, Ma C, Li W, Liu Y, Tong Y, Wu F (2023) Changes in China's river water quality since 1980: management implications from sustainable development. NPJ Clean Water 6(1):45. https://doi.org/10.1038/s41545-023-00260-y
Zulkifli CZ, Garfan S, Talal M, Alamoodi AH, Alamleh A, Ahmaro IYY, Sulaiman S, Ibrahim AB, Zaidan BB, Ismail AR, Albahri OS, Albahri AS, Soon CF, Harun NH, Chiang HH (2022). IoT-based water monitoring systems: a systematic review. Water 14(22):3621. https://www.mdpi.com/2073-4441/14/22/3621

Chapter 8
Comparative Studies on Ground Water Quality of Krishnagiri, Tiruvannamalai, and Vellore Districts of Tamil Nadu, India

Ambrose Paulraj, Narasimman Manickam, Razack Rafi Mohamed, Chandrasekaran Vasumathi, and Rajagopal Bhuvaneswari

Abstract In the current study, groundwater levels were monitored in three districts (Krishnagiri, Vellore, and Tiruvannamalai) in the northwest of Tamil Nadu, India. Water samples were taken at 28 locations, including 8 in Krishnagiri, 13 in Vellore, and 7 in Tiruvannamalai. All twelve borewell samples, twelve open well samples, three lake samples, and one municipal source sample were analyzed. The physical and chemical characteristics were examined and recorded. The three districts of water samples were analyzed for appearance, odor, turbidity, total dissolved solids (TDS), electrical conductivity (EC), pH, total hardness (TH), total alkalinity (TA), calcium (Ca), magnesium (Mg), iron (Fe), sodium (Na), potassium (K), free ammonia ($NH_3{}^-N$), nitrate ($NO_3{}^-$), chloride (Cl^-), heavy metals, sulfides ($S_2{}^-$),

A. Paulraj (✉)
PG and Research Department of Zoology, Government Arts College for Men, Krishnagiri 635001, Tamil Nadu, India
e-mail: ambrose.paulraj@rediffmail.com

N. Manickam
Marine Planktonology and Aquaculture Laboratory, Department of Marine Science, School of Marine Sciences, Bharathidasan University, Tiruchirappalli 620024, Tamil Nadu, India

Crustacean Biology Laboratory, Department of Zoology, Bharathiar University, Coimbatore 641046, Tamil Nadu, India

Department of Zoology, Sree Meenakshi Arts & Science College, Meenakshi Nagar, Eriyur (Po), Pennagaram (Tk), Dharmapuri 636810, Tamil Nadu, India

R. R. Mohamed
PG and Research Department of Zoology, C. Abdul Hakeem College (Autonomous), Melvisharam, Ranipet 632509, Tamil Nadu, India

C. Vasumathi
PG and Research Department of Zoology, Government Arts College, Dharmapuri 636701, Tamil Nadu, India

R. Bhuvaneswari
Fish Disease Diagnostic Laboratory, Department of Zoology, Ayya Nadar Janaki Ammal College (Autonomous), Sivakasi 626124, Tamil Nadu, India

A. K. Yadav et al. (eds.), *Integrated Management of Water Resources in India: A Computational Approach*, Water Science and Technology Library 129,
https://doi.org/10.1007/978-3-031-62079-9_8

oil and greases fluoride, sulfate (SO_4^{2-}), and phosphate (PO_4^{3-}). Higher values of turbidity, TDS, EC, pH, TA, TH, Ca, Mg, Fe, Na, K, NH_3^-N, NO_3^-, Cl^-, F^-, SO_4^{2-} and PO_4^{3-} are found in Krishnagiri district, respectively. For TDS, TH, Ca, Mg, NH_3^-N, NO_3^-, Cl, and F^- exceed the WHO and TWAD permissible limits in Krishnagiri district, In the pH ranges of Vellore district, TSS, TDS, Ca, SO_4^{2-}, BOD, COD, S_2^-, NH_3N, Na, TA, TH, hardness of Ca of hardness and Mg hardness were recorded, respectively. Phenolic compounds, hexavalent chromium, total chromium (Cr), F^-, and oil and greases showed minimum detectable amounts in the Vellore district water samples. Higher and lower values of EC, pH, TSS, TDS, Cl, SO_4^{2-}, BOD, COD, TA, TH, Ca and Mg hardness is recorded in Tiruvannamalai district water samples, respectively, and other parameters are below the minimum detectable limits. Compared to WHO and TWAD standards, the maximum values of the water sample exceed the permissible limits in the present study of three districts. Consequently, public awareness campaigns regarding the indicators of water quality and their significance will be scheduled in addition to education in schools, colleges and the public.

Keywords Physical and chemical parameter · Ground water · Borewell · Open well water · Krishnagiri · Tiruvannamalai · Vellore

8.1 Introduction

Life originated on Earth because it is the only planet with liquid water. Since water is a plentiful resource in nature and the stuff of life, humans have used it more than any other resource. Water is also known as "life." Since it gives the planet the ability to support life, water is revered as a precious gift from God. It makes up nearly 70.9% of the Earth's surface, of which 97.2% of marine water is found in the seas and just 2.8% is found in freshwater environments, including rivers, lakes, and streams (Strahler 1992; Manickam et al. 2015b). A prominent topic in scientific, political, social, and human discourse is water. There is a general fear that our civilization's ability to share the planet's limited water resources will determine future peace and prosperity on a finite planet. Furthermore, it is acknowledged that such considerate water sharing comes with hitherto unseen social, scientific, and human obstacles entwined in intricate ways (Narasimhan 2005; Shirin and Yadav 2014; Shirin et al. 2019). However, water quality is gradually declining since new substances and contaminants (including pesticides and antibiotics) have been discovered in a variety of settings. Understanding the dynamics, composition, and structure of biotic communities is crucial to minimizing negative consequences and maintaining the ecological integrity of freshwater aquatic ecosystems (Manickam et al. 2015b, 2023; Kumar et al. 2024). On a socioeconomic level, man-made ponds, lakes, and reservoirs are crucial to human societies because they enable global irrigation and human consumption, generate electricity, and promote the growth of agriculture, cities, and industries (Manickam et al. 2015b, 2018; Aravinth et al. 2023; Yadav 2014). In terms of their low flow velocity, abiotic characteristics and capacity to hold water, reservoirs are similar to natural lakes compared to rivers. Any water body must first conduct a primary survey of its phytoplankton and zooplankton because this list of organisms is essential for the future management of aquatic resources, conservation efforts, and sustainable exploitation (Manickam et al. 2014, 2015a, 2018; Aravinth et al. 2023).

The sun provides most of the energy for the hydrological cycle. About 40% of the total solar energy collected by the Earth's surface is lost to evaporation as latent heat, and the remaining 18% is returned as sensible heat. Water is essential to the redistribution of energy in the atmosphere through zonal or latitudinal circulation patterns, as well as meridional or longitudinal convection cells. Across the whole surface of the Earth, the average yearly precipitation is estimated to be around 100 cm, with a total volume of approximately 5105 km. Given that this is around 39 times the total amount of water in the atmosphere, the average residence time of water in the atmosphere is estimated to be 9.4 days. The Earth's evolution over billions of years has been impacted by this atmospheric circulation, which is dynamically related to the considerably larger time scales of surface water circulation (years to decades) and groundwater circulation (decades to centuries), as reported by Drever (1997) and ENPHO (1998, 2003).

8.2 Materials and Methods

In the present study, groundwater quality was studied in the Krishnagiri, Vellore and Tiruvannamalai district areas (Tamil Nadu, India). A total of 28 water sample analyzes were carried out. Krishnagiri district represents 8 samples, Vellore district represents 13 samples, and Tiruvannamalai district (T.V. Malai) represents 7 samples. In this study, 12 borewell water samples, 12 open well water samples 3 lake water samples, and 1 municipal water sample are analyzed. The Krishnagiri, Vellore and Tiruvannamalai districts are located northeast of Tamil Nadu, India (Fig. 8.1). Geologically, these districts are surrounded by small hills, except for Tiruvannamalai town. The Pennai and Palarru river systems run in these districts. Two liters of water samples were collected in the plastic bottle from January 2008 to April 2008 in various places. The sampling bottles were thoroughly percleaned with 5% of HNO_3 for the physical and chemical examinations.

Appearance, odor, turbidity, total dissolved solids (TDS), electrical conductivity (EC), pH, total hardness (TH), total alkalinity (TA), calcium, magnesium, iron, sodium, potassium, free ammonia, nitrate, chloride, fluoride, sulfate, and phosphate were analyzed in Krishnagiri district water samples from Tamilnadu Water Supply and Drainage Board (TWAD), District Water Testing Laboratory, Dharmapuri. WaterWater samples were collected from the eight places in Krishnagiri district (Six places of borewell water: Guruvinayanapalli, Guddur, Rengasamudram, Puroshothamapuram, Vinayagapuram and Indra Gandhi Nagar; two places of open well water: Kuppanatham and Mittapalli). pH, total suspended solids, total dissolved solids, chloride, sulfate, BOD, COD, sulphides, ammonical nitrogen,

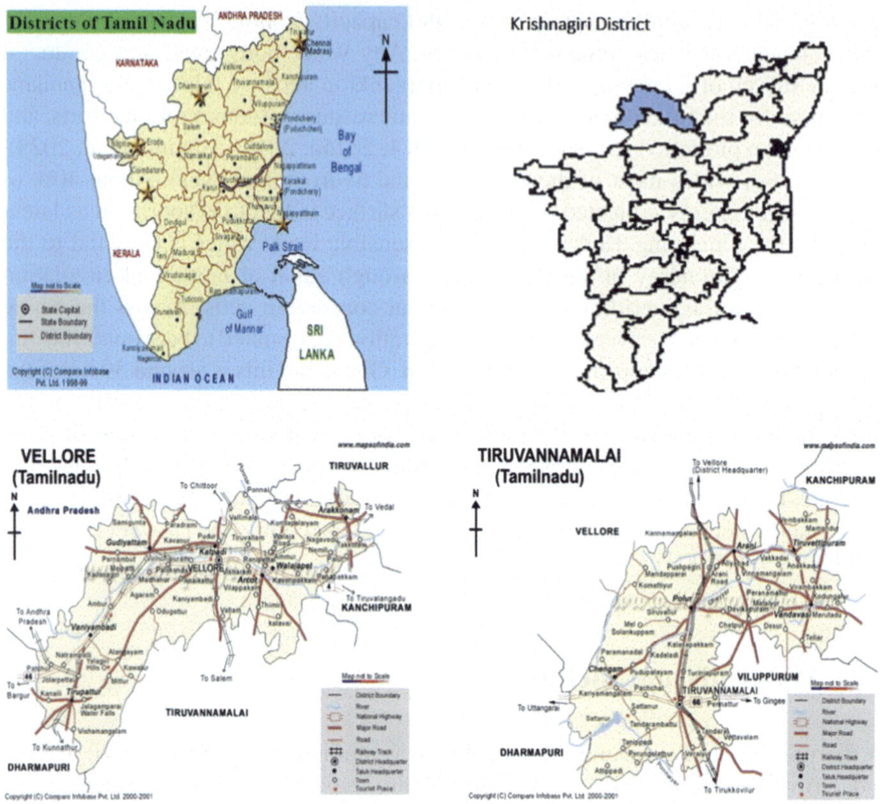

Fig. 8.1 Map showing the sampling areas

phenolic compounds, hexavalent chromium, total chromium, sodium in percentage, alkalinity, total hardness, calcium hardness, magnesium hardness, fluoride, oils and greases were analyzed in Vellore district water samples at the Tamil Nadu Pollution Control Board Laboratory, Vellore. The water samples were collected from the 13 places in the Vellore district: four places of borewell water: Vaniambadi I, Vaniambadi II, Ambur I, and Ambur II; one place of tank water: Vellore Municipal Water; Six places of open-well water: Sathavacheri, Arcot, Tirupattur, Uadapudupet, Melvisharam, and Girisamudram; and two places of lake water: Puliyankannu and Karai.

Color, odor, taste of turbidity, electrical conductivity, pH, Total suspended solids (TSS), Total dissolved solids, chloride, sulfates, biological oxygen demand (BOD), chemical oxygen demand (COD), oil and greases, sulfurides, ammonical nitrogen, phenolic compounds, hexavalent chromium, total chromium, alkalinity, total hardness, calcium hardness, magnesium hardness, and fluoride were analyzed in water samples from Tiruvannamalai district in Tamilnadu Pollution Control Board Laboratory, Vellore. The water samples were collected from the seven places in Tiruvannamalai district: one place of lake water place: Usambodi; Four places of open-well water: Arni, Tiruvannamalai I, Tiruvannamalai II, and Cheyyar; and two places of borewell water: Polur and Arunagiri Chatriam. In addition to this, heavy metals such as mercury, cadmium, lead, zinc, and copper were also analyzed for Tiruvannamalai and Vellore district water samples in the Tamil Nadu Pollution Control Board Laboratory, Vellore. The physical and chemical parameters of various samples are tabulated and statistically analyzed. The mean, standard deviation (SD), standard error (SE), and significance test (t test) were analyzed in SPSS software version 10.0. The results were tabulated.

8.3 Results

This ocean covers 71% of the rest of the planet, which is made up of glaciers and ice caps. Streams and lakes, subsurface soils, and the atmosphere are other places that contain water. There are frozen oceans that contain 97.3% of the world's water, with the remaining 95% of water being chemically bound into rocks. Groundwater, soil water, and atmospheric water vapor are found in the remaining 81.1%, while ice and permanent glaciers remain at 2.1%. The surface and inland water accounts for just 0.8% of the total. Surface water on land, such as lakes and streams, contains the remaining 0.2%. Unfortunately, the careless release of chemical waste and hazardous materials is contaminating this finite supply of clean water. Three districts provided water samples, which are: Krishnagiri, Vellore, and Tiruvannamalai, the three districts in question. All varieties of groundwater, such as lake water, borehole water, and open-well water, are included in this water.

Eight water samples were taken in all from the districts of Krishnagiri. The Bargur Union contains the locations of Guruvinayanapalli, Guddur, Rengasamudram, and Puroshothamapuram; Uthangarai Union contains the locations of Kuppanatham and the Mittapalli; and Hosur Union contains the locations of Vinayagapuram and Indra Gandhi Nagar. Of these eight samples, two are open well waters (Kuppanatham and Mittapalli) and six are borewell fluids (Guruvinayanapalli, Guddur, Rengasamudram, Puroshothamapuram, Vinayagapuram, and Indra Gandhi Nagar). Table 8.1 shows the physical and chemical characteristics of the findings of the eight-place water sample. The physical examination of the water's appearance reveals that it is colorless. All the samples had no taste or smell. The water samples from the Krishnagiri district have turbidities ranging from 0.6 to 2.0. Vinayagapuram and Rengasamudram samples had electrical conductivity (EC) ranging from 1070 to 1620, respectively. The total values of dissolved solids (TDS) in Rengasamudram and Vinayagapuram, Krishnagiri district, are 735 and 1124, respectively (Table 8.1).

In Krishnagiri district's Guruvinayanapalli and Puroshothamapuram had pH values ranging from 7.2 to 7.81, respectively (Table 8.1). In Kuppanathan and Rengasamudram, the total alkalinity ranged from 298 to 376 ppm, respectively.

Table 8.1 Physical and Chemical parameters of water samples in different places of Krishnagiri District

Physical and chemical parameters	Water samples of different places of Krishnagiri District							
	1	2	3	4	5	6	7	8
Appearance	Colourless	Colourless	Colourless	Colourless	Clear	Clear	Clear	Clear
Odour	None	None	None	None	None	None	None	None
Turbidity	2.0	1.3	1.0	0.8	1.2	0.6	0.6	1.2
TDS	991	1063	1124	872	814	905	735	878
EC	1420	1530	1620	1270	1180	1310	1070	1270
pH	7.20	7.31	7.66	7.81	7.46	7.24	7.46	7.72
Alkalinity pH	0	0	0	0	0	0	0	0
Total Alkalinity (ppm)	318	341	376	318	298	318	321	341
Total hardness (ppm)	428	464	580	412	404	448	348	368
Calcium (ppm)	101	106	133	96	94	104	82	85
Magnesium (ppm)	42	48	60	41	40	45	35	37
Iron (ppm)	0.28	0.14	0.28	0.14	0.00	0.14	0.00	0.28
Sodium (ppm)	112	126	112	82	72	76	90	108
Potassium (ppm)	13	15	17	10	8	14	7	7
Free Ammonia (ppm)	0.02	0.03	0.03	0.03	0.04	0.06	0.0	0.0
Nitrate (ppm)	36	31	56	41	54	31	40	33
Chloride (ppm)	228	251	281	166	152	190	127	160
Fluoride (ppm)	1.2	1.3	1.2	1.3	1.2	1.3	1.4	1.3
Sulphate (ppm)	20	34	26	17	14	19	19	19
Phosphate (ppm)	0.04	0.00	0.00	0.01	0.7	0.0	0.0	0.0

In Vinayagapuram and Rengasamudram, the total hardness (TH) varied from 348 to 580 ppm, respectively (Table 8.1). Rengasamudram (133 ppm) and Vinayagapuram (82 ppm) were the locations of the higher and lower calcium readings, respectively. In Vinayagapuram and Regasamudram, the magnesium concentrations were 35 and 60 parts per million, respectively. The potassium and sodium were 7–17 ppm and 72–126 ppm, respectively, in Kuppantham, Guddur, Indra Gandhi Nagar, and Rengasamudram. Groundwater samples from the Krishnagiri district have sulfate concentrations ranging from 14 to 34 parts per million. Phosphate levels were found to be 0.01, 0.04, and 0.7 ppm in Gurvinayanapalli, Puroshothamapuram, and Kuppanatham, respectively. No phosphate was found in the groundwater samples from Guddur, Rengasamudram, Mittapalli, Vinayagapuram, and Indra Gandhi Nagar (Table 8.1).

Table 8.1 shows that the water samples from the Rengasamudram well had higher concentrations of nitrate (56 ppm) and chloride (281 ppm) than the samples from Vinayagapuram and Rengasamudram (127 and 56 ppm, respectively). Not every sample had nitrate. Indra Gandhi Nagar, Guddur, Rengasamudram, Vinayagapuram, Mittapalli, and Guruvinayanapalli, and Indra Gandhi Nagar, respectively, had higher and lower amounts of iron, fluoride, and free ammonia, at 0.14–0.28 ppm, 1.2–1.4 ppm, and 0.06–0.02 ppm, according to Table 8.1. The mean, standard deviation (SD) and SE of the Krishnagiri district samples are shown in Table 8.2. Table 8.3 shows the significant values found in the t-test findings for each water sample in the Krishnagiri district. 13 groundwater samples were collected in Vellore district. This contains samples of water from the wells in Ambur and Vaniambadi. Open-pit wells were used in Sathavacheri, Arcot, Tirupattur, Vadapudupet and Girisamudram were used to collect water samples. Two lake water samples were collected from Puliyankannu and Karai. The municipal water supply tank in Vellore yielded one sample of water.

The values of the Vellore district water samples are presented in Table 8.4. Physical parameters such as appearance and odor turbidity are normal in all water samples. The highest and lowest electrical values were recorded in the water of Girisamudram and Melvisharam, respectively. A higher pH value was recorded in the Puliyankannu lake water sample (8.7). Lower pH values were recorded in Vaniambadi I (7.0) and Girisamudram (7.0) borewell and well water samples, respectively (Table 8.4). The crucial determinants were determined to be 12–116 mg/l, 516–11,868 mg/l, 450–3550 mg/l, 110–1530 mg/l, and 230–2020 mg/l, respectively, in Vaniambadi II-Puliyankannu Lake, Tirupattur-Girisamudram, Vaniambadi I-Girisamudram and Melvisharam-Girisamudram (Table 8.4). The highest concentrations of alkalinity, sulfate and chloride in the Vellore district are 800 mg/l (Girisamudram), 1730 mg/l (Puliyankannu), and 4797 mg/l (Girisamudram), in that order.

Lower values were found in Melvisharam (166 mg/l) Sathavacheri (58 mg/l), and Melvisharam (255 mg/l) for alkalinity, sulfate and chloride, respectively (Table 8.4). The water of Karai Lake (28 mg/l) and Puliyankannu Lake (51 mg/l) had higher BOD (Biological Oxygen Demand) values. The remaining locations, including Melvisharam, Ambur I and II, Sathavacheri, Arcot, Tirupattur, Vadapudupet, and Arcot, registered 3 mg/l. Lower BOD values were found in Table 8.4 for Vaniambadi

Table 8.2 The mean, standard deviation (SD) and standard error (SE), values of water samples in the Krishnagiri District

Physical–chemical parameters	Mean	SD	SE
pH	7.4825	0.2282	8.068 (E–02)
TDS	922.7500	129.1729	45.6695
Nitrate	40.2500	9.8525	3.4834
Chloride	194.3750	53.6947	18.9840
Fluoride	1.2750	7.071	2.500
Sulphate	21.000	6.2335	2.2039
Phosphate	9.375 (E–02)	0.2454	8.867 (E–02)
Total alkalinity	328.8750	23.5581	8.3290
Total hardness	431.5000	71.2481	25.1900
Calcium	100.1250	15.7792	5.5788
Magnesium	43.5000	7.8376	2.7710
Iron	0.1575	0.1168	4.131 (E–02)
Sodium	97.2500	19.826	7.0096
Potassium	11.3750	3.8891	1.3750
EC	1333.7500	181.1816	64.0574

Table 8.3 Test of significance values of physical and chemical parameters of water samples in the Krishnagiri District

Physical chemical parameters	T-test values	Significant values
pH	92.739	0.00
TDS	20.205	0.00
Nitrate	11.555	0.00
Chloride	10.239	0.00
Fluoride	51.00	0.00
Sulphate	9.529	0.00
Phosphate	1.081	0.316
Total alkalinity	39.485	0.00
Total hardness	17.130	0.00
Calcium	7.947	0.00
Magnesium	5.698	0.00
Iron	3.813	0.07
Sodium	13.874	0.00
Potassium	8.273	0.00
EC	20.821	0.000

*Significant at 5% level

Table 8.4 Physical and chemical parameters of water sample in the Vellore District

Parameters	Water sampling for the different places of Vellore District												
	1	2	3	4	5	6	7	8	9	10	11	12	13
pH	7.00	7.7	7.2	7.2	7.70	7.9	7.1	7.6	7.1	7.5	7.0	8.7	8.82
TSS (mg/l)	12	12	16	28	12	12	20	12	56	12	76	116	56
TDS (mg/l)	844	952	724	1404	744	764	1152	516	2740	616	11,868	4248	3212
Chloride (mg/l)	298	248	199	470	243	243	380	253	820	166	4797	1213	850
Sulphate (mg/l)	63	190	230	177	85	58	135	147	272	102	640	1730	1081
BOD (mg/l)	2	2	3	3	2	3	3	3	3	3	2	51	28
COD (mg/l)	24	16	32	24	16	24	32	24	24	24	24	600	344
Sulphides (mg/l)*	< 1.0	< 1.0	< 1.0	< 1.0	< 1.00	< 1.0	< 1.0	< 1.0	< 1.0	< 1.0	< 1.0	< 1.0	< 1.0
Ammonical Nitrogen (mg/l)*	< 5.0	< 5.0	< 5.0	< 5.0	< 5.0	< 5.0	< 5.0	< 5.0	8	< 5.0	8	18.0	14.0
Phenolic compound (mg/l)*	< 0.0005	< 0.0005	< 0.0005	< 0.0005	< 0.0005	< 0.0005	< 0.0005	< 0.0005	< 0.0005	< 0.0005	< 0.0005	< 0.0005	< 0.0005
Hexavalent chromium (mg/l)*	< 0.01	< 0.01	< 0.01	< 0.01	< 0.01	< 0.01	< 0.01	< 0.01	< 0.01	< 0.01	< 0.01	< 0.01	< 0.01
Total chromium (mg/l)*	< 0.003	< 0.003	< 0.003	< 0.003	< 0.003	< 0.003	< 0.003	< 0.003	< 0.003	< 0.003	< 0.003	< 0.003	< 0.003
Sodium (%)	8	22	16	12	Nil	Nil	Nil	12	33	Nil	24	60.0	66

(continued)

Table 8.4 (continued)

Parameters	Water sampling for the different places of Vellore District												
	1	2	3	4	5	6	7	8	9	10	11	12	13
Alkalinity (mg/l)	385	330	205	455	500	500	340	455	680	255	800	375	455
Total hardness (mg/l)	450	520	480	590	600	540	450	540	1050	350	3550	480	580
Calcium hardness (mg/l)	220	270	180	200	310	230	150	170	610	110	1530	220	300
Magnesium hardness (mg/l)	230	250	300	390	290	310	300	370	440	240	2020	260	280
Fluoride (mg/l)*	< 1.0	< 1.0	< 1.0	< 1.0	< 1.0	< 1.0	< 1.0	< 1.0	< 1.0	< 1.0	< 1.0	< 1.0	< 1.0
Oils and Greases (mg/l)*	< 1.0	< 1.0	< 1.0	< 1.0	< 1.0	< 1.0	< 1.0	< 1.0	< 1.0	< 1.0	< 1.0	< 1.0	< 1.0

**LMDL* less than minimum detectable limit

I and II (2 mg/l), Vellore municipal water (2 mg/l), and Girisamudram water (2 mg/l). Water from Karai Lake (344 mg/l) and Paliyankannu (600 mg/l) has higher chemical oxygen demand values. Vaniambadi I, Ambur II, Sathavacheri, Tirupattur, Vadapudupet, Melvisharam and Girisamudram were found to have 24 mg/l COD. COD levels of 32 mg/l were found in Arcot and Ambur I water. Table 8.4 shows that Vellore municipal water (16 mg/l) and Vaniambadi II (16 mg/l) had the lowest COD values.

In the different water samples of the Vellore district, other chemical parameters such as sulfurides, ammoniacal nitrogen, phenolic compounds, hexavalent chromium, and total chromium demonstrated minimum detectable limits (Table 8.4). The minimal detectable limits for heavy metals such as mercury, cadmium, lead, zinc and copper were found to be 0.009 mg/l, 0.8 mg/l, 0.09 mg/l, 1.5 mg/l, and 1.5 mg/l, respectively (Table 8.4). Vaniambadi I and Karai Lake have sodium levels ranging from 8 to 66%. Table 8.5 displayed the mean ± SD and SE values of the water samples from the Vellore district, while Table 8.6 displayed the t-test data. Every water sample showed noteworthy results.

Water was collected from one lake, four wells, two wells and Usambadi, Arni, Tiruvannamalai, Polur, Arunagiri Chatriam, Tiruvannamalai, and Cheyyar for the Tiruvannamalai district. Table 8.7 shows the water values for the Tiruvannamalai district. All samples lacked color and smell. The water from Usambadi Lake was found to be slightly murky, whereas the other samples had a clear look. The water from Usambadi Lake, the Tiruvannamalai well, the Arunagiri Chatriam well, and the Cheyyar well all tasted salty. There was no flavor in the other water samples. The electrical conductivity was highest in Cheyyar (10,525) and Tiruvannamalai well-water (8950). The low EC values (322) in the water of Usambadi Lake are listed in Table 8.7.

Table 8.5 The mean, SD and SE of physical and chemical parameters of water samples in the Vellore District

Physical chemical parameters	Mean	SD	SE
pH	7.5785	0.6022	0.1670
TSS	33.8462	32.8071	9.0991
TDS	2291.0769	3105.1854	861.2235
Chloride	783.0769	1247.3514	345.9530
Sulphate	377.6923	497.9709	138.1123
BOD	8.3077	14.62	4.0583
COD	92.9231	176.2283	48.8769
Sodium	28.111	21.2158	7.0719
Alkalinity	441.1538	161.7245	44.8543
T.H	783.0769	847.5298	235.0624
C.H	348.1538	376.3761	104.3880
M.H	439.9231	479.6246	133.0239

Table 8.6 Test of significance values of physical and chemical parameters of water samples in the Vellore District

Physical and chemical parameters	T-test values	Significant values
pH	45.371	0.00
TSS	3.720	0.03
TDS	2.660	0.021
Chloride	2.264	0.043
Sulphate	2.735	0.018
BOD	2.048	0.063
COD	1.901	0.082
Sodium	3.975	0.004
Alkalinity	9.835	0.000
T.H	3.331	0.006
C.H	3.316	0.006
M.H	3.285	0.007

*Significant at 5% level

The pH values of the water samples from the Tiruvannamalai district varied from 6.2 to 7.6 for the Tiruvannamalai well water and Cheyyar well-water, respectively (Table 8.7). The levels of TSS in Tiruvannamalai well water and the Usambadi lake water varied from 4 to 32 mg/l, as shown in Table 8.7. In the Cheyyar well water, the highest values of TDS, total hardness, calcium hardness, and magnesium hardness were measured (5524 mg/l, 3580 mg/l, 2725 mg/l, and 855 mg/l, respectively). In water samples from the Tiruvannamalai district, the lower values of TDS, total hardness, calcium hardness, and magnesium hardness were found to be approximately 168 mg/l 125 mg/l, 75 mg/l, and 50 mg/l, respectively (Table 8.7).

Table 8.7 shows the range of values for chloride, sulfate, and alkalinity in the Tiruvannamalai district, which are 78–2213 mg/l 32–1102 mg/l, and 155–555 mg/l, respectively. The well water from Arni and Tiruvannamalai did not show alkaline readings. Table 8.7 shows the BOD and COD values. Usambadi lake water has higher BOD (16 mg/l) and COD (172 mg/l) readings. Low BOD levels (24 mg/l) were found in the Arni well, the Polur well, and the Tiruvannamalai water. In every sample taken from the Tiruvannamalai district, the levels of oil and grease, sulfur, ammonia, phenolic compounds, and total chromium were found to be below the lowest detectable limit. The sample values for the mean, SD, SE and t-test for the Tiruvannamalai districts were shown in Tables 8.8 and 8.9. Notable results were obtained from the physical and chemical parameters of each t-test sample.

The Vellore district has higher mean levels of alkalinity (441.15 mg/l) TDS (2291.07 mg/l), and pH (7.5785) than the other two districts. The water samples from the Tiruvannamalai district had higher mean values of total hardness (1340.71 mg/l), sulfate (442.28 mg/l), and chloride (866 mg/l) than those of the Vellore and Krishnagiri districts. Comparing the mean values of TSS (33.84 mg/l) BOD (8.30 mg/l), and

Table 8.7 Physical and chemical parameters of the water sample in the Tiruvannamalai District

S. No	Parameters	Water sampling for the different places of Tiruvannamalai District						
		1	2	3	4	5	6	7
1	Colour	No	No	No	No	No	No	No
2	Odour	No	No	No	No	No	No	No
3	Turbidity	Slight	Clear	Clear	Clear	Clear	Clear	Clear
4	Taste	Salty	No	Salty	No	Salty	No	Salty
5	E C (μmh)	522	322	8950	950	5950	524	10,525
6	pH	6.8	7.3	6.7	7.2	6.3	7.6	6.2
7	TSS (mg/l)	32	16	8.0	4.0	8.0	4.0	8.0
8	TDS (mg/l)	288	168	4475	654	3876	328	5524
9	Chloride (mg/l)	102	78	1278	278	1986	127	2213
10	Sulphate (mg/l)	65	32	842	142	1102	46	867
11	BOD (mg/l)	16	3	9	3	6	3	12
12	COD (mg/l)	172	40	88	24	72	32	128
13	Oils and greases (mg/l)*	< 1.0	< 1.0	< 1.0	< 1.0	< 1.0	< 1.0	< 1.0
14	Sulphide (mg/l)*	< 1.0	< 1.0	< 1.0	< 1.0	< 1.0	< 1.0	< 1.0
15	Ammonical nitrogen (mg/l)*	< 5.0	< 5.0	< 5.0	< 5.0	< 5.0	< 5.0	< 5.0
16	Phenolic compound (mg/l)*	< 0.005	< 0.005	< 0.005	< 0.005	< 0.005	< 0.005	< 0.005
17	Hexavalent chromium (mg/l)*	< 0.01	< 0.01	< 0.01	< 0.01	< 0.01	< 0.01	< 0.01
18	Total chromium (mg/l)*	< 0.003	< 0.003	< 0.003	< 0.003	< 0.003	< 0.003	< 0.003
19	Alkalinity (mg/l)	200	Nil	325	155	300	Nil	555
20	Total hardness (mg/l)	185	125	2580	380	2350	185	3580
21	Calcium hardness (mg/l)	110	75	1800	215	1750	100	2725
22	Magnesium hardness (mg/l)	75	50	780	175	800	85	855
23	Fluoride (mg/l)*	< 1.0	< 1.0	< 1.0	< 1.0	< 1.0	< 1.0	< 1.0

* *LMDL* less than minimum detectable limit

COD (92.92 mg/l) in water samples from the Tiruvannamalai district and Vellore showed higher values. In the current study, the mean values of calcium hardness and magnesium hardness in water samples from the Tiruvannamalai and Vellore districts were found to be higher than those from the other two districts. The t-test analysis revealed the same pattern.

Table 8.8 The mean, SD and SE of physical and chemical parameters of water samples in the Tiruvannamalai District

Physical–chemical parameters	Mean	SD	SE
pH	6.8714	0.5219	0.1973
TSS	11.4280	9.9139	3.7471
TDS	2187.5714	2334.9434	882.5236
Chloride	866.0	942.9	356.40
Sulphate	442.2857	471.3932	178.1699
BOD	7.4286	5.1270	1.9378
COD	79.4286	54.6713	20.6638
T.H	1340.7143	1451.5033	548.6167
Alkalinity	219.2857	196.2445	74.1734
C.H	967.8571	1098.8511	415.326
M.H	402.8571	384.9985	145.5157

Table 8.9 Test of significance values of physical and chemical parameters of water samples in the Tiruvannamalai District

Physical–chemical parameters	T-test values	Significant
pH	34.834	0.00
TSS	3.050	0.00
TDS	2.479	0.00
Chloride	2.430	0.00
Sulphate	2.482	0.00
BOD	3.833	0.00
COD	3.844	0.00
T.H	2.444	0.00
Alkalinity	2.956	0.00
C.H	2.330	0.00
M.H	2.768	0.00

8.4 Discussion

Water scarcity is a common occurrence in the Krishnagiri, Vellore, and Tiruvannamalai districts of Tamil Nadu (India). In addition, lakes and ponds are seasonal, and the last 10 years have seen a failure in rainfall in these areas, resulting in a severe shortage of water for irrigation and human consumption. Most of the residents of these districts relied on open wells and wells for irrigation and residential needs, and the ongoing usage of these groundwater sources revealed various physical and chemical characteristics during the current study. Additionally, tanneries and the leather industry severely pollute the Vellore district (there are several reports to that effect, as do orders from the Supreme Court and the High Court). The findings of the current study revealed that the three districts' physical and chemical factors of the three districts varied. Compared to the WHO drinking water standard and the TWAD standard, TDS, total hardness, calcium, and magnesium in Krishnagiri district water samples were found to be higher than the allowed limits. The same kind of groundwater results were also observed by Datta (2005) in a different area of Delhi, Nepal (Sharma et al. 2005). This region was severely impacted and became much more susceptible to pollution with a wide range of containments, including nitrate (< 22.16 mg/l) and fluoride (< 1.16 mg/l).

According to our current findings, the amount of rainfall and the level of groundwater correspond exactly. Should precipitation naturally decrease, so would the groundwater levels in both borewell and open-well water. Sediment and rock precipitation is released into the water and combined with it, increasing TDS and TSS. When the TDS rises simultaneously with other chemical parameters such as alkalinity and total hardness, nutrients in the soil are also suspended in the water, causing contamination of the groundwater. As a result of additional pollution, such as soil, surface water, and agricultural pollution, the quality of the groundwater in the current study may also deteriorate. If this condition persists, the groundwater becomes unfit for human consumption.

Gneisses and granites from the Achaean period that have been nitrated by dykes can be found in the Krishnagiri district (Balasubramaniam 1980). Their main elements are hornblende, biotite, and scattered gray and plink spells with quart grains (Thangarajan and Singh 1998). Analysis revealed that there were significant differences in the chemical properties of groundwater samples from the districts of Krishnagiri, Vellore and Tiruvannamalai. The TDS and TSS of the three districts were much higher since there was no groundwater recharge zone around the sites where the water samples were taken. Until now, people have relied on open wells and wells for domestic and irrigation purposes. The constant extraction of water from these sources has resulted in a decrease in groundwater levels in the affected areas, and as a result, the TDS and TSS values in groundwater sources have increased in these districts. The states of Rajasthan, Uttar Pradesh, and Gujarat experienced a similar phenomenon (Datta 2005). Compared to WHO (1990) guidelines, Rengasamudram borewell water samples in this investigation had high values of TDS, CE, total alkalinity, total hardness, nitrate, and chloride. Most of the water samples from the Krishnagiri district are above the WHO guidelines. A large amount of water is required for domestic, agricultural, and industrial uses due to the development of the large population. As a result, water is constantly drawn from borewells and open wells and there is no substitute supply of water available. This poses a threat, as climate change and prolonged monsoon failures have had a severe negative impact on the area, raising groundwater values in the Krishnagiri District.

Additionally, several activities have greatly polluted lakes and ponds, no pollution management measures have been implemented, and there is a lack of public knowledge in these places. According to Karthikeyan and Shunmugasundarraj (2000), the Krishnagiri block's high chemical parameter values have the same causes. Similar results were noted in Guwahati, Assam, India's groundwater. The findings of Garg and Hassan (2007) provide a comprehensive explanation of India's water scarcity. According to Shaji et al. (2007), most of the groundwater quality in Kerala's Palghat district exceeds allowable standards. The same threats to most of the various water bodies in the north Indian states were reported by Soni (2007), Raju (2007), and Madhnure et al. (2007). In the present study, the groundwater values of the Burgur Union showed higher levels of physical and chemical values compared to Uthangarai and Hosur Union. Higher levels of sulfate and chloride were found in water samples from Tiruvannamalai, Polur and Cheyyar in the Tiruvannamalai district. This suggests that groundwater in these areas is unfit for human consumption because high levels of TDS inevitably translate into high levels of nutrients. Nutrients and TDS depend on one another. According to Garg and Hassan (2007), the supply of groundwater is not uniform in place or time and is an annual replenishable resource. The occurrence and transport of groundwater is governed by a complex interplay of hydrogeological, hydrological, and climatological elements (Shirin and Yadav 2021; Shirin et al. 2021, 2023). The high values of total hardness, TDS, chloride, and sulfate in the three districts under investigation may be caused by the same reasons.

The maximum allowable limits for fluoride and iron were exceeded in Krishnagiri district, but were within acceptable bounds in Vellore and Tiruvannamalai districts. High levels of fluoride are found in Krishnagiri district groundwater, according to reports by Karthikeyan and Shunmugasundarraj (2000) and Harikumar et al. (2007). According to Harikumar et al. (2007), the villages of Dharmapuri and Krishnagiri had the highest percentage of water sources with 1.00 ppm of fluoride (100%), followed by Vellore and Salem (70–80%) and Erode (17%). The current study found that more than 50% of the water sources in the districts of Krishnagiri (62%), Vellore (60%), Salem (56%), Dharmapuri (53%) and Erode (26%), had 1.00 ppm of fluoride, which is further supported by the categorization of data. The districts of Krishnagiri, Vellore, and Tiruvannamalai have groundwater pH averages and standard deviations for the current inquiry of 7.4 ± 0.2, 7.5 ± 0.60, and 6.8 ± 0.52. Madhnure et al. (2007) found 7.18–8.01 in the Yavatmal district of Maharastra, India's Pandharkawada region; Shaji et al. (2007) noted 7.2–8.2 in the Palghat district of Kerala. The present investigation indicates that the Krishnagiri district has higher than allowed amounts of calcium, magnesium phosphate, and salt; nevertheless, Madhnure et al. (2007) and Shaji et al. (2007) suggested that these nutrients were found to be low in certain areas and high in others.

In the Yavatmal district of Maharashtra, Madhnure et al. (2007) recorded calcium values of 24–206 and 6–96 mg/l in shallow and deep aquifers, respectively. In the same way, the sodium concentration in the shallow and deep aquifers in the Yavatmal district, Maharashtra, varies from 121 to 128 and from 21 to 353 mg/l, respectively. Maximum groundwater samples exhibited unsatisfactory limits in the districts of Krishnagiri, Vellore and Tiruvannamalai. However, nutrients such as calcium, sodium, phosphate and sulphate levels are within the allowed range in all 236 samples from Guwahati, Assam (Das et al. 2003). The present study found that chloride levels in the districts of Krishnagiri, Vellore and Tiruvannamalai varied from 160 to 281 ppm, 166 to 4799 mg/l and 78 to 2213 mg/l, respectively. In their investigations, Madhnure et al. (2007) and Shaji et al. (2007) also noted increased levels of ion content. Although the iron level in samples from Vellore and Tiruvannamalai was not examined, it was found in the Krishnagiri district of our current investigation to be above the allowed limits.

Because of their ecotoxicological effects and the complexity of eliminating them through waste management, heavy metals in the environment in general and subsurface aquatic systems, in particular, are a major source of concern (Basta and McGowen 2004; Naik et al. 2007; Patel et al. 2022). Most of the heavy metal levels in this investigation fall within the LMDL level, or less than the minimum detectable limit. In the current investigation, BOD levels are within allowable limits, but COD levels-especially in lake water samples from the Vellore district-exceed allowable limits. This study has shown that there is a definite relationship between land use and water quality. Anthropogenic activities are the main contributors to the degradation of water quality. Rapid population growth, urbanization, and industrialization have led to a greater demand for an increasingly smaller supply of water. Of the water resources in the country, the majority are consumed in agriculture (70–90%), and the remaining are consumed in industrial activities and for domestic purposes such as drinking water and sanitation (Rao and Mamatha 2004; Manickam et al. 2015b). In most cases, discharges into pits, open ground, or open, unlined drains near factories, allowing it to move to low-lying depressions, resulting in groundwater pollution.

Across India, a wide range of initiatives, campaigns and programs are being implemented to raise public awareness and encourage participation in the management of water resources. Nongovernmental organizations have adopted some of these, while central state governments are putting others into practice. An efficient groundwater management strategy focused on the advancement of justice, sustainability, and efficiency is necessary to protect the aquifers from exploitation. Groundwater resource extraction should be controlled to avoid depleting the available recharge capacity. The state and federal governments must take decisive action to mitigate the damaging effects of groundwater extraction on the environment. Better management of current storage and the construction of new storage through the construction of small, medium, and large dams, while taking economic, environmental, and social issues, is necessary to increase the availability of water resources. Rejuvenating aging ponds, lakes, and tanks, as well as expanding artificial groundwater recharge methods, can improve the availability of water resources.

8.5 Conclusion

Analysis was done on eight Krishnagiri samples, 13 Vellore samples and seven samples from the district of Tiruvannamalai. Higher values of TDS, CE, pH, total alkalinity, total hardness, calcium, magnesium, iron, sodium, potassium, free ammonia, nitrate, chloride, fluoride, sulfate and phosphate are found in the Guruvinayapalli district of Krishnagiri, Rengasamudram, Vinayagapuram, Guddur and Kuppanatham, respectively. Water tests in Rengasamudram, Krishnagiri district, showed higher values for most criteria than in other areas. The WHO and TWAD permissible limits for TDS, TH, calcium, magnesium, free ammonia, nitrate, chloride, and fluoride are exceeded in the Krishnagiri district. In trace levels, phenolic compounds, fluoride, hexavalent chromium, total chromium, oil, and greases were discovered in water samples from the Vellore district. Higher values of CE, pH, TSS, TDS, chloride, sulphate, BOD, COD, alkalinity, total hardness, calcium magnesium, and hardness were found in water samples from the Tiruvannamalai district. The remaining attributes lie in the range of the smallest observable bounds. Three t-test analyzes of district water samples produced informative findings. The maximum values of the water samples used in this investigation are higher than what is allowed compared to the WHO and TWAD recommendations.

Acknowledgements The first author thanks the Principal and Head, PG and Research Department of Zoology, Government Arts College for Men, Krishnagiri-635001, Tamil Nadu, India for providing the necessary laboratory facilities and infrastructure provided for completing the experiment.

References

Aravinth A, Kannan R, Chinnadurai G, Manickam N, Raju P, Perumal P, Santhanam P (2023). Temporal changes in plankton diversity in relation to hydrographical characteristics at Perumal Lake, Cuddalore District, Tamil Nadu, India. J Basic Appl Zool 84:13. https://doi.org/10.1186/s41936-023-00337-7

Balasubramaniam K (1980) Geology of parts of Krishnagiri taluk, Dharmapuri district, Tamilnadu. Progress Report, GSI Tech. Rep., Madras, p 14

Basta NT, McGowen SL (2004) Evaluation of chemical immobilization treatments for reducing heavy metal transport in a smelter-contaminated soil. Environ Pollut 127:73–82

Das B, Talukdar J, Sarma S, Gohain B, Dutta RK, Das HB, Das SC (2003) Fluoride and other inorganic constituents in groundwater of Guwahati, Assam, India. Curr Sci 85(5):657–661

Datta PS (2005) Groundwater ethics for its sustainability. Curr Sci 89(5):812–817

Drever JI (1997) The geochemistry of natural waters, 3rd edn. Prentice Hall, Upper Saddle River, p 436

ENPHO (2003) The South Asian transboundary water quality monitoring programme. Environment and Public Health Organisation Magazine, World Water Day, pp 57–59

ENPHO (1998) Siraha drinking water, health, education and sanitation project (1993–1998). Environment and Public Health Organization, Bijuli Bazzar, Kathmandu, Nepal, vol 8, pp 1–7

Garg NK, Hassan Q (2007) Alarming scarcity of water in India. Curr Sci 93(7):932–941

Harikumar R, Khandare AL, Brahman GNV, Venkiah K, Reddy CG, Sivakumar B (2007) Assessment of current status of fluorosis in North-Western districts of Tamil Nadu using community index for dental fluorosis. J Hum Ecol 21(1):27–32

Karthikeyan G, Shunmugasundarraj A (2000) Isopleth mapping and in-situ fluoride dependence on water qulity on the Krishnagiri block of Tamilnadu in South India. Fluoride 33(3):121–127

Kumar A, Chava R, Gupta S, Shirin S, Jamal A, Yadav AK (2024) Environmental and health effects of fluoride contamination and treatment of wastewater using various technologies. In: Advanced treatment technologies for fluoride removal in water: water purification. Springer Nature Switzerland, Cham, pp 323–341. https://doi.org/10.1007/978-3-031-38845-3_18

Madhnure P, Sirsikar DY, Tiwari AN, Ranjan B, Malpe DB (2007) Occurrence of fluoride in the groundwaters of Pandharkawada area, Yavatmal district, Maharashtra, India. Curr Sci 92(5):675–679

Manickam N, Bhavan PS, Santhanam T, Muralisankar V, Srinivasan S, Radhakrishnan K, Vijayadevan P, Chitrarasu, Ali AJ (2014) Seasonal variations of zooplankton diversity in a Perennial Reservoir at Thoppaiyar, Dharmapuri District, South India. Austin J Aquac Mar Biol 1(1):1–7

Manickam N, Bhavan PS, Santhanam P, Muralisankar T, Srinivasan V, Radhakrishnan S, Vijayadevan K, Bhuvaneswari R (2015a) Biodiversity of freshwater zooplankton and physico-chemical parameters of Barur Lake, Krishnagiri District, Tamil Nadu, India. Malaya J Biosci 2(1):1–12

Manickam N, Bhavan PS, Santhanam P (2015b) Physico-chemical characteristics of lake ecosystem. LAMBERT Academic Publishing, pp 1–80. ISBN: 978-3-659-79409-4

Manickam N, Santhanam P, Bhavan PS, Vijayadevan K, Ashokan V, Bhuvaneswari R, Dhanalakshmi B, Chitra J (2018) Environmental pollution biodiversity conservation and climate change: issues and challenges. In: Guha A, Ray A, Barik KL (eds) Physico-chemical characteristics and zooplankton diversity in the reservoirs (dams) of Dharmapuri District, Tamil Nadu, Southern India. Sparrow Publication, India. ISBN: 978-81-89140-17-7

Manickam N, Santhanam P, Bhavan PS (2023) Biofloc-copefloc: technology in aquaculture. LAP—LAMBERT Academic Publishing, pp 1–61. ISBN: 978-620-6-18441-6

Naik PK, Dehury BN, Tiwari AN (2007) Groundwater pollution around an industrial area in the coastal stretch of Maharashtra state, India. Environ Monit Assess 2007(132):207–233

Narasimhan TN (2005) Water: science and society. Curr Sci 89(5):31–38

Patel D, Pamidimukkala P, Chakraborty D, Yadav A (2022) Bharuch District, Gujarat, India: factor analysis and geographical distribution of water quality characteristics. Environ Nanotechnol Monit Manage 18:100732. https://doi.org/10.1016/j.enmm.2022.100732

Raju NJ (2007) A season-wise estimation of total dissolved solids from electrical conductance and silica in ground waters of upper Gunjanaeru River basin, Kadapa district, Andhra Pradesh. Curr Sci 92(3):371–376

Rao SM, Mamatha P (2004) Water quality in sustainable water management. Curr Sci 87(7):942–947. http://www.jstor.org/stable/24109398

Shaji E, Viju BJ, Thambi DS (2007) High fluoride in groundwater of Palghat district, Kerala. Curr Sci 92(2):240–245

Sharma S, Bajracharya RM, Sitaula BK, Merz J (2005) Water quality in the Central Himalaya. Curr Sci 89(5, 10):774–786

Shirin S, Yadav AK (2014) Physico-chemical analysis of municipal wastewater discharge in Ganga River, Haridwar district of Uttarakhand, India. Curr World Environ 9(2):536–543

Shirin S, Yadav AK (2021) Assessment and suitability for the dumping of flyash in an abandoned mine. Environ Challenges 4:100111. https://doi.org/10.1016/j.envc.2021.100111

Shirin S, Jamal A, Ranjan P, Yadav AK (2019) Study on assessment of slope stability and mixed disposal of overburden in voids of Singrauli Coalfield. Environ Qual Manage 28(3):131–139. https://doi.org/10.1002/tqem.21616

Shirin S, Jamal A, Emmanouil C, Yadav AK (2021) Assessment of characteristics of acid mine drainage treated with fly ash. Appl Sci 11(9):3910. https://doi.org/10.3390/app11093910

Shirin S, Jamal A, Emmanouil C, Singh VP, Yadav AK (2023) Assessment and characterization of waste material used as backfilling in an abandoned mine. Int J Coal Prep Util 43(8):1402–1410. https://doi.org/10.1080/19392699.2022.2118259

Soni V (2007) Three waters—an evaluation of urban groundwater resource in Delhi. Curr Sci 93(6):760–761

Strahler AN (1992) Modern physical geography. Wiley, New York

Thangarajan M, Singh VS (1998) Estimation of parameters of an extensive aquifer a case study. J Geol Soc India 52:477–481

Word Heath Organization (WHO) (1990) Guidelines for drinking water quality, Geneva

Yadav AK, Sahoo SK, Mahapatra S, Kumar AV, Pandey G, Lenka P, Tripathi RM (2014) Concentrations of uranium in drinking water and cumulative, age dependent radiation doses in four districts of Uttar Pradesh, India. Toxicol Environ Chem 96(2):192–200

Chapter 9
An Investigation of Coastal Water Quality and SEMBV Disease in *Penaeus monodon* in Coastal Southern India

Razack Rafi Mohamed, Narasimman Manickam, Ambrose Paulraj, and Rajagopal Bhuvaneswari

Abstract *Penaeus monodon* infected shrimps in various size groups were randomly collected using a clean bucket from cultured ponds in all the sampling stations (Ramnad, Tuticorin, Pattukottai, Nagapattinam, Karaikal, Srikhazhi, Chidambaram, Cuddalore, Pondicherry and Marakkanam, Coastal Southern India) during the study period of four months. Infected water and specimens were also collected from the same location. Water quality data including temperature, salinity, pH, dissolved oxygen (DO), ammonia (NH_3), nitrate (NO_2), and nitrite (NO_3) were measured at the same location as the pond along with the coastline. Infections with mesodermal and ectodermal baculovirus (SEMBV) were found in the middle intestinal gland and other organs of the cultured *P. monodon* at all ten sites under examination. Furthermore, SEMBV infections were found in neighboring populations. The mature grass shrimp that were obtained from Tuticorin, Karaikal, and the vicinity of Chidambaram showed significant damage compared to other regions like Ramnad,

R. R. Mohamed (✉)
PG and Research Department of Zoology, C. Abdul Hakeem College (Autonomous), Melvisharam, Ranipet 632509, Tamil Nadu, India
e-mail: rafimohamed620@gmail.com

N. Manickam
Marine Planktonology and Aquaculture Laboratory, Department of Marine Science, School of Marine Sciences, Bharathidasan University, Tiruchirappalli 620024, Tamil Nadu, India

Crustacean Biology Laboratory, Department of Zoology, Bharathiar University, Coimbatore 641046, Tamil Nadu, India

Department of Zoology, Sree Meenakshi Arts & Science College, Meenakshi Nagar, Eriyur (Po), Pennagaram (Tk), Dharmapuri 636810, Tamil Nadu, India

A. Paulraj
PG and Research Department of Zoology, Government Arts College for Men, Krishnagiri 635001, Tamil Nadu, India

R. Bhuvaneswari
Fish Disease Diagnostic Laboratory, Department of Zoology, Ayya Nadar Janaki Ammal College (Autonomous), Sivakasi 626124, Tamil Nadu, India

A. K. Yadav et al. (eds.), *Integrated Management of Water Resources in India: A Computational Approach*, Water Science and Technology Library 129,
https://doi.org/10.1007/978-3-031-62079-9_9

Pattukottai, Nagapattinam, Sirkhazhi, Cuddalore, Pondicherry, and Marakkanam. The highest recorded percentages were Tuticorin, Sirkhazhi, and Cuddalore (74%) followed by Chidambaram (46%). However, in the Tamil Nadu Coastal Region there was a small but statistically significant variation in the prevalence of SEMBV. The available statistics strongly suggest that white spot disease could pose a significant threat to the coastal Tamil Nadu farming community. The most common main pathogen detected throughout the survey was SEMBV, which was discovered in *P. monodon* samples from each of the ten survey locations. These results suggest that SEMBV is a new infectious agent present in the culture stocks of Indian *Penaeid* species. Many of the shrimp sampled in the survey had a few common Penaeid epicommensals on their gills, appendages, and general body surface before a few days of infection; some of these were associated with illness and even mortality.

Keywords Coastal water · Shrimps · *Penaeus monodon* · SEMBV · *Penaeid*

9.1 Introduction

The farming of aquatic animals in a controlled setting, such as fish, shellfish, mollusks, and aquatic plants, is known as aquaculture. It helps to enhance the production of protein-rich food to the ever-expanding global population. Exports are the lifeline of a country's economy. After fin fish, shell fish, such as crustaceans, hold the major position in production in both import and export values. Among crustacean animals shrimp is the important aquaculture source due to its great demand in the world, as well as aquaculture industry. Shrimps hold the top pride of place among the various economically important organisms in the international market and occupy an unassailable No. 1 position. There is rapid expansion of the culture of penaeid shrimps along both east and west coasts of India, and commercial culture is slowly becoming a major component of India's economy. Corporate companies and individual farmers are showing quite a lot of interest in entering this industry.

Successful aquaculture production has been hampered by diseases of viral and bacterial origin. An estimated USD 2 billion in economic losses were attributed to diseases in Asia's emerging nations in 1990 alone. Losses have escalated since then. According to a 1995 estimate, aquatic animal diseases and environmental issues might cost Asian countries' aquaculture industries more than US $4 billion annually in lost productivity. According to recent statistics, shrimp disease has resulted in losses above USD 5 billion worldwide. The World Bank has allocated $175 million to support studies on diseases affecting shrimp. The numbers illustrate the total financial cost of illnesses that affect aquatic life. Since then, the epidemic plague has come as a damper and its causing concern for the growth of this promising industry. During the past 20 years, viral infections have become a major financial barrier to effective shrimp farming. Viral infections are common in crustaceans, and approximately 20 viruses have been reported infectious (Nadal et al. 1998; Johnson et al. 1999). Some of these are *Penaeus monodon* Denso virus, *Baculovirus penaei*,

Yellow head virus (YHV), infectious hypodermal and hematopoietic virus, Gill-associated virus, White Tail disease, and Taura syndrome virus, the virus that had a detrimental effect on world economics and drastically decreased shrimp farming mortality.

Most of these viruses were discovered because they played havoc with the commercial culture of penaeid shrimps. Knowledge of these viruses and the diseases they cause is essential for the prevention and effective control of the disease. Some of the penaeid viruses have been described solely based on diseases or disease syndromes and because of their morphological characteristics (Shirin and Yadav 2021; Shirin et al. 2021b). The biochemical characteristics of some of the penaeid viruses have also been determined, permitting the taxonomic placement of their agents (Johnson arid Lightner 1988; Couch 1981, 1991; Lightener 1992, 1993; Shirin and Yadav 2014; Brock and Lightner 1990; Francki et al. 1991; Lightener and Redman 1991; Adams and Bonami 1991; Shirin et al. 2019, Yadav et al. 2014). It is known that penaeid shrimp viruses are widespread throughout the world and can infect a variety of host species (Francki 1984; Bell and Lightner 1988; Lightener et al. 1992, 1993; Shirin et al. 2023). In cultivated black tiger shrimp, *Penaeus monodon*, a DNA-type virus known as systemic ectodermal and mesodermal baculovirus (SEMBV) has been reported to induce white spot disease (Pratanpipat et al. 1994). Shrimps that contract the white spot (WS) disease typically die within 3 to eight days due to its acute and deadly nature. In addition to white spot disease, frequent outbreak of epizootics has also been nagging the industry. The baculovirus SEMBV has been recognized during all stages of the culture period. However, there is no concerted effort to document the diseases that affect the industry. The objective of the present study is to examine certain aspects of significance, pathogenesis, possible causes, and the distribution (geographical) of the agents involved in the outbreak of viral disease in the cultured shrimps.

9.2 Study Area

The water and shrimp samples were collected from the grow-out farms located in different places of Ramnad, Tuticorin, Pattukottai. Nagapattinam, Karaikal, Sirkhazhi, Chidambaram, Cuddalore, Pondicherry and Marakkanam, South India, during the study period of four months from November 1995 to February 1996.

9.3 Collection of Water Samples

Water samples were taken from reservoirs, creeks, and grow-out ponds within the sampling regions indicated previously. The water samples were collected in separate fresh plastic containers without any air bubbles. Before sampling, the containers were rinsed and filled with water. After collection, the containers were tightly sealed and labeled. The water samples were brought to the laboratory. Various physico-chemical parameters such as water temperature, pH, salinity, and dissolved oxygen were measured at the farm itself. Other parameters such as alkalinity, hardness, ammonia (NH_3), nitrite (NO_2), and nitrate (NO_3) were measured in the laboratory using standard procedures (Strickland and Parsons 1968).

9.4 Physicochemical Analysis of Water Samples

Using a mercury thermometer with a minimum count of 1 °C and a range of 0–100 °C, the temperature of the water was monitored in the field. The thermometer was submerged in water for a minute to measure the temperature and the reading was noted. Transparency of the water was measured by the Secchi disk. Salinity was measured using a salinometer (Model No-211. HANNA instruments). A single drop of water was tested on the meter's main prism. Before inserting the water sample, distilled water was used to clean the meter's prism and set the scale to 0 ppt. A digital pH meter (Systronics, India) was used to measure the pH of water samples with a precision of 0.01. The pH meter was calibrated using buffer solutions with pH values of 7 and 10 before use. Without going beyond the maximum immersion level, the pH electrode was submerged in the water samples that were to be evaluated. The sample was then gently swirled. The reading was recorded once it had steadied.

Winkler's method was used to measure the amount of dissolved oxygen in the water samples (Strickland and Parson 1968). By titration with standard sulfuric acid (0.02 N) at room temperature and using a methyl orange indicator and phenolphthalein, the alkalinity of the water sample was determined (APHA 1992). Total hardness of the water samples was measured by a titrimetric method using ethylenediamine-tetra acetic acid. The spectral light degree method of phenol disulfonic acid was used to evaluate the amount of nitrate nitrogen (NO_3-N). The colorimetric method Nessler reagent was used to detect ammonia nitrogen (NH_3-N) and naphthalene ethylenediamine was used to measure nitrite nitrogen (NO_2-N).

9.5 Collection of Shrimps

Using a clean bucket, samples were randomly taken from grow-out ponds cultivating Penaeus monodon in various places (Ramnad, Tuticorin, Pattukottai, Nagapattinam, Karaikal, Sirkhazhi, Chidambaram, Cuddalore, Pondicherry, and Marakkanam). The animals were aged between 17 and 133 days of culture (DOC) and varied in size from 0.3 to 26.7 g. They were transported in ice boxes to the lab for examination. On most farms, samples were taken every month for active surveillance. Whenever a disease epidemic occurred, samples were also taken from shrimp farms.

9.5.1 Preparation of samples

Shrimp samples were thawed by keeping them in an ice pack. The samples were dissected out aseptically using sterile scissors and forceps. Before exoskeleton removal, clinical signs, body weight (g) and length (mm) of moribund shrimps were recorded. Pointed sterile dissecting scissors were used to open the exoskeleton to remove infected organs such as the gills, the hepatopancreas (midgut gland), heart and hematopoietic organ. Similarly, abdominal muscle tissue was also removed for diagnosis.

9.5.2 Histopathology

Organs from disease-suspected shrimp, including the hepatopancreas, gills, heart, muscle, and stomach, were removed and preserved in Davidson fixative for a full day in preparation for histological examination. Subsequently, the preserved organ specimens were placed in 70% alcohol and preserved. The tissues were cleared and dehydrated at room temperature to prepare for cutting slices of various tissues into paraffin. The tissues were classified twice in 95% alcohol and absolute alcohol, cleared through a mixture of absolute alcohol and chloroform (1: 1 v/v) and then passed twice in pure chloroform for one hour each. The tissues were first washed in two changes of 70% alcohol for one hour each, dehydrated for two hours, and further dehydrated for one hour each in two changes of 80% alcohol. Because chloroform does not cause the tissue to become brittle and rigid, it was preferred to xylene. After clearance, the tissues were kept overnight at room temperature in a solution of paraffin wax and chloroform (approximately 1: 1). The tissues were impregnated for one hour in each of ceresin-containing three variations of paraffin wax with a melting point of 58–60 °C before embedding. Using a manual rotatory microtome, transverse slices were cut at a thickness of 5 to 7 μm. Sections were deparaffinized in xylene, and then hydrated using a graded sequence of alcohol up to 70%. Harris alum hematoxylin was used for staining and 1% alcoholic eosin was used as a counterstain (Preece 1972). Following the standard protocol, stained slices were mounted with a glass cover-slip in DPX using xylene after being dried using a graded series of alcohol. Histology sections were seen using a Carl Zeiss binocular compound microscope. Using a digital camera (Nikon) coupled to a Carl Zeiss microscope with a 10 X projection eyepiece and objectives 10, 20, 40, and 100 X, photomicrographs were captured. With the aid of an ocular and stage micrometer, the magnification of the enlarged prints was determined.

9.6 Results

SEMBV infections were discovered at 10 distinct places where samples were taken, including the midgut gland of cultivated P. monodon. Additionally, SEMBV infections were discovered in nearby people. One or more eosinophilic occlusions, also known as eosinophilic occlusion bodies, inside the hypertrophied nuclei of hepatopancreatic and midgut epithelial cells, were previously thought to be necessary for the diagnosis of SEMBV diseases. Occlusion bodies were also frequently observed to be free in the lumens of the hepatopancreatic tubules and the midgut lumen, as a result of the cytolysis of necrotic SEMBV-infected hepatopancreatic epithelial cells. The components of these occlusion bodies were crystalline arrays of viral proteins. The diagnosis of SEMBV was accompanied by multifocal necrosis and nuclear inclusion in organs such as the heart, hepatopancreatic tissue, gill epithelium, and tissues of ectodermal and mesodermal origin. These pathognomonic inclusion bodies were typically seen inside the small lesions that were seen in the gills of the animals that were being evaluated as growers and finishers. The occurrence of SEMBV was measured in relation to stocking densities with days of culture in various mentioned study areas mentioned. Adult grass shrimp collected from Tuticorin, Karaikal and in and around Chidambaram were highly affected compared with other areas such as Ramnad, Pattukottai, Nagapattinam, Sirkhazhi, Cuddalore, Pondicherry and Marakkanam (Fig. 9.1). Similarly, the rate of mortality in relation to infection in days were 65, 75, 74, 56, 72, 46, 74, 63 and 74, respectively, in the form of percentage after the post-infection. Tuticorin, Sirkhazhi, and Cuddalore recorded the highest with 74% and the lowest with 46% from Chidambaram (Fig. 9.2).

In the gross pathology after, (1) All the Infected shrimps had broken antennae; (2) In some cases, prior to SEMBV infection, Zoothamnium fouling (ectoparasite) was recorded; (3) A distinct reddish yellow of gills has also been noted; (4) The dorsal cephalothorax area of the infected shrimp and generally all other parts of the body (exoskeleton). (5) The white spots appeared in a variable size from (0.5–3.0 mm in diameter). (6) The midgut gland showing through the translucent carapace in moribund shrimp. Furthermore, there was a noticeable lack of neurotic changes in the parenchyma of multiple organs, including the cardiac, striated, and epicardium of the heart. The hepatopancreatic epithelium exhibited decreased lipid vacuolation and was less columnar in most shrimps. Necrotic hepatopancreas epithelial cells have occasionally been seen alone. In certain samples, the lumen of the digestive system included a comparable quantity of sloughed necrotic cells that were rounded and similar in number to gills. Almost all environmental variables were similar in all the study areas to average levels (Table 9.1). The cytoplasm was still visible in the early stages of infection, but thinned out as the virus developed and eventually became translucent. Later, the nuclei also broke down, resulting in larger, empty cells. Hematocytic infiltration of the gills, hepatopancreatic hemal sinuses, hemocoel, and diffusely localized necrosis in tissues of ectodermal and mesodermal origin are other inclusions.

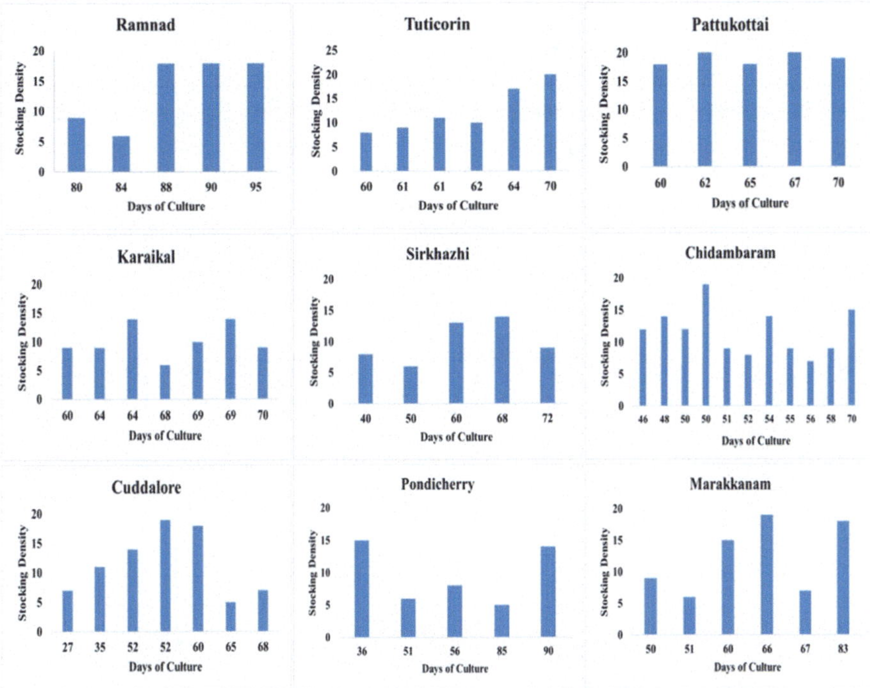

Fig. 9.1 Showing the occurrence of SEMBV in different stocking densities and days of culture in various study areas

9.7 Discussion

The significant negative impact of viral diseases on shrimp farming and the growing number of people searching for them combine to cause the regular discovery of new viruses in cultivated *penaeid* shrimp. According to Lightener et al. (1992), there are already more than a dozen recognized penaeid viruses, which are tentatively categorized into five major families: Parvoviridae, Baculoviridae, Reoviridae, Rhabdoviridae and Togaviridae. The most prevalent pathogen among these viruses has been identified as baculovirus, which is extensively present in cultivated penaeid shrimps. In the present study, the cytopathology and virogenesis of SEMBV were investigated in specific cell types of the hepatopancreas of *P. monodon.* However, some slight, but statistically significant differences in the prevalence of SEMBV were observed in all the coastal line of Tamil Nadu, India. In many farms, the problem was aggravated by water exchanges, both *P. monodon* and *P. indicus* were susceptible. In the small sample size we had, SEMBV was frequently linked to farms that reported low growth and survival, particularly in the early stages of juvenile phases. Light to moderate infections, however, supported previous findings by Lightener et al. (1983)

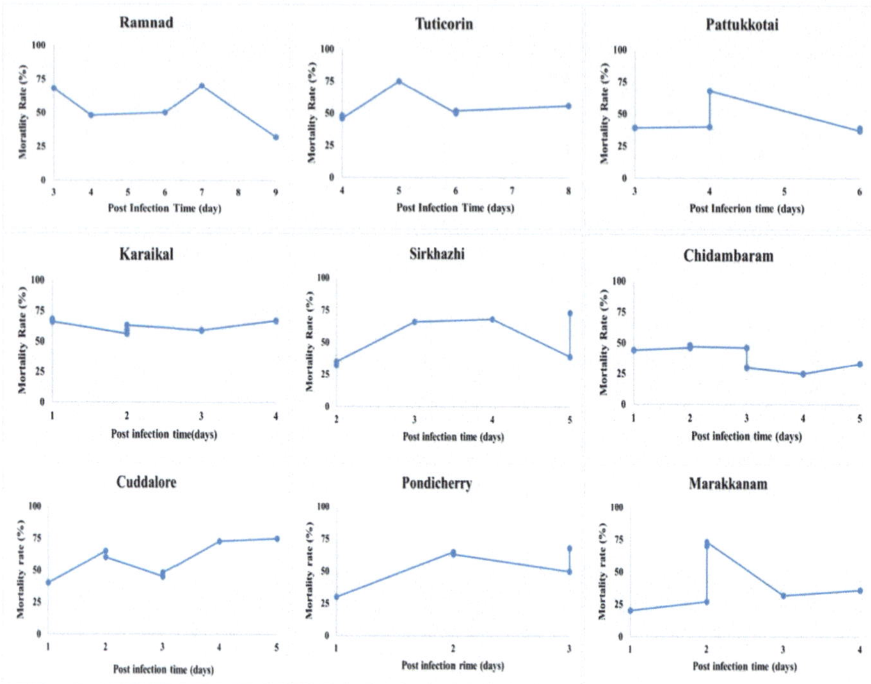

Fig. 9.2 Comparison of mortality rate of *P. monodon* infected with SEMBV

that P. monodon can withstand low to moderate infections if other cultural conditions are ideal, even in areas not reporting this issue.

The available data suggest that white spot disease could pose a significant threat to the agricultural population in the coastal regions of Tamil Nadu (India). The occurrence of the virus in the post-laternal stages would help to explain the rapid spread of the disease throughout the Asian region in 1994. Post larvae are a major limiting factor in shrimp production in several Asian countries, leading to substantial higher prices compared to other countries where the supply/demand situations are more stable. The shipment of post-larves across national borders, legally or illegally, is a common practice, and this could have spread the infection from a country with high production of post larvae (Thailand, Taiwan, Indonesia) to those within the supply (India and Bangladesh). Therefore, when shipping live penaeid shrimp stock from Asian nations, extreme caution should be exercised. Additionally, more research on the disease is desperately needed to understand the characteristics of the virus, its host range, and its geographic distribution, as well as to create efficient protocols for the early identification and management of SEMBV-related disorders.

Recommendations: No treatment for the viral disease is available. Prevention is the only method. Unfortunately, at present, not enough is known about viruses to develop a definite prevention strategy. The following recommendations have been made based

Table 9.1 Water quality parameters for the ponds at study areas

Physiochemical parameters	Stations									
	Ramnad	Tuticorin	Pattukkotai	Nagapattinam	Karaikal	Sirkhazhi	Chidambaram	Cuddalore	Pondicherry	Marakkanam
Temperature										
Average	25.6	30.1	31.0	30.9	29.1	23.8	28.5	28.0	26.0	30.9
SD	2.3	2.3	2.5	1.9	1.8	1.8	1.6	2.0	2.1	2.3
Minimum	19.9	19.5	18.9	15.6	17.7	18.6	18.6	18.6	19.6	20.5
Maximum	30.0	28.7	30.9	30.7	31.3	30.1	29.5	29.1	29.8	20.9
[a]*Salinity* ($^{0}/_{00}$)										
Average	31.9	30.0	29.6	30.9	29.3	31.0	26.1	28.8	27.1	28.0
SD	1.9	1.5	1.8	2.0	1.5	1.7	1.8	1.9	1.5	2.0
Minimum	17.0	18.1	18.5	19.0	19.5	18.2	17.5	19.7	18.0	17.1
Maximum	28.0	30.4	30.0	30.0	27.0	27.5	27.1	29.5	29.2	32.0
[a]*pH*										
Average	7.8	8.1	6.8	7.0	7.8	6.9	6.2	7.4	6.8	7.3
SD	0.3	0.1	0.2	0.3	0.3	0.1	0.1	0.1	0.2	0.3
Minimum	7.1	6.2	6.2	7.1	7.2	7.8	7.2	7.2	7.8	7.2
Maximum	7.4	7.8	7.8	8.0	9.7	9.7	8.1	7.8	7.9	8.2
[a]*DO* (ppm)										
Average	4.8	4.5	4.2	4.1	3.4	3.4	5.0	4.2	4.5	4.1
SD	0.8	0.6	0.9	0.6	0.6	0.6	0.7	0.8	0.8	0.6
Minimum	2.7	3.0	3.0	2.5	2.5	2.5	2.1	1.9	1.8	2.0
Maximum	7.0	6.5	7.1	7.1	7.0	7.0	6.8	6.1	7.0	7.0
[a]NH_3 (ppm)										

(continued)

Table 9.1 (continued)

Physiochemical parameters	Stations									
	Ramnad	Tuticorin	Pattukkotai	Nagapattinam	Karaikal	Sirkhazhi	Chidambaram	Cuddalore	Pondicherry	Marakkanam
Average	0.23	0.25	0.21	0.21	0.21	0.24	0.22	0.22	0.20	0.24
SD	0.34	0.37	0.40	0.32	0.33	0.33	0.32	0.30	0.31	0.35
Minimum	0.01	0.01	0.01	0.01	0.01	0.01	0.01	0.02	0.01	0.01
Maximum	0.30	0.29	0.31	0.27	0.27	0.30	0.31	0.31	0.30	0.28
[a]NO_3 (ppm)										
Average	0.7	0.7	0.6	0.6	1.1	0.6	0.9	0.9	0.4	0.6
SD	0.7	0.6	0.6	0.5	0.9	0.5	0.8	0.9	0.5	0.6
Minimum	0.2	0.2	0.1	0.2	0.1	0.2	0.1	0.2	0.1	0.2
Maximum	3.0	3.1	3.1	2.9	2.5	2.9	2.5	2.8	3.0	2.8
[a]NO_2 (ppm)										
Average	0.13	0.14	0.13	0.17	0.12	0.12	0.17	0.15	0.14	0.11
SD	0.14	0.13	0.13	0.17	0.12	0.12	0.17	0.15	0.12	0.11
Minimum	0.01	0.01	0.02	0.01	0.02	0.02	0.02	0.02	0.01	0.01
Maximum	0.30	0.29	0.25	0.32	0.32	0.31	0.32	0.30	0.30	0.30

[a] ($^0/_{00}$) ppt [a]DO—Dissolved Oxygen [a]NH_3—Ammonia [a]NO_3—Nitrate [a]NO_2—Nitrite

on expensive with yellow head and are subject to change as new information comes to light. Stock the source water sea/estuary in the reservoir and grow-out ponds. Walt for five days to allow the pH to settle to flow 8.5 and apply chlorine 5 ppm and salt for ten days before stocking PL. By increasing the use of lime (100 ppm instead of the chlorine regular 40–60 ppm), the concentration could be lowered to 5 ppm instead of 30 ppm and still as effective in terms of Yellow Head Virus (YHV) disinfection. Groups of farmers can co-operate by closing off their area of the small supply canals and treating with chlorine. The water can then be used as a reservoir for exchange. The use of reduced salinity water has been claimed to prevent or at least reduce the severity of the disease. These claims have remained unproven. The test carried out by NICA (National Institute for coastal Aquaculture, Japan) was found to be table at a variety of salinities. Avoid using water directly from the sea. Set up the reservoir and use the water in the reservoir for exchange. Minimize water exchange. Avoid all use of fresh feeds because the studies are showing that the bivalves, crabs, and crustaceans are the carriers of various to the system. Stocking densities of PL will reduce potential financial losses as well as stress and possibly the spread of disease. Keep the arid gates in good condition to prevent the entry of potential carriers of the virus into the pond.

9.8 Conclusion

During the study period, a survey of the disease issues that negatively impact farmed shrimp farms in coastal Tamil Nadu, South India, was carried out. Shrimp farms that were chosen for inclusion in the survey were dealing with a range of illnesses, such as low growth and death rates in their stocks. Furthermore, samples from a few cultured penaeid stocks that seemed healthy were taken, and any signs of subclinical illness were sought. The predominant cytopathogenic changes associated with SEMBV infection that were observed in the membranes were nuclear hypertrophy and virogenesis, leading to the development of polyhedrin. The hepatopancreatic lumen allowed the virus particles and occlusion bodies that were released from lysed cells to reach the digestive system. In all ten survey locations, samples of P. monodon had SEMBV, the most common major pathogen detected there. These results suggest that SEMBV is a novel infectious agent in India's farmed stocks of penaeid species. Before a few days of infection, many shrimp sampled for the survey had a few seemingly common penaeid epi-commensals on their gills, appendages, and general body surface; some of them were linked to sickness and mortality.

Acknowledgements The first author thanks the Honourable Secretary, Correspondent, and Principal C. Abdul Hakeem College (Autonomous), Melvisharam—632509, Ranipet District, Tamil Nadu, India, for providing the necessary laboratory facilities and infrastructure for completing the experiments.

References

Adams JR, Bonami JR (1991) Preparation of invertebrate virus and tissues for examination. In: Adams JR, Bonami JR (eds) Atlas of invertebrate viruses. CRC Press, Boca Raton, pp 9–30

APHA (1992) Standard methods for the examination of water and wastewater, 18th edn. APHA, Washington

Bell TA, Lightner DV (1988) IHHN virus: infectivity and pathogenecity studies in *Penaeus stylirostris* and *Penaeus vannainei*. Aquaculture 38:185–194

Brock JA, Lightner DV (1990) Diseases of crustacean. Diseases caused by microorganisms. In: Kinne O (ed) Diseases of marine animals, vol 3. Biologische Anstalt Helgolond, Hamburg, pp 245–349.

Couch JA (1991) Baculoviridae. Nuclear polyhedrosis viruses. Part 2. Nuclear polyhedrosis viruses of invertebrates other than Insects. In:Adams JR, Bonami JR (eds) Atlas of invertebrate viruses. CRC Press, Boca Raton, pp 205–226

Couch JA (1981) Viral diseases of invertebrates other than insects. In: Pathogenesis of invertebrate microbial disease. Davidson (EW) Allanheld, Osmun, Totowa, pp 127–160

Francki RIB (1984) Viral diseases of marine invertebrates. Helgo Lander Meenesunters 37:65–98

Francki RIB, Fauquet CM, Knudson DL, Brown F (1991) Classification and nomenclature of viruses Arch viral (Supplementum 2) 1–450

Johnson MC, Maxwell JM, Loh PC, Leong JAC (1999) Molecular characterization of the glycoproteins from two warm water rhabdoviruses: snakehead rhabdovirus (SHRV) and rhabdovirus of penaeid shrimp (RPS)/spring viremia of carp virus (SVCV). Virus Res 64(2):95–106

Johnson PT, Lightener DY (1988) The rod-shaped nuclear viruses of crustaceans: gut-infecting species. Dis Aquatic Org 4:123-141

Lightener DV (1992) Shrimp virus diseases: diagnosis, distribution, and management. In Wyban J (ed) Proceedings of the special session on shrimps farming. World Aquaculture Society, Baton Rouge, pp 238–253

Lightener DV, Redman RM (1991) Hosts, geographic range and diagnostic procedures for the penaeid virus diseases of concern to shrimp culturists in the America. In: Deloach P, Dougherty WJ, Davidson MA (eds) Frontiers of shrimp research. Elsevier, Amsterdam, pp 173–196

Lightener DV, Redman RM, Bell TA (1983) Observations on the geographic distribution, pathogenesis, and morphology baculovirus from Penaeus monodonFabricius. Aqaucultue 32:209–233

Lightener DY, Poulos BI, Bruce L, Redman RM, Mari J, Bonami JR (1992) New development in penaeid virology: application of biotechnology in research and disease diagnosis for shrimp viruses of concern in the America. In: Fulks, W, Main K (ed) Diseases of cultured penaeid shrimp in Asia and the United States. The Oceanic Institute, Honolulu, pp 233–253

Nadal A, Rovira JM, Laribi O, Leon-quinto T, Andreu E, Ripoll C, Soria B (1998) Rapid insulin tropic effect of 17β-estradiol via a plasma membrane receptor. FASEB J 12(13):1341–1348

Pratanpipat P, Nash GL, Withyachumnarnkul B (1994) A preliminary report on shrimp culture problems in India. In: INDAQUA, 22–26 Dec

Preece A (1972) A manual for histologic technicians. Little, Brown, and Company, Boston, p 428

Shirin S, Yadav AK (2014) Physico-chemical analysis of municipal wastewater discharge in Ganga river, Haridwar district of Uttarakhand, India. Curr World Environ 9(2):536–543. https://doi.org/10.12944/CWE.9.2.39

Shirin S, Jamal A, Ranjan P, Yadav AK (2019) Study on assessment of slope stability and mixed disposal of overburden in voids of Singrauli Coalfield. Environ Qual Manage 28(3):131–139. https://doi.org/10.1002/tqem.21616

Shirin S, Yadav AK (2021) Assessment and suitability for the dumping of flyash in an abandoned mine. Environ Challenges 4:100111. https://doi.org/10.1016/j.envc.2021.100111

Shirin S, Jamal A, Emmanouil C, Yadav AK (2021) Assessment of characteristics of acid mine drainage treated with fly ash. Appl Sci 11(9):3910. https://doi.org/10.3390/app11093910

Shirin S, Jamal A, Emmanouil C, Singh VP, Yadav AK (2023) Assessment and characterization of waste material used as backfilling in an abandoned mine. Int J Coal Prep Util 43(8):1402–1410. https://doi.org/10.1080/19392699.2022.2118259

Strickland JDH, Parsons TR (1968) A practical handbook of seawater analysis, 2nd edtion. J Fish Res Board Canada 167:310

Yadav AK, Sahoo SK, Mahapatra S, Kumar AV, Pandey G, Lenka P, Tripathi RM (2014) Concentrations of uranium in drinking water and cumulative, age-dependent radiation doses in four districts of Uttar Pradesh, India. Toxicol Environ Chem 96(2):192–200. https://doi.org/10.1080/02772248.2014.934247

Chapter 10
Heavy Metal Accumulation and Health Risk Assessment in the Puddles of Virudhunagar District, Tamil Nadu, India

Rajagopal Bhuvaneswari, Narasimman Manickam, Ambrose Paulraj, and Razack Rafi Mohamed

Abstract The present investigation examined the levels of lead (Pb), zinc (Zn), copper (Cu), and cadmium (Cd) in various water bodies located in the districts of Rajapalayam, Sivakasi, and Virudhunagar in Tamil Nadu, India. It also examined changes in the hematological and physiological traits of several freshwater fish, such as *Oreochromis mossambicus*, *Catla catla*, *Mystus aor* and *Channa striata*, in addition to the levels of heavy metal accumulation in these fish. An industrial zone in full swing, full of power plants, failed companies, and hundreds of band-aids. Eight randomly chosen locations, both inside and outside the city, were used for the sample. According to the study's conclusions, the water samples from the Sengulam and Vadakarai ponds contained higher levels of lead than were allowed by both national and international rules. Some ponds' water tests revealed lead levels below

R. Bhuvaneswari (✉)
Fish Disease Diagnostic Laboratory, Department of Zoology, Ayya Nadar Janaki Ammal College (Autonomous), Sivakasi, Tamil Nadu 626124, India
e-mail: drbhuvana3@gmail.com

N. Manickam
Marine Planktonology & Aquaculture Laboratory, Department of Marine Science, School of Marine Sciences, Bharathidasan University, Tiruchirappalli, Tamil Nadu 620024, India

A. Paulraj
PG and Research Department of Zoology, Government Arts College for Men, Krishnagiri, Tamil Nadu 635001, India

R. R. Mohamed
PG and Research Department of Zoology, C. Abdul Hakeem College (Autonomous), Melvisharam, Ranipet, Tamil Nadu 632509, India

N. Manickam
Crustacean Biology Laboratory, Department of Zoology, Bharathiar University, Coimbatore, Tamil Nadu 641046, India

Department of Zoology, Sree Meenakshi Arts & Science College, Meenakshi Nagar, Eriyur (Po), Pennagaram (Tk), Dharmapuri, Tamil Nadu 636810, India

A. K. Yadav et al. (eds.), *Integrated Management of Water Resources in India: A Computational Approach*, Water Science and Technology Library 129,
https://doi.org/10.1007/978-3-031-62079-9_10

permissible bounds. The investigation findings demonstrated that the levels of Cd (0.007 ppm) Cu (1.0 ppm), and Zn (15.00 ppm) in each water sample were within allowable limits. The build-up of Pb, Zn, Cu and Cd buildup in the gills, muscles, and bones of *O. mossambicus*, *C. catla*, *M. aor* and *C. striata*. Pb levels were higher than permitted (0.29 ppm) in the muscle, bone, and gills of *O. mossambicus*, *C. catla*, *M. aor*, and *C. striata*. Less zinc than allowed (5.00 ppm) was found in the muscles, gills, and bones of *O. mossambicus*, *C. catla*, *M. aor*, and *C. striata*. Cu concentrations were below the permissible limit (0.5 ppm) in the bones, gills and muscles of *O. mossambicus*, *C. catla*, *M. aor*, and *C. striata*. Multiple sections of *O. mossambicus*, *C. catla*, and *M. aor* had Cd levels over the permissible limit of 0.05 ppm. In *C. striata*, Cd accumulates in the muscle above and the gills and bone below. The natives have used this water for domestic and fishing needs. Given the rate at which heavy metal concentrations increase, it appears that both fishing and water consumption can be avoided. District and government authorities should take the appropriate action in these areas and additional measures aimed at lowering the concentration of heavy metal concentrations through the use of innovative and alternative technologies.

Keyword Heavy metal · Health risk assessment · Freshwater fish · Water

10.1 Introduction

The majority of heavy metals can be found scattered across rock formations. The amount of heavy metals humans have contributed to the biosphere has increased due to industrialization and urbanization. Soil and aquatic habitats contain the highest amounts of heavy metals, while the atmosphere contains relatively smaller amounts of these metals as vapors or particles. It is believed that a variety of heavy metals are essential for plant growth. In a similar vein, numerous heavy metals are vital trace nutrients that both humans and animals require (Wintz et al. 2002). There are more than twenty different heavy metals. According to researches, the heavy metals that have been shown to have detrimental impacts on human health include lead, cadmium, mercury, nickel, copper, chromium, zinc, and lead (Abdullah et al. 1997; Shirin and Yadav 2021; Shirin et al. 2023). Common inorganic pollutants that have a great effect on the environment are toxic heavy metals such as Cu, Cd, Zn, Ni, Fe, Hg, As, and Cr. These dangerous metals, which are present in significant amounts in industrial wastes including pulp from distilleries, paper, textiles, dyes, and printing, are among the primary environmental contaminants (Srivastava and Patil 2006). Micronutrients are some of these heavy metals, including Cu, Zn, Fe, Mn, Mo, Ni, and Co. The biological system is heavily on trace metals like Zn, Cu, manganese, iron, and excess or lack of these metals in the human body can cause a variety of diseases. The poisoning of the aquatic environment by heavy metals has been a global concern in recent years because of their indestructible nature and the dangerous impact they have on species (Farlane and Burchett 2000). The most important heavy metals in terms of water contamination are Zn, Cu, Pb, Ni, and

Cr. Although some elements, such as Pb and Cd, are poisonous, their biological function is unknown (Anderani et al. 2008). A significant environmental issue is the contamination of aquatic ecosystems by heavy metals that endanger human health and harm the environment when they accumulate in the food chain (Chen et al. 2000; Kumar et al. 2024). Heavy metal concentrations in the coastal ecosystem have recently been found to have increased, and this rise has been connected to mining, agriculture, and the disposal of industrial waste (Sankar et al. 2006; Shirin et al. 2019). Aquatic ecosystems are under constant pressure from anthropogenic pollutants, and heavy metal contamination of these environments is a global problem that can harm nontarget populations, such as humans, as well as ecosystem inhabitants through food chains and food webs (WHO 1989). Farmobi et al. (2007) state that heavy metals and chemicals can induce disorders in the reproductive, respiratory, neurological, and hepatic systems of aquatic vertebrates and invertebrates that are either fatal or sublethal.

Heavy metals can have a bioaccumulative, sublethal, or deadly effect on all members of the biota, including fish, birds, benthos, phytoplankton, and zooplankton, as well as people. Due to their consumption of contaminated food or absorption of metals through their gill surface, fish have higher concentrations of heavy metals in their tissues (Miliou et al. 1998). Additionally, the physiological, histological, and biochemical reactions of fish exposed to sublethal levels can be used to gauge the effects of exposure. The fish gills, liver, and kidneys absorb heavy metals in higher quantities than the muscles, which exhibit the lowest level of metal accumulation (Wepener et al. 2001). Fish have been widely used to measure metal contamination bioindicators. Fish is an essential dietary source because it is high in protein, a great source of vitamins and omega-3 fatty acids, and helps reduce the risk of several diseases, such as diabetes, Alzheimer's disease, heart attacks, strokes, and sudden death. Fish intake can improve health and be beneficial to consumers, but it can also pose health risks due to the possibility of contamination from harmful metals found in the contaminated aquatic environment (Kojadinovic et al. 2006). Two heavy metals that are known to be extremely dangerous to both humans and fish are lead (Pd) and copper (Cu). According to Malik et al. (2010), these pollutants in the aquatic environment can negatively impact aquatic species' growth, physiology, and chance of surviving reproduction. Fish are particularly vulnerable to these effects. The accumulation of these hazardous metals in aquatic foods not only disturbs the ecosystem, but also threatens public health (Ubalua et al. 2009). Fish absorb heavy metals through a variety of organ systems (Fatima and Usmani 2013). Cadmium (Cd), among other unnecessary heavy metals, has attracted significant scientific interest due to its potential to harm human health (Tran and Popova 2013).

However, fish exposed to various heavy metal compounds in their natural habitat are typically more hazardous than individual metals because of their additive or more additive (synergistic) actions. Furthermore, aquatic organisms like plankton, aquatic plants, invertebrates, and vertebrates are negatively impacted by heavy metals, even at very low concentrations (Schuurmann and Market 1998). In light of this, the objective of the current study has been to evaluate the levels of heavy metals (Pb, Zn, Cu, and Cd) in a few chosen water bodies in Rajapalayam Taluk (Tamil Nadu,

India). The study examined the accumulation of heavy metals in a variety of freshwater fish species, including *Oreochromis mossambicus*, *Catla catla*, *Mystus aor* and *Channa striata*. Additionally, the fish's hemoglobin content, total red blood cell count, white blood cell count, and oxygen consumption were examined, as well as their physiological and physiological experimental responses to the pollutants in the water.

10.2 Materials and Methods

10.2.1 Distribution of Study Area

Rajapalayam is located in the Virudhunagar District of Tamil Nadu, India. It is an industrial town with hundreds of bandages and dying companies, power looms, and other machinery. Four locations for sampling were chosen in the Rajapalayam area (Fig. 10.1). Karunkulam Pond, which collects sewage waste and industrial effluents, is located near the Rajapalayam New Bus Stand. Senkulam Pond is located close to Chatrapatti in Mudukudi. Samusigapuram is home to Illupankulam Pond. The Chatrapatti and Samusigapuram regions are home to numerous power looms and dying enterprises. The location of Vadakarai Pond is close to Sivalingapuram. It receives fertilizer and pesticide-laden drainage water from agriculture.

10.2.2 Collection of Water Samples

In November and December 2015, when it was raining heavily, water samples were taken from the ponds. Glass labware vials were used to collect water samples and after thoroughly cleaning them with nitric acid, they were rinsed with distilled water.

10.2.3 Collection of Fish Samples and Experimental Animals

Samples of fish are taken from different bodies of water in and around Rajapalayam. with the assistance of knowledgeable local fishermen. *Oreochromis mossambicus* (Tilapia), *Mystus aor* (Catfish), *Catla catla* (Catla), and *Channa striata* (Snake Head Fish) were among the live freshwater fish that were moved to polythene covers and brought to the lab for additional examination.

Fig. 10.1 Study areas in Illuppankulam, Karunkulam, Vadakarai and Senkulam ponds

10.2.4 Acclimatization and Specimen Storage

Fish samples were exposed to dechlorinated tap water in buckets for a whole day to acclimate to laboratory settings. They were slaughtered after one day to obtain tissue and blood samples, which were subsequently used to analyze the bioaccumulation of heavy metals. The dead fish are preserved in the ice boxes and kept at -10 °C until further examination is carried out.

10.2.5 Preparation of Fish Sample and Dissection of Fish Organs

When the fish, precautions were taken to prevent contaminating the samples from the outside. The fish was dissected with a sterile, rust-free stainless steel kit. Place the fish on the dissection pan with its left side facing upwards. Slice the skin from the anal fin to the operculum along its abdomen. Remove the pectoral fin and operculum. The groins, bones, and muscles are cut. After being moved to spotless Petri plates that had been acid-washed, all tissue samples were oven-dried at 80° C until a consistent weight was achieved.

10.2.6 Heavy Metal Analysis in Water

The AA-6300 Atomic Absorption Spectrophotometer was used to analyze water samples to detect elements such as Cu, Cd, Zn and Pb. Air-C2H2 flame was used for the Cu measurement, which was performed at 324.8 nm (1.8 l/min for fuel gas and 15.l/min for support). The air-C_2H_2 flame was used for the Cd measurement, which was performed at 228.8 nm (1.8 l/min for fuel and 15.0 l/min for support gas). The air-C_2H_2 flame was utilized for Zn measurements, which were carried out at 213.9 nm (fuel gas flow rate: 2.0 l/min, support gas flow rate: 15.0 l/min). The Pb measurement was carried out using an Air-C_2H_2 flame at a wavelength of 283.3 nm with a fuel gas flow rate of 2.0 l/min and a support gas flow rate of 15.0 l/min.

10.2.7 Heavy-Metal Analysis in Fishes

To determine whether elements including copper, zinc, lead, and mercury had accumulated in the fish samples, an analysis was performed.

10.2.8 Analytical Procedure

The experimental samples were degraded using a 2:1 v/v mix of perchloric acid and laboratory grade nitric acid. Ten times the sample mass in grams made up the volume of the digesting mixture. After the samples were digested in the mixture for an entire night at room temperature, it was gradually heated to 200° C for three hours in a sand bath. After that, the combination was allowed to digest until no odours were visible and turned pale yellow. After diluting the resulting mixture to a volume of 25 ml using double-distilled water, Swaileh and Sansur (2004) filtered the samples. The sample was run with a standard solution (known volume) and blanks (very pure distilled water, very pure nitric acid and perchloric acid at a 2:1 v/v ratio). In conclusion, the Shimadzu AA-6300 Atomic Absorption Spectrophotometer (AAS), which is kept at the Science Instrumentation Center of Ayya Nadar Janaki Ammal College in Sivakasi, Tamil Nadu, India, was used to determine the concentration of heavy metals, including Pb, Cd, Zn and Cu.

10.2.9 Haematological Studies

To investigate the hematological and physiological parameters, live freshwater fish, including *O. mossambicus*, *M. aor*, *C. catla*, and *C. striata*, were taken from specific water bodies in the Rajapalayam area. Using a cardiac vein puncture technique, blood was extracted from live fish and placed in vials containing ethylene diamine tetraacetic acid (EDTA, Sigma).

10.2.10 Total Count of RBC

Neubauer's hemocytometer was used to calculate the total erythrocyte count. Blood was extracted into an RBC pipette until the 0.5 mark was reached. After that, the blood was well combined in a red pipette and diluted with Hayem fluid (1:200). The blood sample was diluted and placed in a counting chamber after the initial few drops were discarded. The hemocytometer chamber was filled with a single drop of blood. Wintrobe (1967) reported that the RBC count was 106/mm^3.

$$\text{Number of RBCs/mm}^3 = \frac{\text{Number of RBC counted} \times 200 \times 10}{\text{Area of chamber counted}}$$

10.2.11 Total Count of White Blood Cells (WBC)

With Neubauer's counting chamber, the total leukocyte count (TLC) was performed. Using Turk fluid, the blood was diluted to 11 marks after being taken to the 0.5 mark in the WBC pipette. Next, fully spin the pipette while maintaining a horizontal orientation. The blood sample was diluted and placed in a counting chamber after the initial few drops were discarded. At the four corners of a square millimeter, in the middle-ruled zone on both sides of the hemocytometer counting chambers, the total leucocytes were counted. The stain present in the dilution fluid recognizes the leucocytes. Cells that are in contact with the boundary lines are not included in the count. Total leucocytes (TLC) were expressed as total leucocytes × 103/mm3 (Dhayanithi et al. 2013) after calculations were performed with the dilution factor taken into account.

(Dilution = 25, Area counted = 4, Depth of the fluid = 0.1 mm).

$$\text{Number of WBCs/mm}^3 = \frac{\text{Number of WBC} \times \text{Dilution}}{\text{Area counted} \times \text{Depth of the fluid}}$$

10.2.12 Estimation of Hemoglobin and Shali's Acid Haematin Method

Add 0.1 N of HCl to the graduated tube until it reaches the 20 mark (on the percentage scale). Using a hemoglobin pipette, draw 0.02 ml of blood. To ensure that there is no blood left on the pipette's outside, wipe it with a cloth. Pour blood into the Shali tube filled with HCl solution. Once again, a tiny quantity of acid to the pipette and then let it out of the tube, repeating it twice. Using a glass rod, mix the contents briskly but gently for ten minutes. Drop-by-drop, add 0.1 N HCl, stirring in between until the color reaches the standard. Blaxhall and Daisley (1973) state that calibrations are used to determine the hemoglobin concentration once the solution in the graduated tube has been read.

10.2.13 Physiological Studies of Opercular Movement

The fish studied had their number of opercular motions per minute counted using a stopwatch. Three readings were collected for each sample and the mean value was noted.

10.2.14 Oxygen Consumption in Fish

Using Winker's approach, the experimental fish's oxygen consumption was calculated. One milliliter of each of the alkaline iodine and manganous sulphate solution should be carefully added along the sides of the reagent bottle after the top reagent bottle has been filled with sample water. There will be a brown precipitate. To prevent photochemical reactions and give the precipitation time to settle, stop the bottle, shake it thoroughly, and store it in a dark area. Add 1 ml of concentrated H_2SO_4 gradually to dissolve the precipitate in the yellowish solution after 15 min. Once in a conical flask, reduce this sample to 50 ml. Add a few drops of 1% starch solution to serve as a demonstration. The water becomes blue hue. Use a burette containing a 0.025 N sodium thiosulfate solution to titrate this sample. When the blue color fades away, the process is complete, and the liter value is noted. To obtain the concordant value, repeat the titration. Oxygen consumption is determined by the hourly quantity of oxygen in the experimental sample and the quantity of oxygen in the control sample. The oxygen consumption is measured in milligrams per gram or liter per hour. Use the formula to determine how much oxygen is in the sample.

$$\text{Amount of oxygen} = \frac{8 \times 1000 \times 0.025}{\text{Volume of sample}} \times \text{titre value in mg/litre}$$

10.2.15 Statistical Analysis

Heavy metal content in fish tissue was determined using a one-way ANOVA classification. The values are given as the mean ± SD.

10.3 Result

10.3.1 Examination of Heavy Metals in Water Samples

Heavy metals, including Pb, Zn, Cu and Cd, were found in pond water samples taken from the Rajapalayam area and surrounding areas for this investigation (Table 10.1). Pd: The range of Pd concentrations found in the water samples of Karunkulam and Ilupankulam was 0.0862 ppm to 0.884 ppm. Pd levels in these water samples were lower than WHO-permissible levels (0.10 ppm). The water samples from the Senkulam and Vadakarai ponds have Pd concentrations ranging from 0.1437 ppm to 0.1725 ppm. The levels of Pd in these water samples exceeded WHO recommendations. The quantity of Zn ranged from 0.0013 ppm to 0.4880 ppm in the water samples. All water samples had Zn concentrations below the WHO recommended threshold of 15.00 ppm. Cu: The water samples had Cu concentrations ranging from 0.0385 to 0.1243 ppm. Every water sample had less copper than the WHO-permissible threshold of 1.0 ppm. Cd: The water samples had Cd concentrations ranging from 0.0221 ppm to 0.0590 ppm. Each water sample had Cd levels above WHO-permitted thresholds of 0.007 ppm.

Table 10.1 Concentration of heavy metal in four selected water bodies

S. No.	Site of sample collection	Heavy metals			
		Pb (ppm)	Zn (ppm)	Cu (ppm)	Cd (ppm)
	Permissible limits (WHO)	0.10	15	1	0.007
1.	Vadakarai pond water	0.1725*	0.0013*	0.0671*	0.0520*
2.	Karunkulam pond water	0.0862*	0.1814*	0.1243*	0.0590*
3.	Sengulam pond water	0.1437*	0.4176*	0.1129*	0.0451*
4.	Illupankulam pond water	0.0884*	0.0488*	0.0385*	0.0221*
Mean ± SD		0.1227 ± 0.04	0.162 ± 0.19	0.057 ± 0.04	0.045 ± 0.016

* The results are statistically not significant at $p < 0.05$ in one-way ANOVA classification

10.3.2 Examination of Heavy Metals in Fish Samples

The muscle, gills, and bone were shown to accumulate heavy metals such as Pb, Zn, Cu, and Cd (Fig. 10.2; Tables 10.2, 10.3, 10.4, 10.5, and 10.6). Pd: In the current study, Pd has accumulated in the muscles, bones, and gills of *C. catla* and *O. mossambicus* above the allowable limit (0.29 ppm). The fish *O. mossambicus* was observed in many parts, including the gills (0.4025 ppm),, muscle (0.7187 ppm) and bone (0.8337 ppm). The highest concentration of Pd is found in bone. The gills of C. catla have high levels of lead, as do other body tissues such as muscle (0.4350 ppm) and bone (0.3024 ppm). Compared to bone (0.3536 ppm), the gills of *M. aor gills* contain the highest concentration of Pd (0.3733 ppm). Pd accumulates in the muscle (0.1965 ppm) below the allowable level. The accumulation of Pd in the muscle (0.2652 ppm) and gills (0.1866 ppm) of *C. striata* is within the allowable limits. The highest Pd values (0.5305 ppm) were found in bone. Zn: In the gills of *M. aor* muscle (1.1668 ppm) and *C. striata* muscle (1.0571 ppm) (1.2895 ppm), Zn accumulation was below the allowable limit, which is less than 5.0 ppm. Zn concentrations of 6.9552 ppm and 6.2727 ppm were found in the bones of *C. striata* and *M. aor*, respectively. Compared to muscle (4.2087 ppm) and bone (5.7421 ppm), the gills (10.3097 ppm) of *O. mossambicus* were found to have the highest concentration. The same finding was made with *C. catla.* Zn concentrations in the gills (11.1111 ppm) muscle (5.7730 ppm), and bone (1.6824 ppm) are all higher than the allowable limit. Cu: In *O. mossambicus*, Cu was discovered to have accumulated above the allowable limit, or more than 0.5 ppm. The highest concentration was detected in the gills (6.4020 ppm), compared to bone (3.0953 ppm) and muscle (2.4737 ppm) tissue. Cu content in *C. catla* above allowable limits.

Compared to other tissues like the gills (1.5638 ppm), and bone (1.2824 ppm) the concentration of Cu in muscle tissues is the highest (2.1692 ppm). The tissues of *M. aor* muscle (0.0666 ppm), bone (0.1059 ppm), and gills (0.1920 ppm) showed Cu accumulation below the allowable limit. Cu concentrations in *C. striata* muscle (0.0778 ppm) gills (0.0778 ppm), and bone (0.3286 ppm) are below the allowable limit. Cd: The amount of Cd that accumulated in *O. mossambicus* and *C. catla* is above the permissible limit of 0.05 ppm. *O. mossambicus* was found to include muscle (0.1035 ppm) bone (0.0882 ppm), and gills (0.1077 ppm). High levels of Cd have been found in *the* bone (0.0698 ppm) and various bodily tissues, including the muscle (0.0524 ppm) and gills (0.0530 ppm). In the muscles of *M. aor*, the concentration of Cd is below the allowable level (0.02494 ppm), while it is above the limit in the gills (0.0513 ppm) and bone (0.0554 ppm). In *C. striata*, a significant concentration of Cd is found in muscle (0.0676 ppm), whereas the amount of Cd deposited in the gills (0.0367 ppm) and bone (0.0335 ppm) is below the allowable limit.

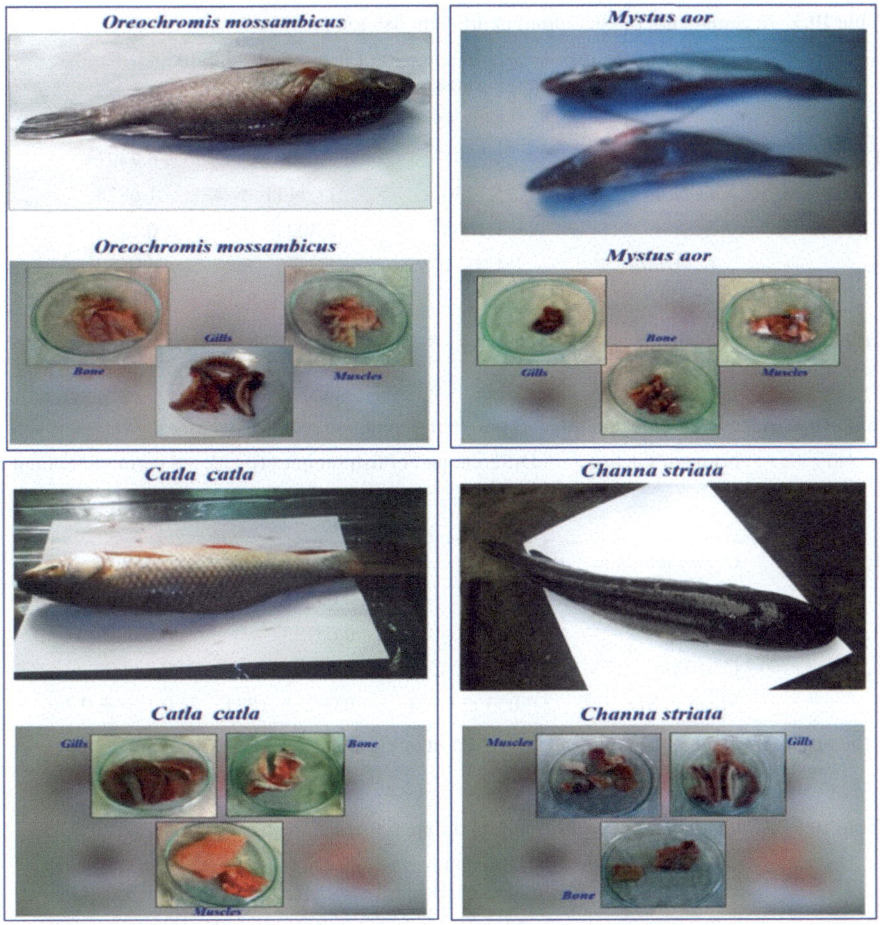

Fig. 10.2 Experimental fishes (*O. mossambicus, M. ao, C. Catla* and *C. striata*)

Table 10.2 Accumulation of Lead (ppm) in different tissues of fish samples

S. No	Name of the fish	Different part of fish samples (ppm) and permissible limit (0.29 ppm)		
		Muscle	Gills	Bone
1.	*O. mossambicus*	0.7187*	0.4025*	0.8337*
2.	*C. catla*	0.4350*	0.5323*	0.3024*
3.	*M. aor*	0.1965*	0.3733*	0.3536*
4.	*C. striata*	0.2652*	0.1866*	0.5305*
	Mean ± SD	0.404 ± 0.23	0.374 ± 0.143	0.505 ± 0.24

Oreochromis mossambicus (O. mossambicus); Catla catla (C. catla); Mystus aor (M. aor); Channa striata (C. striata)

*The results are statistically not significant at $p \leq 0.05$ in one-way ANOVA classification

Table 10.3 Accumulation of zinc (ppm) in different fish sample tissues

S. No.	Name of the fish	Different part of fish samples (ppm) and Permissible limit (5.0 ppm)		
		Muscle	Gills	Bone
1.	*O. mossambicus*	4.2087*	10.3097*	5.7421*
2.	*C. catla*	5.7730*	11.1111*	1.6824*
3.	*M. aor*	1.1668*	2.8024*	6.2727*
4.	*C. striata*	1.0571*	1.2895*	6.9552*
	Mean ± SD	3.05 ± 2.33	6.378 ± 5.05	5.16 ± 2.37

* The results are statistically not significant at $p < 0.045$inone-wayANOVAclassification

Table 10.4 Accumulation of cadmium (ppm) in different fish sample tissues

S. No	Name of the fish	Different part of fish samples (ppm) and permissible limit (0.05 ppm)		
		Muscle	Gills	Bone
1.	*O. mossambicus*	0.1035*	0.1077*	0.0882*
2.	*C. catla*	0.0524*	0.0530*	0.0698*
3.	*M. aor*	0.0294*	0.0513*	0.0554*
4.	*C. striata*	0.0676*	0.0367*	0.0335*
	Mean ± SD	0.063 ± 0.031	0.062 ± 0.031	0.186 ± 0.246

* The results are statically significant at $p < 0.05$ in one-way ANOVA classification

Table 10.5 Accumulation of copper (ppm) in different fish sample tissues

S. No	Name of the fish	Different part of fish samples(ppm) and permissible limit (0.05 ppm)		
		Muscle	Gills	Bone
1.	*O. mossambicus*	2.4737*	6.4020*	3.0953*
2.	*C. catla*	2.1692*	1.5638*	1.2824*
3.	*M. aor*	0.0666*	0.1920*	0.1059*
4.	*C. striata*	0.0778*	0.0778*	0.3286*
	Mean ± SD	1.196 ± 1.30	2.059 ± 2.97	1.202 ± 1.359

* The results are statistically not significant at $p < 0.05$ in one-way ANOVA classification

10.3.3 Investigations in Hematology and Physiology

Red blood cells (RBC): The fish obtained from the water of the pond were subjected to haematological investigation in the current study. The RBC count in O. mossambicus and C. catla was determined to be $5.37 \pm 0.56 \times 10^6$ $mm^3$and $5.56 \pm 1.15 \times 10^6$ mm^3, respectively. The RBC count in *M. aor* and *C. striata* was determined to

Table 10.6 Haematological and physiological changes of fishes collected from selected water bodies of Rajapalayam Taluk

S. no.	Parameter	Experimental fishes			
		O. mossambicus	*C. catla*	*M. aor*	*C. striata*
1.	RBC ($\times 10^6$) mm^3	$5.37 \pm 0.56^*$	$5.56 \pm 1.15^*$	$6.2 \pm 0.62^*$	$6.7 \pm 1.52^*$
2.	WBC ($\times 10^3$) mm^3	$4.99 \pm 2.0^*$	$5.32 \pm 1.52^*$	$4.35 \pm 0.43^*$	$4.27 \pm 1.00^*$
3.	Haemoglobin (g/dl)	$7.0 \pm 0.66^*$	$7.55 \pm 0.33^*$	$8.0 \pm 1.15^*$	$8.23 \pm 0.42^*$
4.	Opercular movement (No/Mint)	$66 \pm 0.53^\#$	$67 \pm 0.53^\#$	$68 \pm 0.47^\#$	$79 \pm 1.16^\#$
5.	O_2 (mg/g/l/h)	$1.5 \pm 0.72^\#$	$1.70 \pm 0.23^\#$	$1.92 \pm 0.34^\#$	$2.0 \pm 1.23^\#$

* The results are statically significant at $p < 0.001$ in one-way ANOVA classification
$^\#$ The results are statically significant in 't' test at (0.01)
The Opercular movement and O_2 consumption is positively correlated (+0.76)

be 6.2 ± 0.62 × 10^6 mm^3 and 6.7 ± 1.52 × 10^6 mm^3, respectively. Low RBC counts were observed in *C. catla* and *O. mossambicus.* White blood cells (WBCs): *O. mossambicus* 4.99 ± 2.00 × 103 mm^3 and *C. catla* 5.32 ± 1.52 × 10^3 mm^3 had high levels of WBC in the current study. The WBC count in *M. aor* and *C. striata* was determined to be 4.35 ± 0.43 × 10^3 mm^3 and 4.27 ± 1.00 × 10^3 mm^3, respectively. Content of hemoglobin: *M. aor* had a hemoglobin content of 8.0 ± 1.15 g/dl, while *C. striata* had a hemoglobin value of 8.23 ± 0.42 g/dl. The hemoglobin content of O. mossambicus and *C. catla* were 7.55 ± 0.33 g/dl and 7.0 ± 0.66 g/dl, respectively. Opercular movement: *C. striata* 79 ± 1.16 (No/mini) showed a greater number of opercular motions. 68 ± 0.47 (No/mini) of opercular movement. In *O. mossambicus* and *C. catla,* opercular movement was measured to be 66 ± 0.53 (No/mini) and 67 ± 0.53 (No/mini), respectively. Oxygen consumption: *C. striata* was found to have an elevated oxygen consumption of 2.0 ± 1.23 mg/g/l. The findings indicate that *M. aor* used 1.92 ± 0.34 mg/g/l of oxygen, *C. catla* consumed 1.70 ± 0.23 mg/g/l, and *O. mossambicus* consumed 1.5 ± 0.72 mg/g/l.

10.4 Discussion

The rapid industrialization and urbanization of today have led to pollution, claim Swaminathan and Manonmani (1997), as a result of the flow of industrial and municipal pollutants into freshwater sources. The current study aims to investigate the concentration of heavy metals in pond water in the Rajapalayam area (Tamil Nadu, India). The findings demonstrated that the Senkulam and Vadakarai pond water samples had elevated concentrations of lead (Pb). Pb concentrations ranged from 0.1437 to 0.1725 ppm. The water tests from the Karunkulam and Illupankulam ponds have Pb levels over the WHO's permissible limit of 0.10 ppm. This might be due to careless discharge of chemicals from human activity and industry, waste materials, and home sewage. Because heavy metals are poisonous and harm aquatic life, industrial heavy metal discharge poses a significant threat to water quality according to a survey by the Central Inland Fisheries Research Institute. Zn, Cu, and Cd are stable in the environment like other metals and are used by the mining, steel and electroplating industries as well as dry and acid manufacturers (CIFRI 1981). According to Hart (1982) assessment, natural freshwater has a Pb content of 0.0003 to 0.03 parts per million. While Chakravarty et al. (1996) discovered that the range of Pb concentration is 0.074 ppm in various creeks of the Hooghly estuary along the Calcutta metropolis, Subramanian (1993) indicated that the concentration of Pb is high in the Cauvery River. Although Pd is used to construct water pipes, most of the lead originates from vehicle emissions. Tetra-ethyl Pb, which is added to gasoline as an antiknock additive, is part of the Pb compound. to aid in the engine's fuel burning evenly. Anemia, brain damage, vomiting, appetite loss, seizures, liver damage, and kidney damage are just a few of the side effects that lead to Kotpal and Bali (1986). Increased concentrations of lead in drinking water lead to disruptions in the synthesis of hemoglobin (Walker 1975). Khurshid and Zaheruddin (2000) claim that the elevated Pb and Zn levels relative to other metals make it dangerous to drink. The Zn concentrations in the water samples used in this investigation ranged from 0.0013 to 0.4880 ppm, which is below the allowable limits. According to Sulthana and Sharief (2005) analysis, Zn concentrations in Chennai's Double Lake range from 0.31 to 0.62 mg/I. The WHO (1984) recommended that the maximum permissible zinc content in drinking water be 15 ppm. Although air does not naturally contain zinc, it exists in the atmosphere around zinc smelters and scrap zinc refineries.

Higher levels of Zn (2.1094 ppm) have been found in the Cauvery River, according to Ayyadurai et al. (1994). According to research done in Lokande and Kelkar (1999) the Ulhas River System is becoming more contaminated with heavy metals like znic1.194 ppm of zinc are found in this watercourse. Safi River sediments have a high concentration of heavy metals, with 3.447 ppm zinc, according to (Saxena et al. 1998). According to reports (Chakravarty et al. 1996), the concentration of Zn in the Hoogly Estuary is 0.263 parts per million. The Zn concentration ranged from 0.31 to 0.62 mg/I in Chennai's Double Lake (Sulthana and Sharief 2005). Pb and Zn concentrations in the Safi River are greater than allowed according to Priyadharshini (1997). Dangerously high values of Pb and Zn were observed. In-depth research was done in the rivers that are highly impacted by human activity. According to Mazharuddin (1989), the distribution of heavy metals, specifically Zn and Pd, in the Great Lakes surface water bodies is greater than what is allowed. Groundwater in the Nagapattinam Districts has 0.0003–0.007 parts per million of cadmium, according to Mathess (1974). According to Zaheeruddin and Khurshid (1998), the groundwater of Faridabad, Haryana, has a Zn concentration of 0.473 parts per million. Zn is essential for human health. It is crucial for the production of proteins, the metabolism of carbohydrates, and enzymes. In a thorough investigation into the hydrochemistry of groundwater in Faridabad, Kakar (1989) found that the highest concentration of Zn was 0.175 parts per million.

The sources of Cu in the atmosphere are both natural and man-made. In rural areas, the average air concentration of Cu is believed to be between 5 and 50 mg, while in metropolitan areas, it is often between 30 and 200 mg. In addition to causing coma, uremia, sporadic fever, and hypertension, Purohit and Rajan (1990). Cu concentrations in the water samples used in this investigation range from 0.0385 to 0.01243 ppm. According to WHO (1984), the concentration of Cu does not exceed allowable limits. According to Lokande and Kelkar (1999), Cu concentrations in Vasai Creek's industrial effluents ranged from 3.1 to 7.8 ppm. According to Sulthana and Sharief (2005), the Cu content of Double Lakes ranged from 0.35 to 0.76 mg/l. Several researchers (Chandra et al. 1991; Bhosle and Patil 2001; Mule and Patil 2000) estimated and examined the Cu content in fresh water, wastewater, groundwater, and other sources. They reported higher levels of Cu content compared to the permissible limit of 1 mg/l of Cu in drinking water. WHO (1984) established a maximum allowable value of 1.00 ppm for copper in drinking water. Cd concentrations in water samples varied from 0.0221 to 0.0590 ppm in the current investigation. Every water sample has more Cd than is allowed (WHO 1984).

Pd builds up in the kidneys and livers of humans, leading to hypertension, emphysema, and kidney disease. For mammals, Purohit and Ranjan (1990) found that it is carcinogenic. Renal impairment may occur if drinking water contains a sizable amount of Pd (Zaheeruddin 1996). Bhatia and Ghosh (1999) found that the presence of Cd in roadside environments in India ranges from 0.844 ppm. According to Forstner and Wittmann (1975), fresh groundwater contains 0.0003–0.007 parts per million of Cd. To protect fish and other aquatic life, the USEPA (1997) recommended a maximum Cd value of 0.04 and 0.02 ppm. This illustrates the Pd concentration found in several parts of the fish in the current investigation. There was no discernible variation in Pd accumulation between organs ($P < 0.05$). According to a related study by Vinodhini and Narayanan (2008), Pd levels in *the* liver and muscle differed significantly. *Oreochromis niloticus*, *Anquilla anguilla*, and *Mugil cephaleus species* were the three species whose organs varied in lead content, according to a different study (Yilmaz 2009). In Mitra (1998), the researcher researched lead (Pb), a hazardous heavy metal that enters coastal waters via industrial wastewater discharges from oil refineries, painting and dyeing businesses and battery manufacturing facilities. Pd is an essential ingredient in antifouling paints, which are used to stop the growth of marine life at the bottom of boats and trawlers. These paints are made to continuously release harmful metals into the water, killing any organisms that might stick to the bottom and make their way into the sediment and aquatic compartments. Microorganisms (*Cereodaphnia cornuta*, *Diaphnosoma sarsi* and *Artemia*) were tested with the two antifouling paints (Mega Shield and Bombay). After experimenting, it was discovered that Mega Shield antifouling paint was considerably more hazardous than Bombay antifouling paint. The toxicity of the paints is assessed using a novel concept called the "surface-to-volume ratio" (Manickam 2008; Manickam et al. 2011).

In the current investigation, Zn accumulates in numerous fish organs in varying amounts. Compared to muscle and bone, the gills of *O. mossambicus* and *C. catla* were found to contain the highest quantity of zinc. Khail and Faragallah (2008) noted similar outcomes. Heavy elements such as Cr, Zn, Cu, Fe, Mn, Cd, and Pb were the most concentrated (52.2%) in the *Tilapia zilli* gills found in the Benue River. With the lowest metal concentration (21.5%), muscular tissues appeared to be the least preferred location for metal bioaccumulation. Similarly, the highest concentration (40.3%) of heavy metals, including Cr, Zn, Cu, Fe, Mn, Cd, and Pb, was found in the gills of *Clarias gariepinus*. However, muscular tissue had the lowest percentage (28.1%). According to Yilmaz (2009), the liver had a higher concentration of Zn than the muscle of *A. anquilla* and *Mugilce phalus*. This was due to the livers high irrigation and extraction capacities of the liver, which make it a target organ that absorbs a lot of heavy metals. The liver function can potentially serve as an indicator of the level of pollution in the ecosystem, according to Yilmaz (2009). According to the findings of Carbonell and Tarazona (1994), various aquatic animal tissues produce and provide immutable binding sites, which in turn Pd to varying amounts of accumulation.

In the current investigation, the accumulation of Cd in various parts of the fish sample is displayed. In tilapia (*O. mossambicus*), the accumulation of Cd exceeds the allowable limit (0.5 ppm). Gills have the highest concentration compared to bone and muscle. Cu is most concentrated in the muscle tissues of *C. catla*, followed by the gills and bones. Cu accumulates in *M. aor* and *C. striata* below the allowable limit. According to the findings of the current study, the concentration of Cd in the muscles of *M. aor* was below the authorized threshold; however, the concentration of Cu in the muscular tissues of *O. mossambicus*, *C. catla* and *C. striata* was greater than the permitted limit (0.05 ppm) (Table 5). The fish of *C. anguillaris* had the lowest Cd content (0.11 mg/g), whereas the fish of *Synodontis budgetti* had the highest concentration (1.03 mg/g). Similar findings were also noted by Sivaperumal et al. (2007). When a person consumes more than 350 mg/g of Cd, they can die from shock and acute renal failure (NAS—NRC 1982). Although Pd is a poisonous substance that accumulates in the human body, it poses a risk to human health. Pb is a dangerous, non-essential metal that can harm humans if breathed in or consumed in large quantities (Pourang 1995). Metal buildup in fish living in contaminated waterways, according to Jezierska and Witeska (2002) demonstrates that a significant amount of different metals can be accumulated in fish tissues without resulting in death. Live fish often have metal levels that are ranked as follows: Fe > Zn > Pb > Cu > Cd > Hg. Zinc levels could be as high as more than 300 $\mu g/g^{-1}$ dry weight Lead and copper have lower maximum values that often do not go above 10 $\mu g/g$ d.w. The fish acquire relatively little mercury and Cd, less than 1 $\mu g/g^{-1}$ dry weight. The build-up of fish metals varies among species that co-exist in the same body of water and is dependent on pollution.

Javed and Usmani (2012a) state that the water of the sewage-fed aquaculture pond in Panethi, Aligarh, which is used as a source of commercial food fish, also contains heavy metals in the following order: Fe > Mn > Zn > Co > Ni > Cu > Cr. A physicochemical evaluation of the sediments was conducted, which included research on the sequential extraction of heavy metals and classification of particle size. Because soluble heavy metals are easier to move and more accessible to aquatic life, they are assumed to be more dangerous. When compared to the gills of the fish species in the sample, the muscle had the lowest concentration of heavy metals, according to Cross et al. (1973). This aligns with earlier research by Ishaq et al. (2011), which showed that muscle does not actively store heavy metals. In *O. mossambicus* and *C. catla*, the current investigation revealed a low total RBC count and hemoglobin concentration (Table 10.6). The fish exposed to *Corydoras punctatus* had a lower hemoglobin content and a lower total RBC count than the control fish, according to Javed and Usmani (2012b). Fish *C. carpio* treated with Cr (VI) showed a significant drop in Hb concentration and RBC counts in two studies by Shaheen and Akhtar (2012) and Javed and Usmani (2013). The results are also in line with earlier studies (Vutkuru 2005; Vinodhini and Narayanan 2009), which, depending on the species, demonstrated a notable drop in RBC and Hb in freshwater fish (such as *L. rohita* and *C. carpio*) exposed to heavy metals (Cr and mixtures of Cd, Pb, Cr and Ni). According to the current study, a low erythrocyte count and a low hemoglobin concentration may be related to the harmful effects of accumulated heavy metals on the kidney.

Since the kidney is the fish's hematopoietic organ, any changes to its structure or function will impact the fish's blood profile. The highest concentration of iron was unquestionably deposited in all organs of *C. punctatus* organs, however, despite this, there was still a drop in RBC and Hb due to altered hematopoiesis or decreased intestinal absorption of iron. Similarly, *C. punctatus* had a low oxygen-carrying capacity, since it depends on the fish's Hb content. Fish such as *Isalvalinus fontinalis* have been observed to have declining red blood cell counts and anemia (Holcombe and Benoit 1976). The present study revealed elevated total WBC levels in *O. mossambicus* and *C. catla*. Although granulocytes and monocytes act as phagocytes to remove debris from injured tissue, lymphocytes produce antibodies (Wedemeyer and McLeay 1981). The concentration showed a more noticeable and clear pattern of change as a result of mercury exposure compared to the control group's erythrocyte levels. Leucocyte concentrations were higher in the blood of all experimental groups than in the control group. Thinning of circulating lymphocytes may cause lymphoid regions to produce more lymphocytes as a compensatory measure. *O. aureus* exposed to mercury showed elevated WBC (leucocyte) counts as reported by Shah and Altindag (2004) and Allen and Wilson (1994).

In the current investigation, *M. aor* and *C. striata* were found to exhibit a greater number of opercular motions. In the same fish, it was discovered that their oxygen consumption had increased. Physiological metrics offer information on how anthropogenic activities influence fish in ways beyond their immediate death and are sensitive indicators of repercussions such as environmental changes. Reproduction, resistance to illness, and growth rates are all adversely affected by physiological changes. When coupled with other stressors, temperature and oxygen stress can be very harmful and cause fish death (Vanlandeghem et al. 2010). Oxygen consumption is a common way to measure the metabolism of fish. A subsequent deficiency in oxygen transfer could be the cause of anemia, according to Robert (1978) and Kutty and Mohamed (1975). Because the fish's immune system has become accustomed to the poison, it is creating more WBC and lymphocytes to combat its effects, which may indicate an infection. Furthermore, there was an exact relationship between platelet count and the toxin concentration. This may lead to increased coagulability or clotting of intravenous blood.

According to Dhanapal et al. (1990), the presence of toxicants in the medium may bind with the fish's hemoglobin fraction, changing the body's physiological functions. Therefore, damage to RBC may be the reason for decreased oxygen consumption. To boost the blood's capacity to carry oxygen per unit volume, red blood corpuscle concentration can be raised (Svobodova et al. 1993). The size and stress level affect both the respiratory rate and ocular beat. After their water's oxygen content was artificially reduced to half of its typical amount, Jones (1952) found that the fish's gill cover motions increased from 95 to 165 per minute. A higher metabolic rate or less efficient oxygen intake and transport could lead to an increase in opercular rate. According to Oke et al. (2013), the fish's reactions are directly correlated with the contamination. Fish's cardiac output decreases when they do not getting enough oxygen, according to Mason (1981). As a result, there was longer time for oxygen uptake and oxygen conservation because less effort required of the muscles. This also slowed the rate at which blood passed through the gills.

10.5 Conclusion

The results of this investigation showed that most of the metal concentrations in the muscle of *M. aor* and *C. striata* were lower than those recommended for human ingestion. However, some of the edible tissues of *O. mossambicus* and *C. catla* have exceeded the allowable limits, therefore they should not be consumed at this time. Controlling the dumping of industrial waste in water bodies and conducting biomonitoring of the trace elements in fish and water are essential. To prevent the potential ingestion of contaminated fish, the practice of trace element detection should be maintained. It is recommended that people are aware of the dangers of drinking contaminated water and the fish that are linked to it. To reduce contamination of this kind, farmers also need to be urged to grow their land away from industrial and contaminated regions, apply pesticides sparingly to prevent wastewater from penetrating the soil, and receive the necessary information. The government should create regulations to ensure that wastewater is properly treated before being released. Promoting the use of heavy metal-absorbing column construction for drainage purposes is a good idea.

Acknowledgements The first author was thankful to the Chairman, Honorable Secretary, Correspondent, Principal and Head, P.G. & Research Department of Zoology, Ayya Nadar Janaki Ammal College (Autonomous), Sivakasi—626124, Tamil Nadu, India for providing the necessary laboratory facilities and infrastructure provided for completing the experiment.

References

Abdallah AT, Wanas MQS, Thompson SN (1997) Effect of heavy metals on survival and growth of *Biomphalaria glabrata* Say (Gastropoda: Pulmonata) and interaction with Schistosome infection. J Moll Stud 63:79–86

Allen GT, Wilson RM (1994) Metals and organic compounds in fish of the Missouri River: Boyd country Nebraska to Kansas City, Missouri USFWS, Manhattan. Kansas, USA

Anderani G, Santoro M, Cottignoli S, Fabbri M, Carpene E, Isani G (2008) Metal distribution and metallothionein in logger head (*Caretta caretta*) and green (*Chelonia mydas*) sea turtles. Sci Total Environ 390:287–294

AyyaduraiK, Swaminathan CS, Krishnasamy V (1994) Studies on heavy metal pollution in the finfish, *Oreochromis mossambicus* from river Cauvery. Indian J Environ Health 36(2):99–103

Bhatia IB, Ghose DK (1999) Heavy metal dispersion along roadside habitats of India. J Environ Pol 6(1):35–42

Bhosle AB, Patil PM (2001) Seasonal variation in the copper content of the river water Godavari at Nandod, Maharashtra. J Ectoxicol Envion Monit 11(1):61–64

Blaxhall PC, Daisley KW (1973) Routine haematological methods for use with fish blood. J Fish Biol 5:771–781

Carbonell G, Tarazona JV (1994) Toxic Kinetics of Cu. Aqua Toxicol 29:213–221

Chakravarty S, Roy US, Vass KK (1996) Heavy metal content in different creeks of Hoogly estuary along Calcutta metropolis. Indian J Environ Pol 16(10):779–783

Chandra H, Modak DP, Gupta BN, Ray RK (1991) Evaluation of drinking water quality during Mahakumbhamela, January–February 1981 at Allahabad. Indian J Environ Prot 10(2):69

Chen CY, Richard S, Stemberger B (2000) Accumulation of heavy metals in food web components across a gradient lake. Limnol Oceanogr 45(7):1525–1536

Cross FA, Hardy LH, Jones NY, Barber RT (1973) Relation between total body weight and concentration of Mn, Fe, Zn, Cu and Hg is in the blue fish (*Pomatonus saltatrix*) and a bathy dermersal fish (*Antimora rostrata*). J Fish Res Bd Can 30:1287–1291

Dhanapal S, Sivakumar AA, Logonkumar K, Aruchami A (1990) T4oxicity of tannery effluent on the fish, *Sarotherodon mossambicus*. Indian J Environ Health 32:153–156

Dhayanithi NB, Ajithkumar TT, Balasubramanian T, Tissera K (2013) A study on the effect of using mangrove leaf extracts as a feed additive in the progress of bacterial infections in marine ornamental fish. J Coast Life Med 1(3):217–224

Farlane GBM, Burchett MD (2000) Cellular distribution of Cu, Pb, and Zn in the Grey Mangrove *Avicemnia marina* (Forsk). Vierh Aquatic Botanica 68:45–59

Farmobi EO, Adelowo OA, Ajimoko YR (2007) Biomarker of oxidative stress and heavy metal levels as induced by environmental pollution in African cat fish *Clarias gariepinus* from Nigeria Ogun river. Int J Environ Res Public Health 4(2):158–165

Fatima M, Usmani N (2013) Histopathology and bioaccumulation of heavy metals (Cr, Ni, and Pb) in fish (*Channa striata* and *Heteropneustes fossilis*) Tissue: a study for toxicity and ecological impacts. Pak Biol Sci 16:412–420

Forstner V, Wittman GTW (1975) Metal pollution in the aquatic environment Berlin Heidelberg, New York. Indian J Environ Prot 20(6):401–406

Hart BT (1982) Trace metals in natural waters. I Speciation. Chem Austraila 49:260–265

Holcombe GW, Benoit DA (1976) Long-term effects of lead exposure on three generations of brook trout, *Salvalinus fontinalis*. J Fish Res Bd Canada 33:1731–1741

Ishaq ES, Ato RS, Annune PA (2011) Bioaccumulation of heavy metals in fish (*Tilapia zilli* and C*larias gariepinus*) organs from River Benue, North–Central Nigeria. Pak J Anal Environ Chem 12:25–31

Javed M, Usmani N (2012a) Uptake of heavy metals by C*hanna punctatus* from sewage-fed aquaculture pond of Panethi, Aligarh. Glob J Res Eng 12:27–34

Javed M, Usmani N (2012b) Toxic effects of heavy metals (Cu, Ni, Fe Co, Mn, Cr, Zn) to the haematology of Mastacembelus armatus thriving in Harduaganj Reservoir, Aligarh, India. Glob J Med Res 12:59–64

Javed M, Usmani N (2013) Haematological indices of *Channa punctatus* as an indicator of heavy metal pollution in waste water aquaculture pond, Panethi, India. Afr J Biotechnol 12:520–525

Jezierska B, Witeska M (2002) Metal toxicity to fish, Wydawnictwo Akademii Podlaskiej, Siedlce, p 318

Jones JRE (1952) The reactions of fish to water of low oxygen concentration. J Exp Bio 29:403–415

Kakar YP (1989) Hydro biological and pollution of ground in Faridabad area, Haryana. C.GWB Min. of Water Resources, pp 1–13

Khail M, Faragallah H (2008) Egypt J Aquat Res 34:1. Khaki C, Patient G, Nelly K, Edorhand PA, Rodrigue A (2011) Evaluation of heavy metals pollution of Nokoue Lake. Afr J EnvironSci Technol 5(3):255–261

Khurshid S, Zaheruddin A (2000) Heavy metal pollution and its toxic effect on water quality in part of Hindon river basin. Indian J Environ Prot 20(6):401–406

Kojadinovic J, Potier M, Lecorre M (2006) Mercury content in commercial pelagic fish and its risk assessment in the Western Indian Ocean. Sci Total Environ 366:688–700

Kotpal RL, Bali NP (1986) Concepts of ecology. Tata Mc Graw Hill Publishing Company Limited, New Delhi., p 225

Kumar A, Chava R, Gupta S, Shirin S, Jamal A, Yadav AK (2024) Environmental and health effects of fluoride contamination and treatment of wastewater using various technologies. In: Advanced treatment technologies for fluoride removal in water: water purification. Springer Nature Switzerland, Cham, pp 323–341. https://doi.org/10.1007/978-3-031-38845-3_18

Kutty MN, Mohamed MP (1975) Metabolic adaptations of mullet, *Rhinomugil cersula* (Hamilton) with special reference to energy utilization. Aquacult 5(3):253–270

Lokande RS, Kelkar N (1999) Studies on heavy metals in water of Vasai Creek, Maharashtra. Indian J Environ Prot 20(6):441–446

Malik N, Biswas AK, Qureshi TA, Borana K, Virha R (2010) Bioaccumulation of heavy metals in fish tissues of a freshwater lake of Bhopal. Environ Monitor Asess 160:267–267

Manickam N, Ali AJ, Bhavan PS (2011) Aquatic Toxicology. Toxicity of antifouling paints in selected zooplanktons. LAP—LAMBERT Academic Publishing, pp 1–80. ISBN-13: 978-3-8454-2369-2; ISBN-10: 3845423692

Manickam N (2008) A cost-effective acute toxicity screening test to determine the toxicity of antifouling paints. Master of Philosophy, Thesis, University of Madras, India, pp 1–73

Mason CF (1981) Biology of freshwater pollution, Chap. IV, 11 pp.

Mathess G (1974) Heavy metals as trace constituents in natural and polluted ground water. Goel Mii Bouw 53:149–155

Mazharuddin KM (1989) National seminar on pollution control environs. In: Mang., Nagpal proceeding, pp 98–102

Miliou H, Zaboukas N, Apostolopoulou MM (1998) Biochemical composition, growth and survival of the guppy. *Poecilia reticulata*, during chronic sub lethal exposure to cadmium. Arch Environ Contam Toxicol 35:58–63

Mitra A (1998) Status of coastal pollution in West Bengal with special reference to heavy metals. J Indian Ocean Stud 5(2):135–135

Mule MB, Patil RK (2000) Study of metal content in water resources from Radhanagari forest. Indian J Environ Prot 20(1):55–59

NAS–NRC (National Academy of Science–National Reachers Council) (1982) Drinking water and health. National Academic Press, Washington DC

Okey IB, Gabriel UU, Keremah RI (2013) Acute toxicity of *Lepidagathis alopecuriodes* to *Heterobranchus bidorsalis* fingerlings. J Environ Chem Ecotoxicol 5(8):208–215

Pourang N (1995) Heavy metal bioaccumulation in different tissue of two fish species with regards to their feeding habits and trophic levels. Environ Monit Assess 35:207–219

Priyadharshini N (1997) Trace elements in the sediments of Safi River in Bachra area of North Karanupura coal field on Hazaribagh Dirst. Indian J Environ Prot 18(7):511–515

Purohit SS, Ranjan R (1990) Ecology, environment andpollution. Agrobios Publications, Jodhpur, pp 682–683

Robert RJ (1978) Fish pathology. Cassel Ltd London, 1–77 pp

Sankar TV, Zgnubheen AA, Mandan R, Nair PGV (2006) Distribution of organo chlorine pesticides andheavy metals residues fish and shell fish from Calicut region, Kerala, India. Chemosphere 65:583–590

Saxena A, Shrivastava P, Swarup A (1998) Heavy metal pollution in a tropical wetland. Indian J Environ Prot 18(7):505–510

Schuurmann G, Markert B (1998) Ecological fundamentals, chemical exposure and biological effects. In: Ecotoxicology, vol 4. Ohn Wiley & Sons, Inc and Spectrum, pp 45–56

Shah SL, Altindag A (2004) Haematological parameters of tench (*Tinca tinca L.*) after acute and chronic exposure to lethal and sublethal mercury treatments. Bull Environ Contam Toxicol 73:911–918

Shaheen T, Akhtar T (2012) Assessment of chromium toxicity in *Cyprinus carpio* through haematological and biochemical blood markers. Turk J Zool 36:682–690

Shirin S, Yadav AK (2021) Assessment and suitability for the dumping of flyash in an abandoned mine. Environ Chall 4:100111. https://doi.org/10.1016/j.envc.2021.100111

Shirin S, Jamal A, Ranjan P, Yadav AK (2019) Study on assessment of slope stability and mixed disposal of overburden in voids of Singrauli Coalfield. Environ Qual Manage 28(3):131–139. https://doi.org/10.1002/tqem.21616

Shirin S, Jamal A, Emmanouil C, Singh VP, Yadav AK (2023) Assessment and characterization of waste material used as backfilling in an abandoned mine. Int J Coal Prep Util 43(8):1402–1410. https://doi.org/10.1080/19392699.2022.2118259

Sivaperumal P, Sankar JV, Viswantahan PGN (2007) Heavy metals concentrations in fish, shellfish and fish products from internal markets of India visa–vis international standards. Food Chem 102(3):612–620

Srivastava VS, Patil SS (2006) Impact of hazardous metals on surrounding environment. Indian J Environ Prot 17(11):1021–1024

Subramanian V (1993) Transport and distribution of heavy metals in Cauvery river. Water Air Soil Pollut 71:13–28

Sulthana M, Sharief D (2005) Heavy metal contamination of the Double lake, Chennai, Tamil Nadu, India. J Aqua Biol 20(6):673–679

Svobodova Z, Richard L, Jana M, Blanka V (1993) Water quality and fish health EIFAC Technical paper 54

Swaileh KM, Sansur R (2004) Monitoring urban heavy metal pollution using the sparrow. J Environ Monit 8:209–213

Swaminathan K, Manonmani K (1997) Studies on toxicity of viscose rayon factory effluents. J Environ Biol 18(1):73–78

Tran TA, Popova LP (2013) Functions and toxicity of cadmium in plants: recent advances and future prospects. Tur J Bot 37:1–13

Ubalua AO, Chijioke VC, Ezeronye OV (2009) Determination and assessment of heavy metal content in fish and shell fish in Aba river, Abia state, Nigeria KMIL. Sci Tech 7(1):56–79

USEPA (1997) Water quality criteria—1972. EPA Washington D.C. Pub. EPAR3—73–33. Guidelines for surface water quality, vol 1. Inorganic chemical substance, Inland water, Directorate water quality branch, Ottawa, Canada

Vanlandeghem MM, Wahl DH, Suski CD (2010) Physiological responses of largemouth bass to acute temperature and oxygen stressors. Fisheries Manage Ecol 17:414–425

Vinodhini R, Narayanan M (2008) Bioaccumulation of heavy metals in organs of freshwater fish *Cyprinus carpio* (common carp). Zoology 5(2):179–182

Vinodhini R, Narayanan M (2009) The impact of toxic heavy metals on the haematological parameters in common carp, *Cyprinus carpio* L. Iran. J Environ Health Sci Eng 6:23–28

Vutkuru SS (2005) Acute effect of Hexavalent chromium on survival, oxygen consumption, hematological parameters and some biochemical profiles of the Indian major carps, *Labeo rohita*. Int J Environ Res Public Health 2:456–462

Walker C (1975) Environmental pollution by chemicals. Hutchinson and Co. Pub. Ltd., London

Wedemeyer GR, Mcleay DJ (1981) Methods of determining the tolerance of fish to environmental stressors in Pickering AD) stress and fish. Academic Press, New York USA, pp 247–275

Wepener W, Vurenvan JHJ, Preezdu HH (2001) Up-take and distribution of a copper, iron and zinc mixture in gill, live rand plasma of a freshwater teleost. Tilapia Sparrmanii, Water SA 27(1):99–108

WHO (1984) International standard for drinking water. Control Fed 51:2810–2822

WHO (1989) Heavy metals environmental aspects. Evironment Health Criteria No. 85. Geneva, Switzerland

Wintrobe MM (1967) Clinical hematology, 6th edn. Lea and Febiger, Philadelphia. In: Library of congress, Print USA

Wintz H, Fox T, Vulpe C (2002)Functional genomics and gene regulation in bio metals research. Biochem Soc Trans 30:766–768

Yilmaz F (2009) The comparison of heavy metal concentrations (Cd, Cu, Mn, Pb and Zn) in tissues of three economically important fish (*Anguilla anguilla*; *Mugil cephathus* and *Oreochromic nilotiucus*) inhabiting Koycegiz Lake-Mugla (Turkey). Turk J Sci Tech 4(1):7–12

Zaheeruddin A (1996) Heavy metal pollution in parts of Delhi. Indian J Environ Prot 16(11):828–830

Zaheeruddin A, Khurshid S (1998) Dispersion of heavy metals and its toxic effect on water quality in the parts of Faridabad, Haryana. J Ecotoxicol Environ Monit 8:9–14

Chapter 11
Water Quality Evaluation, Reservoir Management, and Zooplankton Monitoring in the Dharmapuri District, Tamil Nadu, India

Narasimman Manickam, Razack Rafi Mohamed, Ambrose Paulraj, Rajagopal Bhuvaneswari, Ayyanar Shenbaga Devi, Venugopal Ashokan, Kandasamy Vijayadevan, Albert Anne Rebecca, Shanmugam Gunabal, Palani Sridhar, Annamalai Aravinth, Balan Mohan, Perumal Santhanam, and Periyakali Saravana Bhavan

Abstract The current study examined the hydrobiology of the three perennial reservoirs that were chosen: Panchapalli, Nagavathy, and Thoppaiyar in the district of Dharmapuri, Tamil Nadu, India. Over one year (March 2022 to February 2023) at monthly intervals, seasonal variations in water quality evaluation and zooplankton

N. Manickam (✉) · A. S. Devi · S. Gunabal · P. Sridhar · P. Santhanam
Marine Planktonology and Aquaculture Laboratory, Department of Marine Science, School of Marine Sciences, Bharathidasan University, Tiruchirappalli 620024, Tamil Nadu, India
e-mail: nmanickam5@gmail.com

N. Manickam · K. Vijayadevan
PG and Research Department of Zoology, Government Arts College, Dharmapuri 636705, Tamil Nadu, India

N. Manickam · P. S. Bhavan
Crustacean Biology Laboratory, Department of Zoology, Bharathiar University, Coimbatore 641046, Tamil Nadu, India

N. Manickam
Department of Zoology, Sree Meenakshi Arts & Science College, Meenakshi Nagar, Eriyur (Po), Pennagaram (Tk), Dharmapuri 636810, Tamil Nadu, India

R. R. Mohamed
PG and Research Department of Zoology, C. Abdul Hakeem College (Autonomous), Melvisharam, Ranipet 632509, Tamil Nadu, India

A. Paulraj
PG and Research Department of Zoology, Government Arts College for Men, Krishnagiri 635001, Tamil Nadu, India

R. Bhuvaneswari
Fish Disease Diagnostic Laboratory, Department of Zoology, Ayya Nadar Janaki Ammal College (Autonomous), Sivakasi 626124, Tamil Nadu, India

A. K. Yadav et al. (eds.), *Integrated Management of Water Resources in India: A Computational Approach*, Water Science and Technology Library 129,
https://doi.org/10.1007/978-3-031-62079-9_11

monitoring were examined, encompassing four distinct seasons: summer, before the monsoon, monsoon, and postmonsoon. For the present invitation, different evaluations of water quality parameters were analyzed. Based on current monitoring and the diversity of zooplankton (for example, Rotifera > Copepoda > Cladocera > Ostracoda), thirty species were found in the Panchapalli reservoir, twenty-nine in the Nagavathy reservoir, and forty-four in the Thoppaiyar reservoir. The zooplankton population was favourably associated with reservoir management and water quality monitoring. During the study period, a total of 56 species of zooplankton were observed in the three reservoirs. The current investigation concluded that rotifera dominated the population, distribution and diversity in each of the three reservoirs. Summer was the season with the highest population, while the monsoon was the season with the lowest occurrence. Therefore, in addition to teaching in schools, colleges, universities, and the general public, public awareness campaigns will be planned about markers of hydrobiological inquiry, zooplankton diversity, and their significance.

Keywords Hydrobiology · Zooplankton · Reservoir · Panchapalli · Thoppaiyar · Nagavathy

11.1 Introduction

Water is an essential natural resource for the ecosystem, humans, and others. Water was always thought to be an endless resource, but as it becomes more widely used for all purposes, the rate at which the planet naturally produces water is beginning to exceed it. Most water quality issues arise from industrial production, untreated wastewater, household and urban trash, and intensive agriculture. The most common chemical contaminants found in freshwater resources worldwide are excesses of phosphate and nitrogen, mostly from fertilizer applications. However, the quality of water is gradually declining, though, as new materials and pollutants (including pesticides and antibiotics) have been discovered in several habitats. To reduce adverse effects and preserve the ecological integrity of freshwater aquatic ecosystems, it is

V. Ashokan
Institute of Advanced Study in Education (Autonomous), Saidapet, Chennai 600015, Tamil Nadu, India

A. A. Rebecca
Department of Zoology, Thiruvalluvar University, Vellore 632115, Tamil Nadu, India

A. Aravinth
Department of Marine Science, School of Marine Sciences, Bharathidasan University, Tiruchirappalli 620024, Tamil Nadu, India

B. Mohan
Department of Environmental Sciences, Bharathiar University, Coimbatore 641046, Tamil Nadu, India

essential to understand the dynamics, composition, and structure of biotic communities (Manickam et al. 2023; Pinto et al. 2023). Man-made reservoirs are important for human societies on a socioeconomic level because they support global irrigation and human consumption, contribute to electrical power generation, and foster agricultural, urban, and industrial development (Manickam et al. 2018; Yadav et al. 2014; Aravinth et al. 2023). Reservoirs are comparable to natural lakes when it comes to their ability to store water, low flow velocity, and abiotic features compared to rivers. Any water body must conduct a primary survey of its phytoplankton and zooplankton because these data are essential for sustainable management of aquatic resources, exploiting the organisms present in the water and promoting conservation (Manickam et al. 2014, 2018; Kowthaman et al. 2019; Malik et al. 2020; Aravinth et al. 2023).

Phytoplankton are the food source for zooplankton, which are tiny organisms that float or roam. In an aquatic ecosystem, zooplankton is essential for the trophic structure and is involved in energy transfer. A significant barrier to a deeper understanding of the life processes of freshwater bodies is our limited understanding of plankton and its dynamics (Manickam et al. 2017a). Many health stresses that severely reduce biodiversity have an impact on aquatic ecosystems. It is expected that in the future, aquatic environments will experience a higher loss of biodiversity and its effects than terrestrial ecosystems (Manickam et al. 2020a). Most zooplankton, a heterotrophic group of organisms in the water column, are filter feeders that eat sediment made up of organic detritus, bacteria, and phytoplankton. Although physicochemical methods can help identify how pollution affects water quality, alterations in the trophic conditions of the water can also be seen in the structure of the biotic community, which includes species distribution, pattern, and diversity (Akindele and Olutona 2014; Manickam et al. 2017a).

The physicochemical characteristics and nutritional state of water bodies have a major impact on this process (Manickam et al. 2014; Shirin and Yadav 2014). Because of their short life history, they are used as indicators of the overall health or condition of their habitats, since they respond quickly to changes in aquatic environments, including those related to pH, color, odor, taste, etc. In many reservoirs, lakes, and shallow water bodies, zooplankton communities have been studied (Manickam 2015; Manickam et al. 2015b; Kumar et al. 2023). Plankton productivity, the biology of cultured organisms, and final yields are all significantly influenced by the quality of the water (Dhawan and Karu 2002; Manickam et al. 2017a). Zooplankton is the term for microscopic organisms suspended in water. These include a diverse range of protozoans, microcrustaceans, and planktonic micro-invertebrates. Freshwater zooplankton is necessary for the aquatic environment, food chain and ecosystem ponds, lakes, and reservoirs (Manickam et al. 2014; Sharma and Sharma 2020; Anyanwu et al. 2022).

The great sensitivity of zooplankton to alterations in its surroundings makes it an excellent indicator of ecosystem health. Studies on the connection between environmental conditions, basin age, and zooplankton shape have been conducted extensively in the past. In particular, zooplankton provides many carnivorous and omnivorous finfish and shellfish with an essential source of live food. It also provides the high-nutrient protein required for larval growth. Zooplankton serves as a helpful tool and bioindicator, making it crucial for assessing the degree of water pollution (Manickam et al. 2017a; Guermazi et al. 2023). Plankton is important for the trophic structure and ecosystem processes of reservoirs due to their drifting tendency and fast turnover period. The extensive plankton community seen in Indian reservoirs is a unique trait that results from abiotic factors and variations in nutrient loads. The main objective was to assess, through a case study methodology, the monitoring of the zooplankton community and water quality indicators in the reservoirs of the Dharmapuri District (Tamil Nadu, South India).

11.2 Materials and Methods

11.2.1 Study Area

Three reservoirs in Dharmapuri District, Tamil Nadu, Southern India, were chosen for the study. Panchapalli reservoirs (Station I: Lat. 12°48′ N; Long. 77°92′ E), Nagavathy reservoirs (Station II: Lat. 12°3′ N; Long. 78°1′ E), and Thoppaiyar reservoirs (Station III: Lat. 11°57′ N; Long. 78°6′ E). As seen in Figs. 11.1 and 11.2, three sampling sites (S-I or P-Dam, S-II or N-Dam, and S-III or T-Dam) were chosen for each reservoir and sampling was conducted between March 2022 to February-2023.

11.2.2 Physico-chemical Characteristics

The samples were taken between 5 and 7 AM in the early morning and delivered to the lab under ice cold conditions for the examination of the same day physico-chemical characteristics. A standard centigrade thermometer was used to measure the temperatures of the surface water and the atmosphere on the spot. They measured pH, salinity, electrical conductivity (EC), total dissolved solids (TDS) and dissolved oxygen (DO) using a "μP Based Water and Soil Analysis Kit Model-1160." A three-day incubation period (using a Sanyo BOD incubator) was required at 27 °C to measure the biological oxygen demand (BOD). Meanwhile, the titration method was used to measure additional parameters, including chemical oxygen demand (COD), total alkalinity, chloride, phosphate, sulfate, and ammonia (APHA 2005; Manickam et al. 2015b). To assess the levels of dissolved inorganic nitrite, nitrate, and reactive silicate in water samples, the Millipore filtering system was utilized. The results were reported in micrograms, in accordance with standard protocols outlined by Strickland and Parsons (1972). Four seasons were identified from the recorded data: summer (March-2022), premonsoon (June–August 2022), monsoon (September–November 2022) and post-monsoon (December-2022 to February-2023).

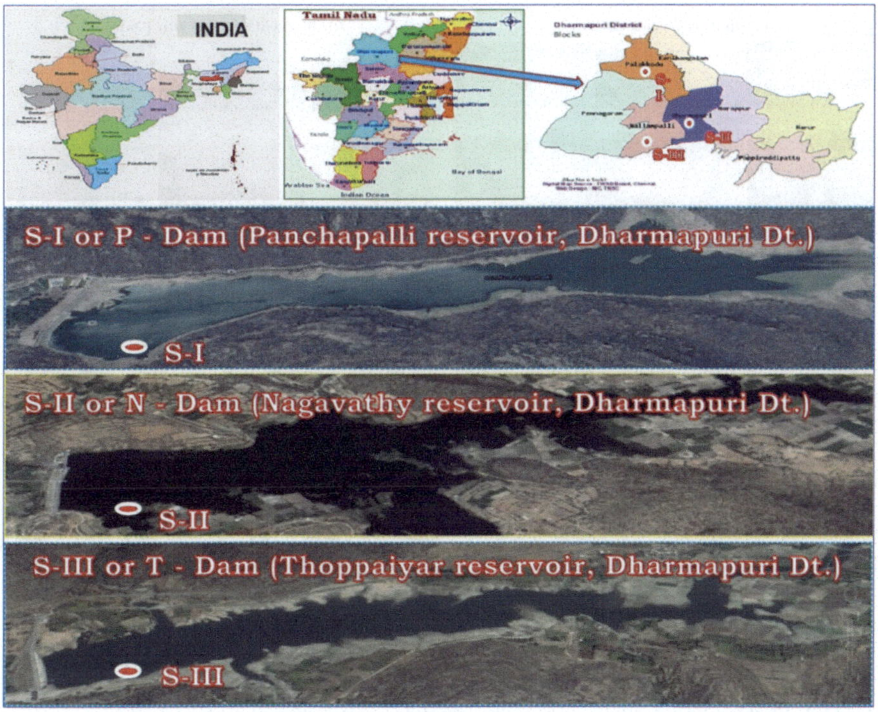

Fig. 11.1 Satellite view of study area in the reservoirs of Dharmapuri District, Tamil Nadu, South India

11.2.3 Qualitative and Quantitative Analysis of Zooplankton

Standard plankton net (150 μm mesh) from Towing-Henson was used to collect samples from reservoirs for the qualitative examination of zooplankton. With the boat running at a steady speed for five to ten minutes, the net was moved horizontally in a *zig-zag* pattern horizontally at a depth of 0.50–1.00 m (Fig. 11.2). A 5-l plastic container was used to filter 100 l of water through a plankton net made of bolting silk (No. 10, mesh size: 150 m) to perform a quantitative analysis of the zooplankton. The specimen vials with 4% formalin were filled with the plankton biomasses, which were then exposed to stereomicroscopic examination after the sample was quickly filtered from the water. The zooplankton was divided into groups, including copepods, ostracoda, rotifera, and cladocera. They were separated under a binocular stereo-zoom dissecting microscope (Magnus, Japan Technology) using a fine needle and brush. After the plankton species on small slides with a drop of 20% glycerine, the stain was done.

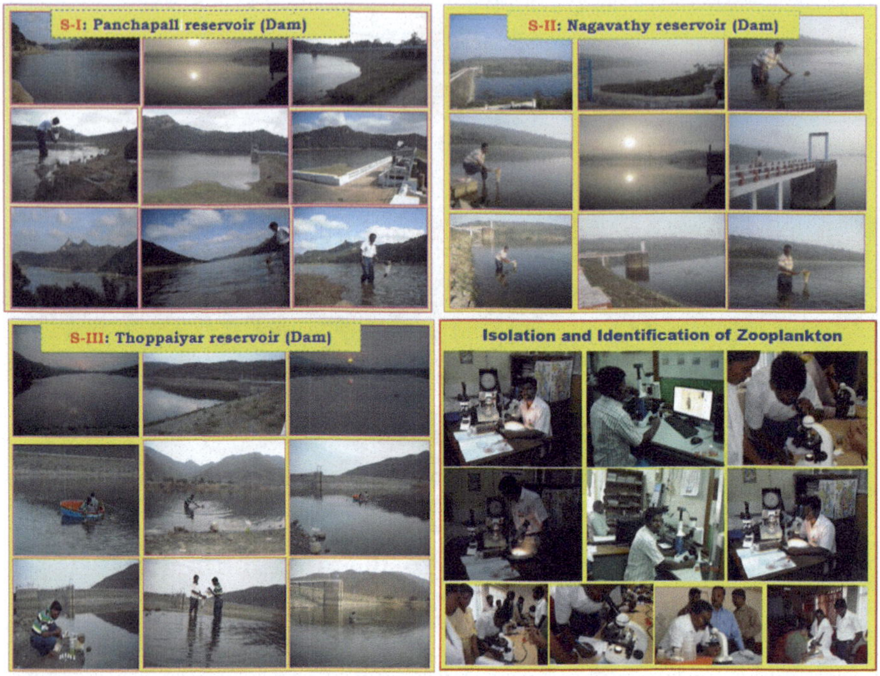

Fig. 11.2 Photograph of study area with sampling sites and isolation and identification of zooplankton

11.2.4 Biological Analysis and Identification of the Zooplankton Sample

The zooplankton was identified using reference materials such as standard textbooks, manuals, and bibliographies on freshwater zooplankton (Battish 1992; Murugan et al. 1998; Altaff 2004; Manickam et al. 2019). The inverted biological microscope (Model Number INVERSO 3000 (TC-100) CETI) attached to a camera (Model IS 300) and a compound microscope were also used. The Sedgwick Rafter counting cell received 1 ml of zooplankton samples that were collected using a wide mouth pipette. We allowed them some time to acclimate before counting them. At least three to five counts of this type were carried out for each species, gender, and stage of plankton growth. We calculated the mean of the data. The following formula was used, according to Santhanam et al. (1989) and Manickam et al. (2019), to calculate the total number of plankton in a 1-l water sample: N is equal to $n \times v/V$, where v is the milliliter of concentrated plankton, V is the liter of total filtered water, and n is the average number of plankton in 1 ml of filtered water.

11.2.5 Statistical Analysis and Diversity Indices

For each group of zooplankton, the average number of individuals per liter (ind./l or ind./m^3) was recorded. The total number of species in the zooplankton, the individual diversity indices, the Shannon diversity index (H′), the evenness of the species, and the richness were determined using the PAST software program (PAST, v2.02). The statistical analysis took advantage of these computations.

11.3 Results

11.3.1 Temperature of the Atmosphere and Surface Water Temperature (°C)

Figures 11.1 and 11.2 show the physicochemical properties of a few chosen reservoirs. The air temperature was measured for the current investigation (Fig. 11.3). The monsoon season of three reservoirs recorded the lowest temperature, while the summer season of all reservoirs (S-I, S-II, and S-III) recorded the highest temperature. Similarly, the lowest air and surface water temperatures were observed during the monsoon and the highest during the summer in research regions S-I, S-II, and S-III.

11.3.2 pH

As seen in Fig. 11.3, the pH ranged from 5.83 to 7.41 at S-I. The post-monsoon season had the lowest pH, whereas the summer and premonsoon seasons had higher pH values. At S-II, the pH ranged from 5.99 to 7.18. The pH of S-III was found to be between 6.59 and 7.19. The monsoon season produced the lowest pH readings, while the summer season produced the highest. The pH in the research region was found to be very low because rain and wastewater enter the reservoir during the monsoon season.

11.3.3 Salinity

The total salinity value at S-I in the current investigation ranged from 0.66 to 0.94 (ppt). The monsoon season saw the lowest salinity levels, while the summer months saw the highest (Fig. 11.3). At S-II, the salinity was measured between 0.83 and 1.88 (ppt). The post-monsoon season had the lowest salt levels, and the summer season

Fig. 11.3 Seasonal variations of physico-chemical parameters in the three reservoirs during March-2022 to February-2023

had the highest. Salinity ranged from 0.75 to 1.41 (ppt) as measured in S-III. The summer months had the highest values, and the monsoon season had the lowest.

11.3.4 Electrical Conductivity (EC)

The current EC varied from 0.78 to 1.23 mg/l in S-I. The most recent CE peaked in the summer after being observed during the monsoon season (Fig. 11.3). At S-II, the EC was found to be between 1.00 and 2.06 mg/l. Summer saw the greatest readings, while the monsoon season saw the lowest. It was discovered to be between 0.83 and 1.66 mg/l of S-III. The summer months had the highest values, and the monsoon season had the lowest.

11.3.5 Total Dissolved Solids (TDS)

The summer season registered the highest TDS and the postmonsoon season the lowest, according to S-I measurements of the research region's total TDS, which ranged from 101.57 to 188.11 mg/l (Fig. 11.3). The TDS level at S-I varied between 97.41 and 174.73 mg/l. The TDS was highest during the summer and lowest during the post-monsoon season. At S-III, TDS was measured within the range of 88.68–173.80 mg/l.

11.3.6 Dissolved Oxygen (DO)

The range of S-I in this study is 5.80–7.76 mg/l. The summer had the highest DO, whereas the monsoon season had the lowest DO (Fig. 11.3). The DO concentration in S-II was reported to range from 6.03 to 7.73 mg/l. Monsoon season reports of DO were the lowest, and summer season reports of DO were the highest. It was found to be in the range of 4.61 and 7.21 mg/l S-III. The monsoon season yielded the lowest DO readings, while the summer season produced the highest. According to current research, the DO in three reservoirs was lowest during the monsoon and highest during the summer. The reason for the summertime greatest DO reported may be the flora's increased photosynthetic activity of the flora, while the monsoon months recorded the lowest DO.

11.3.7 Biological Oxygen Demand (BOD)

Between 14.64 and 16.70 mg/l of BOD were found in S-I in the current observation. According to Fig. 11.3, the summer season produced the most, and the premonsoon season produced the least. From 13.64 to 18.30 mg/l, the BOD concentration of S-II ranged. It was observed that the summer season had the highest BOD, while the post-monsoon season had the lowest. In S-III, the range of 14.82–18.92 mg/l was observed. Summer was the season of maximum record, and the monsoon was the season of minimum.

11.3.8 Chemical Oxygen Demand (COD)

In the current study, COD in S-I ranged from 35.47 to 42.40 mg/l (Fig. 11.3). The monsoon season was when summer production peaked, and the rest of summer saw its lowest levels. The S-II measured COD levels within the range of 35.85–46.39 mg/l. The monsoon season was when summer production peaked, and the rest of the summer saw its lowest levels. The measurement range for S-III was 34.08–37.44 mg/l. Readings were highest during the summer and lowest during the monsoon season. According to the current study, COD levels were highest during the summer and lowest during the monsoon and postmonsoon seasons. However, winter COD is lowest due to low temperatures, minimal decomposition activity, and diluting effects.

11.3.9 Total Alkalinity (TA)

In the S-I reservoir, TA fluctuated between 112.88 and 170.12 mg/l during the current observation (Fig. 11.3). The summer was when the maximum was gained, while the post-monsoon season was when the minimum was acquired. TA in S-II was measured to be between 134.85 and 176.38 mg/l. Before the monsoon was the season with the highest TA, while after the monsoon it was the season with the lowest. It was found to be between 130.62 and 166.77 mg/l in S-III. The values were highest before the monsoon season and lowest during the monsoon season.

11.3.10 Chloride (Cl^-)

According to Fig. 11.3, the total Cl^- at S-I was found to be between 116.13 and 384.18 mg/l. The maximum Cl^- was reported in the months leading up to the monsoon, and the minimum was reported in the months following. In S-II, the Cl^- concentration was measured and found to be between 166.81 and 403.09 mg/l. The highest Cl^- values were observed during the premonsoon season, and the lowest during the postmonsoon. S-III was discovered to have a range of 207.98–451.14 mg/l. The summer months had the highest values, and the monsoon season had the lowest.

11.3.11 Phosphate (PO_4^{3-})

In the current observation, PO_4^{3-} in S-I varied in concentration from 1.25 to 2.60 mg/l. The post-monsoon season produced the least amount of PO_4^{3-}, while the summer produced the maximum (Fig. 11.3). PO_4^{3-}—at S-II ranged in concentration from 1.06 to 1.49 mg/l. PO_4^{3-} levels were highest during the summer and lowest during the monsoon season. It was found to be in the range of 1.25–2.33 mg/l in S-III. PO_4^{3-} was measured at its lowest point in the post-monsoon season and at its highest point in the summer.

11.3.12 Sulfate (SO_2)

The SO_2 content in S-I fluctuated between 19.24 and 56.48 mg/l at the time of the current observation. Figure 11.3 shows that the highest levels of SO_2 occurred during the pre-monsoon season, while the lowest levels occurred during the postmonsoon season. The postmonsoon season had the lowest concentration in S-II, while the pre-monsoon season had the highest concentration, ranging from 19.35 to 46.67 mg/l. S-III was found to be in the 26.61–58.12 mg/l range. The summer season recorded the maximum, while the postmonsoon season recorded the minimum.

11.3.13 Ammonia (NH_3)

The current observation shows that the NH_3 content fluctuated between 3.45 and 5.13 mg/l. According to Fig. 11.3, the summer season saw the highest levels of NH_3 and the monsoon season the lowest (Fig. 11.3). At S-II, NH_3 concentrations varied from 1.55 to 3.78 mg/l, peaking in the summer and falling throughout the monsoon seasons. In S-III, the range was 2.09–3.85 mg/l. Records for the monsoon season indicate the lowest levels of NH_3, while records for the summer season indicate the highest levels.

11.3.14 Nitrite (NO_2)

In the current study, the NO_2 content in I was determined to be between 0.15 and 0.46 mg/l. The summer months had the highest NO_2 levels, while the post-monsoon season had the lowest (Fig. 11.3). NO_2 measured NO_2 at S-II ranged from 0.17 to 0.24 mg/l, with the premonsoon season having the lowest values and the summer months having the highest. S-III was found to have NO_2 values ranging from 0.14 to 0.36 mg/l. NO_2 levels peaked during the pre-monsoon season and decreased throughout the monsoon season.

11.3.15 Nitrate (NO_3^-)

The total amount of NO_3^- in the S-I reservoir was found to fluctuate from 27.53 to 51.32 mg/l, with summertime being the lowest value and monsoon season being the highest (Fig. 11.3). According to reports, the NO_3 content in the S-II reservoir ranges from 19.94 to 40.54 mg/l, with the summer months showing the lowest concentration and the monsoon months showing the highest. The monsoon season had the highest level of S-III, measuring between 32.66 and 43.37 mg/l, while the mid-summer season had the lowest.

11.3.16 Silicate (Si_2O_3)

The observed concentration of Si_2O_3 in the S-I reservoir varied from 0.90 to 2.07 mg/l, with the summer having the highest value and the monsoon having the lowest (Fig. 11.3). Measured in milligrams per liter, the Si_2O_3 concentration at S-II Reservoir peaked in the summer and decreased during the monsoon season. The results ranged from 0.57 to 1.77. At S-III, silicate was detected in the 0.63–2.03 mg/l range. The postmonsoon season saw the lowest, and the summer season the highest.

11.3.17 Composition of Species (Quantitative Analysis)

During the current examination, 30 species of zooplankton were found in the S-I reservoir. These species include 10 rotifera species, 9 cladocera species, 7 copepoda species, and 4 ostracoda species (Tables 11.1 and 11.2). At the S-II reservoir, 29 different species of zooplankton were seen, comprising 11 rotifera species, 7 cladocera species, 7 copepoda species, and 4 ostracoda species. 54 zooplankton species, including 19 rotifera species, 13 cladocera species 14 copepoda species, and 8 ostracoda species, were found in the S-III reservoir.

11.3.18 Percentage Composition

Based on the percentage composition of the current observation, rotifera is the most common form of zooplankton in the S-I, S-II, and S-III reservoirs (Fig. 11.4). In the S-I reservoir, copepoda (30%), cladocera (27%), and ostracoda (9%) were found to make up the remaining 32% of the entire population. According to the findings, copepoda (29%), followed by ostracoda (6%), cladocera (26%), and rotifera (39%), comprised the population of the S-II reservoir. According to the findings, the

Table 11.1 Check of zooplankton recorded in the three reservoirs, during March-2022 to February-2023

Zooplankton	Family	Genus	Species
Rotifera (20)	Brachionidae (Ehrenberg, 1838)	*Anuraeopsis* Lauterborn, 1900	*Anuraeopsis fissa*
			Anuraeopsis navicula
		Brachionus Pallas, 1776	*Brachionus angularis*
			Brachiounus bidentata
			Brachionus budapestinesis
			Brachionus calyciflorus
			Brachionus caudatus personatus
			Brachionus diversicornis
			Brachionus falcatus
			Brachiounus forficula
			Brachionus quadridentatus
			Brachionus rubens
		Keratella Bory de St. Vincent, 1822	*Keratella cochlearis*
			Keratella tropica
		Notholca Gosse, 1886	*Notholca lebis*
		Platyias Harring, 1913	*Platyas quadricornis*
	Lecanidae (Bartos, 1959)	*Lecane* Koste, 1972	*Lecane papuana*
	Asplanchnidae (Harring & Myers, 1933)	*Asplanchna* Gosse, 1850	*Asplanchna brightwelli*
			Asplanchna intermedia
	Filinidae (Bartos, 1959)	*Filinia* Bory and Vincent, 1824	*Filinia longiseta*
Cladocera (13)	Sididae (Baird, 1850)	*Diaphanasoma* Fischer, 1850	*Diaphanasoma sarsi*
			Diaphanasoma excisum
		Daphnia Muller, 1785	*Daphnia carinata*
			Daphnia magna
	Daphnidae (Straus, 1850)	*Ceriodaphnia* Dana, 1853	*Ceriodaphnia cornuta*
			Ceriodaphnia reticulata
	Moinidae (Goulden, 1968)	*Moina* Baird, 1850	*Moina brachiata*
			Moina flagellate

(continued)

Table 11.1 (continued)

Zooplankton	Family	Genus	Species
			Moina micrura
			Moina macrocopa
		Moinodaphnia Herrick, 1887	*Moinodaphnia macleayi*
	Chydoridae (Stebbing, 1902)	*Alona* Baird, 1843	*Alona quadrangularis*
			Alona rectangular
Copepoda (15)	Diaptomidae (Baird, 1850)	*Heliodiaptomus* Kiefer, 1932	*Heliodiaptomus viduus*
			Heliodiaptomus cinctus
		Allodiaptomus Kiefer, 1936	*Allodiaptomus raoi*
		Neodiaptomus Kiefer, 1932	*Neodiaptomus lindbergi*
			Neodiaptomus schmakeri
		Phyllodiaptomus Kiefer, 1936	*Phyllodiaptomus annae*
		Sinediaptomus Kiefer, 1937	*Sinodiptomus indicus*
	Cyclopidae (Rafinesque, 1815)	*Eucyclops* Claus, 1893	*Eucyclops speratus*
			Eucyclops serrulatus
		Mesocyclops Claus, 1893	*Mesocyclops hyalinus*
			Mesocyclops leuckarti
		Thermocyclops Kiefer, 1927	*Thermocyclops hyalinus*
			Thermocyclops decipiens
		Apocyclops Lindberg, 1942	*Apocyclops dengizicus*
		Acanthocyclops Kiefer, 1927	*Acanthocyclops vernalis*
Ostracoda (8)	Cyprididae (Baird, 1845)	*Cypris* Muller, 1776	*Cypris protubera*
		Eucypris Vavra, 1891	*Eucypris bispinosa*
		Strandesia Stuhlmann, 1888	*Strandesia elongate*
		Cyprinotus Brady, 1886	*Cyprinotus nudus*
		Heterocypris Claus, 1892	*Heterocypris dentatomargintus*
			Hemicypris anomala

(continued)

Table 11.1 (continued)

Zooplankton	Family	Genus	Species
		Candonocypris Vavra, 1891	*Candonocypris dentatus*
		Cypretta Vavra, 1895	*Cypretta fontinalis*

most prevalent species in the S-III reservoir were rotifera (41%), copepoda (29%), cladocera (21%) and ostracoda (9%).

11.3.19 Population Density

The population density of the zooplankton research region in the current study is 88,596 ind./m^3 in total within the three reservoirs (Fig. 11.5). The density of the zooplankton population in the S-I reservoir (Table 11.3) ranged from 2152 to 8013 ind./m^3 (with an overall population density of 23,771 ind./m^3). The maximum was seen during the premonsoon season, and the minimum during the monsoon season. Table 11.4 reports a population density in the S-II reservoir that falls between 1441 and 9299 ind./m^3 (totaling 24,021 ind./m^3). Readings were highest during the summer and lowest during the monsoon season. Within an overall S-II1 of 40,804 ind./m^3, the population density at the S-III reservoir (Table 11.5) was observed to range from 3585 to 16,693 ind./m^3, peaking in the pre-monsoon season and throwing in the post-monsoon season.

11.3.20 Species Diversity

The general diversity of zooplankton in the reservoir waters of S-I, S-II, and S-III is shown in the table of three to five. It was discovered that the premonsoon and monsoon seasons had the maximum species diversity in the S-I reservoir, while the post-monsoon season had the lowest (Table 11.3). In the S-II reservoir, the premonsoon season produced the maximum species diversity, whereas the monsoon season produced the lowest (Table 11.4). The variety of the S-III reservoir was shown to peak during the pre-monsoon and peaked during the monsoon season (Table 11.5).

11.3.21 Species Evenness

The overall level of zooplankton species in the waters of the S-I, S-II, and S-III reservoirs was recorded in a table with values ranging from 3 to 5. According to

Table 11.2 The occurrence of zooplankton species during the study period (March 2022 to February 2023) at three reservoirs (S-I. Panchapalli: P-Dam, S-II. Nagavathy: N-Dam, and S-III. Thoppaiyar: T-Dam)

S. No	Zooplankton	Species	S-I	S-II	S-III
1	Rotifera	*Anuraeopsis fissa* Gosse, 1851	–	–	+
2		*Anuraeopsis navicula* Rousselet, 1892	–	–	+
3		*Brachionus angularis* Gosse, 1851	+	+	–
4		*Brachiounus bidentata* Anderson, 1889	–	–	+
5		*Brachionus budapestinesis* Daday, 1885	–	+	+
6		*Brachionus calyciflorus* Pallas, 1776	+	+	+
7		*Brachionus caudatus personatus* Ahlstrom, 1948	+	+	+
8		*Brachionus diversicornis* Daday, 1883	+	+	+
9		*Brachionus falcatus* Zacharias, 1898	+	+	+
10		*Brachiounus forficula* Sudzuki, 1995	–	–	+
11		*Brachionus quadridentatus* Hermann, 1783	+	+	+
12		*Brachionus rubens* Ehrenberg, 1838	–	+	+
13		*Keratella cochlearis* Gosse, 1851	–	–	+
14		*Keratella tropica* Apstein, 1907	+	+	+
15		*Notholca lebis* Gosse, 1887	–	+	+
16		*Platyas quadricornis* Ehrenberg, 1882	–	–	+
17		*Lecane papuana* Murray, 1913	–	–	+
18		*Asplanchna brightwelli* Gosse, 1850	+	–	+
19		*Asplanchna intermedia* Hudson, 1886	+	–	+
20		*Filinia longiseta* Ehrenberg, 1834	+	+	+
21	Cladocera	*Diaphanasoma sarsi* Richard, 1894	+	+	+
22		*Diaphanasoma excisum* Sars, 1885	+	+	+
23		*Daphnia carinata* King, 1853	+	–	+
24		*Daphnia magna* Straus, 1820	+	+	+
25		*Ceriodaphnia cornuta* Sars, 1885	+	+	+
26		*Ceriodaphnia reticulata* Jurine, 1820	+	–	+
27		*Moina brachiata* Jurine, 1820	+	+	+
28		*Moina flagellate* Hudendroff, 1876	–	–	+
29		*Moina micrura* Kurz, 1875	+	+	+
30		*Moina macrocopa* Straus, 1820	_	–	+
31		*Moinodaphnia macleayi* King, 1853	–	+	+
32		*Alona quadrangularis* Sars, 1862	–	–	+
33		*Alona rectangular* Muller, 1776	+	–	+
34	Copepoda	*Heliodiaptomus viduus* Gurney, 1916	+	+	+
35		*Heliodiaptomus cinctus* Gurney, 1907	–	–	+

(continued)

Table 11.2 (continued)

S. No	Zooplankton	Species	S-I	S-II	S-III
36		*Allodiaptomus raoi* Kiefer, 1936	–	–	+
37		*Neodiaptomus lindbergi* Brehm, 1951	–	–	+
38		*Neodiaptomus schmakeri* Poppe & Richard, 1892	–	–	+
39		*Phyllodiaptomus annae* Apstein, 1907	–	–	+
40		*Sinodiptomus indicus* Kiefer, 1934	+	+	+
41		*Eucyclops speratus* Lilljeborg, 1901	+	+	+
42		*Eucyclops serrulatus* Fischer, 1851	–	+	–
43		*Mesocyclops hyalinus* Claus, 1857	+	+	+
44		*Mesocyclops leuckarti* Claus, 1857	+	+	+
45		*Thermocyclops hyalinus* Rehberg, 1880	+	+	+
46		*Thermocyclops decipiens* Kiefer, 1929	+	–	+
47		*Apocyclops dengizicus* Lepeschkin, 1900	–	–	+
48		*Acanthocyclops vernalis* Fischer, 1853	–	–	+
49	Ostracoda	*Cypris protubera* Victor & Fernando, 1978	+	+	+
50		*Eucypris bispinosa* Victor & Michael, 1975	–	–	+
51		*Strandesia elongate* Hartmann, 1964	+	+	+
52		*Cyprinotus nudus* Brady, 1885	+	–	+
53		*Heterocypris dentatomargintus* Baird, 1859	–	–	+
54		*Hemicypris anomala Furtos, 1993*	–	–	+
55		*Candonocypris dentatus* Victor and Michael, 1975	–	+	+
56		*Cypretta fontinalis* Hartmann, 1964	+	+	+

Note + Present; – absent (Occurrence of zooplankton species: S-I: 30; S-II: 29 and S-III: 54)

Table 11.3, the S-I reservoir had the highest species evenness in the premonsoon season and the lowest during the postmonsoon season. For the S-II reservoir, the uniformity of the species peaked in the summer and decreased during the monsoon (Table 11.4). There was a low in the post-monsoon season and a maximum in the pre-monsoon season for species evenness in the S-III reservoir (Table 11.5).

11.3.22 Species Richness

The total species richness of zooplankton was shown in Tables 11.3, 11.4 and 11.5 using the reservoir waters of S-I, S-II and S-III reservoir waters. The richness was noted in the S-I reservoir, with a minimum in the pre-monsoon season and a high in the post-monsoon season (Table 11.3). The species richness of the S-II reservoir peaked in the monsoon and declined over the summer (Table 11.4). In the S-III

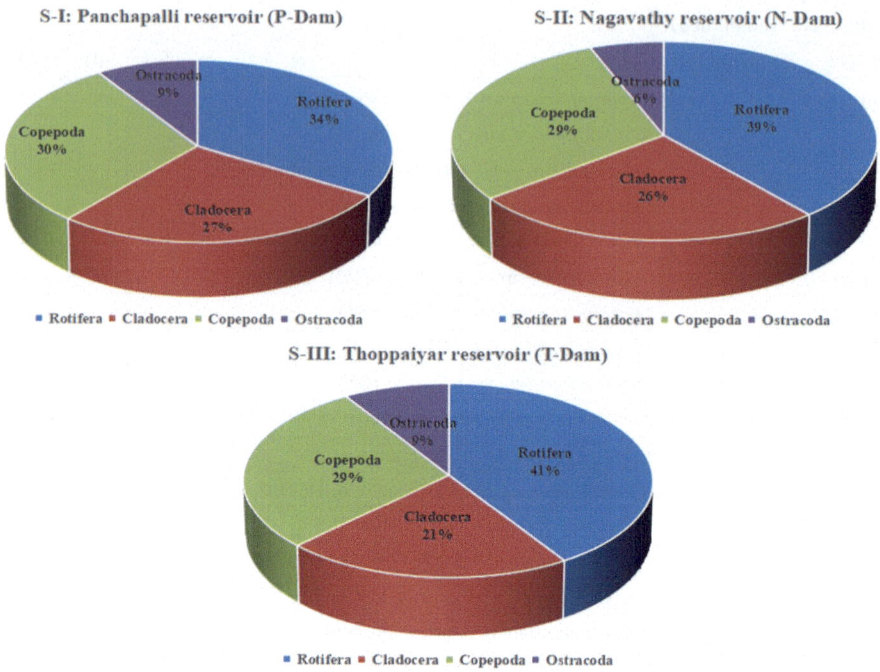

Fig. 11.4 Percentage composition of zooplankton species in the in the three reservoirs

reservoir, the species richness reaches its peak in the postmonsoon season and its minimum in the premonsoon season (Table 11.5).

11.4 Discussion

Significant variations in the physicochemical characteristics suggest the existence of different environmental conditions. These fluctuations could be connected to trends in water use, precipitation, and temperature (Atobatele and Ugwumba 2008; Manickam et al. 2014). An important aspect influencing reservoir primary production of the reservoir is temperature. The amount of plankton in the ecosystem varies with the temperature of the water. Temperature affects every physiological and metabolic activity in aquatic life, including feeding, moving, reproducing, and dispersion (zooplankton and phytoplankton). The carbonic system and the relationship between carbonates and bicarbonates are the primary determinants of water's pH. It is commonly known that pH has a direct relationship with CO_3 and an inverse relationship with free carbon dioxide (Zafar 1964). The overall pH values varied from 7.5 to 8.5 (alkaline range), with a minimum in December and a maximum in May to July in the reservoir, as agreed by earlier reporters (Manickam et al. 2012,

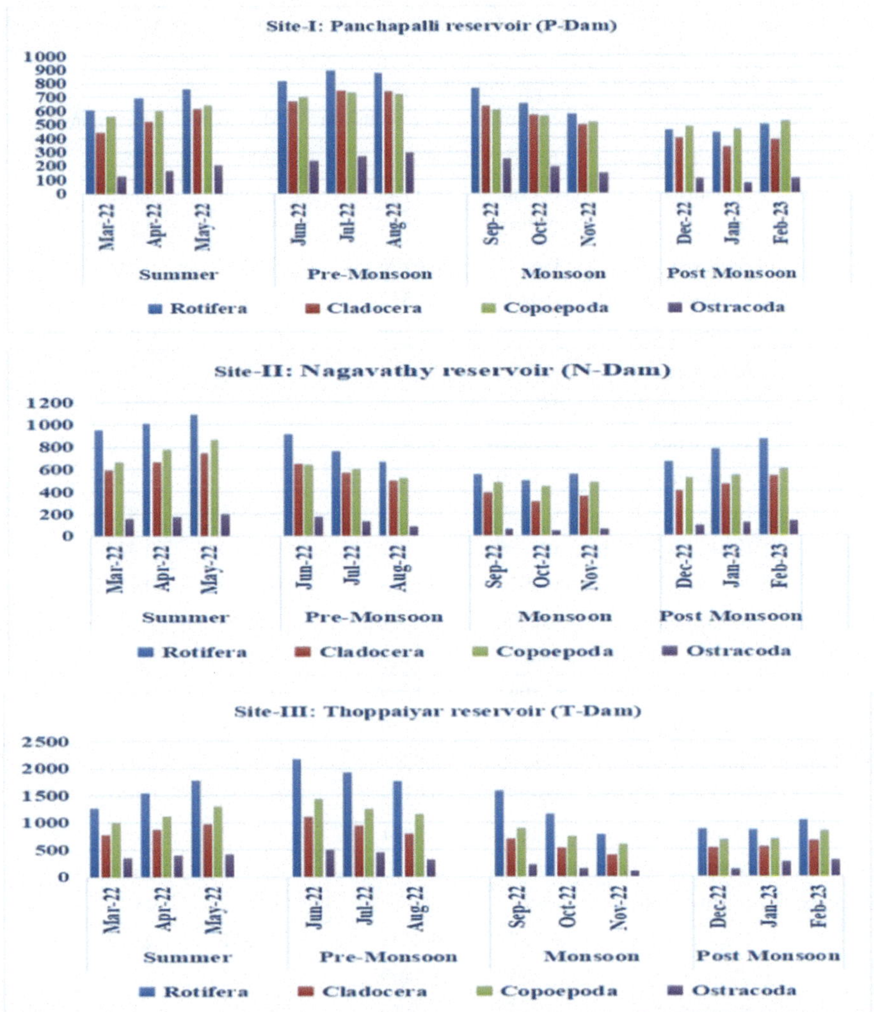

Fig. 11.5 Seasonal variations and population density (ind./m^3 or ind./l) of zooplankton species in the three reservoirs

2017a; Manickam 2015). Salinity is a significant ecological element that regulates brackish and freshwater phytoplankton and zooplankton populations, which fluctuate in size based on salinity conditions (Manickam et al. 2015a, 2017b). According to this study, the salinity range is highest in the summer and lowest during the wet season. Between the monsoon and summer seasons, salinity was observed to change greatly (Kubley 1982; Manickam 2015; Manickam et al. 2017a). The combination of high freshwater intake and low temperatures can be used to explain why salinity is higher during the summer and lower during the monsoon. Plankton production

Table 11.3 Seasonal variation of zooplankton diversity indices in the Panchapalli reservoir (S-I) during March 2022 to February 2023

Zooplankton diversity indices		Seasonal variation			
		Summer	Pre-monsoon	Monsoon	Post monsoon
Rotifera	Density (ind./l)	686	864	665	465
	Dominance_D	0.1183	0.1100	0.1108	0.1182
	Shannon_H	2.2133	2.2533	2.2500	2.2080
	Simpson_1-D	0.8817	0.8900	0.8892	0.8818
	Evenness_e^H/S	0.9147	0.9521	0.9489	0.9100
	Menhinick (R2)	0.3829	0.3405	0.3897	0.4640
	Margalef (R1)	1.3790	1.3310	1.3863	1.4653
Cladocera	Density (ind./l)	526	713	567	374
	Dominance_D	0.1250	0.1258	0.1305	0.1280
	Shannon_H	2.1320	2.1280	2.1007	2.1170
	Simpson_1-D	0.8750	0.8742	0.8695	0.8720
	Evenness_e^H/S	0.9369	0.9332	0.9080	0.9229
	Menhinick (R2)	0.3950	0.3373	0.3796	0.4667
	Margalef (R1)	1.2790	1.2177	1.2633	1.3513
Copepoda	Density (ind./l)	603	713	560	489
	Dominance_D	0.1647	0.1700	0.1782	0.1727
	Shannon_H	1.8710	1.8530	1.8223	1.8423
	Simpson_1-D	0.8353	0.8300	0.8218	0.8273
	Evenness_e^H/S	0.9277	0.9118	0.8838	0.9015
	Menhinick (R2)	0.2854	0.2621	0.2962	0.3169
	Margalef (R1)	0.9376	0.9133	0.9485	0.9692
Ostracoda	Density (ind./l)	165	267	194	95
	Dominance_D	0.2528	0.2290	0.2351	0.2725
	Shannon_H	1.4857	1.5400	1.5237	1.4300
	Simpson_1-D	0.7472	0.7710	0.7649	0.7275
	Evenness_e^H/S	0.8840	0.9332	0.9180	0.8358
	Menhinick (R2)	0.3950	0.3072	0.3668	0.5178
	Margalef (R1)	0.7876	0.7168	0.7651	0.8818

density and diversity may decrease due to the water's increased salinity (Horne and Goldman 1994).

Since most dissolved salts in water are ionic, water can conduct electricity. The earlier researcher agreed with the current findings, which indicated that in the freshwater ecosystem, CE is highest in the summer and lowest in the monsoon (Manickam et al. 2015a, 2017b). During the monsoon and postmonsoon seasons,

Table 11.4 Seasonal variation of zooplankton diversity indices in the Nagavathy reservoir (S-II) during March-2022 to February-2023

Zooplankton diversity indices		Seasonal variation			
		Summer	Pre-monsoon	Monsoon	Post monsoon
Rotifera	Density (ind./l)	1018	780	530	771
	Dominance_D	0.1018	0.1016	0.1083	0.1123
	Shannon_H	2.3397	2.3383	2.2913	2.2807
	Simpson_1-D	0.8982	0.8984	0.8917	0.8877
	Evenness_e^H/S	0.9434	0.9423	0.8994	0.8903
	Menhinick (R2)	0.3451	0.3965	0.4781	0.3979
	Margalef (R1)	1.4440	1.5043	1.5947	1.5060
Cladocera	Density (ind./l)	668	571	352	468
	Dominance_D	0.1562	0.1576	0.1577	0.1565
	Shannon_H	1.8950	1.8903	1.8900	1.8953
	Simpson_1-D	0.8438	0.8424	0.8423	0.8435
	Evenness_e^H/S	0.9507	0.9460	0.9459	0.9509
	Menhinick (R2)	0.2718	0.2942	0.3741	0.3252
	Margalef (R1)	0.9233	0.9464	1.0240	0.9772
Copepoda	Density (ind./l)	768	589	467	554
	Dominance_D	0.1831	0.1867	0.2049	0.1945
	Shannon_H	1.7973	1.7850	1.7177	1.7567
	Simpson_1-D	0.8169	0.8133	0.7951	0.8055
	Evenness_e^H/S	0.8620	0.8517	0.7962	0.8278
	Menhinick (R2)	0.2537	0.2893	0.3241	0.2979
	Margalef (R1)	0.9042	0.9415	0.9763	0.9502
Ostracoda	Density (ind./l)	175	128	54	114
	Dominance_D	0.2583	0.2668	0.2703	0.2692
	Shannon_H	1.3697	1.3533	1.3453	1.3487
	Simpson_1-D	0.7417	0.7332	0.7297	0.7308
	Evenness_e^H/S	0.9836	0.9674	0.9600	0.9630
	Menhinick (R2)	0.3031	0.3642	0.5506	0.3778
	Margalef (R1)	0.5814	0.6255	0.7565	0.6356

an abrupt increase in conductivity in water suggests the presence of some contaminants (Trivedy and Goel 1984). Anthropogenic pollutants that have accumulated and compromised water quality may be the cause of the highest average value of TDS (Manickam et al. 2012, 2017a; Dhanasekaran et al. 2017). In the current study, TDS was found to be highest in summer in reservoirs and lowest in the post-monsoon season. Summertime temperatures were shown to have higher TDS, most likely as a result of water stagnation. When the products of the decomposing matter are released

Table 11.5 Seasonal variation of zooplankton diversity indices in the Thoppaiyar reservoir (S-III) during March-2022 to February-2023

Zooplankton diversity indices		Seasonal variation			
		Summer	Pre-monsoon	Monsoon	Post monsoon
Rotifera	Density (ind./l)	1525	1951	1166	923
	Dominance_D	0.0715	0.0638	0.0731	0.0831
	Shannon_H	2.7627	2.8313	2.7443	2.6563
	Simpson_1-D	0.9286	0.9362	0.9269	0.9169
	Evenness_e^H/S	0.8347	0.8931	0.8215	0.7501
	Menhinick (R2)	0.4899	0.4313	0.5741	0.6272
	Margalef (R1)	2.4597	2.3773	2.5683	2.6380
Cladocera	Density (ind./l)	873	949	542	578
	Dominance_D	0.0969	0.0940	0.1013	0.1027
	Shannon_H	2.4270	2.4477	2.3970	2.3863
	Simpson_1-D	0.9031	0.9060	0.8987	0.8973
	Evenness_e^H/S	0.8711	0.8893	0.8457	0.8381
	Menhinick (R2)	0.4412	0.4248	0.5690	0.5428
	Margalef (R1)	1.7733	1.7537	1.9163	1.8887
Copepoda	Density (ind./l)	1143	1276	739	741
	Dominance_D	0.0857	0.0854	0.0944	0.0947
	Shannon_H	2.5643	2.5657	2.4923	2.4920
	Simpson_1-D	0.9144	0.9145	0.9055	0.9053
	Evenness_e^H/S	0.8663	0.8675	0.8062	0.8059
	Menhinick (R2)	0.4455	0.4211	0.5569	0.5530
	Margalef (R1)	1.9903	1.9590	2.1250	2.1207
Ostracoda	Density (ind./l)	388	422	154	231
	Dominance_D	0.1354	0.1375	0.1489	0.1410
	Shannon_H	2.0390	2.0310	1.9790	2.0140
	Simpson_1-D	0.8646	0.8625	0.8511	0.8590
	Evenness_e^H/S	0.9602	0.9528	0.9049	0.9367
	Menhinick (R2)	0.4069	0.3948	0.6643	0.5489
	Margalef (R1)	1.1750	1.1627	1.4053	1.3043

back into the water in May, the majority of vegetation experiences conditions similar to those of decay, which could increase the concentration of dissolved solids (Verma et al. 1978; Manickam et al. 2014, 2017b). Anthropogenic waste buildup may be the cause of the highest average TDS value, degrading the water quality. Inorganic salts such as NaCl, KCl, $CaCl_2$, etc. are the primary sources of chlorides. These salts are typically found in soil, naturally occurring layers of chloride salts, industrial and municipal wastewater, and animal waste (Gopalkrushna 2011; Manickam et al. 2015b). Water's solvent effect of salts from contaminated materials, like sewage and trade waste, causes chlorides to enter the water. Although Smitha et al. (2007) reported different results, Swaranlatha and Rao (1998) and Manickam et al. (2015b) also reported similar results.

The main sources of phosphates are human waste, leaching of phosphorus-rich bedrock, and decaying organic matter (Girija et al. 2007; Manickam et al. 2015b). High concentrations of phosphates and nitrates have been linked to eutrophication, which raises the bloom and eventually lowers dissolved oxygen in the water, according to studies by Murdoch et al. (2001) and Manickam et al. (2015b). High phosphate concentrations are evident in all three reservoirs at the sampling sites, according to our results. Although sulfate is often found in natural streams, industrial contamination with sulphuric acid, bisulphate, and aluminium sulphate from water treatment plants can raise levels. The study found that water samples taken from agricultural field sites had the highest level of sulfate. Increased sulfate levels in drainage waters result in diarrhea and laxative effects (Guru Prasad 2003; Manickam 2015; Manickam et al. 2017b). In soil and water, nitrogenous organic molecules and biota break down spontaneously to form ammonia. According to Manickam et al. (2015b), natural NH_3 for seasonal oscillations can result from the decomposition and mortality of bacteria, phytoplankton, and other aquatic species. Additional man-made sources of NH_3 are industrial emissions, sewage discharges into natural rivers, and fertilizer runoff. Sewage and residential discharges could be the cause of high quantities of NH_3 and nitrite in general (Mohanraj et al. 2000; Manickam et al. 2015b).

Phosphates and nitrates are reliable markers of eutrophication. According to Kudesia et al. (1986), nitrites and a few other harmful aromatic chemicals give water a brown color and an unpleasant odor, rendering it unsuitable for drinking, irrigation, or fish culture. Nitrate is a pollutant found both in natural water sources and the atmosphere. It is also an essential component of the biological nitrogen cycle (Manzoori et al. 1998; Manickam 2015; Manickam et al. 2015b). Fertilizers, decomposing plant and animal matter, industrial and household wastewater, and atmospheric fallout are important sources of nitrates. As a result, nitrate levels in a body of water must be preserved. Nitrate is a nutrient for plants that enters the system as an inorganic fertilizer from septic systems, animal feed, agricultural fertilizers, manure, industrial effluent and sanitary landfills, according to Saxena (1998), Manickam (2015), Manickam et al. (2015a) and other sources. The highest nitrate value currently reported during the monsoon season may have its origins in organic materials derived from the catchment area. Silicate levels have remained more constant over time, despite increases in N loading and P concentrations caused by anthropogenic activity and detergent use (Gilpin et al. 2004; Manickam et al. 2015b). Microorganisms, invertebrates, and plants all take silica from the water to use in their growth and as building components. In the present result, Si_2O_3 concentration was found to be highest during summer and late monsoon due to runoff and seasonal opening of upstream channels of Thoppaiyar, Nagavathy, and Panchapalli reservoirs.

The species composition (qualitative analysis) of zooplankton is the ecological indicator of water bodies, according to Gajbhiye and Desai (1981). The dissolved oxygen content, light intensity, food availability, and predation all have an impact on the dynamics of the zooplankton population. Low pH and increased salinity have been shown to decrease plankton variety and density (Horn and Goldman 1994, 2015b, 2017a; Manickam et al. 2014). Previous research in the perennial freshwater lake (Haledharmapuri Lake) supported the current findings. 33 species of zooplankton-13 rotifera, 7 cladocera, 8 copepoda and 5 ostracoda-were discovered out of the 55 species of zooplankton that were discovered in Dharmapuri Town, Tamil Nadu, India, according to research (Manickam et al. 2012, 2017b; Dhanasekaran et al. 2017). However, 55 other species found in the area. 36 species of zooplankton have been identified in Sulur Lake (Coimbatore District, Tamil Nadu) by Bhavan et al. (2015). These species include 11 rotifera, 10 cladocera, 7 copepoda, and 6 ostracoda. Manickam et al. (2015a) found 47 species of zooplankton at Barur Lake, Krishnagiri District (Tamil Nadu, India). These species included 18 rotifera, 11 cladocera, 11 copepoda, and 7 ostracoda species. In the Sulur and Ukkadam lakes, early researchers discovered 33 species of zooplankton (Manickam et al. 2017a). The cladocera and copepoda makeup that is currently known is comparable to several other researchers' findings (Sivakumar and Altaff 2004; Manickam et al. 2015a). By altering metabolic rate and activity level, water temperature limits the life cycle features of zooplanktonic organisms, which has a direct impact on growth and reproduction (Burns 1969). Moreover, Mergeay et al. (2006) linked high population density to environmental factors such as water temperature and others that are necessary for resting cladoceran egg hatching in natural water. Nonetheless, Wright (1965) discovered that when the water body's nutrient concentration increased, so did the zooplankton densities.

According to the earlier findings of Bhavan et al. (2015) and Manickam et al. (2015b), the present result was similar. The prevailing physical and chemical attributes of the surroundings dictated the dispersion and density of zooplankton. Rotifera were discovered to be the most common group among zooplankton. Since these species are considered indicators of eutrophication, human activity in watershed areas must be limited to reduce water pollution (Manickam et al. 2012, 2015b, 2017b). The physicochemical properties of the area around the reservoir dictated the density and dispersion of the zooplankton population. The presence of food (phytoplankton) in reservoir ecosystems and good environmental conditions may be the reasons for the large zooplankton population observed during the summer. Furthermore, as demonstrated by Manickam et al. (2014) from the Thoppaiyar reservoir in Dharmapuri, Tamil Nadu; Ramakrishna (2014) from Yelahanka Lake in Banglore, India; Gayathri et al. (2014) from Doddavoderhalli Lake in Banglore; Dede and Deshmukh (2015) from the Bhima river near Ramwadi village in Solapur District, Maharashtra, India; Bhavan et al. (2015) from the Sulur lake (Coimbatore District, Tamil Nadu); Manickam (2015) from Barur Lake (Krishnagiri District, Tamil Nadu). As previously mentioned by Jayabhaye (2010), Manickam et al. (2012, 2014, 2015b, 2017a), Ezhili et al. (2013), Manickam (2015) and Dhanasekaran et al. (2017) also discovered that high turbidity, low light intensity, and cloudy skies, in addition to high rainfall, may be the cause of the overall low population density of zooplankton during the monsoon season in all stations. In the Thoppaiyar, Nagavathy, and Panchapalli reservoirs (Tamil Nadu), zooplankton diversity was highest in the summer and lowest in the monsoon and premonsoon seasons, according to Kannan and Job (1980) and Manickam et al. (2017b). Diverse species According to Odum (1983), there are two main methods for analyzing species diversity in various dominant scenarios: diversity, relative abundance curves, and diversity indices, which are ratios or other mathematical representations of the relationships between species' importance. In stressed and contaminated ecosystems, species diversity is typically low (Bass and Harrel 1981). On the contrary, the monsoon season is characterized by many factors that affect the diversity and density of copepods, rotifera, cladocera, and ostracoda, such as water temperature, turbidity, and transparency (Edmondson 1965). Some of the factors that contribute to the summertime zooplankton population minimum and maximum may be bacteria, nanoplankton, and suspended detritus.

During the rainy season, equitability (evenness) was comparatively high, suggesting a decrease in plankton diversity at that time. Peet (1974) defined species diversity as "richness and evenness in the number of species, as well as equity for the distribution of individuals among the species." The current analysis supports the findings of Odum (1983), who proposed an inverse relationship between the dominance value and the values of species richness, evenness, and diversity. Primary crops do best in Thoppaiyar, Nagavathy, and Panchapalli reservoirs during the research period, when the freshwater flow from monsoon rainfall accumulates nutrients. The plankton population and seasonal fluctuations in environmental conditions support this. The lentic water system may have produced physicochemical conditions that were detrimental to zooplankton growth, as seen by the sudden fall in the zooplankton population during the monsoon season (Manickam 2015; Manickam et al. 2017b). Low temperatures may be associated with the lowest zooplankton population ever seen during the monsoon (Marshall and Orr 1972). According to Bias and Agrawal (1995), a high-standing crop of primary producers, higher temperatures, and improved transparency were all associated with the summertime zooplankton population, which in turn improved food availability. The number of zooplankton grew along with the gradual increase in alkalinity (Manickam et al. 2015b). Rotifera species found in *B. calyciflorus* are believed to be useful eutrophication indicator organisms (Manickam et al. 2012, 2014, 2015b, 2017a; Bhavan et al. 2015; Dhanasekaran et al. 2017). Some species of copepoda (*Heliodiaptomus viduus*, *Mesocyclops hyalinus* and *Thermocyclops hyalinus*), ostracoda (*Cypris protubera* and *Hemicypris anomala*), rotifera (*B. calyciflorus*, *B. falcatus*, and *Filinia longiseta*), and cladocera (*D. sarsi* and *C. cornuta*) are among species that can withstand pollution and exhibit signs of organic matter buildup in the lake (Goel and Charan 1991; Manickam et al. 2012, 2014, 2015b, 2017a, b; Bhavan et al. 2015; Dhanasekaran et al. 2017). Because of the dilution impact and lower primary producer photosynthetic activity, the monsoon months were traditionally linked to a low zooplankton population. Manickam (2015) and Bais and Agrawal (1993) have both produced results that are comparable. Due to a diluting effect, the population decreases during the monsoon. In this reservoir environment, throughout the summer, the coexistence of hard water and DO also facilitate the growth of zooplankton. Multiple researchers have similarly observed similar results

(Manickam 2015; Manickam et al. 2015a, 2017a, b). Numerous species still have to contend with the limitations imposed by dissolved oxygen, temperature, salinity, and other physical or chemical elements on the spectrum of ambient circumstances in which zooplankton can live. Five rotifera species were present, indicating a gradual and steady organic pollution process that was pushing the reservoirs toward eutrophication. The diversity of species and population density demonstrated their abundance in response to favorable conditions; hence, they vanished during unfavorable conditions and resurfaced when favorable conditions in the reservoir environment returned.

11.5 Conclusion

The physical and chemical properties of the reservoirs at Thoppaiyar, Nagavathy, and Panchapalli in the Dharmapuri District of Tamil Nadu, South India, are found to be within tolerance limits, according to the study's findings. This is because no abnormal values were found during the allotted study period. The physical and chemical environment has an impact on the distribution and population density of zooplankton species. Therefore, steps must be taken to reduce the pollution of freshwater by prohibiting bathing, laundry, and other human activities inside the ecology of the reservoir. Additionally, data created in the form of memoirs is crucial for decision makers to effectively utilize and conserve water bodies within reservoir ecosystems.

Acknowledgements The authors are grateful to the Head, Department of Marine Science, School of Marine Sciences, and the authorities of Bharathidasan University for the necessary laboratory facilities provided.

References

Akindele EO, Olutona GO (2014) Water physicochemistry and zooplankton fauna of Aiba reservoir head water streams, Iwo, Nigeria, pp 1–11

Altaff K (2004) A manual of zooplankton. Department of Zoology, The New College, Chennai, Tamil Nadu, pp 19–145

Anyanwu ED, Orjikwe CJ, Chinasa PO (2022) Water quality and zooplankton assessment of Iyiakwu river, Southeast Nigeria. J Inst Landsc Ecol Slovak Acad Sci Ekologia 41(1):9–16

APHA (2005) American Public Health Association, standard methods for the examination of water and wastewater, 21st edn. Washington D.C., pp 10–132

Aravinth A, Kannan R, Chinnadurai G, Manickam N, Raju P, Perumal P, Santhanam P (2023) Temporal changes in plankton diversity in relation to hydrographical characteristics at Perumal Lake, Cuddalore District, Tamil Nadu, India. J Basic Appl Zool 84:13. https://doi.org/10.1186/s41936-023-00337-7

Atobatele OE, Ugwumba OA (2008) Seasonal variation in the physico-chemistry of a small tropical reservoir (Aiba Reservoir, Iwo, Osun, Nigeria). Afr J Biotecnol 7:1962–1971

Bais VS, Agrawal WL (1993) Seasonal variation of nutrient content in *Hydrilla verticollata*. J Freshwat Biol 3:259–265
Bais VS, Agrawal NC (1995) Comparative study of the zooplanktonic spectrum in the Sagar Lake and Military engineering Lake. J Environ Biol 16(1):27–32
Bass D, Harrel RC (1981) Water quality of South East Texas Stream. Hydrobiology 76:69–79
Battish SK (1992) Freshwater zooplankton of India, vol 6. Oxford and IBH Publication. Co. Pvt. Ltd, Calcutta, p 233.
Bhavan PS, Selvi A, Manickam N, Srinivasan V, Santhanam P, Vijayan P (2015) Diversity of zooplankton in a perennial Lake at Sulur, Coimbatore, India. Int J Ext Res 5:31–44
Burns CW (1969) Relation between filtering rate, temperature, and body size in 4 species of *Daphnia*. Limnol Oceanogr 14:693–700
Dede AN, Deshmukh AL (2015) Study on zooplankton composition and seasonal variation in Bhima River Near Ramwadi Village, Solapur District (Maharashtra), India. Int J Curr Microbiol Appl Sci 4(3):297–306
Dhanasekaran M, Bhavan PS, Manickam N, Kalpana R (2017) Physico-chemical characteristics and zooplankton diversity in a perennial lake at Dharmapuri (Tamil Nadu, India). J Entomol Zoo Stud 5(1):285–292
Dhawan A, Karu S (2002) Pig dung as pond manure: effects on water quality, pond productivity and growth of carps in polyculture system. Naga ICLARM Q 25:11–14
Edmondson WT (1965) Reproductive rate of planktonic rotifers as related to food and temperature. Ecol Monogra Orginat 35:61–111
Ezhili N, Manikandan R, Ilangovan R (2013) Diversity and seasonal variation of zooplankton in Ukkadam Lake, Coimbatore, Tamil Nadu, India. Int J Cur Res 8(5):2091–2094
Gajbhiye SN, Desai BN (1981) Zooplankton variability in polluted and unpolluted waters of Bombay. Mahasagar Bull Natl Inst Oceanogr 4:173–182
Gayathri S, Latha N, Mohan MR (2014) Studies on population dynamics and seasonal abundance of zooplankton community in Doddavoderahalli Lake, Bangalore, India. Int J Emerg Trends Eng Dev 4(1):50–55
Gilpin LC, Davidson K, Roberts E (2004) The influence of changes in nitrogen: silicon ratios on diatom growth dynamics. Elsevier J Sea Res 51:21–35
Girija TR, Mahanta C, Chandramouli V (2007) Water quality assessment of an untreated effluent impacted urban stream: the Bharalu Tributary of the Brahmaputra River, India. Environ. Monit Assess 169–671
Goel PK, Charan VR (1991) Studies on the limnology of polluted fresh water tank. In: Gopan B, Asthana V (eds) Aquatic sciences in India. Ind. Ass. Lim. Oce, pp 51–64
Gopalkrushna MH (2011) Assessment of physico-chemical status of ground water samples in Akot city. Res J Chem Sci 1(4):117–124
Guermazi W, El-khateeb M, Abu-Dalo M, Sallemi I, Al-Rahahleh B, Rekik A, Belmonte G, Ayadi H, Annabi-Trabelsi N (2023) Assessment of the zooplankton community and water quality in an artificial freshwater lake from a semi-arid area (Irbid, Jordan). Water 15:2796. https://doi.org/10.3390/w15152796
Guru Prasad B (2003) Physico-chemical and bacteriological quality of ground water at Tadepalli Mandal of Guntur District, Andhra Pradesh. Nat Environ Pollut Tech 2:173–178
Horne AJ, Goldman CR (1994) Limnology, 2nd edn. McGraw-Hill, New York
Jayabhaye UM (2010) Studies on zooplankton diversity of river Kayadhu, Near Hingoli City, Hingoli, Dist. Maharashtra. India. Int Res J 11(12):47–49
Kannan V, Job SV (1980) Diurnal, seasonal vertical study of primary production in Sathiar Reservoir. Hydrobiology 70:171–178
Kowthaman R, Rebecca AA, Manickam N (2019) Diversity and distribution of Planktonic communities in Krishnampathy Lake, Coimbatore, Tamil Nadu, India. Environ Ecol 37(4):1230–1239
Kubley DM (1982) Physical and chemical features of Playa Lakes in Southeastern California U.S.A. Arch Hydrobiol Suppl 26(3–2):491–525

Kudesia VP, Verma SP, Sangh KP, Saniv P (1986) Pollution studies of drinking water quality of environmentally degreded village Kamalpur (Dist. Meerut) and their remedial measures. Ind J Environ Agric 1(1):38–44

Kumar A, Chava R, Gupta S, Shirin S, Jamal A, Yadav AK (2023) Environmental and health effects of fluoride contamination and treatment of wastewater using various technologies. In: Advanced treatment technologies for fluoride removal in water. Water science and technology library, vol 125. Springer, Cham. https://doi.org/10.1007/978-3-031-38845-3_18

Malik DS, Sharma SK, Sharma AK (2020) Status of zooplankton diversity, abundance in relation to hydrography of Ganga River and its Tributaries at Rishikesh. Indian J Ecol 47(3):739−745

Manickam N, Bhavan PS, Santhanam P, Chitrarasu P, Ali AJ (2012) Zooplankton diversity in a perennial freshwater Lake. In: Desai PV, Roy R (eds) Diversity and physiological processes. Goa University, India, pp 25–37

Manickam N, Bhavan PS, Santhanam P, Muralisankar T, Srinivasan V, Radhakrishnan S, Vijayadevan K, Chitrarasu P, Ali AJ (2014) Seasonal variations of zooplankton diversity in a perennial reservoir at Thoppaiyar, Dharmapuri District, South India. Austin J Aquac Mar Biol 1(1):1–7

Manickam N, Bhavan PS, Santhanam P, Muralisankar T, Srinivasan V, Radhakrishnan S, Vijayadevan K, Bhuvaneswari R (2015a) Biodiversity of freshwater zooplankton and physico-chemical parameters of Barur Lake, Krishnagiri District, Tamil Nadu, India. Malaya J Biosci 2(1):1–12

Manickam N, Bhavan PS, Santhanam P (2015b) Physico-chemical characteristics of lake ecosystem. LAMBERT Academic Publishing, pp 1–80. ISBN: 978-3-659-79409-4

Manickam N, Santhanam P, Bhuvaneswari R, Bhavan PS, Vijayadevan K, Ashokan V (2017a) Seasonal variations in zooplankton diversity and physico-chemical characteristics of the Nagavathi reservoir, Dharmapuri District, Tamil Nadu, India). J Ter Mar Res1(1):28–35

Manickam N, Bhavan PS, Santhanam P (2017b) Evaluation of nutritional profiles of wild mixed zooplankton in Sulur and Ukkadam Lakes of Coimbatore, South India. Turk J Fish Aqu Sci 17:509–517

Manickam N, Santhanam P, Bhavan PS, Vijayadevan K, Ashokan V, Bhuvaneswari R, Dhanalakshmi B, Chitra J (2018) Physico-chemical characteristics and zooplankton diversity in the reservoirs (Dams) of Dharmapuri District, Tamil Nadu, Southern India. In: Guha A, Ray A, Barik KL (eds) Environmental pollution biodiversity conservation and climate change: issues and challenges. Sparrow Publication, India. ISBN: 978-81-89140-17-7

Manickam N, Santhanam P, Bhavan PS (2019) Techniques in the collection, preservation and morphological identification of freshwater zooplankton. In: Santhanam P, Begum A, Perumal P (eds) Basic and applied zooplankton biology. Springer Nature, pp 139–195. https://doi.org/10.1007/978-981-10-7953-5_5

Manickam N, Bhavan PS, Santhanam P, Muralisankar T, Dinesh Kumar S, Balakrishnan S, Ananth S, Shenbaga Devi A (2020a) Phytoplankton biodiversity in the two perennial lakes of Coimbatore, Tamil Nadu, India. Acta Ecol Sin 40:81–89. https://doi.org/10.1016/j.chnaes.2019.05.014

Manickam N, Santhanam P, Bhavan PS (2023) Biofloc-Copefloc: technology in aquaculture. LAP—LAMBERT Academic Publishing, pp 1–61. ISBN: 978-620-6-18441-6

Manickam N (2015) Biodiversity of plankton in two perennial lakes of Coimbatore, India and suitability of wild mixed zooplankton as live feed for rearing of the giant freshwater prawn *Macrobrachium rosenbergii* early post larvae. Ph. D., Thesis, Bharathiar University, Tamil Nadu, India, pp 1–145

Manzoori JL, Sorouraddin M, Haji-Shabani AM (1998) Spectrophotometric determination of nitrite based on its catalytic effect on the oxidation of carminic acid by bromate. Talanta 46:1379

Marshall SM, Orr AP (1972) The charophytes of the DulLake. Kashmir. In: Abstract proceeding. 21st Land Congress, p 295

Mergeay J, Declerck S, Verschuren D, Meester L (2006) *Daphnia* community analysis in shallow Kenyan Lakes and ponds using dormant eggs in surface sediments. Freshw Biol 51:399–411

Mohanraj R, Sathishkumar M, Azeez PA, Sivakumar R (2000) Pollution status of wetlands in urban Coimbatore, Tamil Nadu, India. Bull Environ Contamin Toxicol 64:638–643

Murdoch T, Cheo M, O'Laughlin K (2001) Streamkeeper's field guide: watershed inventory and stream monitoring methods. Adopt-a-stream foundation. Everett, WA, pp 1–297

Murugan N, Murugavel P, Kodarkar MS (1998) Cladocera: the biology, classification, identification and ecology. Indian Assoc Aq Biol Hyderabad 5:1–55

Odum EP (1983) Basic ecology. Holt-Saunders international editions. Saunders College Publishing, Japan, p 558

Peet RK (1974) The measurement of species diversity. Ann Rev Ecol Syst 5:285–307

Pinto I, Nogueira S, Rodrigues S, Formigo N, Antunes SC (2023) Can zooplankton add value to monitoring water quality? A case study of a meso/eutrophic Portuguese reservoir. Water 15:1678. https://doi.org/10.3390/w15091678

Ramakrishna S (2014) Zooplankton seasonal abundance in relation to physicochemical feature in Yelahanka Lake, Bangalore, India. Global J Res Anal 3(6):218–219

Santhanam R, Velayutham P, Jegatheesan G (1989) A manual of freshwater ecology. Daya Publishing House, Delhi, pp 1–109

Saxena A (1998) Primary productivity studies in a sewage polluted Lake with special reference to phytoplankton. Ph. D., Thesis, Barkatullah University, Bhopal, India

Sharma BK, Sharma S (2020) Zooplankton diversity of a subtropical reservoir of Meghalaya, northeast India with remarks on spatial and temporal variations. Opusc Zool 51(1):67–86. https://doi.org/10.18348/opzool.2020.1.67

Shirin S, Yadav AK (2014) Physico chemical analysis of municipal wastewater discharge in Ganga River, Haridwar District of Uttarakhand, India. Curr World Environ 9(2):536

Sivakumar K, Altaff K (2004) Ecological indices of freshwater Copepods and Cladocerans from Dharmapuri District, Tamil Nadu, India. J Zoos' Print 19(5):1466–1468

Smitha PG, Byrappa K, Ramaswamy SN (2007) Physico-chemical characteristics of water samples of Bantwal Taluk, south-western Karnataka, India. Natl Environ Pollut Technol 28(3):591–595

Strickland JDH, Parsons TR (1972) A manual of sea water analysis. Can Fisher Res Board Bull 167:310

Swaranlatha S, Rao AN (1998) Ecological studies of Banjara Lake with reference to water pollution. J Environ Biol 19:179–186

Trivedy RK, Goel PK (1984) Chemical and biological methods for water pollution studies. Environ Publ 1–215

Verma SR, Tyagi AK, Dalela RC (1978) Physico-chemical and biological characteristics of Kadarbad Drain in Uttar Pradesh, India. Ind J Environ Health 20(1):1–13

Wright JC (1965) The population dynamics and production of *Daphnia* in Canyon Ferry Reservoir, Montana. Limnol Oceanogr 12:492–502

Yadav AK, Sahoo SK, Mahapatra S, Kumar AV, Pandey G, Lenka P, Tripathi RM (2014) Concentrations of uranium in drinking water and cumulative, age-dependent radiation doses in four districts of Uttar Pradesh, India. Toxicol Environ Chem 96(2):192–200

Zafar AR (1964) On the ecology of algae in certain fish ponds of Hyderabad, India. I Physico Chem Compl Hydrobiol 23:176–196

Chapter 12
Exploratory Data Analysis of Groundwater Physio-chemical Parameters in R Software Program: A Case Study from Jammu Himalayas, India

Iftikhar Hussain Beigh and Saba Riyaz

Abstract Water as an essential natural resource vital for sustaining life, plays a pivotal role in the socioeconomic advancement of nations. Its quality is paramount, influencing its suitability for human consumption, agricultural irrigation, and industrial processes. Groundwater, among the myriad available water sources, serves as a crucial reservoir for drinking, household use, and agricultural activities. The protection of groundwater from the dangers of agricultural and industrial pollution is imperative to preserve ecological balance, safeguard public health, and ensure the integrity of drinking water sources. This study offers a comprehensive examination of groundwater samples collected from the Jammu Himalayan region, presenting an exploratory data analysis of several physiological parameters outlined in the Groundwater Directive (GWD). These parameters include fluoride, iron, nitrate, pH, total hardness, calcium, chloride, magnesium, bicarbonate, sodium, sulfate, potassium, and total dissolved solids present in the samples. The discussion covers the utilization of latest R packages, mentioning the codes employed for generating plots. The analytical insights gained from this study are relevant and applicability across diverse fields beyond groundwater analysis.

Keywords Groundwater quality · Box plot · Pair plot · Correlation analysis · Heat map · Scatter plot · R codes

---

I. H. Beigh (✉)
Department of Civil Engineering, National Institute of Technology, Srinagar, J&K 190006, India
e-mail: iftikharbeigh@nitsri.ac.in

S. Riyaz
Department of Statistics, University of Kashmir, Srinagar, J&K 190006, India

A. K. Yadav et al. (eds.), *Integrated Management of Water Resources in India: A Computational Approach*, Water Science and Technology Library 129,
https://doi.org/10.1007/978-3-031-62079-9_12

12.1 Introduction

Groundwater is a very important natural resource that is present underground in saturated zones below the land surface (van der Gun 2021). It refers to the water that seeps into the soil and rocks getting stored below the ground in the rocks called aquifers. Most of the substances that are soluble in groundwater are minerals that are soluble in soil and sedimentary rocks (Committee 2002). Providing a clean and renewable supply of drinking water is one of the most significant drivers of a country's growth. The largest source of nonfrozen freshwater on earth is groundwater (Gleeson et al. 2015). According to the International Association of Hydrogeologists (2005), it is used for drinking by most of the world population. Understanding the availability and quality of groundwater is crucial for assessing the present and future health of many ecosystems. Today, groundwater quality is threatened by agricultural practices, urbanization, industrial activity, and climate change (Lee 2006), thus a growing field of research (Hartmann 2022). Suhag (2016) presented an analysis of the groundwater scenario in India. Swain et al. (2022) has offered a comprehensive analysis of the relationship between groundwater, the climatic system, and anthropogenic activity in India. Romshoo et al. (2017) evaluated the physiochemical parameters in the groundwater of the plains of Jammu and Kashmir, India. Positive and negative ionic species make up most of the salinity, also known as total dissolved solids (TDS). Sundaram et al. (2009) have provided in detail the basic information to plan and undertake field groundwater sampling, ensuring the high integrity and representativeness of samples collected and submitted for lab analysis. However, Jammu and Kashmir have vast supplies of surface and groundwater, which are mostly derived from rainfall and snowfall, snow melting, and spring discharge. Recent climate changes have influenced water resources in the region, reducing the extent of glaciers, changing precipitation patterns, and reducing river and stream flow. Due to all these factors, surface water resources are diminishing, and home and industrial water demand is increasingly reliant on groundwater resources (Arora 2011; Jasrotia and Kumar 2014). Poor water quality threatens plant growth, human health, social prosperity, and economic development (WHO 2011). Yadav et al. (2014) have determined uranium concentrations in drinking and groundwater samples. Shirin and Yadav (2014) have analyzed the water quality parameters of municipal wastewater discharge from various sources. Shirin et al. (2019) have done a study on assessment of slope stability and mixed disposal of overburden in voids of Singrauli Coalfield. Shirin and Yadav (2021) have also shown how fly ash may be effectively used for sustainable neutralization of coalfields. Shirin et al. (2021) have suggested mitigation measures and how to control the pollutants so that the ecosystem of the area gets least disturbed by the associated mining activity. Kumar et al. (2023) have demonstrated a novel potential for fluoride removal. Patel et al. (2023) have shown that fluoride can be effectively removed from water using iron-based materials. Shirin et al. (2023) have studied how some opencast mines are not properly backfilled which causes harmful environmental effects, such as leaching of acid mine drainage Further work has been done by Kulwant et al. (2023) on the practice of artificial water fluoridation.

To ensure the long-term viability of portable groundwater, appropriate management methods must be used to reduce the negative effects of anthropogenic and industrial contamination of groundwater resources and to prevent additional contamination to the point where it becomes unsuitable for human use. (Lee 2006) offers a mathematical framework for assessing water quality, which is divided into five categories of the CCME WQI range (0–44): Poor; (45–59): Marginal; (60–79): Fair; (80–94): Good; (95–100): Excellent. Exploratory data analysis (EDA) or Data Visualization is a fundamental early step of any scientific process after collecting and preprocessing it, where the data is to be visualized, manipulated, plotted to check its features, the distribution like outliers, setting hypothesis, and building models (Komorowski et al. 2016). EDA has earned a large following as the gold standard for analyzing a data set since the initial work by Tukey in 1977. Bondu et al. (2020) provided exploratory data analysis of naturally occurring contaminants such as arsenic, fluorine, manganese, barium, etc., in groundwater from a region of Canada. Researchers utilize many statistical tools such as Microsoft Excel, SPSS, S-Plus or programming languages such as SQL, JavaScript, Scala, GENSTAT, and Python to analyze their data, while some create their computer programs in FORTRAN, C++, or C for the same purpose, but some computer programs lack the detailed exploratory data analysis ability or to incorporate environmental correlations (Omuto and Gumbe 2009). For statistical analysis, graphic representation, and reporting in this work, we employ the R programming language. Robert Gentleman and Ross Ihaka (University of Auckland in New Zealand) developed the R programming language. Currently being worked on by the R Development Core Team, it initially appeared in 1993. It is a high-level language that is more efficient, fast, accurate, and executable as free-source software (Ihaka and Gentleman 1996). Pradhan et al. (2021) used the R language to visualize the evolution and improvement of the Olympic Games over the years. Too et al. (2022) recently utilized the capabilities of R to analyze some models. Arunkumar and Velmurugan (2021) have attempted to use exploratory data analysis as a method to assess groundwater quality using the Python programming language. In this study, we have attempted to perform an exploratory data analysis of the chemical parameters of groundwater from the Jammu Himalayan region, a northwestern Himalayan region. Parameters considered for analysis are Potential Hydrogen (pH), Electrical conductivity (EC), Iron (Fe), Potassium (K), Sodium (Na), Calcium (Ca), Magnesium (Mg), Chloride (Cl), Bicarbonate (HCO_3), Fluoride (F), Nitrate (NO3), Sulfate (SO_4), Total Dissolved Solids (TDS) and Total Hardness (TH), using standard methods as suggested by APHA, 2005. This study serves as a guide for exploratory data analysis of groundwater datasets using the R programming language.

12.2 Study Area

Groundwater data have been gathered from the Jammu region of the state of J&K, which lies between 32.28 and 37.06° latitude and 72.53–80.32° longitude, in the northwest of the Himalayas in India (Romshoo et al. 2020) (Fig. 12.1). This region

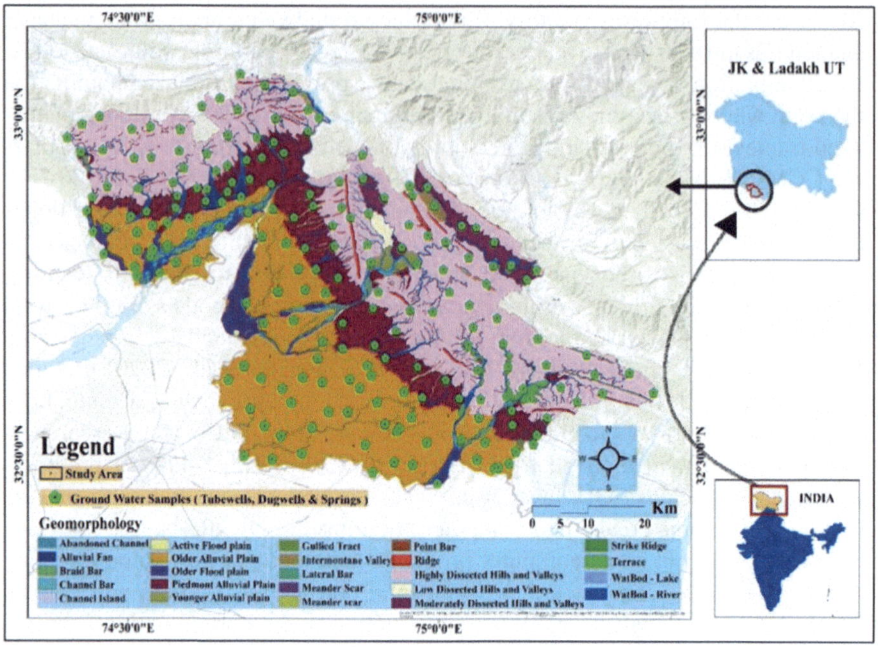

Fig. 12.1 Geomorphological map of the Jammu Himalaya, J&K with groundwater sampling locations

experiences a subtropical to moist temperate climate, contributing to regular surface recharge of groundwater through annual rainfall.

12.3 Datasets and Methodology

A total of 207 tube wells and wells were meticulously sampled to ensure representativeness (Beigh 2013). The data set comprises 22 attributes, 3 being non-numerical—Well. no., District, and Type—and the remaining 19 being numerical—F, Fe, NO_2, NO_3, RSC, pH, Alkalinity, Cl, TDS, TH, HCO_3, Ca, Mg, Na, K, SAR, BeI, SO_4 and CO_3. Our objective in this study is to visualize and comprehensively analyze our groundwater dataset, specifically focusing on the premonsoon data.

Following data collection and input into a spreadsheet, sifting through a plethora of figures or an entire dataset can be daunting. Extracting meaningful insights from raw numerical data can be tedious and overwhelming. Therefore, the necessity for exploratory data analysis (EDA) and visualization arises. Stem-and-leaf plots, histograms, box plots, and scatter plots stand out as prevalent tools within the realm of EDA.

12.3.1 Exploratory Data Analysis

To effectively address any problem and achieve desired outcomes, it is imperative to adopt a systematic approach. The ensuing diagram delineates the procedural flow applicable to various scientific data analysis endeavours.

Following data collection and input into a spreadsheet, sifting through a plethora of figures or an entire dataset can be daunting. Extracting meaningful insights from raw numerical data can be tedious and overwhelming. Therefore, the necessity for exploratory data analysis (EDA) and visualization arises. Stem-and-leaf plots, histograms, box plots, and scatter plots stand out as prevalent tools within the realm of EDA. Ma et al. (2017) exemplifies the significance of EDA in formulating hypotheses within a data science framework. The sequential steps involved in EDA are elucidated as depicted in Fig. 12.2.

Data Preprocessing

- *Explore* the raw dataset—attempt to understand the content and structure of the dataset, variable types, and check for errors in order to make the data useful.
- *Cleaning the data*/dealing with missing data—check for any incomplete or missing data. If missing data/null values exist, we eliminate it or replace it with valid values, which addresses the problem and leads to producing accurate results.

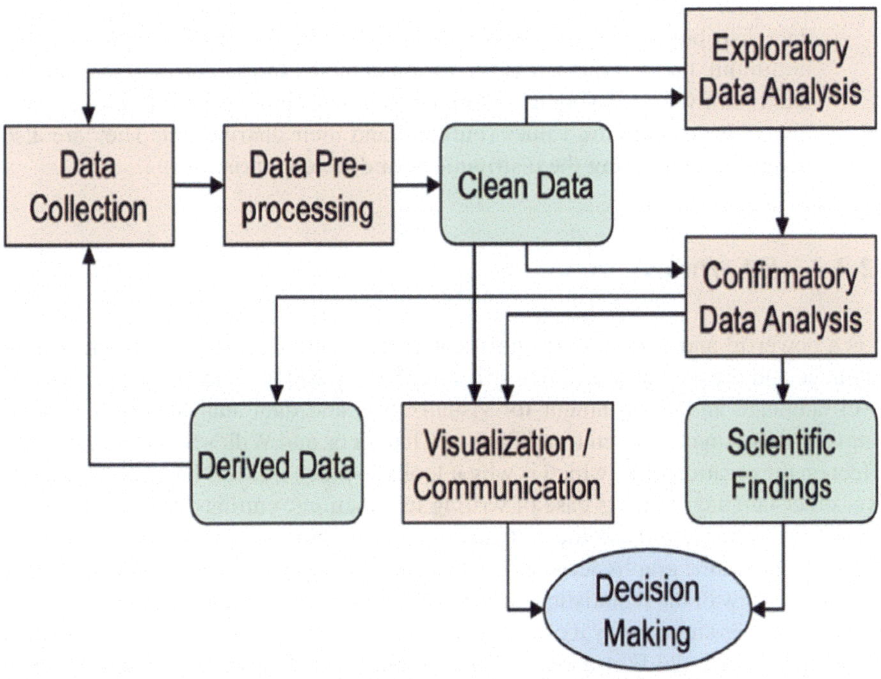

Fig. 12.2 Data analysis process flow (adapted from Schutt and O'Neil 2013)

A. **Describe and summarize.**

- Spread/range of data—to see the minimum and maximum value for each variable, the range it covers, and the Inter-Quartile range.
- Centrality—median and mean of each variable.
- Variance—spread of values of variable from the center.
- *Summary statistics* to see summary of whole dataset at a glance.
- *Correlation matrix*—this investigates the dependence between multiple variables with help of a table of correlation coefficients. The coefficient indicates the direction and the strength of a relationship.

B. **Visualize**

- ***Heatmap***—is a graphical representation in which each colored square shows the relationship between the factors on every pivot. The correlation ranges from − 1 to + 1. If correlation is near 1 it depicts a closer association; that is, as one increases, so does the other, and it displays a darker colored square in the heatmap. A correlation closer to − 1 means a strong negative association. Diagonals show the value 1 since that is the correlation of a variable with itself. We choose strongly correlated variables for further analysis.
- ***Pair plots***—It is a matrix of scatterplots that helps in understanding the pairwise relationship between selected variables in the dataset once the correlations between various variables are recognized.
- ***Box plot*** (or box and whisker plot) is a common method for displaying data distribution and skewness that is based on a five-point summary of the minimum, the first quartile (Q1), the median, the third quartile (Q3), and the maximum values. Using the interquartile range (IQR) criterion, it can reveal the values of extreme values (outliers) and their distribution. They are also helpful in comparing the distributions of different variables.

12.3.2 R Software

R is a powerful and widely used statistical analysis and data science programming language and is executable as a free software (Ripley 2001; Team 2019). It is a high-level language and environment for visualization and data analysis. Two existing languages, Sussman's Scheme and Becker, Chambers, and Wilk's S, had a significant effect on the creation of R, which is why it looks so much like S. The advantages of R over other languages are its ease of writing its functions, cutting-edge applications, unrivaled coverage, and a dedicated back-up and support system. It gives options to choose layout of the graphs according to the needs of the user. See Grunsky (2002) for a great overview of the R statistical aspects of the software package. Beyond its statistical capabilities and the way it offers a lovely programming environment that permits connecting to external Fortran, C, or C++ (Schlather and Huwe 2004), promoted an important philosophy of R. RStudio is an integrated development environment (IDE) for R that supports direct code execution (Verzani 2011). The first port of call for everything to do with R is the 'Comprehensive R Archive Network' (CRAN) (Hornik 2012). One may download and install R from CRAN, as well as access all the packages, instructions, and information on the newest advancements. Collections of data, R functions, and code compiled in a well-defined format are called packages, which are stored in the directory called library. All standard packages are provided with R. Others are available for download and installation. Currently, the CRAN package repository features 18,966 packages available. Tidyverse is a group of code packages that work together to clean, process, model, and visualizing the data. The tidyverse, when compared to the traditional base R, is better to use as it emphasizes readability and flexibility, which is great when needed to write scalable code that others can easily read. Base R is what you get when you open R for the first time and was the only dialect for a long time. The tidyverse, in contrast, are add-ons (via packages) to the language. It drastically improves certain aspects of R, though there are merits to coding only in base (Wickham 2010). The core tidyverse includes code packages that we use in regular analysis. In tidyverse 1.3.0, library(tidyverse) will load the following main tidyverse packages:

- readr, for data import (fast and friendly way to read comma-separated values (CSV) and tab-separated values (TSV))
- tidyr, for tidying the data.
- dplyr, for manipulating the data;
- ggplot2, for visualizing the data.
- forcats, for factors.
- stringr, for strings.
- purrr, for functional programming.
- tibble, for tibbles, a modern re-imagining of data frames.

The packages we used for the analysis are described as follows:

- **tidyverse**—main package for data import, tidying, manipulation, visualization, and programming.
- **readxl**—This package is designed to work with tabular data. It gets data from Excel and into R. readxl supports both the.xls format and the xml based.xlsx format. The easiest way to install its latest version from CRAN is to install the whole tidyverse and then call it with library(readxl).
- **ggplot2**—was released in 2005; this is a code package used to visualize different types of data in any format in beautiful graphs. It is based on *The Grammar of Graphics* (Wilkinson 2012). The **ggcorrplot ()** function can be used to easily visualize a correlation matrix using the ggplot2 package, instead of the heatmap () function in base R.
- **GGally**—'GGally' is an extension to 'ggplot2' adding more functions to reduce the complexity. It lets us obtained the Pearson correlation coefficient between variables and a pairwise plot matrix of the selected variables with the help of **ggpairs ()** from the GGally library.

12.3.2.1 Installation of Packages and Execution of Codes

- Install all the code packages in the tidyverse by running the command **install.packages ("tidyverse")**. Run **library(tidyverse)** to load the tidyverse core into your current R session. Then we install further packages, load the libraries, and run the commands as follows:
- **# Input/Read the data into R**
- install.packages('readxl')
- library(readxl)
- df < -read_excel("C:/groundwater.xlsx") #These inputs our groundwater dataset into R environment and names it as 'df', for convenience.
- **# Explore the dataset**
- View(df) # To view full dataset table.
- dim(df) # To see dimensions of our dataset.
- head(df) # To see first 6 rows of the dataset.
- tail(df) # To see last 6 rows of the dataset.
- names(df) # To see the names of variables as mentioned in data table.
- str(df) # To see overall structure of dataset.
- sum(is.na(df)) # Count missing values if any.
- **# Describe and summarize the dataset**
- summary(df) # Summary of full data set—
- gw<-df %>% # drop the categorical variables in the table and name it as a new table 'gw'
- select(-name, -Type)
- correlation_matrix < -round(cor(gw),
- digits = 2 # correlation matrix for all variables rounded to 2 decimals
-)
- View(correlation_matrix) # to view the correlation matrix created.
- **# Visualize the dataset**
- ***# Heatmap***
- install.packages('ggplot2')
- install.packages("ggcorrplot")
- library(ggplot2)
- library(ggcorrplot)
- # To display the lower matrix and add the correlation coefficient to the plot and white borders
- ggcorrplot(correlation_matrix, hc.order = TRUE,
- type = "lower", lab = TRUE, outline.color = "white")
- ***# Pairs plot***
- # Select the variables for which we want to do further analysis and make a new table
- newtable < - gw[,c('Cl','TH','Ca')]
- head(newtable) # Displays first 6 rows of selected variables.
- # using ggplot2 package
- install.packages('GGally')

- library(GGally)
- ggpairs(newtable) # Most informative plot as it has scatterplot + density + correlation values.
- ***# Boxplot***
- # For plotting multiple boxplots side by side using the ggplot2 package, we need to first reshape the data frame from wide to long format using melt() function
- install.packages("reshape2")
- library(reshape2)
- newtable_long < - melt(newtable)
- ggplot(newtable_long, aes(x = variable, y = value)) + # ggplot function
- geom_boxplot()

12.4 Results and Discussions

In the present study, Table 12.1 shows the statistical summary of selected groundwater elements. We get the following results from the **summary** () command. We can understand that pH of water samples is 6.5–8.45, i.e., acidic to slightly alkaline water, which is in range according to (WHO 2011). A heatmap or correlation plot, also referred as a correlogram or corrgram (Friendly 2002) highlights the variables that are correlated in most positively and negatively way. Darker boxes indicate stronger correlation. Although the heatmap presents the same information as in the correlation matrix, the heatmap is a visual representation, allowing us to quickly observe which variables are correlated and which are not. We get the following better version of the heatmap **ggcorrplot** () command in R.

In the present groundwater analysis, moderate correlation of calcium and Chloride was found with Total Hardness. pH showed a mild correlation with total hardness. The correlation of chloride with sodium absorption ratio (SAR) exists as well. NO_2 shows a high positive correlation with Alkalinity and a low negative high correlation with NO_2, as shown in Fig. 12.3.

Once we study the correlations, we can select variables of our interest for further study. The pair plots represent the relationship between two individual attributes in pairwise order. We have used the **ggpairs** () function of R for this plot. Scatter plots of each pair of numeric variables are visible on the left part of the figure and we can see Pearson's correlation on the right. The diagonal shows the variable distribution. Using **ggpairs** () over the base R function pairs () helps to obtain more information about the variables. Specifically, we can clearly see the correlation coefficient between each variable as a pairwise combination and also a density plot for each (Fig. 12.4).

Thus, we are able to produce a relatively complex matrix of scatterplots with just one line of code.

The way to interpret this matrix is as follows.

Table 12.1 Summary of variables of the dataset

	F	Fe	NO_2	NO_3	RSC	pH	Alkalinity	Cl	TDS	TH
Min	0.0100	0.0300	0.0000	5.0000	*5.3100*	6.5000	0.0000	2.8000	30.0000	10.0000
Ist Qu	0.0140	0.1000	0.0000	20.0000	*0.4000*	7.1250	0.0000	10.0000	48.0000	20.0000
Median	0.2700	0.1000	0.0000	40.0000	0.0000	7.5000	0.0000	10.0000	48.0000	47.0000
Mean	0.2825	0.4498	0.1110	36.6100	*0.1466*	7.4440	6.8500	19.2000	48.7100	128.0000
3rd Qu	0.4000	0.3850	0.2000	48.5000	0.0000	7.6500	10.0000	21.0000	60.0000	250.0000
Max	0.8800	9.7500	0.5000	81.0000	4.9000	8.4500	20.0000	92.0000	60.0000	565.0000

	HCO_3	Ca	Mg	Na	K	SAR	BeI	SO_4	CO_3
Min	36.0000	10.0000	6.1000	2.0000	0.2000	0.0000	*5.0600*	5.0000	0.0000
Ist Qu	214.0000	30.0000	16.0000	15.0000	1.0000	0.0000	*0.2300*	10.0000	0.0000
Median	245.0000	40.0000	21.0000	29.0000	2.0000	0.0000	0.0000	20.0000	0.0000
Mean	277.7000	47.2300	24.5600	29.9400	3.3390	0.4514	*0.2940*	23.7500	0.3228
3rd Qu	354.0000	60.0000	32.5000	36.0000	6.0000	0.5417	0.0000	32.0000	0.0000
Max	531.0000	98.0000	75.0000	98.0000	9.0000	5.2700	1.0000	77.0000	24.0000

Note The value in italic means a negative value

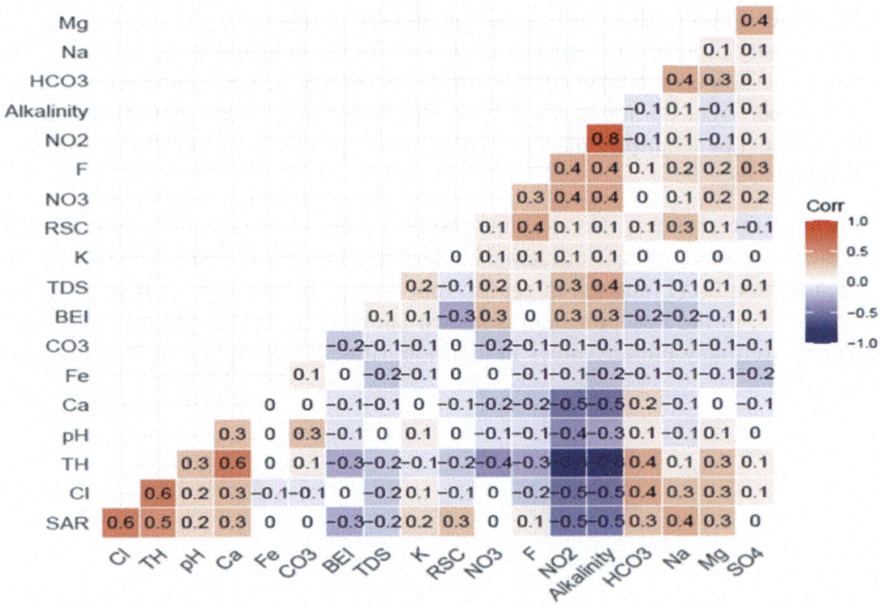

Fig. 12.3 Heatmap displaying the correlation values

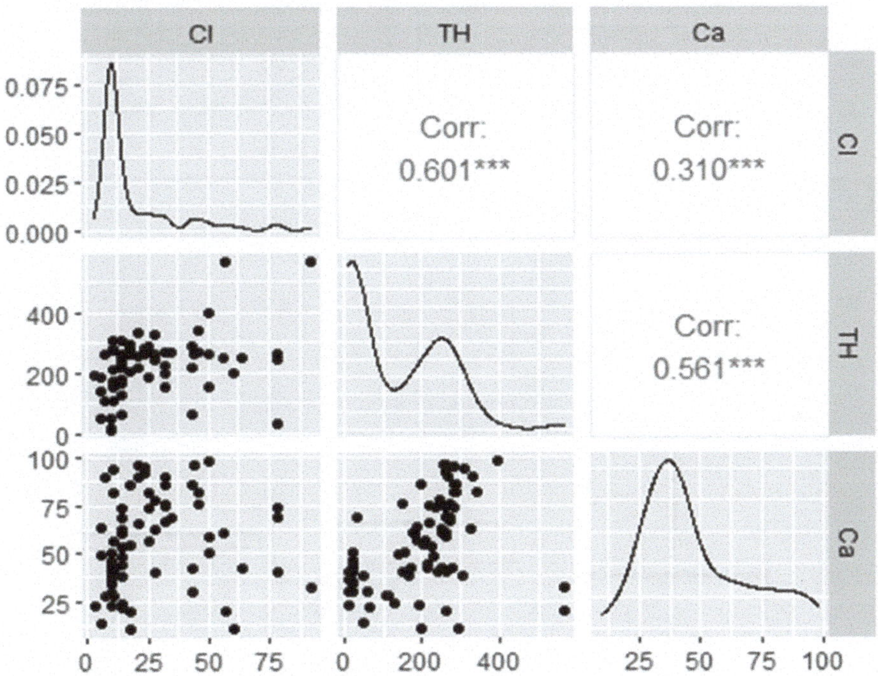

Fig. 12.4 Pairs plot of desired variables

- The outer edges of the matrix show the variable names.
- The density plot for each variable is seen along the diagonals.
- The lower left corner boxes display the scatterplot between each variable.
- The upper right corner boxes display the correlation coefficients between the variables. For example, the correlation between chloride and Total Hardness is **0.601**.

Furthermore, we can also examine the five-value summary and any particular variable for extreme values / outliers with the help of boxplots. For this we have used the **geom_boxplot** () function in ggplot from the ggplot2 package of R (Fig. 12.5).

Moreover, in the box plot figures shown above (Fig. 12.5), we can understand that there are many outlier values in chloride. Effective completion of exploratory

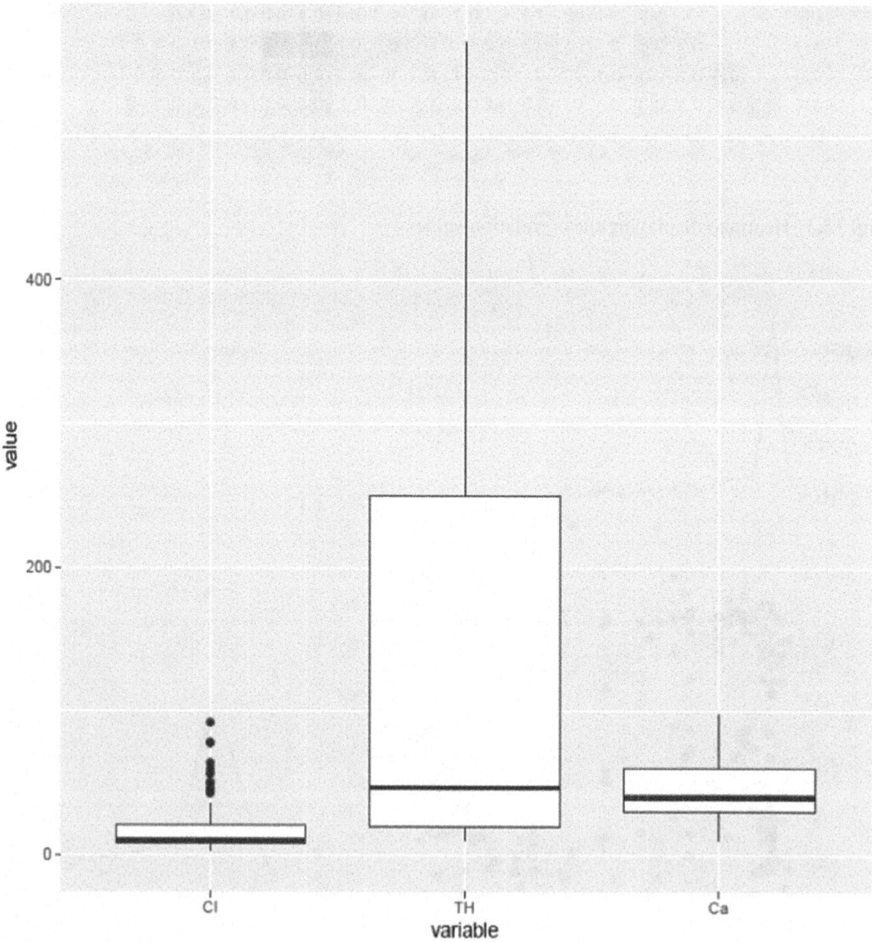

Fig. 12.5 Box plot of chloride, total hardness, and calcium

information examination measure leads to specific methods to investigate work. This EDA cycle helped the groundwater quality dataset deal with considerably more without any problem.

12.5 Conclusions

Water quality degradation is an ever-growing problem as new contaminants emerge. The research work aims to visualize the variables that make up the composition of groundwater in a graphical format. We have used R programming language, which is a free versatile software that contains packages to easily visualize and analyze the data. We have performed exploratory data analysis on the groundwater quality dataset after which we derived correlations between different parameters present in the dataset and were immediately viewed with the help of a heatmap. Pair plots of the groundwater variables Chloride, Calcium, and Total Hardness have then been constructed to represent the detail of the relationship and distribution of the variables. The outliers are examined with the help of box plots. Understanding the exploratory information gives an idea for additional investigation. The future scope of this work is to calculate various water quality indices and create predictive models that can predict the quality of groundwater in the Himalayas, particularly in the Jammu Himalayan region.

Acknowledgements The authors are grateful to Dr Munmun Chakarvorty for her encouragement and prompt support. In addition, the authors would like to thank everyone who assisted during this study. Lastly, the authors would also like to thank anonymous reviewers for their insightful remarks and suggestions.

References

Arora S (2011). Sustainable water resource management in the foothills of the Siwaliks, Northwest India. Groundw Manage Pract 103–120

Arunkumar R, Velmurugan T (2021) An exploratory data analysis process on groundwater quality data. J Maharaja Sayajirao Univ Baroda 54(2)

Beigh IH (2013) Groundwater quality mapping of Jammu District. Unpublished Master's Dissertation, University of Kashmir

Bondu R, Cloutier V, Rosa E, Roy M (2020) An exploratory data analysis approach for assessing the sources and distribution of naturally occurring contaminants (F, Ba, Mn, As) in groundwater from southern Quebec (Canada). Appl Geochem 114:104500

Committee WWAS (2002) Waterwatch Australia National technical manual. Environment Australia, Canberra, ACT. www.Waterwatch.Org.Au/Publications/Index.Html

Friendly M (2002) Corrgrams: exploratory displays for correlation matrices. Am Stat 56(4):316–324

Gleeson T, Befus KM, Jasechko S, Luijendijk E, Cardenas MB (2015) The global volume and distribution of modern groundwater. Nat Geosci 9(2):161–167

Grunsky EC (2002) R: a data analysis and statistical programming environment–an emerging tool for the geosciences. Comput Geosci 28(10):1219–1222

Hartmann A (2022) The hydrology of groundwater systems—from recharge to discharge. Ref Module Earth Syst Environ Sci 324–330

Hornik K (2012) The comprehensive r archive network. Wiley Interdisc Rev Comput Stat 4(4):394–398

Ihaka R, Gentleman R (1996) R: a language for data analysis and graphics. J Comput Graph Stat 5(3):299–314

International Association of Hydrogeologists (2005) Groundwater—more about the hidden resource, https://iah.org/education/general-public/groundwater-hidden-resource

Jasrotia AS, Kumar A (2014) Groundwater quality mapping based on the geographical information system (GIS) of Jammu District, Jammu and Kashmir India. J Spat Hydrol 12(1):1–21

Komorowski M, Marshall DC, Salciccioli JD, Crutain Y (2016) Exploratory data analysis. In: Secondary analysis of electronic health records. Springer International Publishing, pp 185–203

Kulwant M, Patel D, Shirin S, Sharma SN, Yadav AK (2023) Future frameworks for fluoride and algorithms for environmental system. In: Advanced treatment technologies for fluoride removal in water. Water science and technology library, vol 125. Springer, Cham. https://doi.org/10.1007/978-3-031-38845-3_19

Kumar A, Chava R, Gupta S, Shirin S, Jamal A, Yadav AK (2023) Environmental and health effects of fluoride contamination and treatment of wastewater using various technologies. In: Advanced treatment technologies for fluoride removal in water. Water science and technology library, vol 125. Springer, Cham. https://doi.org/10.1007/978-3-031-38845-3_18

Lee G (2006) A sensitivity analysis of the Canadian Water Quality Index. A report for the Canadian Council of Ministers of the Environment prepared by Gartner Lee Limited Markum, ON, Canada

Ma X, Hummer D, Golden JJ, Fox PA, Hazen RM, Morrison SM, Downs RT, Madhikarmi BL, Wang C, Meyer MB (2017) Using visual exploratory data analysis to facilitate collaboration and hypothesis generation in cross-disciplinary research. ISPRS Int J Geo Inf 6(11):368

Omuto CT, Gumbe LO (2009) Estimating water infiltration and retention characteristics using a computer program in R. Comput Geosci 35(3):579–585

Patel D, Kulwant M, Shirin S, Varshney R, Pandey G, Yadav AK (2023) Fluoride removal from aqueous solution using iron-based materials: preparation, characterization, and applications. In: Advanced treatment technologies for fluoride removal in water. Water science and technology library, vol 125. Springer, Cham. https://doi.org/10.1007/978-3-031-38845-3_4

Pradhan R, Agrawal K, Nag A (2021) Analyzing evolution of the olympics by exploratory data analysis using R. IOP Conf Ser Mater Sci Eng. 1099(1):012058

Ripley BD (2001) The R project in statistical computing, MSOR Connections. Newslett LTSN Maths Stats OR Netw 1(1):23–25

Romshoo SA, Dar RA, Murtaza KO, Rashid I, Dar FA (2017) Hydrochemical characterization and pollution assessment of groundwater in Jammu Siwaliks, India. Environ Monit Assess 189(3):1–19

Romshoo SA, Rashid I, Altaf S, Dar GH (2020) Jammu and Kashmir state: an overview. Biodiversity of the Himalaya: Jammu and Kashmir State, pp 129–166

Schlather M, Huwe B (2004) The use of the language interface of R: two examples for modelling water flux and solute transport. Comput Geosci 30(2):197–201

Schutt R, O'Neil C (2013) Doing data science: straight talk from the frontline, O'Reilly Sebastopol, CA

Shirin S, Yadav AK (2014) Physico chemical analysis of municipal wastewater discharge in Ganga River, Haridwar District of Uttarakhand, India. Curr World Environ 9(2):536

Shirin S, Yadav AK (2021) Assessment and suitability for the dumping of flyash in an abandoned mine. Environ Chall 4:100111

Shirin S, Jamal A, Ranjan P, Yadav AK (2019) Study on assessment of slope stability and mixed disposal of overburden in voids of Singrauli Coalfield. Environ Qual Manage 28(3):131–139

Shirin S, Jamal A, Emmanouil C, Yadav AK (2021) Assessment of characteristics of acid mine drainage treated with fly ash. Appl Sci 11(9):3910

Shirin S, Jamal A, Emmanouil C, Singh VP, Yadav AK (2023) Assessment and characterization of waste material used as backfilling in an abandoned mine. Int J Coal Prep Util 43(8):1402–1410

Suhag R (2016) Overview of ground water in India. PRS Legisl Res 9504:12

Sundaram B, Feitz A, de Caritat P, Plazinska A, Brodie R, Coram J, Ransley T (2009) Groundwater sampling and analysis—a field guide. Geosci Aust Rec 27(95):104

Swain S, Taloor AK, Dhal L, Sahoo S, Al-Ansari N (2022) Impact of climate change on groundwater hydrology: a comprehensive review and current status of the Indian hydrogeology. Appl Water Sci 12(6):1–25

Team RC (2019) A language and environment for statistical computing. R Found Stat Comput 2

Too VK, Omuto CT, Biamah EK, Obiero JP (2022) Analysis of soil water retention characteristics (SWRC) models using R computer program. Model Earth Syst Environ 8(2):2581–2590

Tukey JW (1977) Exploratory data analysis, vol 2. Reading, MA

van der Gun J (2021) Groundwater resources sustainability. In: Global groundwater. Elsevier, pp 331–345

Verzani J (2011) Getting started with RStudio. O'Reilly Media, Inc

WHO G (2011) Guidelines for drinking-water quality. World Health Org 216:303–304

Wickham H (2010) A layered grammar of graphics. J Comput Graph Stat 19(1):3–28

Wilkinson L (2012) The grammar of graphics. In: Handbook of computational statistics. Springer, pp 375–414

Yadav AK, Sahoo SK, Mahapatra S, Kumar AV, Pandey G, Lenka P, Tripathi RM (2014) Concentrations of uranium in drinking water and cumulative, age-dependent radiation doses in four districts of Uttar Pradesh, India. Toxicol Environ Chem 96(2):192–200

Chapter 13
Studies on the Level of Fluoride in Ground Water, Particularly in Rural Regions Located in the Krishnagiri District, Tamil Nadu, India

N. Gowrappan and D. Manivelu

Abstract The current research investigation aims to illustrate the significant impact that groundwater plays in rural as well as urban environments. Excessive levels of fluorine in household groundwater causes serious health problems in many places of the world. For drinking and irrigation, groundwater is essential for more than 90 of the inhabitants of rural regions. Millions of people around the world face a risk of health problems due to excessive levels of fluoride in groundwater. Depending on the mineral composition of the rocks and minerals that occur naturally that carry it at particular depths, fluoride has been documented that fluoride is found mainly in groundwater in greater quantities. High levels of fluorine increase the incidence of both skeletal and fluorosis of the teeth. The purpose of this study was to determine the fluoride content in a few chosen rural places in the Krishnagiri district. Groundwater samples have been taken across 25 villages in the district of Krishnagiri. The fluoride level of groundwater ranged from 0.1 to 14.8 mg/l. Among those, the fluoride concentration was considerably greater than 10.8 mg/l in eighteen different localities. According to the research findings, fluoride concentrations were above acceptable in the majority of areas. The excessive amount of fluoride in the Krishnagiri district has been caused by the deterioration of rocks and the absorption of groundwater. Therefore, to provide fluoride-free water in Krishnagiri, it has become essential that people learn about the causes of fluorosis and support the capture of rainwater and the decontamination processes.

Keyword Groundwater · Fluoride · Fluorosis

N. Gowrappan · D. Manivelu (✉)
PG and Research Department of Zoology, Government Arts College for Men, Krishnagiri 635001, Tamil Nadu, India
e-mail: prof.dmani@gmail.com

A. K. Yadav et al. (eds.), *Integrated Management of Water Resources in India: A Computational Approach*, Water Science and Technology Library 129,
https://doi.org/10.1007/978-3-031-62079-9_13

13.1 Introduction

Water is a vital natural resource for sustaining life and the environment, but its overuse has caused in declining water quality in recent decades. When the primary goal is sustainable development that keeps humanity at the center of gravity, water quality is a crucial parameter that needs to be studied. All living things depend on water as an amazing gift from nature to survive. The chemical makeup of the surface and subsurface determines whether or not water is suitable for use in household, agricultural, and industrial settings. Approximately a billion people still lack knowledge about hygienically safe drinking water, despite progress in almost every part of the world. Safe drinking water is essential for people and other living beings. Water is an essential natural resource that has a unique place in life (Bhat 2010, Gupta et al. 2024). The nation's groundwater, surface water, and other resources are essential for a variety of uses, including forests, fisheries, hydroelectric generation, agriculture, cattle production, industrial processes, and leisure activities (Saxena 2012).

The quality and biological resources of rivers, lakes, reservoirs, and other bodies of water are steadily declining as a result of poor management, rapid growth in industry and agriculture, and the disposal of untreated wastewater into agricultural fields, as well as other human and animal wastes (Elmaci et al. 2008; Yadav et al. 2014).

In India, among sixty-five million people drink groundwater contaminated with fluoride, as there are few enough drinkable water resources, particularly in areas with hard rock. Drinking water accounts for most of the daily intake of fluoride, however it can also enter its system through meals, water, industrial exposure, drugs, cosmetics, and other sources. With a frequency of occurrence of 17, it makes up between 0.06% and 0.09% of the crust of the earth. A tiny amount of fluoride is necessary for proper mineralization of bone and the synthesis of tooth enamel. Other factors that influence the harmful effects of fluoride include age, sex, occupation, growth, and bone remodeling of bone in relation to the Ca, Mg, P, and Al content of food. High levels of fluoride in foods and vegetables, including sorghum, ragi, and bajra, are found in fluoride-endemic locations, aggravated by the disease (Kodalirama and Sasikira 2013, Patel et al., 2022).

To create a groundwater development strategy, it is necessary to comprehend how groundwater levels fluctuate in relation to artificial or natural recharge in both space and time. Rainfall is the main source of groundwater replenishment and plays an significant role in the water cycle. However, because each location in India has a unique physiographic and climatic context, the overall pattern of rainfall also differs (Sarkar et al. 2009). India is one of the nations where hydrofluorosis, which affects 18 of the 28 states and 8 union territories that make up the nation, is a serious public health issue. Ten out of 38 districts in Tamil Nadu, a state in the South, have been impacted by fluorosis. Consumption of excessive fluoride is the main cause of Fluorosis, primarily through contaminated drinking water (Kumar et al. 2024, Kumari et al., 2024). In many regions of the world, there is considerable concern about the high concentration of fluoride found in surface and groundwater.

A correlation has been established between fluoride, sodium, and calcium (Dibal and Lar 2005, Shirin & Yadav, 2014).

According to Thivya et al. (2017), the lesser EC concentration is where the highest fluoride ion concentration is found, and the isotopic investigation indicates that fluoride is naturally occurring. Using a stratified random sample approach, a cross-sectional study was conducted in 13 chosen villages in five consecutive North Western districts of Tamil Nadu. Vellore, Dharmapuri, Krishnagiri, Salem, and Erode, to determine the prevalence of fluorosis. 8700 people in all, including 1745 children aged 5–14, had their dental mottling analyzes from 2800 households. According to the study, the primary supply of drinking water in villages was their communal wells. In the urban areas of Dharmapuri (24%), Krishnagiri, which (36%), and Salem (33%), the entire population had an elevated incidence of dental mottling (DM), with mean fluoride levels of 2.2, 2.7, and 1.2 ppm, respectively. Skeletal abnormalities were less common. In the three areas mentioned above, the percentage of DM cases in children aged 5 to 14 remained high at 43%, 53% and 42%, respectively.

The amount of fluoride in the water used for drinking and the degree to which residents in the Dharmapuri district are exposed to varying levels of fluoride contamination. Most people living in these areas are affected by dental and skeletal fluorosis, which includes persistent discomfort in the joints, osteosclerosis of the pelvis and spinal column, and mottling of the teeth. According to this study, water in a small number of villages has excessive levels of fluoride and hardness. Therefore, drinking water from the ground needs to undergo appropriate processes, such as defluoridation and softening (Muniyan and Thirupathi 2018; Patel et al., 2024). Because surface waters in this part of Krishnagiri district are being depleted, the underground is the only source of potable water. Because of this, people who use groundwater to drink put their health at risk. According to this study, the average fluoride concentration is 1.79 ppm, with variations between 0.37 and 3.47 ppm. The following are the areas' predominant cations and anions, while high levels of fluoride are discovered to be caused by ion leaching from biotite-rich rock segregations, especially shear zones (Shanmugasundaram et al. 2015).

Fluorosis is important for public health in most villages, according to an evaluation of the Community-wide Assessment for Dental Fluorosis (CIDF), where > 42% of the population is considered to be in good health. The most successful future strategy for managing and avoiding fluorosis is to distribute fluoride-free water of the surface from a river to communities, as this is the only way to address the high prevalence of the disease. Using isopleths mapping, Karthikeyan and Shunmugasundarraj (2000) documented the dependence of fluoride on the quality of water in the Krishnagiri block of Tamil Nadu, South India. The subject of groundwater flow has been the subject of several studies conducted throughout the world. The main health issues that fluoride causes in humans include mouth fluorosis, tooth mottling, skeletal Fluorosis, and bone deformation in both children and adults. Overfluoride has an impact on both plants and animals. The reduction of plant metabolism that results in necrosis, needle scrape, and tip burn diseases was another clear impact on agriculture (Susheela et al. 1993). According to the standards, maximum desirable level of fluoride in potable water is 0.6 mg/l while maximum level is 1.5 mg/l. This is a very alarming situation, as most people are unaware of the situation and take water without treatment. In the general population of the research region, chronic consumption of drinking water containing fluoride concentrations over permissible limits can result in dental and skeletal fluorosis (Srinivasulu et al. 2020).

13.1.1 Sources of Fluoride

The water compounds and fluoride emission from fertilizers and metal producers are the main sources of fluoride contaminants in drinking water. Dental fluorosis is caused by an excessive intake of fluoride as a result of the use of toothpaste that contains fluoride insufficiently. Over thirty percent of the toothpaste used for each tooth brushing by children under the age of five is swallowed (Villene 2000). Due to the widespread availability of numerous sources of fluoride, fluoride treatment is thought to play a significant role in the avoidance and treatment of dental caries. Fluoride has been shown to decrease dental caries and increase the frequency of dental fluorosis in both populations with and without fluoridated drinking water (Tabari et al. 2000).

13.1.2 Fluoride Content Present in Tamil Nadu

The groundwater of Krishnagiri, Dharmapuri, Thiruvannamalai, and Salem districts in Tamil Nadu had the greatest concentration of fluoride, closely followed by those of Coimbatore, Madurai, Trichy, Dindukal, and Chidambaram Districts. Table 13.1 shows the fluoride level in Tamil Nadu by a few districts.

Table 13.1 Fluoride content in bore well water

S. No.	Districts	Status of fluoride
1	Krishnagiri, Dharmapuri, Salem, Thiruvannamalai	Severe
2	Coimbatore, Madurai, Trichy, Dindukal, Chidambaram	Moderate
3	Thirunelveli, Pudukottai, Vellore, Ramnad	Less

13.2 Materials and Methods

13.2.1 Location of the Study Area

The study area is located in Tamil Nadu, India's Krishnagiri district, and a span around 20 square km. Bangaluru is 100 km away from it. The South Indian states of Karnataka, Andhra Pradesh, and Tamil Nadu come together in the Krishnagiri district. The district is located in Tamil Nadu's northwest region. One of Tamil Nadu's most significant districts is Krishnagiri. The study region, which spans an area of roughly 5143 km^2, is located within North latitudes 12° 16′ N and 12° 88′ N and East longitudes 77° 50′ E and 78° 55′ E. The administrative divisions of the Krishnagiri district consist of 6 Taluks, 11 blocks and 626 villages. The district is drained by the main river Thenpennaiyar. In the district, wells, tanks, canals, and dug are the main irrigation sources. The Uthangarai block has the highest rate of dug well irrigation, followed by Kaveripattinam. The Taluks of Krishnagiri and Kaveripattinam have the highest levels of tank and canal irrigation, respectively.

13.2.2 Sampling Procedure

In Krishnagiri District, 25 groundwater samples have been collected from hand pumps, bore wells, and open wells to evaluate the quality of the groundwater. Samples were taken using two-liter sterile polythene bottles, and standard procedures were followed for the analysis of acidity, electricity conductivity of electricity (Ec), nitrate (NO_3) and fluoride (F) (APHA, 1998). The Bureau of Indian Criteria (2009) and drinking water quality criteria were followed in the evaluation of the results. A digital pH meter (Model 802 Systronics) was used to test the pH, a conductivity meter (Model 304 Systronics) was used to measure the EC, a spectrophotometer was used to estimate the nitrate, and an Orion ion detector with a fluorine selective electrode was used to detect the fluoride concentration.

13.2.3 Collection of Water Samples

The purpose of this study was to evaluate the fluoride concentration of drinking water samples taken between March 2023 and November 2023 from a variety of sampling points. Using a routine sampling approach, water samples were taken monthly around 25 bore well sampling areas (Hand Pumps) at depths ranging from 750 to 1400 ft. Table 13.2 shows the sample that was gathered from the Krishnagiri union and district in Tamil Nadu.

Table 13.2 Fluoride concentration in bore water in twenty five zones of Krishnagiri district

S. No.	Panchayat	Village	Houses	Population	Purpose
1	Pachakanapalli	Bolparaikottai	40	182	Drinking
2	Pachakanapalli	Karathappanagar	15	118	Drinking
3	Pachakanapalli	Mandhrigoundankottai	35	188	Drinking
4	Periyamuthur	Solakottai	50	225	Drinking
5	Sokkadi	Sokkadi	345	1642	Drinking
6	Gangalari	Moongilpudur	10	68	Drinking
7	Gangalari	Marigampalli	25	120	Drinking
8	Moramadugu	Moramadugu	427	1800	Drinking
9	Moramadugu	Gandhi nagar	10	49	Drinking
10	Moramadugu	Ramarkoil	56	322	Drinking
11	Bellarampalli	Venkelikanapalli	40	256	Drinking
12	Sokkadi	Chinna sokkadi	26	86	Drinking
13	Sokkadi	Nadar thottam	25	187	Drinking
14	Sokkadi	Boma kottai	15	78	Drinking
15	Sokkadi	Nagappan kottai	54	187	Drinking
16	Sokkadi	Dhuduganahalli	156	489	Drinking
17	Sokkadi	Kovil kottai	54	190	Drinking
18	Sokkadi	Thimmarayanahalli	44	168	Drinking
19	Sokkadi	Muthuran kottai	34	89	Drinking
20	Pachakanapalli	Rice mill	33	259	Drinking
21	Moramadugu	Barathi nagar	20	59	Drinking
22	Moramadugu	Usathalahalli	83	300	Drinking
23	Moramadugu	Kodugoor	43	123	Drinking
24	Alapatti	Alapatti	212	500	Drinking
25	Alapatti	Poovathi	46	167	Drinking

13.3 Result and Discussion

Twenty-five communities were chosen at random for the sampling. The samples were taken from wells that were often known as public hand pumps. The physical characteristics revealed that they are odorless and colourless by nature. There are variations in the area's groundwater fluorine concentration. This could be due to variations in the availability and amount of fluorine-containing minerals in flowing water as well as weathering and leaching processes.

There is a range of 0.2–2.8 mg per milliliter of fluoride. The village of Sokkadi recorded the highest content of 2.8 mg/l, while the villages of Bomakottai and Kovilkottai, while Gandhi Nagar samples had the lowest concentration of 0.4 mg/l. The maximum desired limit for Indian Drinking Water Standards is IS: 10500, 1993 (i.e., 1.0 mg/l), although seven sample sites were found to be above the allowable range. The results of a nine-month sample were displayed in Tables 13.2 and 13.3 and Figs. 13.1 and 13.2.

According to Hussain et al. (2013) offered five categories that showed the distribution of fluoride levels among all habitations.

- Category-1 (Green): Less than 1.0 mg/l of fluoride
- Category-2 (Yellow): fluoride concentrations ranging from 1.0 to 1.5 mg/l
- Category-3 (Red): Fluoride content in 1.5 mg/l to 3.0 mg/l
- Category-4 (Brown): Fluoride content in 3.0 mg/l to 5.0 mg/l
- Category-5 (Black): Fluoride content greater than 5.0 mg/l.

Elevated fluoride levels in drinking water, especially in dry and semi-arid areas, can lead to serious health problems. This might be due to the climate that is common there, which makes individuals drink more water than they would in temperate places. When establishing national or local fluoride criteria or assessing the potential health effects of fluoride exposure, it is crucial to take into account the average water intake per day of the target population as well as fluoride intake from other sources (such as food and air). Setting an established or local recommendation at an amount less than 1.5 mg/l would be appropriate when the intakes will probably reach or beyond 6 mg/day.

13.3.1 Soils

Many different factors influence the amount of fluoride present in both surface and subsurface fluids and parent fluoride minerals' primary solubility and availability in the waters they are in touch with.

13.3.2 Routes of Fluoride Intake

Fluorosis can be extremely abundant in some food items, beverages, and cosmetics, even if fluoride can be consumed through drinking water. However, drinking water has been found to be the most significant source of fluoride exposure. Drinking water has been estimated to account for 60% of the overall amount of fluoride consumed. In particular in children, low doses protect against tooth cavities. Up to 2 mg/l of water to consume has a higher concentration of this protective action. It takes about 0.6 mg/l of fluorine in drinking water as the minimum concentration to have the desired effect. Series 1 to 9 (Mar'2023 to Nov'2023) show the levels of fluorine ion concentrations in the groundwater of the research area. Throughout the study period, there were fluctuations in total fluoride concentration values. At site 5 (Sokkadi), the fluoride content varied between 2.4 and 2.8 mg/l, with May, June and july is having

Table 13.3 Fluoride content present in drinking water from March'2023 to November'2023 from in and around 25 villages (mg/l)

Name of Villages	Mar' 2023	Apr'2023	May'2023	Jun'2023	Jul'2023	Aug'2023	Sep'2023	Oct'2023	Nov'2023
Bolparai Kottai	1.3	1.6	1.8	1.8	1.7	1.5	1.4	1.4	1.4
Karathappa Nagar	0.9	1.0	1.0	1.0	0.9	0.8	0.7	0.7	0.7
Mandhrigoundan Kottai	1.0	1.1	1.0	0.9	0.8	0.8	0.8	1.0	1.0
Solakottai	1.2	1.2	1.0	1.1	1.2	1.0	0.9	0.6	0.4
Sokkadi	2.5	2.5	2.8	2.8	2.8	2.5	2.4	2.6	2.5
Moongilpudur	1.1	1.0	1.0	0.9	0.8	0.8	0.7	0.5	0.7
Marigampalli	1.7	1.5	1.5	1.4	1.4	1.0	1.0	0.9	0.8
Moramadugu	1.2	1.2	1.3	1.1	0.9	0.8	0.9	0.8	0.7
Gandhi Nagar	0.5	0.3	0.4	0.4	0.5	0.9	0.7	0.8	0.9
Ramarkoil	1.0	1.2	1.4	1.1	1.2	1.3	1.4	1.1	1.0
Venkelikanapalli	2.4	2.4	2.5	2.4	2.4	2.3	2.2	2.2	2.1
Chinna Sokkadi	0.8	1.2	1.2	1.2	1.1	1.2	1.1	1.0	1.1
Nadar Thottam	1.4	1.3	1.1	1.2	1.3	1.2	1.1	1.1	0.9
Boma Kottai	0.9	0.2	0.6	0.7	0.8	0.6	0.8	0.6	0.4
Nagappan Kottai	1.2	0.6	1.1	0.8	0.6	0.8	0.9	0.6	0.6
Dhuduganahalli	1.5	1.4	1.5	1.4	1.3	1.2	1.5	1.5	1.2
Kovil Kottai	1.0	0.2	0.8	1.1	0.6	1.2	1.1	1.2	0.8
Thimmarayanahalli	0.5	0.6	0.4	0.6	0.4	0.8	0.9	0.9	0.9
Muthuran Kottai	1.2	1.2	1.2	1.1	1.2	1.0	1.1	1.0	1.0
Rice Mill	1.4	1.1	1.2	1.4	1.4	1.6	1.2	1.4	1.2
Barathi Nagar	1.2	1.3	1.4	1.1	0.9	0.8	0.6	0.5	0.8
Usathalahalli	1.6	1.5	1.4	1.6	1.8	1.7	1.7	1.8	1.7

(continued)

Table 13.3 (continued)

Name of Villages	Mar' 2023	Apr'2023	May'2023	Jun'2023	Jul'2023	Aug'2023	Sep'2023	Oct'2023	Nov'2023
Kodugoor	2.1	2.1	2.3	2.4	2.2	2.4	2.2	2.1	2.0
Alapatti	0.4	0.8	0.6	0.7	0.8	0.5	0.6	0.4	0.6
Poovathi	0.7	0.7	0.5	1.0	1.2	1.3	1.1	1.2	1.4

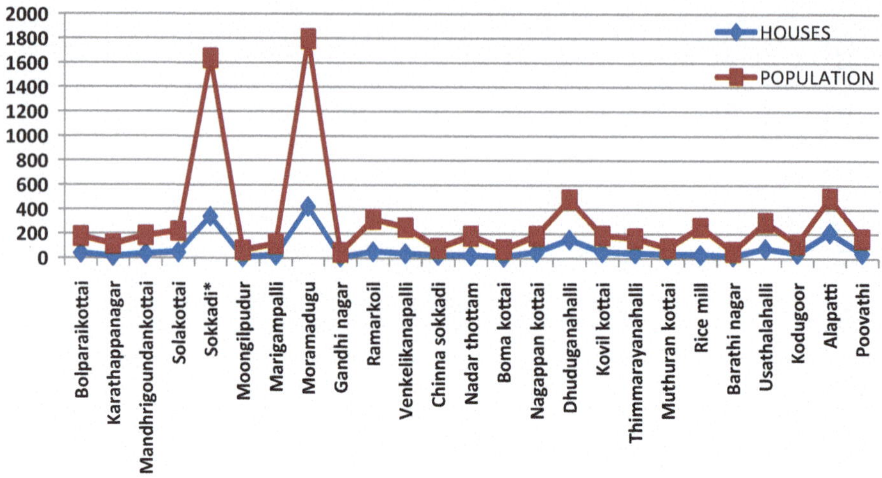

Fig. 13.1 Fluoride content present in drinking water from March'2023 to November'2023 from in and around 25 villages

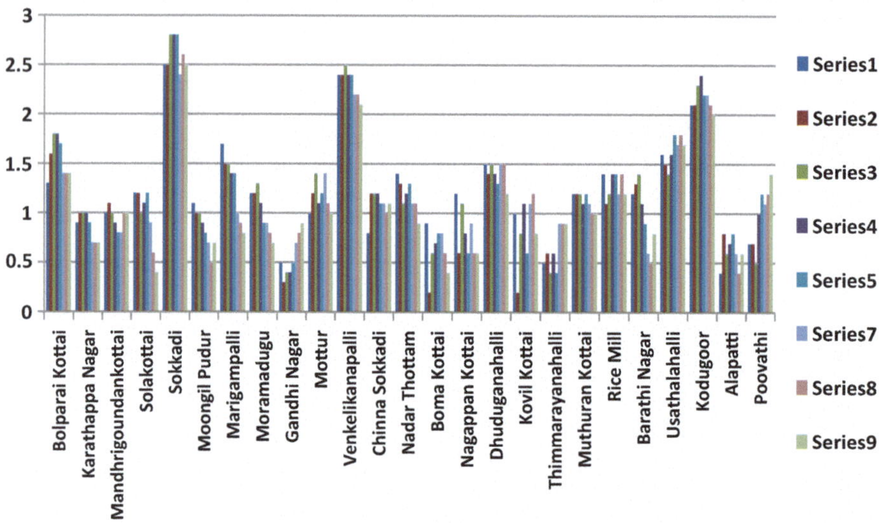

Fig. 13.2 Fluoride occurrences in the months (March'2023 to November'2023 – Series 1 to Series 9) of different study sites. *Note* The graph showed the highest and lowest levels of fluoride present at the sampling locations

the highest fluoride level (2.8 mg/l) and September having the lowest fluoride level (2.4 mg/l).

The large number of sites showed contradictory results in terms of fluoride concentration, which ranged from 0.2 to 2.8 mg/l, with some sites had concentrations much below the acceptable level. Although all samples from hand pumps and bore wells had fluoride concentrations below the allowable limit (< 1.5 mg/l), most of the water sample tests showed fluoride amounts that were adequate compared to the recommended permitted levels (< 1.5 mg/l). It was found that there was extremely high dental caries contamination in each of these distinct settlements. In this study, the health hazards associated with reduced fluoride concentrations in drinking water are also examined. Residents of the study area saw deformities in their mottling teeth, which is indicative of the effects of fluoride concentration.

For the majority of the population, especially young people, high concentration causes skeletal and dental fluorosis and dental caries, while low concentration protects against these conditions. According to 2.0 mg per liter of water for consumption was the concentration at which this impact increases. At the very least, about 0.6 mg/l fluoride must be present in drinking water to have the desired effect. An increasingly important component in the fight against tooth caries is fluorine. Fluoride has two opposing effects on health: It is both good and harmful. Drinking water is the primary way that fluoride enters the human body (Singh et al. 2007). According to the results of the study, the fluoride content of the water samples obtained from the wells varies between 0.2 and 2.8 mg/l at various sampling points. Dental cavities in both adults and children are significantly reduced when fluoride is added to the water. The amount of fluoride in drinking water is inversely correlated with the severity of periodontal disease. This has frequently been associated with fluorosis in the skeleton and teeth caused by drinking water (Pandey and Pandey 2011; Sudhir et al. 2012). In water, it has traditionally been preferred if the permitted level is less than 0.6 mg/l. If the limit exceeds the threshold limitations, the source of drinking water cannot be thrown away as is nevertheless; some health considerations must be accompanied to enhance the quality of water.

The fluoride concentration was determined to be considerably above the minimum permissible level in the current study, exceeding the acceptable limit. In addition to minerals found in rocks that can weather and add to the fluorine content of ground water, phosphoric fertilizers used in agriculture and industrial processes can also increase the concentration of fluoride in groundwater (Bhupinder 2011). When groundwater has a low concentration of fluoride, fluoridation may be advised (Khadsan 2007). Fluorosis was thought to be a dental-specific issue. However, it seems to represent a serious threat to someone's health. It may have a major impact on bones, and their well-known effect includes joint and muscle pain. A person is not only physically affected by it, but also become socially and culturally disabled (Bharati 2013).

The health risk assessment indicates that children are especially vulnerable to fluoride exposure. Therefore, to rate the water as being of excellent quality, additional treatments must be started and government agencies must choose and supply suitable groundwater to the general population (Kalaivanan Kaliyan et al. 2022). However, dental and bone disorders that interfere with daily living are the main health risks associated with fluoride. Compared to bore well sources, the content of open bore sources was generally better (Chicas et al. 2022). The government's main responsibility is to provide the public with clean, safe drinking water, lower the fluoride concentration of that water using modern technologies, and raise public awareness through initiatives like health camps and documentaries. This study highlights the issue of fluoride in the Krishnagiri District.

13.3.3 Mitigation of the Problem for a Sustainable Future

It is true that drinking water with high levels is causing major health problems. Consequently, the following mitigation actions are recommended to address this issue and raise people's awareness of the negative consequences of quality factors in rural areas. The proper awareness of the awareness to use the defluoridation methods of drinking water in endemic areas at the community level in those particular areas by the methods mentioned below.

- Steer clear of the open defecation habit.
- Avoid directly dumping sewage into groundwater directly.
- Reverse osmosis, ion exchange, and electro-dialysis facilities should be installed in affected areas.
- Steer clear of the careless application of nitrogen fertilizers and encourage the use of biofertilizers and biopesticides.
- Building wells and tube wells in regions with low levels of fluoride and nitrate.
- The use of cosmetics, high-end products, and edibles containing fluoride is prohibited.
- Encouragement of breastfeeding for a baby at least six months of age. Fluoride intrusion is reduced by the Aquifer Protection Policy (APP), which was implemented to prevent excessive groundwater extraction.

13.4 Conclusion

The aforementioned findings make it abundantly obvious that untreated effluent water has contaminated the groundwater surrounding the wells. This study, which examines the fluoride content of the well water, found that the Krishnagiri district has a serious fluoride problem. The presence of a high fluoride rock salt system is one of the favorable factors contributing to the growth of fluorine in ground water. The evaluation of the study's data leads to the conclusion that the Krishnagiri district ground water is contaminated with varying levels of fluoride, which could cause dental Fluorosis. Hence, the researchers should concentrate on defluoridation methods in real situations and on raising community awareness of safe drinking water practices in the endemic areas.

Acknowledgements The authors thank our greatest parents, prestigious institute and other dedicated supporters who are involved in this study for their direct and indirect contributions.

References

Ahmad K, Shah I.A, Ali S, Khan M.T, Qureshi M.B.A, Shah Sha, Ali A, Rashid W, Gul H.N. (2021) Synthesis and evaluation of Ca-doped ferrihydrite as a novel adsorbent for the efficient removal of fluoride. *Environ Sci Pollut Res Int. 2022 Jan;29(4)*:6375–6388. https://doi.org/10.1007/s11356-021-16105-5. Epub 2021 Aug 27. PMID: 34449025.

APHA (1998) Standard methods for the examination of water and wastewater. APHA, Washington, pp 1–46

Arumugam T, Kunhikannan S, Radhakrishnan P (2019) Assessment of fluoride hazard in groundwater of Palghat District, Kerala: a GIS approach. Int J Environ Pollut 66(1/2/3). https://doi.org/10.1504/ijep.2019.104533

Bharati KT (2013) Potable groundwater quality in jorve village of sangamner, Mahrashtra, India: focus on fluoride. Rasa J Chem 6(1):20–23. http://www.rasayanjournal.co.in

Bhat S (2010) Natural resources conservation law. SAGE Publications India, p 564

Bhupinder S (2011) Assessment of groundwater quality with respect to fluoride. Univ J Env Res Tech 1:45–50

Bureau of Indian Standards (2006) Manak bhavan, 9-Bahadur shah, New Delhi

Chicas SD, K. Omine , , M. Prabhakaran , T.G. Sunitha , V. Sivasankar (2022) High fluoride in groundwater and associated non-carcinogenic risks at Tiruvannamalai region in Tamil Nadu, India, Published by Elsevier Inc. Ecotoxicol Environ Saf 233:113335, 1–12. https://doi.org/10.1016/j.ecoenv.2022.113335

Dasaiaha S, Rama Mohan Kurakalva B, Pavan Kumar Pindi A (2020). Data on fluoride concentration profile in groundwater of rural habitats in Mahabubnagar district, Telangana, India. Science Direct, Data in Brief 32:106165, 1–6. https://doi.org/10.1016/j.dib.2020.106165

Dhiman SD, Keshari AK (2001) Genesis of fluoride contamination in Western Indian Aquifers. In: *Proc. Int. Conf. On Future Groundwater Resources at Risk, FGR 01, June 25–27, Lisbon, Portugal, Themes 1–3*, 1–8. https://doi.org/10.31018/jans.v4i1.234

Dibal HU, Lar UA (2005) Preliminary survey of fluoride concentrations in the groundwater of Kaltungo area, Gombe State, northeastern Nigeria. J Environ Sci 9(2):41–52

Elmaci A, Topac FO, Ozengin N, Teksoy A, KurtogluS,Baskaya HS.(2008). Evaluation of physical,chemical and microbiological properties of lake Ulubat, Turkey. *J Env Biol.29(2):*205–10. www.jeb.co.in

Gupta, S., Mishra, N., Kumar, A., Yadav, A.K. (2023). Fluoride Pollution in Subsurface Water: Challenges and Opportunities. In: Yadav, A.K., Shirin, S., Singh, V.P. (eds) Advanced Treatment Technologies for Fluoride Removal in Water. Water Science and Technology Library, vol 125. Springer, Cham. https://doi.org/10.1007/978-3-031-38845-3_2

Jiawei Han and Micheline Kamber, (2008) "Data mining concepts and Techniques",*Second Edition, Morgan Kaufmann Publishers, second edition.* www.mkp.comorwww.books.elsevier.com

Hussain J, Husain I, Arif M (2013) Fluoride contamination in groundwater of central Rajasthan, India and its toxicity in rural habitants. Toxicol Environ Chem 95(6):1048–1055. https://doi.org/10.1080/02772248.2013.832545

Karthikeyan G, Shunmugasundarraj A (2000) Research Report-Fluoride, 33: 121–127. E-mail: gricc@md3.vsnl.net.in

Keller E.A. (1979). Environmental geology, *Charles&Merril, Columbus, OH*, 548.

Khadsan RE (2007). Analysis of fluoride in borewells water of Chikhli City, District Buldana (M.S.). *Journal of Ultra Chemistry 3(1)* 93–95. http://journalofchemistry.org/paper/798.

Kodalirama K, Kiran S (2013) Fluorosis-an update. Int J Res Pharm Biomed Sci. 4(4):1084–1088

Kumar, A., Chava, R., Gupta, S., Shirin, S., Jamal, A., Yadav, A.K. (2023). Environmental and Health Effects of Fluoride Contamination and Treatment of Wastewater Using Various Technologies. In: Yadav, A.K., Shirin, S., Singh, V.P. (eds) Advanced Treatment Technologies for Fluoride Removal in Water. Water Science and Technology Library, vol 125. Springer, Cham. https://doi.org/10.1007/978-3-031-38845-3_18

Kumari, S., Dhankhar, H., Abrol, V., Yadav, A.K. (2023). Bioaccumulation of Fluoride Toxicity in Plants and Its Effects on Plants and Techniques for Its Removal. In: Yadav, A.K., Shirin, S., Singh, V.P. (eds) Advanced Treatment Technologies for Fluoride Removal in Water. Water Science and Technology Library, vol 125. Springer, Cham. https://doi.org/10.1007/978-3-031-38845-3_15

Malago J (2017) Fluoride levels in surface and groundwater in Africa: a review. Am J Water Sci Eng 3(1):1–17. https://doi.org/10.11648/j.ajwse.20170301.11

National Research Council, (2006). Fluoride in Drinking Water. https://doi.org/10.17226/11571.

National Water Policy, (2000).*Ministry of Water Resources.*

Pandey J, Pandey U (2011) Fluoride contamination and fluorosis in rural community in the vicinity of a phosphate fertilizer factoryin India. Bull Environ ContamToxicol. 87(3):245–249. https://doi.org/10.1007/s00128-011-0344-6

Patel D, Pamidimukkala P, Chakraborty D, Yadav A (2022) Bharuch District, Gujarat, India: factor analysis and geographical distribution of water quality characteristics. Environmental Nanotechnology, Monitoring & Management 18:100732. https://doi.org/10.1016/j.enmm.2022.100732

Patel, D., Kulwant, M., Shirin, S., Varshney, R., Pandey, G., Yadav, A.K. (2023). Fluoride Removal from Aqueous Solution Using Iron-Based Materials: Preparation, Characterization, and Applications. In: Yadav, A.K., Shirin, S., Singh, V.P. (eds) Advanced Treatment Technologies for Fluoride Removal in Water. Water Science and Technology Library, vol 125. Springer, Cham. https://doi.org/10.1007/978-3-031-38845-3_4

Ramesh K, Bhuvana Jagadeeswari P (2012). Hydro chemical Characteristics of Ground Water for Domestic and Irrigation Purpose In Periyakulam Taluk of Theni District, Tamil Nadu, *I. Res. J. Environment Sci.Vol. 1(1),* 19–27. www.isca.in

Saxena M (2012) Water quality and trophic status of Raipur reservoir in Gwalior, Madhya Pradesh. J Nat Sci Res 2(8):82–97

Sendesh KK, Ramasubramanian V (2011) Assessment of Fluoride Contamination In Groundwater Using GIS, Dharmapuri District, Tamilnadu, India., *International Journal of Engineering Science and Technology*. Vol. 3 No. 2 :1077–1085.

Shanmugasundharam A, Saubhagya Ranjan Mahapatra, G.Kalpana and M.Jayaprakash(2015). High Fluoride concentration in the Groundwater of Uttangarai Taluk, Krishnagiri District, Tamil Nadu, *EnviroGeoChemica Acta, 2(1)*: 1- 9. https://www.researchgate.net/publication/279195298.

Shirin S, Yadav A. K. Physico Chemical Analysis of Municipal Wastewater Discharge in Ganga River,Haridwar District of Uttarakhand, India. Curr World Environ 2014;9 (2) https://doi.org/10.12944/CWE.9.2.39

Singh B, Gaur S, Garg VK (2007) Fluoride in drinking water and human Urine in Southern Haryana, India. J Hazard Mater 144:147–151. https://doi.org/10.1016/j.jhazmat.2006.10.010

Sudhir KM, Suresh S, Prashant GM, Reddy VV, Shafiulla M, Chandu GN (2012) Distribution patterns of enamel fluorosis in permanentdentition. Oral Health Prev Dent 10(2):167–174

Susheela AK (1991) Prevention and control of fluorosis, Technical information training cum awareness camp for doctors, public health engineers and other officers. Published by National Techonology Mission of drinking water, New Delhi

Tabari ED, Ellwood R, Rugg-Gunn AJ, Evans DJ, Davies RM (2000) Dental fluorosis in permanent incisor teeth in relation to water fluoridation, social deprivation and toothpaste use in infancy. Br Dent J 189(4):216–220. https://doi.org/10.1038/sj.bdj.4800726

Thirupathi S and Muniyan M (2018) Evaluating fluoride contamination in Ground Water samples of Dharmapuri District, Tamilnadu, India. *International Journal of Recent Scientific Research Vol. 9, Issue, 1(G)*, pp. 23306–23314. https://doi.org/10.24327/ijrsr.2018.0901.1442.

Thivya, Chidambaram S, Rao MS, Thilagavathi R, Prasanna MV, Manikandan (2017) Assessment of fluoride contaminations in groundwater of hard rock aquifers in Madurai district, Tamil Nadu (India), Appl Water Sci 7:1011–1023. https://doi.org/10.1007/s13201-015-0312-0

Tiwari KK, Raghav R, Pandey R (2023) Recent advancements in fluoride impact on human health: a critical review. Environ Sustain Indic 20:1–11. https://doi.org/10.1016/j.indic.2023.100305

Venkatayogi S (2015) Geochemistry of fluoride bearing groundwater in parts of Telangana State, India. Journal of Water Resource and Hydraulic Engineering 4(4):380–387. https://doi.org/10.5963/jwrhe0404009

Vennila S (2023) Fluoride Contamination in Drinking Water in Palacode Region, Tamil Nadu. Int J Res Chem Envion 2(1):116–123. http://ijrce.org

Villena R.S. (2000). An investigation of the transverse technique of dentifrice application to reduce the amount of fluoride dentifrice for young children. *Pediatr Dent.22(4):*312–7. https://pubmed.ncbi.nlm.nih.gov/10969439.

Yadav AK, Sahoo SK, Mahapatra S, Kumar AV, Pandey G, Lenka P, Tripathi RM (2014) Concentrations of uranium in drinking water and cumulative, age-dependent radiation doses in four districts of Uttar Pradesh, India. Toxicol Environ Chem 96(2):192–200. https://doi.org/10.1080/02772248.2014.934247

Chapter 14
Evaluation of Water Quality Using Soft Computing Techniques

Kajal Chaudhary and Rashmi Bhardwaj

Abstract Water plays a very important role in our daily lives as it is used in various purposes. The quality of water is a concern due to its usage for drinking, cleaning, and other daily purposes. There are different parameters, such as pH, temperature, dissolved oxygen, salinity, nitrogen, phosphorous, etc., on which the quality of water depends. Industrial processes, irrigation, chemical, and other wastes pollute the water resources and hence affect the water management services. Quality water is a basic need for our health and our lives, and it is required to protect our water resources and further better assessment of water quality. Different methods and techniques are used for assessing water quality. In this work, the different parameters are analyzed with respect to their importance and role in the assessment of water quality. Then different soft computing techniques such as neural networks and fuzzy interface systems are used to predict water quality indexes by taking the most affecting parameters as input to the network.

Keywords Water parameters · Water quality index · Correlation · Neural networks · Fuzzy interface

14.1 Introduction

Water is one of the most important sources for our existence on this earth. The main concern is the consumption of healthy and safe water, that is, the water consumed should be free of harmful chemicals and microbes that can cause severe diseases (Gaur et al. 2022). Different activities of industries, people, and agriculture are continuously polluting water resources (Singh et al. 2021). This water pollution is above the allowable limit in many areas that contain water bodies. Polluted water bodies degrade the quality of water that frightens our lives and imbalance the economic

K. Chaudhary · R. Bhardwaj (✉)
University School of Basic and Applied Sciences, Guru Gobind Singh Indraprastha University, Dwarka, Delhi 110078, India
e-mail: rashmib@ipu.ac.in

A. K. Yadav et al. (eds.), *Integrated Management of Water Resources in India: A Computational Approach*, Water Science and Technology Library 129,
https://doi.org/10.1007/978-3-031-62079-9_14

development and growth of our society (Gaur et al. 2022). Rivers are one of the main sources of water. They are a large stream of water flowing naturally to any lake, ocean, or different water bodies. Rivers are the main transporter of water and its nutrients to various parts of the world, as about 0.0001% of water is contained by them at any time (Kumar and Dua 2009).

The quality of water is determined by the nature of its physiochemical and biological parameters. The main idea of traditional methods for assessing water quality is based on analyses of parameter values, which are computed experimentally under the guidelines. These traditional methods help to identify the polluting sources of water but do not provide a complete analysis of temporal and spatial trends in the quality of water in various water bodies (Gupta et al. 2019). Educating people about the quality of water is very crucial before its use for various purposes. Water experts face a big problem while incorporating or interpreting the complexity of environmental data into understandable information, which can be beneficial to policy makers and also for common people (Gupta et al. 2019). Different researchers have carried out many studies to fill this gap between complex environmental data and user-friendly information. For productive planning and management of water resources, a proper assessment of these water resources is of primary concern (Gupta et al. 2019). Both anthropogenic (domestic chores, agricultural purposes, industrial activities) and natural (geology, geography of water bodies, precipitation, and atmosphere) activities determine the biological, chemical and physical composition of ground and surface water (Gaur et al. 2022). The process of precipitation helps any river in collecting its within water through the recharge of groundwater, surface runoff, and discharge of water stored in natural waterbodies (Kumar and Dua 2009). In past years, severe damages have been faced due to excess amount of fluorides in water bodies which directly impacts human health (damage to teeth and bones) and the environment. Therefore. it is required to limit the consumption of harmful fluoride by improving safety measures in industries so that flowing of fluoride compounds to the environment can be reduced (Kulwant et al. 2023). Kumar et al. (2023), studied the effects of fluoride on contamination of drinking water and environment. Different methods and technologies are also examined in this study for the treatment of fluorosis. In Patel et al. (2023), it is observed that fluoride content can be removed effectively from water bodies with the help of iron based materials.

In the late eighteenth century, world has encountered with many new indicators of pollution after the industrial revolution (Omer 2020). Based on origin, water can be classified into two categories: surface water and groundwater (Singh et al. 2021; Omer 2020). Both categories can get polluted due to industrial, agricultural, and household activities including various pollutants such as hazardous chemicals, fertilizers, heavy metals, oils, pesticides, etc. Uranium is found in groundwater, rivers, surface water, lakes, and seas because of its nature to get dissolved in water containing more oxygen. Uranium can be very toxic for humans on the basis of its solubility in drinking water. Yadav et al. (2014), has performed the research to estimate the amount of uranium in drinking water and further determination of absorbed uranium dose in different districts of Uttar Pradesh. Indirectly, rivers provide water to all parts of earth for different purposes and therefore become a centre point of concern for all environmental issues in society. Therefore, they are studied by many specialists, including engineers, geomorphologists, hydrologists, and geologists (Kumar and Dua 2009). Many environmental issues are faced by mining activities involving drilling, dumping, hauling, mineral handling processes, removing of topsoil, and overloaded transportation. These issues are responsible for soil, air, and water pollution, our biological communities also get disturbed by opencast mining (Shirin et al. 2023). Recently mined regions containing the pyritic substances in shafts used for mining and in the spoil piles has created one of the major pollutant, Acid mine drainage (AMD). AMD is a hazardous byproduct of chemical reactions takes place in industries (Shirin et al. 2021). Shirin et al. (2021), has proposed some reduction plans to control the disturbance created in ecosystem due to the pollutants from mining. The main and common method used for the assessment of water quality is the Water Quality Index (WQI). WQI is just a way to encapsulate a large dataset of water quality in a simplified manner for the further use by policy makers, management, and the public. It is very beneficial for categorizing water quality and hence classify the health status of different water reservoirs (Singh et al. 2021). Similar to other indices like air quality index (AQI), the WQI can inform us if the water quality of different water bodies can cause any possible threat to irrigation water for agriculture, livestock, drinking water supply, cleaning purposes, habitat for aquatic life, and other uses. The first WQI was introduced by Horton (1965), which is basically a method to compute a single value from different data sets and test results (Kumar and Dua 2009). Then, other researchers have also developed different WQI indices by rating the parameters of water quality. The WQI is a set of standard values used to measure any change in quality of the water at any station of river over time (Kumar and Dua 2009). The WQI provides a procedure to present a derived numerical expression that interprets a particular level of water quality (Miller et al. 1986). After Horton (1965), Brown et al. (1970) has proposed method for the computation of WQI. Using the concept of WQI, Kannel et al. (2007), analyzed the spatial and climatic changes of the Bagmati River. By incorporating nine physiological parameters of water quality, Debels et al. (2005) calculated a WQI for Chill'an river.

To address the non-linearity of water quality data, many soft computing techniques have been used by different workers. They are used to model water quality quickly and have the great ability to minimize the computation time and errors (Bhagat et al. 2019). The most commonly used soft computing techniques are adaptive neuro fuzzy interference system (ANFIS), random forest (RF), a generalized regression neural network (GRNN), Gaussian process (GP), M5P model tree, support vector machine (SVM) and artificial neural network (ANN) (Singh et al. 2021). In Gao et al. (2015), particle swarm optimization is combined with a back-propagation neuronal network to predict the quality of water. Balraj et al., have compared the ANN, GRNN, and ANFIS techniques to predict WQI in three different sub-water sheds in Iran. In Kumar and Dua 2009, the water quality of the Ravi River is evaluated using this WQI technique. Gupta et al. (2019), has also made an attempt to generate a WQI on the basis of all physiological and chemical parameters and ANN.

This chapter first introduces the physio-chemical and biological parameters. The working and basic architecture of some soft computing techniques are also presented.

Then, a research work is introduced, which analyses the water quality of the Cauvery River and Tamiraparani river using WQI. The research area considers only four stations of these rivers as follows: Pallipalayam, Sirumugai, Cheranmadevi, and Murapanadu. First, the correlation coefficient of different parameters with WQI is computed to figure out which parameters are mostly affecting the value of WQI. Then an ANFIS model is used to predict the water quality of the stations considered.

14.2 Water Quality Parameters

Basically, the parameters of water quality are divided into three categories as: physical, chemical, and biological parameters. Some of the required parameters (for research work) are briefly explained below:

1. Physical parameters

Turbidity - It is the amount of cloudiness present in water. It is used to check whether the light has power to pass through the water. It happens because of the presence of suspended matter and other particulate matter in the water. A nephelometric turbidimeter is used for the measurement of turbidity. Its unit is NTU, which means that 1 NTU is equal to 1 mg/L of silica in the suspension (Omer 2020). Normally, much less turbidity is observed in groundwater due to the penetration of water through the soil.

Solids - Water contains solids basically in the form of suspension or liquid (Omer 2020; Debels et al. 2005). A fiber filter of glass is used to detect these forms of solids. After passing through this filter, the suspended form of water remains, while the other form passes the filter along with water. After the filtered quantity of water is evaporated, the residual solids are known as total dissolved solids.

$$\text{Total solids} = \text{total dissolved solids (TDS)} + \text{total suspended solids (TSS)}.$$

Electrical conductivity - is used to measure the power of any solution for conducting electrical current through it. For any water solution, an increase in electrical conductivity (EC) is observed with an increase in concentration of ions. Therefore, this parameter informs about the appropriateness of water for agricultural practices. Its unit is μS/cm.

2. Chemical parameters

pH – it is the negative logarithm of the concentration of hydrogen ions in water (Omer 2020). It is dimensionless and used to determine the capability of any basic or acidic solution.

Dissolved oxygen (DO) - oxygen is soluble in water slightly. There is a variation in the amount of dissolved oxygen with temperature, salinity, and pressure of water. Water with dissolved oxygen does not directly affect our health, but water with less or no oxygen is not suitable for drinking.

Nitrogen - Four forms of nitrogen are found in water: ammonia, nitrite, nitrate, and organic nitrogen. The contaminated water with sewage contains ammonia and organic nitrogen, which result in nitrates and nitrites by microbes. There is a reduction in the ability of blood to control oxygen as a result of the reaction of haemoglobin with nitrate ions. This results in a disease called methemoglobinemia or blue baby (Omer 2020).

3. Biological parameters

A healthy water system is formed with the presence of some species in a balanced manner. To check the presence of pollutants in water, these organisms can be used on the basis of their allowance for any particular impurity. Microbes are present in wastewater in millions (per millilitre), they are harmful when this waste water comes from infected people (Omer 2020).

14.3 Methodology

14.3.1 Correlation

Correlation is used to represent any statistical relationship between two variables. It examines and measures the way and magnitude of connection between variables. It is also used to test whether the relationship between variables is statistically significant or not. If there is correlation, then it is linear and we can present this relationship with a straight line. The correlation coefficient, r, is used to measure the extent to which the variables are related. Consider two variables X and Y, if X changes in a particular direction, Y also changes in same direction (r is positive means positive correlation) but if Y changes in opposite direction (r is negative means negative correlation). Zero correlation means that there is no relationship between the variables under study. The correlation coefficient is computed as:

$$r = \frac{\sum_{i=1}^{n}(X_i - \overline{X})(Y_i - \overline{Y})}{\sum_{i=1}^{n}\sqrt{(X_i - \overline{X})^2}\sqrt{(Y_i - \overline{Y})^2}} \tag{14.1}$$

where, r is correlation coefficient between variables X and Y,

X_i is the ith observation of variable X, Y_i is the ith observation of variable Y,

$\overline{X}$ and $\overline{Y}$ are the arithmetic mean values of variable X and Y respectively.

14.3.2 Computation of WQI

This method of computing WQI is proposed by Brown et al. (1970), which is completely based on assigning particular weight factors to different parameters of water quality. The standard values (s_i) of the parameters are given in Table 14.1. The steps to calculate WQI are as follows:

Step 1: For each parameter, calculate the unit weight (w_i) factors using the formula given in Eq. (14.2).

$$w_i = \frac{c}{s_i}; i = 1,2, \ldots, n \quad (14.2)$$

where, $c = \frac{1}{\frac{1}{s_1}+\frac{1}{s_2}+\cdots+\frac{1}{s_n}}$

n is the total number of parameters,

s_i is the standard value of ith parameter.

Step 2: For each parameter, sub-index (Q_i) value is computed by using formula defined in Eq. (14.3).

$$Q_i = 100 \times \frac{(v_i - v_o)}{(s_i - v_o)} \quad (14.3)$$

Where, v_i is the observed value of ith parameter,

Table 14.1 Standard values of parameters (CPCB 1974; Gao et al. 2015)

Water quality parameters	Standard values (s_i)
pH	7.5
BOD	30
D.O	5
Electrical conductivity	500
Total Coliform	250
Total alkalinity	200
Chloride	250
COD	250
Ammonia	5
Turbidity	5
Calcium	75
Magnesium	50
Sulphate	250
Sodium	200
TDS	300
TSS	100
Potassium	10
Fluoride	1.9

All values in mg/L except pH (on a scale), electrical conductivity in µS/cm, turbidity in NTU, and total coliform in MPN/100 mL

v_o is the actual value of ith parameter in pure water and it is taken as zero for all parameters except for ph, $v_o = 7$.

Step 3: Overall, WQI is computed as defined in Eq. (14.4).

$$WQI = \frac{\sum w_i Q_i}{\sum w_i} \tag{14.4}$$

.

14.3.3 Adaptive Neuro-Fuzzy Interference System (ANFIS)

It is nothing more than a type of ANN that works on the principle of the Takagi–Sugeno fuzzy inference system (Sihag et al. 2019). By combining both ANN and fuzzy logic systems, this hybrid model can learn from data for the decision-making process and can also deal with incomplete data, continuously varying data or incorrect data. In fuzzy logic, membership functions are used to characterize the fuzziness in an effective manner. The truth of fuzzy logic is represented with the help of these functions. The architecture of ANFIS with eight inputs X_1, X_2, …, X_8, n-rules and one output Y, is presented in Fig. 14.1. This model consists of five layers (Horton 1965) and their working is as follows:

1st Layer: The first layer measures the degree of given inputs. The membership degrees are the outputs of this layer, which assures the membership function μ_{Akj} (X_k), A_{kj} are the non-linear parameters; j = 1,2,…, n; k = 1,2,…, 8.

2nd Layer: The product of the incoming values is the output of layer 2. The firing strength T_j of the jth rule as an output to each node is defined in Eq. (14.5).

$$T_j = \prod_{k=1}^{8} \mu_{A_{kj}}(X_k);\ \mathrm{j} = 1,2,\ldots,\mathrm{n} \tag{14.5}$$

3rd Layer: In the third layer, every node calculates the ratio of jth rule firing strength to the sum of all rules firing strengths, as explained in Eq. (14.6).

$$\overline{T_j} = \frac{T_j}{\sum_{j=1}^{n} T_j} \tag{14.6}$$

.

4th Layer: In the fourth layer, every node is an adaptive node with output as in Eq. (14.7).

$$\overline{T_j} X_k^{(j)} = \overline{T_j}(\theta_{j1}X_1 + \theta_{j2}X_2 + \cdots + \theta_{j8}X_8 + \theta_{j0}) \tag{14.7}$$

.

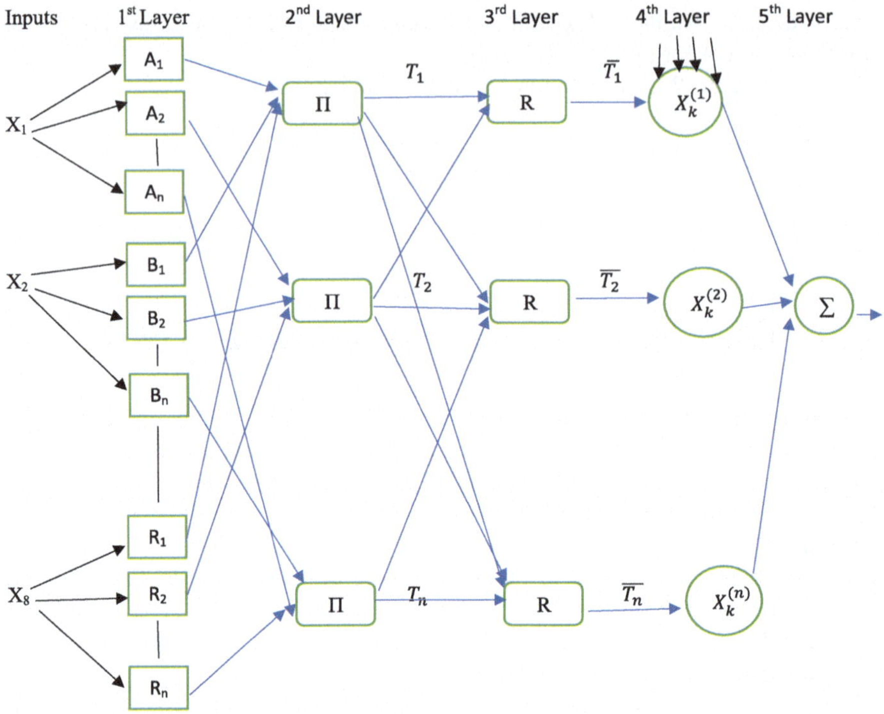

Fig. 14.1 Architecture of the ANFIS model (Tarno et al., 2018)

, where $X_k^{(j)} = \theta_{j0} + \sum_{k=1}^{8} \theta_{jk} X_k$ as consequent section if $X_k is A_{kj}$ as premise section; θ_{jk} and θ_{j0} are the linear parameters.

5th Layer: The fifth layer has fixed nodes that just sum up all the inputs and the final output is:

$$Y = \sum_{j=1}^{n} \overline{T_j}(\theta_{j1}X_1 + \theta_{j2}X_2 + \cdots + \theta_{j8}X_8 + \theta_{j0}) \tag{14.8}$$

.

14.3.4 Root Mean Square Error (RMSE)

It is used mostly to work on the fitness of any model. It calculates error terms and then takes the square root of mean of these terms. It is defined in Eq. (14.9).

RMSE

$$= \sqrt{\frac{1}{n}\sum_{i=1}^{n}(A_i - P_i)^2} \tag{14.9}$$

.

where P_i is the predicted value of ith parameter.
A_i is the actual value of ith parameter.

14.3.5 Mean Absolute Percentage Error (MAPE)

MAPE

$$= \frac{100}{n}\sum_{i=1}^{n}\left|\frac{A_i - P_i}{A_i}\right| \tag{14.10}$$

.

14.4 Research Work

14.4.1 Research Area

This work was carried out using quality data from the water at four stations on the Cauvery River and Tamiraparani river. These stations are: Pallipalayam and Sirumugai in Cauvery River, Murapanadu, and Cheranmadevi in the Tamiraparani River. The data of the water quality parameters were taken from the website of CPCB (Central Pollution Control Board) (CPCB 1974), from january 2018 to December 2019.

14.4.2 Data Description

Data were collected for different physio-chemical parameters including pH, electrical conductivity, turbidity, BOD, DO, total coliform, total alkalinity, chloride, COD, ammonia, calcium, magnesium, sulphate, sodium, TDS, TSS, fluoride, and potassium. After the computation of correlation coefficients (discussed in Sect. 14.4.3), the number of input variables for ANFIS model got reduced, due to low correlation with WQI. The 70% of the dataset was used to train the model and 30% was used for testing of model.

14.4.3 Correlation Plot

The plots were drawn between different physiological parameters (input variable) and the WQI (output variable). Figure 14.2 presents the correlation plots of different training and testing datasets. The correlation coefficients were calculated to evaluate the effect of the input variables on the output variable.

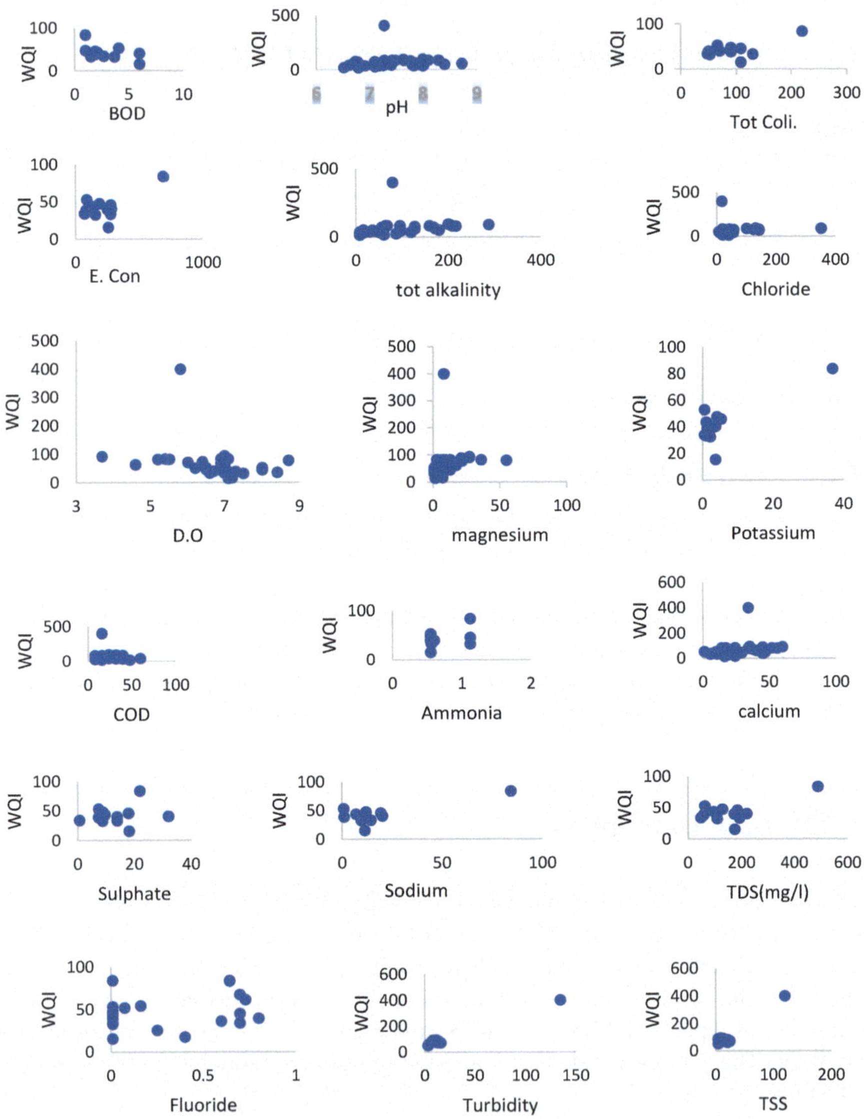

Fig. 14.2 Correlation plot of parameters with WQI

The correlation coefficients of parameters Ph, electrical conductivity, total coliform, turbidity, sodium, TDS, TSS, potassium with WQI were obtained as 0.54, 0.63, 0.61, 0.99, 0.75, 0.63, 0.95, 0.79 respectively. Therefore, these parameters are highly and significantly correlated with WQI; due to this reason they can be used as an input to the model. The rest of the parameters are less correlated with WQI, as their correlation coefficients were below 0.4.

14.5 Results

The soft computing technique, ANFIS with two membership functions, ANFIS_trimf and ANFIS_gaussianmf, was used in this study. The execution of this model was performed in MatLab. The accuracy of the model can be increased or decreased by changing the values of the regulator parameters. The determination of optimal values of parameters were done by trial and error, and thereafter the modelling was performed to increase the accuracy of model. The total number of non-linear parameters introduced while training the dataset with ANFIS_trimf and ANFIS_gaussianmf were 72 and 48 respectively. The total number of epochs were set to 50.

The error values obtained by RMSE and MAPE are summarized in Table 14.2. The error values suggest that ANFIS model performed quite better with the trimf membership function in both training and testing datasets. Figure 14.3 represents the prediction performance of both membership functions in training and testing datasets. The performance of both membership functions is quite well with the training data set, but trimf is more close to the predict the WQI values.

Table 14.2 Error analyses of model

Error for model fitness	ANFIS_trimf	ANFIS_gaussianmf
Training dataset		
RMSE	0.1363457	0.192778501
MAPE	0.331567	0.86964933
Testing dataset		
RMSE	0.3243403	0.5271698
MAPE	1.236319	2.1130768

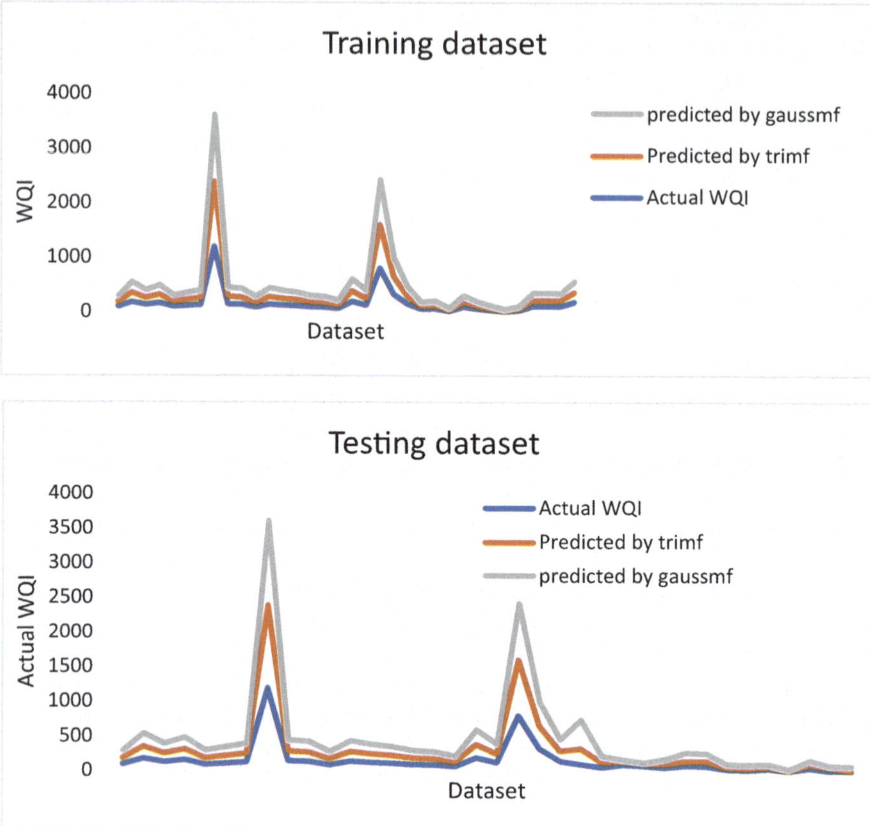

Fig. 14.3 Performance of WQI with the ANFIS model

14.6 Conclusion

Firstly, various water quality parameters were introduced, which forms the basis for the assessment of water quality. Then different methods were discussed, including correlation, ANFIS architecture and its working, WQI calculation, and errors for model performance. Basically, correlation is used to determine whether there is any linear relationship between the variables. Then some research work was proposed, which used the ANFIS model with Gaussian and trimf membership function for the prediction of WQI. The correlation coefficients were determined to know the highly affected parameters against WQI. It is observed that the ANFIS model performs quite better with the trimf membership function.

Acknowledgements The authors are really so thankful and grateful to the "Guru Gobind Singh Indraprastha University" for every possible support financially and all facilities to perform the research.

References

Bhagat SK, Tung TM, Yaseen ZM (2019) Development of artificial intelligence for modeling wastewater heavy metal removal: state of the art, application assessment and possible future research. J Clean Prod 250:119473

Brown RMNI, McClelland RA, Deininger RG, Tozer A (1970) Water quality index—do we dare

Central Pollution Control Board (CPCB) (1982) Water quality standards. https://cpcb.nic.in (1974), accessed 31/Jan/2024–13, Environmental Standards, https://scclmines.com › env › linkfile2

Debels P, Figueroa R, Urrutia R, Barra R, Niell X (2005) Evaluation of water quality in the Chillán River (Central Chile) using physicochemical parameters and a modified water quality index. Environ Monit Assess 110(1–3):301–322

Gao F, Feng MQ, Teng SF (2015) On the way for forecasting the water quality by BP neural network based on the PSO. J Saf Environ 15:338–341

Gaur N, Sarkar A, Dutta D, Gogoi BJ, Dubey R, Dwivedi SK (2022) Evaluation of water quality index and geochemical characteristics of surface water from Tawang India. Sci Rep 12:11698. https://doi.org/10.1038/s41598-022-14760-3

Gupta R, Singh AN, Singhal A (2019) Application of ANN for water quality index. Int J Machine Learn Comput 9(5)

Horton RK (1965) An index number system for rating water quality. J Water Pollut Control Fed 37(3):300–306

Kannel PR, Lee S, Lee YS, Kanel SR, Khan SP (2007) Application of water quality indices and dissolved oxygen as indicators for river water classification and urban impact assessment. Environ Monit Assess 132(1–3):93–110

Kulwant M, Patel D, Shirin S, Sharma SN, Yadav AK (2023) Future frameworks for fluoride and algorithms for environmental system. In: Advanced treatment technologies for fluoride removal in water. Water science and technology library, vol 125. Springer, Cham. https://doi.org/10.1007/978-3-031-38845-3_19

Kumar A, Dua A (2009) Water quality index for assessment of water quality of river ravi at madhopur (India). Global J Environ Sci 8(1):49–57

Kumar A, Chava R, Gupta S, Shirin S, Jamal A, Yadav AK (2023) Environmental and health effects of fluoride contamination and treatment of wastewater using various technologies. In: Advanced treatment technologies for fluoride removal in water. Water science and technology library, vol 125. Springer, Cham. https://doi.org/10.1007/978-3-031-38845-3_18

Miller WW, Joung HM, Mahannah CN, Garrett JR (1986) Identification of water quality differences in Nevada through index application. J Environ Quality 15:265–272

Omer NH (2020) Water quality parameters, In: Summers K (ed) Water quality—science, assessments and policy. IntechOpen

Patel D, Kulwant M, Shirin S, Varshney R, Pandey G, Yadav AK (2023) Fluoride removal from aqueous solution using iron-based materials: preparation, characterization, and applications. In: Advanced treatment technologies for fluoride removal in water. Water science and technology library, vol 125. Springer, Cham. https://doi.org/10.1007/978-3-031-38845-3_4

Shirin S, Yadav AK (2014) Physico chemical analysis of municipal wastewater discharge in Ganga River, Haridwar District of Uttarakhand, India. Current World Environment 9(2):536

Shirin S, Jamal A, Ranjan P, Yadav AK (2019) Study on assessment of slope stability and mixed disposal of overburden in voids of Singrauli Coalfield. Environ Qual Manage 28(3):131–139

Shirin S, Jamal A, Emmanouil C, Yadav AK (2021) Assessment of characteristics of acid mine drainage treated with fly ash. Appl Sci 11(9):3910

Shirin S, Jamal A, Emmanouil C, Singh VP, Yadav AK (2023) Assessment and characterization of waste material used as backfilling in an abandoned mine. Int J Coal Prep Util 43(8):1402–1410

Sihag P, Tiwari NK, Ranjan S (2019) Prediction of unsaturated hydraulic conductivity using adaptive neuro-fuzzy inference system (ANFIS). ISH Journal of Hydraulic Engineering 25(2):132–142

Singh B, Sihag P, Singh VP, Sepahvand A, Singh K (2021) Soft computing technique-based prediction of water quality index. Water Supply 21(8):4015–4028

Tarno, Rusgiyono A, Sugito (2018) Adaptive neuro fuzzy inference system (ANFIS) approach for modeling paddy production data in Central Java. In: The 8th international seminar on new paradigm and innovation on natural science and its application. Journal of Physics: Conference Series, 1217, Central Java, Indonesia, 26 Sep 2018

Yadav AK, Sahoo SK, Mahapatra S, Kumar AV, Pandey G, Lenka P, Tripathi RM (2014) Concentrations of uranium in drinking water and cumulative, age-dependent radiation doses in four districts of Uttar Pradesh, India. Toxicol Environ Chem 96(2):192–200

Chapter 15
Intercomparison of Extreme Rainfall Estimates Using Normal, Gamma and Pearson, and Extreme Value Families of Probability Distributions

N. Vivekanandan

Abstract Estimation of extreme rainfall for a given return period is considered as one of the important parameters for planning, design, operation and management of civil and hydraulic structures. This can be achieved by performing extreme value analysis (EVA) of rainfall that consists of fitting probability distribution to the annual maximum series of the observed rainfall data. In this book chapter, the EVA of rainfall for the Dharamshala and Kangra sites in Himachal Pradesh was carried out by adopting the Log Normal, Gamma, Pearson Type-3, Log Pearson Type-3 (LP3), Extreme Value Type-1 (EV1), Extreme Value Type-2 (EV2), Generalized Extreme Value (GEV) and Generalized Pareto (PRT) belong to normal, gamma and pearson and extreme value families of probability distributions. The parameters of the distributions were determined by Method of Moments (MoM), Maximum Likelihood Method (MLM) and L-Moments (LMO), and are used for rainfall estimation. The adequacy of fitting distributions adopted in EVA of rainfall was evaluated by Goodness-of-Fit (GoF) (viz., Chi-square and Kolmogorov–Smirnov) and diagnostic (viz., D-index) tests, and the fitted curves of the estimated rainfall. The findings of the study indicated that the distributions other than LP3 (MoM and MLM) and EV2 (MoM and MLM) for Kangra whereas PRT (MoM, MLM and LMO) for Dharamshala are found to be suitable for EVA of rainfall. The outcomes of EVA results weighed with GoF and diagnostic tests indicated that the EV1 (LMO) is the most suitable distribution for estimating the rainfall in Dharamshala and GEV (LMO) for Kangra.

Keywords Chi-square · D-index · Extreme value type-1 · Generalized extreme value · Kolmogorov-Smirnov · L-moments · Rainfall

N. Vivekanandan (✉)
Central Water and Power Research Station, Pune, Maharashtra 411024, India
e-mail: vivek.n@cwprs.gov.in

A. K. Yadav et al. (eds.), *Integrated Management of Water Resources in India: A Computational Approach*, Water Science and Technology Library 129,
https://doi.org/10.1007/978-3-031-62079-9_15

15.1 Introduction

In hydrological studies, estimation of rainfall for a given period is considered as one of the important parameters for planning, design, operation and management of civil and hydraulic structures such as dams, bridges, barrages, culverts, storm water drainage systems, etc. (Vivekanandan and Mathew 2010; Yadav et al. 2014; Shirin and Yadav 2021; Kumar et al. 2023). This can be carried out through extreme value analysis (EVA) of rainfall by fitting probability distributions belong to normal, gamma and Pearson, and extreme value families to the series of observed annual 1-day maximum rainfall data.

15.1.1 Background

A number of probability distributions such as 2-parameter Log Normal (LN2), Gamma (GAM), Pearson Type-3 (PR3), Log Pearson Type-3 (LP3), Extreme Value Type-1 (EV1), Extreme Value Type-2 (EV2), Generalized Extreme Value (GEV) and Generalized Pareto (PRT) distributions were generally adopted in EVA. The LN2 was a member of normal family of whereas GAM, PR3 and LP3 were the members of gamma family. The EV1, EV2, GEV and PRT distributions were grouped under extreme value family (Cho et al. 2004; Kamal et al. 2017; Baghel et al. 2019; Sasireka et al. 2019; Kaur et al. 2021). Standard parameter estimation procedures like Method of Moments (MoM), Maximum Likelihood Method (MLM) and Method of L-Moments (LMO) were commonly applied for determination of parameters (Hosking and Wallis 1993; Dingre and Shahi 2006; Iresh 2019; Vivekanandan and Srishailam 2020) of the distributions. During the past, several studies on EVA rainfall have been carried out by various researchers for different regions. AlHassoun (2011) estimated the intensity using LP3 for Riyadh region while Mani Sabarish et al. (2015) carried out the EVA of rainfall for Tiruchirappalli region by applying PR3, LP3 and EV1. Mukherjee (2013) applied GEV to develop a mathematical model to estimate peak flood discharge for a given return period. The study by Afungang and Bateira (2016) revealed that EV1 is better suited for the estimation of rainfall in the Bamenda mountain region, Cameroon. Amin et al. (2016) applied normal, LN2, LP3, and EV1 (also known as Gumbel) distributions for modelling annual maximum rainfall in the northern regions of Pakistan and identified the best fit distribution through Goodness-of-Fit tests. Arvind et al. (2017) analyzed annual and monthly rainfall data for Musiri region, Tiruchirappalli, Tamil Nadu, using EV1. Chifurira and Chikobvu (2017) applied LN2, GAM and EXP distributions to assess the effect of data length on the selection of suitable distribution for rainfall estimation of Zimbabwe. Esberto (2018) made an attempt to determine the best-fit frequency of rainfall pattern for event forecasting in order to address potential disasters using sixty distributions. Sharma and Sharma (2019) evaluated the Normal, LN2 and GAM applied in EVA of the rainfall for the Shimla region through Chi-square (χ^2), Kolmogorov–Smirnov

(KS) and Anderson Darling qualities. Rosmaini and Saphira (2019) applied normal, GAM and LN2 distributions for the estimation of monthly rainfall of Tuntungan, Tanjung Selamat and Medan Selayang stations in Medan City, Indonesia. Hasan et al. (2020) evaluated the GAM, LN2 and Weibull distributions for predicting rainfall in ungauged catchments of Australia. The study by Vivekanandan (2020) revealed that LP3 is the best-fit distribution for the estimation of the peak flood at the Haora site. Vaidya et al. (2020) carried out the study on evaluation of rainfall analysis of distinct rainfall intensity for various stations in Gujarat. Kaur et al. (2021) applied Gumbel, LP3, LN2, and Ven Te Chow distributions for the prediction of annual 1-day, 2-day, and 3-day maximum rainfall values of Roorkee. Van derSpuy and Du Plessis (2022) stated that GEV is a preferred choice at the lower probabilities amongst three distributions applied in flood frequency analysis for South Africa. However, there was no general agreement on applying particular distribution for EVA of rainfall for any region.

In this book chapter, a study on EVA of rainfall was presented by applying MoM, MLM and LMO of eight distributions (viz., LN2, GAM, PR3, LP3, EV1, EV2, GEV and PRT) belongs to the normal, gamma and Pearson and extreme value families for the rain gauge sites of Dharamshala and Kangra of Himachal Pradesh. Goodness-of-Fit (GoF) (viz., χ^2 and *KS*) and diagnostic (viz., D-index) tests, and qualitative evaluation using the fitted curves of the estimated values were applied to select the best-fit distribution for EVA of rainfall. The findings of the results obtained through the methodology adopted in EVA of rainfall by three different parameter estimation methods of eight probability distributions belong to the normal, gamma and pearson and extreme value families were described in the following sections.

15.2 Methodology

15.2.1 Probability Distributions Adopted in EVA

Table 15.1 presents the cumulative distribution function (CDF) and quantile estimator (*x(T)*) for a given return period (*T*) of eight probability distributions (Bobee and Askhar 1991; Rao and Hamed 2000; Samantaray and Sahoo 2020; Yue 2001) adopted in EVA of rainfall. The empirical equations involved in determining the parameters of eight distributions by MoM, MLM, and LMO are presented in Table 15.2.

In Table 15.1, x is the random variable, N is the number of samples, ξ is the location parameter, α is the scale parameter, k is the shape parameter, $\mu(y)$ and $\sigma(y)$ are the average and standard deviation of the log transformed series of the observed data, *F(x)* is the CDF of x (i.e., annual 1 day maximum rainfall (AMR)), *K(T)* is the frequency factor for a return period *(T)* corresponding to the probability of exceedance *(P)* and the coefficient of skewness (i.e., $C_s = 0.0$ for LN2 and C_s of the observed AMR data for PR3), φ^{-1} is the inverse of the standard normal distribution, *G(…)* is the incomplete gamma integral and *x(T)* is the estimated extreme rainfall

Table 15.1 CDF and quantile estimator of eight probability distributions adopted in EVA

Family of distribution	Probability distribution	Cumulative distribution function *F(x)* of *F(y)*	Quantile estimator (*x(T)*)
Normal	LN2 ($\mu(y)$, $\sigma(y)$)	$F(y)=\varphi\left(\frac{y-\mu(y)}{\sigma(y)}\right),\ 0<y<\infty,\ \sigma(y)>0$	$x(T)=\exp\ (\mu(y)+\sigma(y)\,\varphi^{-1}(1/T))$ with y=ln(x)
Gamma and Pearson	GAM (α, k)	$F(x)=\begin{cases} G\left(k,\frac{x}{\alpha}\right) & ,\alpha>0 \\ 1-G\left(k,\frac{x}{\alpha}\right) & ,\alpha<0 \end{cases}$	No explicit expression ofquantile function is available
	PR3 (ξ, α, k)	$F(x)=\begin{cases} G\left(k,\frac{x-\xi}{\alpha}\right) & ,\alpha>0 \\ 1-G\left(k,\frac{x-\xi}{\alpha}\right) & ,\alpha<0 \end{cases}$	No explicit expression of the quantile function is available
	LP3 (ξ, α, k)	$F(x)=\begin{cases} G\left(k,\frac{\ln(x)-\xi}{\alpha}\right) & ,\alpha>0 \\ 1-G\left(k,\frac{\ln(x)-\xi}{\alpha}\right) & ,\alpha<0 \end{cases}$	No explicit expression of quantile function is available
Extreme Value	EV1 (ξ, α)	$F(x)=\exp\left(-\exp-\left(\frac{x-\xi}{\alpha}\right)\right), \alpha>0, -\infty<x<\infty$	$x(T)=\xi-\alpha\ln[-\ln(1-(1/T))]$
	EV2 (α, k)	$F(x)=\exp\left(-\left(\frac{x}{\alpha}\right)^{-k}\right),\ \alpha>0,\ k>0,\ 0<x<\infty$	$x(T)=\alpha\ [-\ln\ (1-(1/T))]^{-(1/k)}$
	GEV (ξ, α, k)	$F(x)=\exp\left(-\left[1-\frac{k(x-\xi)}{\alpha}\right]^{1/k}\right),\ -\infty\le x\le \xi+(\alpha/k)\ for\ k>0, \alpha>0$	$x(T)=\xi+\frac{\alpha}{k}(1-[-\ln(1-(1/T))]^{k})$
	PRT (ξ, α, k)	$F(x)=1-\left(1-\frac{k(x-\xi)}{\alpha}\right)^{1/k}, \xi\le x\le \xi+(\alpha/k)\ for\ k>0, \alpha>0$	$x(T)=\xi+\frac{\alpha(1-(1/T)^{k})}{k}$

Table 15.2 Determination of parameters of eight distributions using MoM, MLM and LMO

Distribution	Parameters of the distribution		
	MoM	MLM	LMO
LN2 $(\mu(y), \sigma(y))$	$\mu(y) = \frac{1}{N}\sum_{i=1}^{N} y(i) - \frac{1}{2N}\sum_{i=1}^{N} y(i)^2 + \frac{1}{2}\left(\frac{1}{N}\sum_{i=1}^{N} y(i)\right)^2$ $\sigma(y) = \left[\frac{1}{N}\sum_{i=1}^{N} y(i)^2 - 2\left(\frac{1}{N}\sum_{i=1}^{N} y(i)\right)\right]^{1/2}$	$\mu(y) = \frac{1}{N}\sum_{i=1}^{N} y(i)$ *where* $y(i) = \ln(x(i))$ $\sigma(y) = \left(\frac{1}{N}\sum_{i=1}^{N} (y(i) - \mu(y))^2\right)^{1/2}$	$\mu(y) = \lambda_1 = b_0 = \frac{1}{N}\sum_{i=1}^{N} y(i)$ $\lambda_2 = 2b_1 - b_0 = \sigma(y)/\sqrt{\pi}$ $b_1 = \frac{1}{N(N-1)}\sum_{i=2}^{N} (i-1)y(i)$
GAM (α, k)	$\mu(x) = \alpha k$ $\sigma(x)^2 = \alpha^2 k$	Parameters can be obtained by the Newton–Raphson algorithm	$\lambda_1 = \alpha k; \lambda_2 = \frac{\alpha}{\sqrt{\pi}}\frac{\Gamma(k+0.5)}{\text{sgn}(\tau_3)\Gamma k}$ $\tau_3 = \begin{cases} 6I_{1/3}(k, 2k) - 3, \alpha > 0 \\ -6I_{1/3}(k, 2k) + 3, \alpha < 0 \end{cases}$
P3 (ξ, α, k)	Can be determined through Eqs. (1 and 2)	Can be determined through Eq. (3)	Not defined
LP3 (ξ, α, k)	Can be determined through Eqs. (1 and 2) by replace x(i) with ln(x(i))	Can be determined through Eq. (3) by replacing x(i) with ln(x(i))	Not defined

(continued)

Table 15.2 (continued)

Distribution	Parameters of the distribution		
	MoM	MLM	LMO
EV1 (ξ, α)	$\xi = \mu(x) - (0.5772157)\ \alpha$ $\alpha = \left(\frac{\sqrt{6}}{\pi}\right)\sigma(x)$	$\xi = -\alpha \ln\left[\sum_{i=1}^{N} \exp(-x(i)/\alpha) \Big/ N\right]$ $\alpha = \mu(x) - \left[\sum_{i=1}^{N} x(i) \exp(-x(i)/\alpha) \Big/ \sum_{i=1}^{N} \exp(-x(i)/\alpha)\right]$	$\xi = \lambda_1 - (0.5772157)\ \alpha$ $\alpha = \frac{\lambda_2}{\ln(2)}$
EV2 (α, k)	Using logarithmic transformation of observed data, the EV1 parameters are initially obtained by MoM, MLM, and LMO; and used to determine the EV2 parameters of EV2 from $\alpha = \exp(\xi)$ and k = 1/(scale parameter of EV1)		

(continued)

Table 15.2 (continued)

Distribution	Parameters of the distribution		
	MoM	MLM	LMO
GEV (ξ, α, k)	$\mu(x) = \xi + \frac{\alpha(1-\Gamma(1+k))}{k}$ $\sigma(x) = \frac{\alpha}{k}\left(\Gamma(1+2k) - \Gamma(1+k)^2\right)^{1/2} C_s =$ $(sign\, k)\frac{\Gamma(1+3k)+3\Gamma(1+k)(1+2k)-2\Gamma^3(1+k)}{\{\Gamma(1+k)-\Gamma^2(1+k)^2\}^{1/2}}$	Parameters can be obtained by the modified Newton–Raphson algorithm	$z = [2/(3+\tau_3) - \ln(2)/\ln(3)]$ $\tau_3 = [2(1-3^{-k})/(1-2^{-k})] -$ 3 $k = 7.817740z + 2.930462z^2 +$ $13.641492z^3 + 17.206675z^4$ $\alpha =$ $\lambda_2 k/(1-2^{-k})\Gamma(1+k)$ $\xi = \lambda_1 + [\alpha(\Gamma(1+k) - 1)/k]$
PRT (ξ, α, k)	$\mu(x) = \xi + [\alpha/(1+k)]$ $\sigma(x) = \left[\alpha^2/(1+2k)(1+k)^2\right]^{1/2} C_s =$ $2(1-k)(1+2k)^{1/2}/(1+3k)$	$\sum_{i=1}^{N} \frac{(x(i)-\xi)/\alpha}{1-(k(x(i)-\xi)/\alpha)} = \frac{N}{1-k}$ $\sum_{i=1}^{N} \ln[1 - (k(x(i) - \xi)/\alpha)] = -Nk$ $\xi \leq$ lowest value of observed data	$\xi = \lambda_1 - [\alpha/(1+k)]$ $k = (1 - 3\tau_3)/(\tau_3 + 1)$ $\alpha = (1+k)(2+k)\lambda_2$

(i.e., 1 day maximum) for a return period *(T)*. In Table 15.2, *Ix(…)* τ_3 as a function of k that are rational function approximations (Hosking and Wallis 1993).

15.2.2 MoM of PR3 Distribution

The MoM estimators of PR3 distribution (Bobee and Askhar 1991) can be determined by solving the system of equations, which is given below:

$$M_r' = \frac{\exp(r\xi)}{(1 - r\alpha)^k} \text{ wherein } 1 - \text{r}\alpha > 0, \text{r} = 1, 2, 3 \tag{15.1}$$

$$\mu(x) = M_1', \ \sigma(x)^2 = M_2' - (M_1\prime)^2 \text{ and } \gamma(x) = M_3' - 3M_2'M_1' + 2(M_1')^3 \tag{15.2}$$

where, $M_r\prime$ is the rth moment of x about the origin and $\gamma(x)$ is the C_s of the observed data (x).

15.2.3 MLM of PR3 Distribution

The MLM estimators of PR3 distribution (Bobee and Askhar 1991) can be determined by solving the following system of equations:

$$\left.\begin{array}{l} \sum_{i=1}^{N}(x(i) - \xi) = N\alpha k \\ \sum_{i=1}^{N}\frac{1}{\alpha}(x(i) - \xi) = N\psi(k) \\ \sum_{i=1}^{N}\frac{1}{(x(i)-\xi)} = \frac{N}{\alpha(k-1)} \end{array}\right\} \tag{15.3}$$

Here, $\psi(k)$ is the digamma function of estimator of the scale parameter *(k)*.

15.2.4 Goodness-of-Fit Tests

GoF (viz., χ^2 and KS) tests were applied to check the adequacy of probability distributions adopted in EVA. Theoretical descriptions of GoF tests statistic (Zhang 2002) are given by:

$$\chi^2 = \sum_{j=1}^{NC}\frac{\left(O_j(x) - E_j(x)\right)^2}{E_j(x)} \tag{15.4}$$

$$KS = \underset{i=1}{\overset{N}{Max}}(F_e(x(i)) - F_D(x(i))) \tag{15.5}$$

Here, $O_j(x)$ is the observed frequency value of x in j^{th} class, $E_j(x)$ is the expected frequency value of x in j^{th} class, NC is the number of frequency classes, $x(i)$ is the observed value of x for i^{th} sample, $F_e(x(i)) = i/(N+1)$ is the empirical CDF of $x(i)$ and $F_D(x(i))$ is the derived CDF of $x(i)$ by probability distribution. If the computed values of the GoF tests statistic given by the probability distribution are lower than its theoretical values at the desired significance level (1 or 5%), then the distribution is acceptable for estimation of rainfall at that level.

15.2.5 Diagnostic Test

The selection of a suitable distribution for rainfall estimation was made through a diagnostic test (say, D index) (USGS 2019), which is given as follows:

$$\mathrm{D-index} = (1/\mu(\mathrm{x}))\sum_{i=1}^{6}\left|\mathrm{x}(i) - \mathrm{x}^*(i)\right| \tag{15.6}$$

Here, $\mu(x)$ is the average of the observed data (x), $x(i)(i = 1 \text{ to } 6)$ and $x^*(i)$ are the six highest observed and the corresponding estimated rainfall by the probability distribution pertaining to i^{th} sample. The distribution with the lowest value of the D-index was considered as better suited distribution compared to the other distributions adopted in the EVA for rainfall estimation.

15.3 Study Area and Data Used

In the present study, an intercomparison of extreme rainfall estimates of Dharamshala and Kangra in Himachal Pradesh was carried out using eight distributions belong to the normal, gamma and pearson and extreme value families. The Dharamshala rain-gauge station is located between the latitude 32° 13′ 12″ N and longitude 76° 18′ 36″ E in Hamirpur district. The Kangra rain-gauge station is located between the latitude 32° 06′ 00″ N and longitude 76° 16′ 12″ E in Kangra district. The daily rainfall observed in Dharamshala (1970 to 1977, 1980 to 1998, 2002 to 2005 and 2007 to 2018) and Kangra (1970 to 2006, 2011 to 2015 and 2017) was used. Considering the importance of the hydrologic extremes, the missing data of Dharamshala (1978 to 1979, 1999 to 2001 and 2006) and Kangra (2007 to 2010 and 2016) were not considered in EVA. The AMR series was derived from daily rainfall data and used in the EVA. The descriptive statistics of the AMR series is given in Table 15.3.

Table 15.3 Descriptive statistics in AMR series of Dharamshala and Kangra

Rain-gauge site	Observed				Log-transformed			
	AVG (mm)	SD (mm)	C_s	C_k	AVG (mm)	SD (mm)	C_s	C_k
Dharamshala	175.8	61.8	0.881	1.926	5.107	0.367	− 0.611	1.570
Kangra	140.2	45.5	− 0.174	− 0.669	4.881	0.381	− 1.079	1.290

AVG—average; SD—standard deviation; C_s—coefficient of skewness; C_k—Coefficient of Kurtosis

15.4 Results and Discussion

15.4.1 Extreme Value Analysis of Rainfall

By applying the procedures, as described above, extreme (i.e., 1-day maximum) rainfall were estimated using normal, gamma and pearson, and extreme value families of distributions for the Dharamshala and Kangra sites. The EVA results are presented in Tables 15.4 and 15.5, and the fitted curves are delineated in Figs. 15.1 and 15.2. From the EVA results of Dharamshala and Kangra, it was observed that (i) the estimated rainfalls by EV2 (MLM) are comparatively higher than the values of other distributions; (ii) the estimated rainfalls for the return periods from 50-year to 100-year by PRT (MLM) are lower than the values of other distributions; and (iii) there is no appreciable difference between the estimated rainfalls by three methods (viz., MoM, MLM, and LMO) of LN2. In the present study, the EVA results of GAM (LMO), PR3 (LMO) and LP3 (LMO) were not computed and also not given in Tables 15.4 and 15.5 due to the non-feasibility (NF) of the fitting LMO of GAM, PR3 and LP3 distributions to the AMR series. The GoF and diagnostic tests values of GAM (LMO), PR3 (LMO) and LP3 (LMO) were not computed and also not presented in Tables 15.6, 15.7 and 15.8. From Figs. 15.1 and 15.2, it could be seen that the fitted lines of the estimated rainfall using LN2 and EV1 are linear while the GAM, EV2, GEV, PR3 and LP3 curves are in exponentially upward and PRT curves are in exponentially downward.

15.4.2 Evaluation of Results Using GoF Tests

The adequacy of the fitting probability distributions adopted in EVA was evaluated by GoF (viz., χ^2 and *KS*) tests. For computing the χ^2 statistic values by eight distributions for Dharamshala and Kangra, the number of frequency classes was considered as 6 while the degree of freedom was considered as 3 for 2-parameter distributions (viz., LN2, GAM, EV1 and EV2) and 2 for 3-parameter distributions (viz., PR3, LP3, GEV and PRT). Using the results of the GoF tests (Tables 15.6 and 15.7), it was observed that (i) χ^2 and *KS* tests results confirmed the applicability of the

Table 15.4 Estimated 1-day maximum rainfall by eight distributions for Dharamshala

Distribution	Method	Maximum rainfall (in mm) for different return periods (in year)							
		2	5	10	20	25	50	75	100
LN2	MoM	165.8	221.0	256.8	290.7	301.4	334.3	353.4	366.9
	MLM	165.2	224.2	263.0	300.0	311.8	348.0	369.2	384.2
	LMO	165.2	223.3	261.4	297.8	309.3	344.7	367.9	380.1
GAM	MoM	168.6	224.6	258.1	288.2	297.4	324.6	339.8	350.4
	MLM	168.7	224.2	257.4	287.1	296.2	323.1	338.2	348.6
	LMO	NF	NF	NF	NF	NF	NF	NF	NF
PR3	MoM	166.8	223.4	258.5	290.4	300.2	329.6	346.2	357.8
	MLM	169.8	223.5	257.0	287.7	297.2	325.6	341.7	352.9
	LMO	NF	NF	NF	NF	NF	NF	NF	NF
LP3	MoM	159.2	221.5	269.1	319.8	337.0	393.6	429.1	455.5
	MLM	160.3	232.1	286.7	344.5	364.0	427.9	467.9	497.5
	LMO	NF	NF	NF	NF	NF	NF	NF	NF
EV1	MoM	165.6	220.2	256.4	291.1	302.1	336.0	355.7	369.7
	MLM	166.6	225.9	265.1	302.8	314.7	351.5	372.9	388.0
	LMO	165.5	220.8	257.4	292.5	303.7	338.0	357.9	372.0
EV2	MoM	160.6	205.7	242.2	283.4	297.9	347.3	379.6	404.4
	MLM	159.5	258.5	355.8	483.5	532.9	719.0	855.8	968.0
	LMO	156.8	209.0	252.9	303.6	321.7	384.6	426.7	459.2
GEV	MoM	167.1	222.5	257.6	290.1	300.1	330.5	347.7	359.7
	MLM	169.1	224.3	257.9	288.0	288.0	324.1	339.1	349.3
	LMO	168.3	223.3	257.1	287.6	297.0	324.7	340.0	350.6
PRT	MoM	161.6	228.1	266.9	297.8	306.4	329.4	340.7	347.8
	MLM	125.0	148.1	158.5	165.1	166.7	170.3	171.8	172.7
	LMO	167.0	232.6	263.8	284.7	289.8	302.2	307.5	310.5

seven distributions other than PRT (MoM, MLM and LMO) for EVA of rainfall for Dharamshala; (ii) χ^2 test didn't confirm the adequacy of the fitting LP3 (MoM and MLM) and EV2 (MoM and MLM) distributions for Kangra; and (iii) the *KS* test supported the use of the eight distributions adopted in EVA of rainfall for Kangra.

15.4.3 Evaluation of Results Using Diagnostic Tests

The selection of the most suitable distribution for the estimation of rainfall was performed through diagnostic test using D-index and the results are presented in Table 15.8. From the results of the diagnostic tests, it was found that the D-index

Table 15.5 Estimated 1-day maximum rainfall by eight distributions for Kangra

Distribution	Method	Maximum rainfall (in mm) for different return periods (in year)							
		2	5	10	20	25	50	75	100
LN2	MoM	133.4	174.1	200.1	224.5	232.1	255.5	269.0	278.5
	MLM	131.7	180.9	213.5	244.8	254.8	285.6	300.6	316.0
	LMO	131.7	180.1	212.2	242.9	252.6	282.8	302.5	318.2
GAM	MoM	135.3	176.4	200.8	222.5	229.1	248.7	259.6	267.2
	MLM	134.5	179.0	205.7	229.5	236.8	258.4	270.5	278.9
	LMO	NF	NF	NF	NF	NF	NF	NF	NF
PR3	MoM	138.9	179.2	198.7	214.2	217.9	230.3	236.4	240.1
	MLM	140.2	178.4	198.3	214.8	219.6	233.3	240.7	245.7
	LMO	NF	NF	NF	NF	NF	NF	NF	NF
LP3	MoM	123.1	175.2	219.6	270.8	289.0	351.7	393.2	425.1
	MLM	219.4	248.7	271.3	295.1	303.0	328.6	344.4	355.9
	LMO	NF	NF	NF	NF	NF	NF	NF	NF
EV1	MoM	132.7	173.0	199.6	225.2	233.3	258.3	272.8	283.1
	MLM	133.6	183.6	216.7	248.5	258.6	289.6	307.7	320.4
	LMO	132.2	175.3	203.8	231.2	239.9	266.6	282.1	293.1
EV2	MoM	129.2	163.0	190.2	220.4	231.0	266.9	290.2	308.0
	MLM	121.2	193.0	262.6	352.9	387.6	517.3	611.8	688.9
	LMO	125.5	166.4	200.5	239.7	253.7	302.2	334.5	359.4
GEV	MoM	141.4	180.4	199.2	213.2	217.0	227.1	231.9	234.9
	MLM	127.3	162.4	179.3	191.9	195.3	204.4	208.7	211.4
	LMO	141.5	181.2	200.3	214.5	218.3	228.4	233.2	236.3
PRT	MoM	143.4	187.9	200.2	205.5	206.4	208.2	208.7	208.9
	MLM	136.2	178.5	190.2	195.2	196.1	197.8	198.3	198.5
	LMO	142.3	188.0	201.6	207.7	208.9	211.1	211.7	212.0

values given by EV2 (LMO) for Dharamshala and GEV (LMO) for Kangra are lower than those values of other distributions adopted in EVA.

15.4.4 *Selection of Probability Distribution*

The best fit for rainfall estimation was re-assessed through qualitative assessment together with GoF and diagnostic tests results, and accordingly final selection was made.

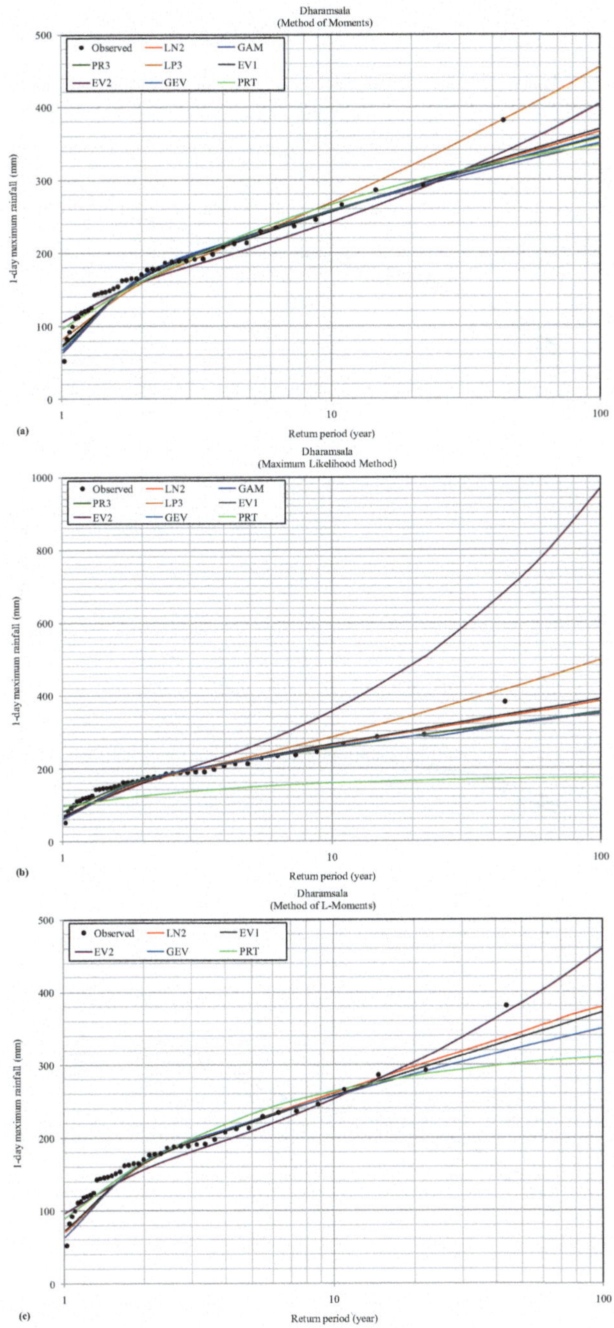

Fig. 15.1 Estimated 1-day maximum rainfall by MoM, MLM and LMO estimators of normal, gamma and Pearson, and extreme value families of distributions for Dharamshala

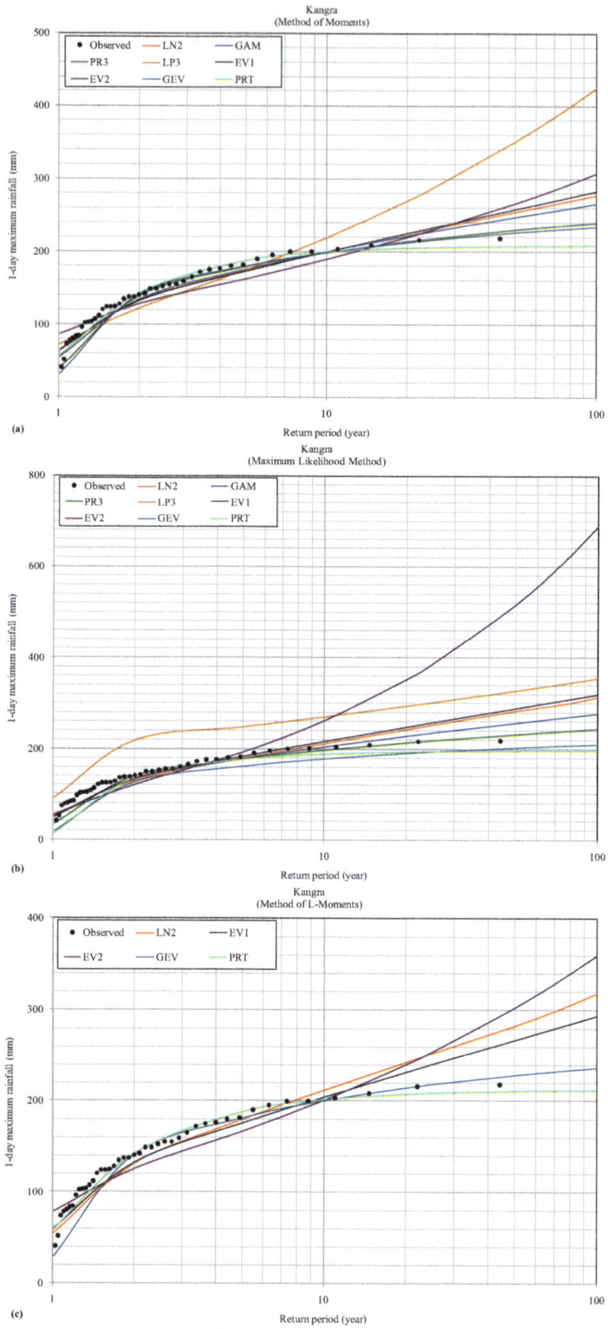

Fig. 15.2 Estimated 1-day maximum rainfall by MoM, MLM and LMO estimators of normal, gamma and Pearson, and extreme value families of distributions for Kangra

Table 15.6 Computed values of GoF tests by eight distributions for Dharamshala

GoF tests/method	Computed values of GoF tests							
	LN2	GAM	PR3	LP3	EV1	EV2	GEV	PRT
Chi-square (χ^2) (Theoretical values of χ^2 test statistic are $\chi^2_{0.05,2} = 5.991$ and $\chi^2_{0.05,3} = 7.815$)								
MoM	4.302	3.744	3.744	3.186	4.302	6.814	4.302	8.767
MLM	4.305	3.748	3.748	3.192	4.302	6.954	3.605	7.651
LMO	4.302	NF	NF	NF	4.302	7.093	2.907	6.535
Kolmogorov–Smirnov (KS) (Theoretical value of $KS_{0.05,\,43}$ is 0.203)								
MoM	0.079	0.074	0.082	0.126	0.077	0.123	0.075	0.515
MLM	0.080	0.076	0.081	0.128	0.078	0.127	0.073	0.412
LMO	0.079	NF	NF	NF	0.080	0.132	0.070	0.310

Table 15.7 Computed values of GoF tests by eight distributions for Kangra

GoF tests/method	Computed values of GoF tests							
	LN2	GAM	PR3	LP3	EV1	EV2	GEV	PRT
Chi-square (χ^2) (Theoretical values of χ^2 test statistic are $\chi^2_{0.05,2} = 5.991$ and $\chi^2_{0.05,3} = 7.815$)								
MoM	2.070	1.512	4.302	7.651	2.345	11.844	0.395	1.512
MLM	2.075	1.521	4.503	7.809	2.221	8.771	0.398	1.233
LMO	2.071	NF	NF	NF	2.070	5.698	0.395	0.953
Kolmogorov–Smirnov (KS) (Theoretical value of $KS_{0.05,\,43}$ is 0.203)								
MoM	0.090	0.071	0.058	0.169	0.098	0.161	0.043	0.105
MLM	0.091	0.075	0.061	0.185	0.096	0.162	0.041	0.113
LMO	0.092	NF	NF	NF	0.094	0.164	0.039	0.120

Table 15.8 D-index values given by eight distributions for Dharamshala and Kangra

Site/method	D-index values given by							
	LN2	GAM	PR3	LP3	EV1	EV2	GEV	PRT
Dharamshala								
MoM	0.457	0.514	0.486	0.463	0.446	0.620	0.478	0.569
MLM	0.452	0.531	0.509	1.199	0.456	5.201	0.517	4.208
LMO	0.531	NF	NF	NF	0.441	0.265	0.514	0.711
Kangra								
MoM	0.470	0.389	1.265	1.832	0.503	0.704	0.163	0.229
MLM	0.999	0.543	0.225	3.309	1.141	4.804	1.438	0.656
LMO	0.682	NF	NF	NF	0.622	1.003	0.140	0.153

(i) For Dharamshala, it was found that

   (a) The fitted curves of the estimated rainfall using EV2 (LMO) indicated that there is no line of agreement between the observed and estimated values, though its D-index is minimum than those values of other distributions adopted in EVA.
   (b) After eliminating the value of the D-index of EV2 (LMO) from the selection, it was identified that the D-index of EV1 (LMO) is the second minimum next to EV2 (LMO).

(ii) For Kangra, it was found that the D-index of GEV (LMO) is a minimum value of other distributions adopted in EVA.

Based on quantitative and qualitative assessments, it was identified that EV1 (LMO) is identified as better suited distribution for the estimation of rainfall at Dharamshala whereas GEV (LMO) for Kangra. Figure 15.3 presents the estimated maximum rainfall by the selected distribution (viz., EV1 (LMO) for Dharamshala and the GEV (LMO) for Kangra) and observed AMR for the Dharamshala and Kangra sites respectively. From Fig. 15.3, it can be seen that approximately 75% of the observed AMR are within the confidence limits of the estimated rainfall by the selected distribution.

15.5 Conclusion

The chapter presented a study on the intercomparison of normal, gamma and Pearson, and extreme value families of probability distributions (viz., LN2, GAM, PR3, LP3, EV1, EV2, GEV and PRT) adopted in the EVA of rainfall for Dharamshala and Kangra in Himachal Pradesh. Goodness-of-Fit (viz., viz., χ^2 and *KS*) and diagnostic (viz., D-index) tests, and fitted curves of the estimated values were applied to check the adequacy of fitting distributions used in the EVA of rainfall. Based on the evaluation of the EVA results, the conclusions drawn from the study were summarized and are presented below:

- χ^2 and *KS* tests results confirmed the applicability of the MoM, MLM and LMO of LN2, GAM, PR3, LP3, EV1, EV2 and GEV distributions for the EVA of rainfall for Dharamshala.
- χ^2 test didn't support the use of LP3 (MoM and MLM) and EV2 (MoM and MLM) distributions for Kangra.
- The *KS* test results witnessed all eight distributions for EVA of rainfall for Kangra.
- The outcomes of the EVA results was weighed with GoF and diagnostic tests, and accordingly EV1 (LMO) is identified as a better suited distribution for the estimation of rainfall in Dharamshala while GEV (LMO) for Kangra.

The outcome of the results would be helpful to decision makers to arrive at a rainfall depth for the design of civil and hydraulic structures, as also for carrying out

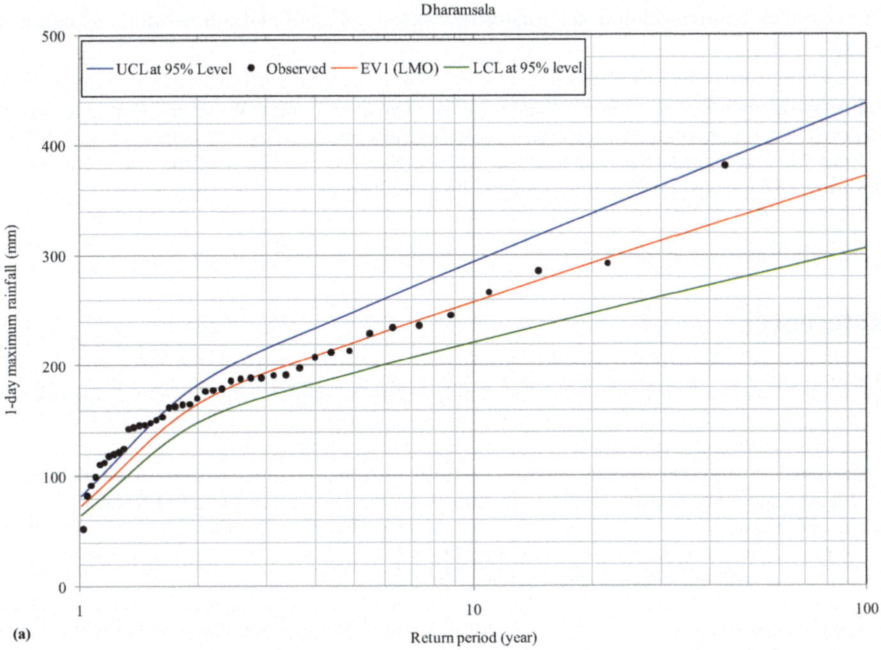

(a)

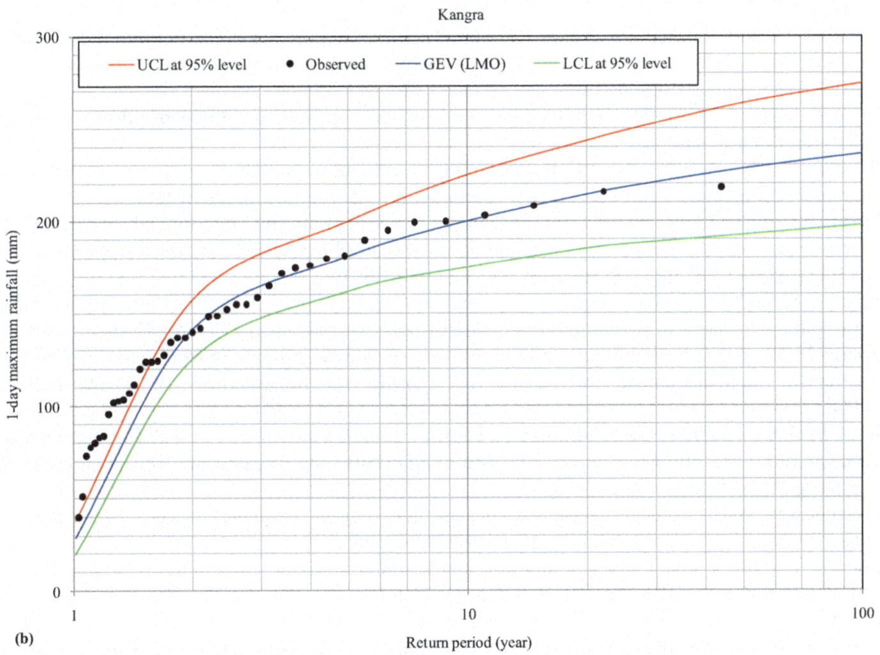

(b)

Fig. 15.3 Estimated 1-day maximum rainfall with 95% confidence limits by EV1 (LMO) for Dharamshala and GEV (LMO) for Kangra, and observed AMR

the related activities related to disaster management at the Dharamshala and Kangra sites.

Acknowledgements The author is thankful to the Director, Central Water and Power Research Station, Pune, for providing the research facilities to carry out the study. The author is also thankful to India Meteorological Department, Pune, and the Executive Engineer of the Water Resources Department, Himachal Pradesh for providing the rainfall data.

References

Afungang R, Bateira C (2016) Statistical modelling of extreme rainfall, return periods and associated hazards in the Bamenda Mountain, NW Cameroon. Geogr Spat Plann J 1(9):5–19. https://doi.org/10.17127/got/2016.9.001

AlHassoun SA (2011) Developing empirical formulae to estimate rainfall intensity in Riyadh region. J King Saud Univ-Eng Sci 23(1):81–88. https://doi.org/10.1016/j.jksues.2011.03.003

Amin MT, Rizwan M, Alazba AA (2016) A best-fit probability distribution for the estimation of rainfall in northern regions of Pakistan. Open Life Sci, Special Issue on CleanWAS 2015, 11(1):432–440. https://doi.org/10.1515/biol-2016-0057

Arvind G, Kumar PA, GirishKarthi S, Suribabu CR (2017) Statistical analysis of 30 years rainfall data: a Case Study. Proc IOP Conf Ser: Earth Environ Sci. https://doi.org/10.1088/1755-1315/80/1/012067

Baghel H, Mittal HK, Singh PK, Yadav KK, Jain S (2019) Frequency analysis of rainfall data using probability distribution models. Int J Curr Microbiol Appl Sci 8(6):1390–1396. https://doi.org/10.20546/ijcmas.2019.806.168

Bobee B, Askhar F (1991) The Gamma family and derived distributions applied in hydrology. Water Resources Publications

Chifurira R, Chikobvu D (2017) Extreme rainfall: candidature probability distribution for mean annual rainfall data. Res J Pharm, Biol Chem Sci 8(3):237–248. Google Scholar

Cho HK, Bowman KP, North GR (2004) A comparison of Gamma and Lognormal distributions for characterizing satellite rain rates from the tropical rainfall measuring mission. J Appl Meteorol 43(11):1586–1597. https://www.jstor.org/stable/26186015

Dingre S, Shahi NC (2006) Consecutive days maximum rainfall prediction from one day maximum rainfall for Srinagar in Kashmir valley. Indian J Soil Conserv 34(2):153–156. Google Scholar

Esberto MDP (2018) Probability distribution fitting of rainfall patterns in Philippine regions for effective risk management. Environ Ecol Res 6(3):178–186. https://doi.org/10.13189/eer.2018.060305

Hasan MM, Croke BFW, Liu S, Shimizu K, Karim F (2020) Using mixed probability distribution functions for modelling non-zero sub-daily rainfall in Australia. Geosciences 10(2):1–11. https://doi.org/10.3390/geosciences10020043

Hosking JRM, Wallis JR (1993) Some statistics useful in regional frequency analysis. Water Resour Res 29(2):271–281. https://doi.org/10.1029/92WR01980

Iresh ADS (2019) Flood frequency analysis for Kelani river basin: case study in Sri Lanka. In: Annual sessions of IESL, pp 571–578. Google Scholar

Kamal V, Mukherjee S, Singh P, Sen R, Vishwakarma CA, Sajadi P, Asthana H, Rena V (2017) Flood frequency analysis of Ganga river at Haridwar and Garhmukteshwar. Appl Water Sci 7(4):1979–1986. https://doi.org/10.1007/s13201-016-0378-3

Kaur L, Anvesha Kumar M, Verma SL Kumar P (2021) Annual maximum rainfall prediction using frequency analysis for Roorkee, Uttarakhand, India. MAUSAM 72(2):359–372. https://doi.org/10.54302/mausam.v72i2.623

Kumar A, Chava R, Gupta S, Shirin S, Jamal A, Yadav AK (2023) Environmental and health effects of fluoride contamination and treatment of wastewater using various technologies" in "Advanced treatment technologies for fluoride removal in water. Water Sci Technol Lib 125. https://doi.org/10.1007/978-3-031-38845-3_18

Mani Sabarish R, Narasimhan R, Chandhru AR, Suribabu CR, Sudharsan J, Nithiyanantham S (2015) Probabilistic analysis of consecutive-day maximum rainfall for Tiruchirappalli city (South India, Asia). Appl Water Sci. https://doi.org/10.1007/s13201-015-0307-x

Mukherjee MK (2013) Flood frequency analysis of River Subernarekha, India, using Gumbel's extreme value distribution. Int J Comput Eng Res 3(7):12–19. Google Scholar

Rao AR, Hamed KH (2000) Flood frequency analysis. CRC Publications, USA

Rosmaini E, Saphira YY (2019) Probability distribution of rainfall in Medan. J Res Math Trends Technol 1(2):39–45. https://doi.org/10.32734/jormtt.v1i2.2835

Samantaray S, Sahoo A (2020) Estimation of flood frequency using statistical method: Mahanadi River basin, India. H_2 Open J 3(1):189–207. https://doi.org/10.2166/h2oj.2020.004

Sasireka K, Suribabu CR, Neelakantan TR (2019) Extreme rainfall returnrainfall return periods using Gumbel and Gamma distributions. Int J Recent Technol Eng 8(4):27–29. https://doi.org/10.35940/ijrte.D1007.1284S219

Sharma NK, Sharma S (2019) Frequency analysis of rainfall data of Dharamshalaregion. Int J Sci Res 8(2):886–892. https://doi.org/10.1051/matecconf/20165703013

Shirin S, Yadav AK (2021) Assessment and suitability for the dumping of flyash in an abandoned mine. Environ Challenges 4:100111. https://doi.org/10.1016/j.envc.2021.100111

USGS (2019) Guidelines for determining flood flow frequency Bulletin 17C. In: Hydrologic analysis and interpretation, United States Geological Survey (USGS), Washington

Vaidya VB, Suvarna D, Kulshreshtha MS (2020) Evaluation of frequency analysis of distinctive rainfall intensity for various stations of Gujarat. Int J Innov Sci, Eng Technol 7(12):420–435. Google Scholar

Van derSpuy D, Du Plessis JA (2022) Flood frequency analysis—part 1: review of the statistical approach in South Africa. Water SA 48(2):110–119. https://doi.org/10.17159/wsa/2022.v48.i2.3848.1

Vivekanandan N, Srishailam C (2020) Selection of best fit probability distribution for extreme value analysis of rainfall. CBIP Water Energy Int 63(10):13–19. Google Scholar

Vivekanandan N, Mathew FT (2010) Probabilistic modelling of annual d-day maximum rainfall. ISH J Hydraul Eng 16(1):122–133. https://doi.org/10.1080/09715010.2010.10515021

Vivekanandan N (2020) Evaluation of estimators of probability distributions for frequency analysis of rainfall and river flow data. Int J Sci Res Civ Eng 4(4):67–75. http://ijsrce.com/IJSRCE204412

Yadav AK, Sahoo SK, Mahapatra S, Kumar AV, Pandey G, Lenka P, Tripathi RM (2014) Concentrations of uranium in drinking water and cumulative, age-dependent radiation doses in four districts of Uttar Pradesh, India. Toxicol Environ Chem 96(2):192–200. https://doi.org/10.1080/02772248.2014.934247

Yue S (2001) A bivariate gamma distribution for use in multivariate flood frequency analysis. Hydrol Process 15(6):1033–1045. https://doi.org/10.1002/hyp.259

Zhang J (2002) Powerful goodness-of-fit tests based on the likelihood ratio. J R Stat Soc (ser b) 64(2):281–294. https://doi.org/10.1111/1467-9868.00337

Chapter 16
Application of Machine Learning Model for Assessing Water Quality Index

Vinita Sangwan and Rashmi Bhardwaj

Abstract Predicting water quality is essential for safeguarding public health, protect the environment, ensuring sustainable resource management, and meet regulatory requirements. The complexity of forecasting water quality across a catchment area arises from the inherent non-linear interactions among water quality parameters and their temporal and spatial variability. This chapter introduces various machine learning models to improve water quality. In this chapter, the author delves into the foundational concepts of machine learning models and introduce an approach based on 8 Machine Learning Models for assessing Water Quality Index (WQI). The predictive performance of the ML models was assessed using different metrics such as accuracy, classification error, recall, precision, specificity, balanced accuracy and F1 score. The model considers physical and chemical parameters such as pH, EC, turbidity, total hardness, Total alkalinity, TDS, Mg, Ca, Cl^-, F^-, NO_3^- and HCO_3^- to predict the water quality index (WQI). As a real-world application, a case study of Farrukhabad district in Uttar Pradesh, where ML models were successfully used to predict groundwater quality. Given precise predictions of water quality, the findings could contribute to enhancing the National Environmental Policy on water resources through ongoing improvements in water quality.

Keyword Water quality index · Machine learning models · Electrical conductivity · Total dissolved solids

16.1 Introduction

Water is an important invaluable resource that all life needs to survive. Contamination of water leads to a decline in its quality, adversely affecting the well-being of marine organisms and consequently posing risks to human health, particularly those

V. Sangwan · R. Bhardwaj (✉)
University School of Basic and Applied Sciences, Guru Gobind Singh Indraprastha University, Dwarka, Delhi 110078, India
e-mail: rashmib@ipu.ac.in

A. K. Yadav et al. (eds.), *Integrated Management of Water Resources in India: A Computational Approach*, Water Science and Technology Library 129,
https://doi.org/10.1007/978-3-031-62079-9_16

reliant on aquatic resources. Consequently, monitoring of water quality becomes imperative to safeguard the survival and survival of marine life (Jain et al. 2011). Understanding water quality issues is essential for effective reduction and control of water pollution. Many governments around the world have initiated ecological water management programs to assess the health of aquatic ecosystems. Shockingly, approximately one billion people lack access to clean drinking water, resulting in two million deaths annually due to polluted water and inadequate sanitation. Therefore, preserving freshwater quality is of the utmost importance (Clark et al. 2011; Kumar et al. 2023; Yadav et al. 2014). However, surface water bodies face significant pressure due to increased human activities, including industrialization, urbanization, and the upsurge of the population (Ambade and Sethi 2021; Ambade et al. 2022). The quality of water plays a vital role in the sustainability of a diversion plan. Inferior quality water can incur significant costs, as resources need to be redirected to repair the water delivery infrastructure whenever challenges arise. There has been a growing demand for improved water governance and quality assurance to ensure the provision of affordable potable water. Addressing these challenges requires systematic evaluations of potable water supplies, waste disposal infrastructure, and administrative surveillance methods (Hu et al. 2019; Shirin et al. 2019). Furthermore, inadequate management of water availability and purity, coupled with effects of climate change, has resulted in a decline in water quality over recent decades, contributing to water pollution (Asadollah et al. 2021; Singha et al. 2021). The use of AI and ML technologies is presently crucial to address security risks, emphasizing mapping correlation between system data inputs and results rather than relying solely on intricate operational schemes (Waqas et al. 2022; Halim et al. 2020). Hence, the assessment and determination of the water quality status are currently a significant focal point.

The WQI serves as a tool used to assess water quality for various purposes. It helps in determining whether water is fit for drinking, industrial purposes, aquatic environments, and other uses. A higher WQI indicates better water quality (Bhardwaj and Verma 2017). Water Quality Classification distinguishes water into mildly contaminated or clean based on the range of WQI values it encompasses. The index developed by Brown et al. is calculated using physicochemical parameters of water, such as the concentrations (pH, turbidity, total alkalinity, total hardness, electrical conductivity, total dissolved solids (TDS), Ca, Mg, Cl, F, NO_3, and HCO_3) of other pollutants. It aims to assess the water quality level (Brown et al. 1970). The WQI provides policy makers and strategists with quantitatively meaningful insights to manage water resources. However, the calculations involved in formulating the WQI are extensive and require considerable time and effort (Singha et al. 2021). The formulations for calculating the WQI often lack consistency, as they typically rely on different equations. The WQI presents significant statistical insights for policy makers. However, a universally applicable method to predict and categorizing WQI remains elusive (Bui et al. 2020).

Over the last few years, ML techniques have been widely employed to assess river water, encompassing estimation of WQI (Tiyasha et al. 2020). Modeling using such techniques eliminates the need for sub-index calculations and expedites the generation of a WQI value. Furthermore, ML approach boasts advantages such as insensitivity to null entries and ability to manage intricate mathematical calculations involving vast amounts of data and structures which are not linear (Nearing et al. 2020). In terms of modelling and forecasting water quality, the effectiveness of many techniques consisting of Ada Boost, Gradient Boosting, Decision Tree, Random Forest, Extra Tree, K-Nearest Neighbors, Naïve Bayes, SVM has been documented (Nayan et al. 2020; Naloufi et al. 2021; Gazzaz et al. 2012; Hameed et al. 2017; Bowes et al. 2022; Prasad et al. 2022). Despite the availability of numerous ML algorithms, researchers continue to grapple with challenges, including determining which ML techniques are most suitable or best suited appropriate for a particular issue.

Numerous studies have been conducted to tackle issues related to water quality. Many studies use traditional laboratory testing and statistical methods to help monitor water quality. On the contrary, other research endeavour's leverage ML techniques to achieve enhanced approaches for addressing water quality concerns. (Wang et al. 2019) used the LSTM network to forecast drinking water quality. (Khullar and Singh 2022) introduced a Bi-LSTM model based on deep learning (DLBL-WQA) to predict water quality variables of the Yamuna River in India. Comparative analysis revealed that the proposed method outperformed all methods in terms of errors and predicting accuracy. (Hassan et al. 2021) utilized ML algorithms that include MLR, neural network, random forest, SVM to categorize a water quality in various locations in India based on various factors such as pH, CE, BOD, TC, Nitrate, DO are all influential in determining water quality. (Shams et al. 2021; Shirin and Yadav 2014) introduced a predictive regression model that incorporates saltwater intrusion indicators and AI methods to regulate groundwater salinization. (Abba et al. 2020) investigated various ML approaches of MLP, ANFIS, SVR, BPNN to predict WQI and obtained results illustrated the effectiveness of the intelligent methods in predicting WQI at 3 stations, taking advantage of the superior modelling outcomes of NNE. (Garabaghi et al. 2021) introduced ensemble learning approaches-based ML algorithms, which include Ada Boost, Random Forest, XGBoost, LogitBoost, for water quality categorization. (Elbeltagi et al. 2022) applied standalone techniques like the SVM, AR, RSS, M5P tree model to predict WQI based on the variable elimination strategy. A review of the literature highlights that each machine learning algorithm possesses unique strengths and weaknesses. Furthermore, its performance is dependent on the input variables related to water quality within different study regions. The objective of this research paper is to evaluate the accuracy of different machine learning algorithms, which comprise Ada Boost, Gradient Boosting, Decision Tree, Random Forest, Extra Tree, K-Nearest Neighbors, Nave Bayes, SVM to forecast water quality.

Predicting water quality using various ML techniques proves to be a more reliable method when contrasted with evaluating it using any single technique, which is more reliable than assessing it with any single technique. Thus, in pursuit of this objective, a model is constructed to predict water quality and its performance is evaluated against well-established ML models, namely different machine learning algorithms, which comprise Ada Boost, Gradient Boosting, Decision Tree, Random Forest, Extra Tree, K-Nearest Neighbors, Naïve Bayes, SVM. This chapter is divided into four sections. Section 16.2 outlines the materials and methods used for the analysis in this study.

Section 16.3 contains a case study, and Sect. 16.4 contains the results and discussion of the experiment, while the chapter concludes in Sect. 16.5.

16.2 Materials and Methodology

16.2.1 Water Quality Parameters

Water quality is influenced by its physical, chemical, and biological components. Alterations in parameter values such as pH, EC, Turbidity, total hardness, Total alkalinity, TDS, Mg, Ca, Cl^-, F^-, NO_3^- and HCO_3^- result in variations in water quality.

- **pH**: It stands as one of the crucial attributes in evaluating water quality. According to WHO standards, the acceptable pH value falls within the range of 6.5–8.5. If the pH value drops below 6.5, water loses its ability to facilitate the absorption of vitamins and minerals in the human body. On the contrary, if the pH rises above 8.5, it can cause skin irritation and impart a salty taste to the water. Aquatic life cannot thrive when the pH of the water falls within the range of 3.5–4.5.
- **Electrical Conductivity**: It is an assessment of water's ability to conduct electricity. As the concentration of dissolved solids and inorganic chemicals increases, the conductivity also increases. The highest allowable value is 250 μS/cm
- **Turbidity**: Turbidity refers to the degree of cloudiness in water, primarily caused by suspended particles.
- **Total Hardness**: The hardness of water is a trait that prevents soap from lathering and increases the boiling point of the water.
- **Total Alkalinity**: Total alkalinity refers to the water's capacity to neutralize acids. Alkalinity is due to minerals dissolved in water, often derived from mineral-rich soils. In rare cases, ammonia or hydroxide may also contribute to alkalinity.
- **Total Dissolved Solids**: TDS serves as a gauge to determine the combined amount of organic and inorganic substances present in a liquid.
- **Magnesium**: Magnesium plays a vital role in shaping the quality of water. Magnesium is carried away from rocks through erosion and eventually finds its way into the water. Magnesium serves various functions, and consequently, it can find its way into water through multiple pathways. Magnesium is a vital mineral absorbed by living beings, except insects. It is found in rivers and rainwater and in saltwater, enabling its organic dispersion throughout the ecosystem. Elevated levels of magnesium contribute to hard water.
- **Calcium**: Calcium occurs naturally in water, primarily originating from the dissolution of rocks such as marble, calcite, limestone, apatite, fluorite, gypsum, and dolomite in aquifer. The concentration of Ca rises when water encounters porous rocks containing mineral deposits such as limestone or dolomite. Calcium primarily contributes to the hardness of water and can potentially affect the toxicity of other chemicals.

- **Fluoride**: Fluorine is the lightest halogen and among elements that react the most in the universe. In water, fluorine exists as negatively charged ion, occurring in either small amounts or large amounts as a major ion. Long-term exposure to drinking water containing more than 1.5 ppm fluoride can lead to health problems, as described by the WHO. The Bureau of Indian Standards specifies that the maximum allowed F^- content in drinking water is 1.5 mg/l.
- **Bicarbonates**: A potential source of bicarbonate is the presence of organic materials in the aquifer, which undergo oxidation to produce CO_2, thus facilitating mineral solubilization. Fifty percent of HCO_3^- ions originate from old carbon present in calcite and dolomite within the aquifer. This process of weathering introduces Ca, Mg, and HCO_3^- ions into groundwater. HCO_3^- ions can also be generated by weathering of silicate minerals.
- **Nitrates**: Naturally, NO_3^- concentrations in groundwater are low, usually below 10 mg/L. However, human activities such as industry, agriculture, combustion engine emissions, and residential effluents contribute to an increase in NO_3^- concentrations. Pollution of groundwater can directly impact human health, as elevated levels of NO_3^- in drinking water might make people feel unwell.
- **Chloride**: The concentration of chloride in natural water varies widely. Chlorides are a common form of water contamination. The abundance of Cl^- in water in excessive amounts is main reason it is not suitable for drinking. Chloride contamination in surface water may arise from nearby storage of salt, the mixing of freshwater with the decomposition of saline industrial waste, among other factors. The migration of salts from the soil into the well water reservoir leads to presence of Cl^-, a common source of contamination in well water. The Cl^- limit deemed acceptable by the BIS is 250 mg/l. Cl^- increases the EC of water, rendering it corrosive.

16.2.2 WQI

For assessing comprehensive vulnerabilities present in water quality, WQI is used. Furthermore, it helps to delineate water quality to pinpoint potential problems related to its suitability for consumption, thereby enhancing the efficacy of protective measures. In this chapter, the physicochemical factors that include pH, EC, Turbidity, total hardness, Total alkalinity, TDS, Mg, Ca, Cl^-, F^-, NO_3^- and HCO_3^- are considered in calculation of WQI. Initially, a weight (w_i) is allocated to every physicochemical parameter using a scale ranging from 1 to 5 as per Table 16.1, considering its significance for human health and appropriateness for drinking purposes.

The sub-index of every parameter is calculated using the following equation:

$$SI_i = W_i \times q_i, \tag{16.1}$$

here q_i denotes quality rating scale of every parameter and W_i denotes relative weight of every parameter and computed using formula below:

Table 16.1 Displays the weight and the relative weight of the parameters

Parameters	Bureau of Indian Standards	Weight (w_i)	Relative weight (W_i)
pH	8.5	4	0.148
Total hardness	200	2	0.074
TDS	500	4	0.148
Magnesium	30	2	0.074
Calcium	75	2	0.074
Fluoride	1	4	0.148
Bicarbonates	200	1	0.037
Nitrates	45	5	0.185
Chloride	250	3	0.111
		$\sum_{i=1}^{n} w_i = 27$	$\sum W_i = 1$

$$W_i = \frac{w_i}{\sum_{i=1}^{n} w_i} \tag{16.2}$$

The WQI is calculated using the following equation:

$$\text{WQI} = \sum SI_i \tag{16.3}$$

According to Chaurasia et al. (2018), if WQI is less than 50 then the type of water is excellent, if WQI lies between 50 and 100, then it is considered as good, if it lies within 100.1–200 then it is poor water, if it lies in between 200.1 and 300, then it is considered in the category of very poor water and finally if WQI is greater than 300.1 then that water is not suitable for consumption.

Groundwater samples were partitioned into two sets: 80% for training and 20% for testing. This division ratio is commonly employed in data-driven modelling (Bui et al. 2020; Nouraki et al. 2021). To improve the predictive accuracy and training speed of the ML models, groundwater quality data were normalized to the 0–1 range before the modeling process, using the following equation:

$$\mathrm{x_i}' = \frac{\mathrm{x_i} - \mathrm{x_{min}}}{\mathrm{x_{max}} - \mathrm{x_{min}}} \tag{16.4}$$

where $\mathrm{x_i}$ and $\mathrm{x_i}'$ represent original value and normalised value of pH, EC, Total Dissolved Solids, etc., while $\mathrm{x_{min}}$ and $\mathrm{x_{max}}$ denote min. and max. values of that specific variable.

16.2.3 Machine Learning Models

As previously stated, this chapter employed different machine learning models to predict SQI which comprise Ada Boost, Gradient Boosting, Decision Tree, Random Forest, Extra Tree, K-Nearest Neighbors, Naïve Bayes, SVM.

- **Adaptive Boosting and Gradient Boosting**:

The boosting algorithm serves as a meta-ensemble algorithm designed to enhance predictive capabilities of multiple underlying base learners. Its primary objective is to mitigate bias and variance in the context of labelled dataset challenges. The fundamental concept of the boosting technique involves initial construction of a model using training data. Subsequently, another model is created based on the preceding one, with aim of minimizing bias errors that may emerge when the initial model fails to capture relevant patterns in the provided data. Upon the incorporation of a new learning algorithm, there is a weight adjustment of the data, a process commonly referred to as 're-weighting.' The sequential addition of these models continues until the ensemble model adequately predicts the training data or reaches the maximum specified number of learners. In the present study, two distinct boosting algorithms, namely adaptive boosting (AdaBoost), gradient boosting (GBM), were employed. Comprehensive information about these boosting algorithms is provided in (Wu et al. 2020) work.

- **Decision Tree, Random Forest, and Extra Trees**:

The decision tree, along with its numerous variations, represents an alternative class of decision-making algorithms that partition input space into distinct regions, each with its own set of parameters. They fall under the category of nonparametric supervised learning methods extensively utilized for tasks such as classification and regression. Additionally, they graphically and clearly depict decisions and processes involved in decision making, and standard configuration of this model follows a flowchart shaped like a tree, reflecting its nomenclature. In this structure, every decision node signifies a 'test' conducted on an attribute, each branch signifies the result of the test, and each leaf node denotes a class label or prediction, representing the final decision made after evaluating all attributes. Moreover, the routes from root to leaf nodes symbolize classification rules. In this study, 3 models based on decision trees, employing diverse training algorithms: decision tree, random forest, and extra trees were employed. (Ahmad et al. 2018) provide comprehensive details on these decision tree-based algorithms.

- **Naïve Bayes**:

Naïve Bayes model is widely utilized in ML applications due to its simplicity, allowing each attribute to contribute equally to the final decision. This straightforwardness not only enhances computational efficiency but also renders the Naïve Bayes technique appealing and applicable across diverse fields. The key components of the Naïve Bayes Classifier encompass three aspects: prior probability, posterior probability, and class conditional probability (Wibawa et al. 2019).

- **(SVM) Support Vector Machine**:

SVM—supervised ML model used primarily for classification assignments. Although SVM algorithms can be used for regression analysis, they are commonly employed in practice for classification purposes, such as categorizing binary data into two separate categories. SVM is centred around fundamental concept of identifying optimal decision boundary, which is a line plotted on graph that effectively divides data points into two separate categories. This decision boundary, also known as a decision plane, a hyperplane, or a boundary front, serves as a key element in SVM. SVM works by determining the nearest data point called the 'support vector' from both classes to the boundary lines. It aims to optimize a hyperplane to maximize margin, which signifies greatest distance between supporting data points. The term 'support vectors' in SVM refers to the coordinates of individual data points in the plotted graph, where the value of each feature in a data point corresponds to a specific coordinate (Gupta et al. 2021).

- **K-Nearest Neighbors**:

The KNN algorithm serves as a supervised ML approach, primarily used for classification tasks. In this algorithm, there is a flexible factor called 'k,' which indicates the count of 'nearest neighbors. In operation, the KNN algorithm identifies closest data points from a training dataset based on proximity to a given query. The process of finding the nearest data points depends on how close they are to the query point. After identifying the 'k' closest data points, the algorithm uses a rule in which it looks at which class appears most frequently to determine the final classification. Then, this most prevalent class is assigned as the final classification for the query (Uddin et al. 2022).

16.2.4 Performance of ML Models

The effectiveness of proposed models is evaluated against various existing models. To assess the classification models, performance metrics including accuracy, recall, precision, specificity, classification error, balanced accuracy, and F1 score were utilized.

- **Accuracy**: Accuracy is the predominant measure employed in classifier evaluation. It gauges algorithm's overall effectiveness by estimating probability of accurately predicting class label's actual value, and it is represented by the following equation:

$$\text{Accuracy} = \frac{TP + TN}{TP + FP + FN + TN}, \tag{16.5}$$

where TP = True Positive, TN = True Negative, FP = False Positive and FN = False Negative

- **Recall/ Sensitivity**: Sensitivity, also known as recall, computes proportion of true positive outcomes accurately predicted by model and defined as follows:

$$\text{Recall} = \frac{TP}{TP + FN} \tag{16.6}$$

- **Precision**: It is defined as follows:

$$\text{Precision} = \frac{TP}{TP + FP} \tag{16.7}$$

- **Specificity**: It quantifies the accuracy of true negative outcomes accurately predicted by the model and is defined as follows:

$$\text{Specificity} = \frac{TN}{TN + FP} \tag{16.8}$$

- **Classification Error (CE)**: It represents an approximation of likelihood of mischaracterisation derived from model predictions and defined as follows:

$$\text{CE} = 1 - \text{Accuracy} \tag{16.9}$$

- **Balanced Accuracy**: Balanced accuracy serves as a valuable metric, particularly in scenarios where data are in imbalance. Consider both positive and negative outcome classes and remains robust even in the presence of imbalanced datasets. It is defined as the average of recall and Specificity.

$$\text{Balanced Accuracy} = \frac{Recall + Specificity}{2} \tag{16.10}$$

- **F1 Measure or Score**: Harmonic mean of precision and sensitivity provides more comprehensive evaluation of accuracy, with values 0–1, a higher score indicates better accuracy. It can be computed as follows:

$$\text{F1 Score} = \frac{2 \times Recall \times Precision}{Recall + Precision} \tag{16.11}$$

16.3 Case Study

16.3.1 Study Area

The Farrukhabad district of Uttar Pradesh covers a total geographical area of 228,830 hectares, consisting of three tehsils, Ampritpur, Farrukhabad and Kaimganj. The map in Fig. 16.1 displays location map of Farrukhabad. However, the region examined by the researcher within this district includes 7 sections namely Rajepur, Barhpur, Nawavganj, Kamalganj, Shamsabad, Mohamdabad, Kaimganj, situated in Farrukhabad district. This area spans latitudes between 26°46' and 27°43'N and longitudes between 79°7' and 80°2'E. The climate is characterized by hot and dry summers, along with pleasantly chilly winter seasons, falling under the category of subtropical humid. During the monsoon season, an excess of water is attainable for thorough infiltration into the soil. Usually, the coldest it gets during the day is about 8 °C, and the warmest it gets is around 23 °C. In May, it is typically the hottest, with an average utmost and least temp. of 41 °C and 26 °C, respectively. The maximum monthly temperature recorded at 32.2 °C, with the minimum monthly temperature averaging 19.5° C. Morning relative humidity averages 65 percent monthly, while evening relative humidity stabilizes around 47 percent. Winds are light, blowing at an average speed of 9.6 km per hour and water-vapour loss potential measures 1464.5 mm. The region receives an average rainfall of 810 mm. The tropical forests in this area typically feature short shrubs and thorn-shaped plants. In 2011, Farrukhabad recorded a population of 18.87 lakhs, consisting of 53.36% males and 46.63% females. Farrukhabad is located on the fertile plains of the Ganges River, where the river serves as a vital component of the district's irrigation system. Rajepur block is situated in the Ganga and Ramganga rivers. The primary sources of irrigation for the district are groundwater and canals. Most areas within the district have the capacity to produce three crops annually, boasting exceptionally high acreage productivity. The administrative division (tahsil) of Kaimganj, located approximately 25 km from Farrukhabad, is famous for its production of mangoes, guavas, and various other fruits. Tobacco cultivation is also of significant importance in the region, with Kaimganj serving as the center for the chewing tobacco industry. Additionally, Kaimganj also houses a sugar mill. In recent times, several small-scale agro industries, such as rice plants and oil mills, have been established in the district.

16.3.2 Samples Collection

Groundwater samples were collected at 14 distinct locations around Furrakhabad and represent them as Site 1 to Site 14 as shown in Table 16.2 and the samples from 14 different locations underwent examination for 12 physicochemical parameters: pH, EC, turbidity, total hardness, Total alkalinity, TDS, Mg, Ca, Cl^-, F^-, NO_3^- and HCO_3^-.

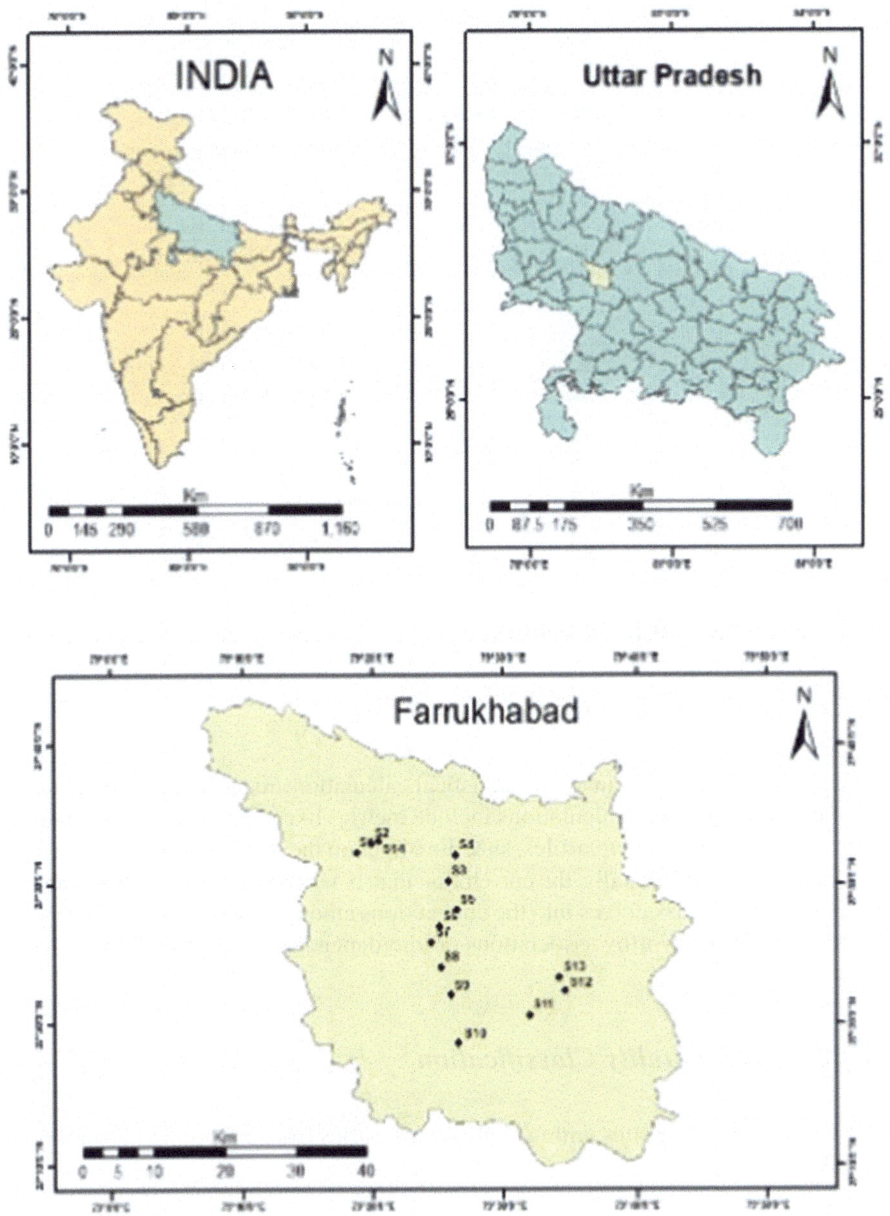

Fig. 16.1 Location map of study area and sample sites

Table 16.2 Sampling sites

Site	Coordinates
1	25.54° N,79.31° E
2	27.55° N,79.33° E
3	27.50° N,79.43° E
4	27.53° N,79.43° E
5	27.47° N,79.44° E
6	27.45° N,79.42° E
7	27.43° N,79.41° E
8	27.40° N,79.42° E
9	27.37° N,79.43° E
10	27.31° N,79.44° E
11	27.34° N,79.53° E
12	27.97° N,79.58° E
13	27.38° N,79.57° E
14	27.55° N,79.34° E

16.4 Results and Discussions

16.4.1 Statistical and Correlation Analysis

In Table 16.3, the analysis involved statistical calculations regarding the characteristics of the data set. These calculations include metrics like mean, standard deviation, minimum, maximum, and quartiles, shedding light on the distribution and characteristics of the data. Additionally, the correlation matrix was scrutinized, as illustrated in Table 16.4. This matrix delves into the connections among various features, aiding in the detection of noteworthy associations or interdependencies among the variables.

16.4.2 Water Quality Classification

The WQI for sampling sites within study region varies from 51.2 to 129.5, as shown in Table 16.5.

16.4.3 Performance of ML Models

The outcomes of the prediction using 8 ML models, including Ada Boost, Gradient Boosting, Decision Tree, Random Forest, Extra Tree, K-Nearest Neighbors, Naïve Bayes, SVM were outlined. These classification algorithms utilized pH, CE, turbidity,

Table 16.3 Statistical analysis of water parameters

Parameters	Min	Max	Mean	Std. dev	CV	Skewness	Kurtosis
pH	7.1	7.79	7.4	0.23	0.05	0.18	−1.33
Turbidity	0	4	1.36	1.34	1.78	0.58	−0.74
Total alkalinity	252	638	395.43	111.85	12,511.19	0.92	0.50
Total hardness	22	556	316.43	154.44	23,851.19	−0.018	−0.57
EC	442	1429	815.36	326.43	106,555.5	0.50	−1.13
TDS	213	599	403.93	131.43	17,274.84	0.13	−1.38
Ca	72	174	112	32.25	1039.85	0.68	−0.47
Mg	42	92	68	15.75	248.15	0.18	−1.01
C^-	6	210	70.96	73.21	5359.68	1.15	−0.19
F^-	0.16	2.26	0.64	0.56	0.32	2.28	5.32
NO_3^-	2.6	22	8.56	5.92	35.08	1.25	0.73
HCO_3^-	224	408	291.57	58.26	3394.88	0.73	−0.43

total hardness, Total alkalinity, TDS, Mg, Ca, Cl^-, F^-, NO_3^- and HCO_3^-. Performance metrics of ML models were scrutinised in Table 16.6.

Table 16.4 Correlation analysis of water parameters

	Ph	Turbidity	TA	TH	EC	TDS	Ca	Mg	C^-	F^-	NO_3^-	HCO_3^-
Ph	1											
Turbidity	−0.21	1										
TA	0.22	0.54	1									
TH	−0.57	0.06	0.26	1								
EC	−0.34	0.31	0.03	0.016	1							
TDS	−0.68	0.42	0.06	0.51	0.09	1						
Ca	−0.56	0.09	−0.08	0.68	−0.21	0.71	1					
Mg	0.07	0.11	0.002	−0.16	0.24	0.39	0.22	1				
C^-	−0.19	0.3	0.22	0.24	0.54	0.02	0.1	0.18	1			
F^-	−0.3	0.1	−0.09	−0.02	0.39	0.32	0.23	0.63	0.62	1		
NO_3^-	−0.3	0.02	−0.37	−0.04	0.39	0.11	−0.2	−0.09	0.29	0.2	1	
HCO_3^-	−0.33	0.11	−0.11	0.16	0.68	0.06	−0.02	0.17	0.72	0.65	0.51	1

Table 16.5 WQI values and water type

Sampling sites	WQI	Water type
Site 1	62.6	Good water
Site 2	60	Good water
Site 3	56.8	Good water
Site 4	88.1	Good water
Site 5	71.3	Good water
Site 6	51.2	Good water
Site 7	64.7	Good water
Site 8	64	Good water
Site 9	68.8	Good water
Site 10	80.5	Good water
Site 11	73.4	Good water
Site 12	109.1	Poor water
Site 13	129.5	Poor water
Site 14	78.8	Good water

Table 16.6 Performance metrics of ML models

ML models	Adaptive boosting	Gradient boosting	Decision tree	Extra tree	Random forest	Naïve Bayes	SVM	k−nearest neighbor
Accuracy	93.37	94.90	86.22	86.27	93.88	90.31	92.86	91.84
Classification error	6.63	5.10	13.78	13.73	6.12	9.69	7.14	8.16
Recall	82.50	80	70	70	80	77.50	72.50	70
Precision	94.62	94.12	65.12	66.34	88.89	75.61	90.62	87.50
Specificity	96.15	98.72	90.38	91.44	97.44	93.59	98.08	97.44
Balanced accuracy	89.33	89.36	80.19	80.72	88.72	85.54	85.29	83.72
F1 score	83.54	86.49	67.47	68.12	84.21	76.54	80.56	77.78

16.5 Conclusion

In this study, various ML models were investigated to predict water quality status, which delineates distinct classes according to WQI. The methodology proposed in this study utilized pH, CE, turbidity, total hardness, Total alkalinity, TDS, Mg, Ca, Cl^-, F^-, NO_3^- and HCO_3^- parameters to evaluate water quality along with 8 ML models, including Ada Boost, Gradient Boosting, Decision Tree, Random Forest, Extra Tree, K-Nearest Neighbors, Naïve Bayes, SVM. Analysis revealed that the gradient boost model outperformed among the 8 algorithms in efficiently predicting water quality status and exhibited the lowest Classification Error along with highest Recall, Precision, Specificity, Balanced Accuracy, and F1 Score. Random Forest stands after Gradient Boosting, while Decision Tree demonstrated poorest performance, exhibiting the lowest among in case of every metric. Therefore, relying on precise quality of water predictions, findings could contribute to ongoing improvements in quality of water and inform the formulation of national environmental policies regarding water resources. Additionally, they could aid in the preservation of the diversity and vitality of the water. In future research, it is recommended that attention be given to the challenge of imbalanced data in water quality datasets, as it has an impact on prediction accuracy.

Acknowledgements The authors are grateful to the "Guru Gobind Singh Indraprastha University" for financial support and research facilities.

References

Abba SI, Pham QB, Saini G, Linh NTT, Ahmed AN, Mohajane M, Khaledian M, Abdulkadir RA, Bach QV (2020) Implementation of data intelligence models coupled with ensemble machine learning for prediction of water quality index. Environ Sci Pollut Res 27(33):41524–41539

Ahmad MW, Reynolds J, Rezgui Y (2018) Predictive modelling for solar thermal energy systems: a comparison of support vector regression, random forest, extra trees, and regression trees. J Clean Prod 203:810–821

Ambade B, Sethi S (2021) Health risk assessment and characterization of polycyclic aromatic hydrocarbon from the hydrosphere. J Hazard, Toxic, Radioactive Wast 25(2):05020008

Ambade B, Sethi SS, Giri B, Biswas JK, Bauddh K (2022) Characterization, behavior, and risk assessment of polycyclic aromatic hydrocarbons (PAHs) in the estuary sediments. Bull Environ Contam Toxicol 108(2):243–252

Asadollah S, Sharafati A, Motta D, Yaseen ZM (2021) River water quality index prediction and uncertainty analysis: a comparative study of machine learning models. J Environ Chem Eng 9(1):104599

Bhardwaj D, Verma N (2017) Research paper on analysing impact of various parameters on water quality index. Int J Adv Res Comput Sci 8(5):2496–2498

Bowes BD, Wang C, Ercan MB, Culver TB (2022) Reinforcement learning-based real-time control of coastal urban stormwater systems to mitigate flooding and improve water quality. Environ Sci: Water Res Technol 8(12):2065–2086

Brown RM, McClelland NI, Deininger RA, Tozer RG (1970) A water quality index: do we dare? Water Sewage Works 117(10):339–343

Bui DT, Khosravi K, Tiefenbacher J, Nguyen H, Kazakis N (2020) Improving prediction of water quality indices using novel hybrid machine-learning algorithms. Sci Total Environ 721:137612

Chaurasia AK, Pandey HK, Tiwari SK, Prakash R, Pandey P, Ram A (2018) Groundwater quality assessment using water quality index (WQI) in parts of Varanasi District, Uttar Pradesh, India. J Geol Soc India 92:76–82

Clark RM, Hakim S, Ostfeld A (2011) Handbook of water and wastewater systems protection. In: Protecting critical infrastructure. Springer, New York, pp 1–29. https://doi.org/10.1007/978-1-4614-0189-6

Elbeltagi A, Pande CB, Kouadri S, Islam ARM (2022) Applications of various data-driven models for the prediction of groundwater quality index in the Akot basin, Maharashtra, India. Environ Sci Pollut Res 29:17591–17605

Garabaghi FH, Benzer S, Benzer R (2021) Performance evaluation of machine learning models with ensemble learning approach in classification of water quality indices based on different subset of features. Res Square 1:1–35

Gazzaz NM, Yusoff MK, Aris AZ, Juahir H, Ramli MF (2012) Artificial neural network modeling of the water quality index for Kinta River (Malaysia) using water quality variables as predictors. Mar Pollut Bull 64(11):2409–2420

Gupta BM, Dhawan SM, Mamdapur GM (2021) Support vector machine (SVM) research in India: a scientometric evaluation of India's publications output during 2002–19. J Indian Lib Assoc 57(3):12–25

Halim Z, Waqar M, Tahir M (2020) A machine learning-based investigation utilizing the in-text features for the identification of dominant emotion in an email. Knowl-Based Syst 208:106443

Hameed MM, Sharqi S, Yaseen ZM, Afan HA (2017) Application of artificial intelligence (AI) techniques in water quality index prediction: a case study in tropical region, Malaysia. Neural Comp Appl 28(1):893–905

Hassan MM, Akter L, Rahman MM (2021) Efficient prediction of water quality index (WQI) using machine learning algorithms. Human-Centric Intelligent System 1(3–4):86–97

Hu Z, Zhang Y, Zhao Y, Xie M, Zhong J, Tu Z, Liu J (2019) A water quality prediction method based on the deep LSTM network considering correlation in smart mariculture. Sensors 19(6):1420

Jain J, Shah S, Mehta H, Lodaria A, Kurup L (2011) A machine learning approach to analyze marine life sustainability. In: Proceedings of International Conference on Intelligent Computing, Information and Control Systems. Advances in Intelligent Systems and Computing, Vol 1272. Springer, Singapore. https://doi.org/10.1007/978-981-15-8443-5_53

Khullar S, Singh N (2022) Water quality assessment of a river using deep learning Bi-LSTM methodology: forecasting and validation. Environ Sci Pollut Res 29(9):12875–12889

Kumar A, Chava R, Gupta S, Shirin S, Jamal A, Yadav AK (2023) Environmental and health effects of fluoride contamination and treatment of wastewater using various technologies. In: Advanced treatment technologies for fluoride removal in water. Water Science and Technology Library, 125, Springer, Cham. https://doi.org/10.1007/978-3-031-38845-3_18

Naloufi M, Lucas FS, Souihi S, Servais P, Janne A, Abreu MD (2021) Evaluating the performance of machine learning approaches to predict the microbial quality of surface waters and to optimize the sampling effort. Water 13(18):2457

Nayan AA, Kibria MG, Rahaman MO, Saha J (2020) River water quality analysis and prediction using GBM. In: Proceedings of the 2020 2nd International Conference on Advanced Information and Communication Technology (ICAICT), Dhaka, Bangladesh, 28–29 November, IEEE: New York, NY, USA, pp 219–224

Nearing GS, Kratzert F, Sampson AK, Pelissier CS, Klotz D, Frame JM, Prieto C, Gupta HV (2020) What role does hydrological science play in the age of machine learning? Water Resour Res 57(3). https://doi.org/10.1029/2020WR028091

Nouraki A, Alavi M, Golabi M, Albaji M (2021) Prediction of water quality parameters using machine learning models: a case study of the Karun River, Iran. Environ Sci Pollution Res 28(40):57060–57072

Prasad DVV, Venkataramana LY, Kumar PS, Prasannamedha G, Harsana S, Srividya SJ, Harrinei K, Indraganti S (2022) Analysis and prediction of water quality using deep learning and auto deep learning techniques. Sci Total Environ 821(1):153311

Shams MY, Nosair A, AbouElmagd L, Hassanien AE (2021) Predictive model for progressive salinization in a coastal aquifer using artificial intelligence and hydrogeochemical techniques: a case study of the Nile Delta aquifer, Egypt. Environ Sci Pollution Res 29(7):9318–9340

Shirin S, Yadav AK (2014) Physico chemical analysis of municipal wastewater discharge in Ganga River, Haridwar District of Uttarakhand, India. Current World Environ 9(2):536

Shirin S, Yadav AK (2021) Assessment and suitability for the dumping of flyash in an abandoned mine. Environ Challeng 4:100111. https://doi.org/10.1016/j.envc.2021.100111

Shirin S, Jamal A, Ranjan P, Yadav AK (2019) Study on assessment of slope stability and mixed disposal of overburden in voids of Singrauli Coalfield. Environ Qual Manage 28(3):131–139

Singha S, Pasupuleti S, Singha SS, Singh R (2021) Prediction of groundwater quality using efficient machine learning technique. Chemosphere 276(4):130265

Tiyasha T, Yaseen ZM (2020) A survey on river water quality modelling using artificial intelligence models: 2000–2020. J Hydrol 585:124670

Uddin S, Haque I, Lu H (2022) Comparative performance analysis of K-nearest neighbour (KNN) algorithm and its different variants for disease prediction. Sci Rep 12(1):6256

Wang J, Liu P, Sangaiah AK, Xie Y, Yin X (2019) Analysis and prediction of water quality using LSTM deep neural networks in IoT environment. Sustainability 11(7):2058

Waqas M, Tu S, Halim Z, Rehman SU, Abbas G, Abbas ZH (2022) The role of artificial intelligence and machine learning in wireless networks security: principle, practice, and challenges. Artif Intell Rev 55(7):5215–5261

Wibawa AP, Kurniawan AC, Murti DMP, Adiperkasa RP, Putra SM, Kurniawan SA, Nugraha YR (2019) Naïve Bayes classifier for journal quartile classification. Int J Recent Contrib Eng, Sci & IT 7(2):91–99

Wu T, Zhang W, Jiao X, Guo W, Hamoud YA (2020) Comparison of five Boosting-based models for estimating daily reference evapotranspiration with limited meteorological variables. PLoS ONE 15(6):e0235324

Yadav AK, Sahoo SK, Mahapatra S, Kumar AV, Pandey G, Lenka P, Tripathi RM (2014) Concentrations of uranium in drinking water and cumulative, age-dependent radiation doses in four districts of Uttar Pradesh, India. Toxicol Environ Chem 96(2):192–200

Part III
Water Resources Management Strategies and Technologies

Chapter 17
An Innovative Technique for Collecting Subsurface Rainwater in Agriculture Fields with IoT

K. Umapathi, S. Nishanth, and S. Shobana

Abstract Rainwater management and harvesting constitute critical elements for the advancement of agricultural fields. The predominant challenge faced by most agricultural landscapes is the accumulation of excess rainwater, particularly during heavy rain seasons and instances of water stagnation. Effortless approach to the removal of this excess rainwater stored within agricultural fields during such periods remains a formidable task. While there are numerous methodologies to address excess surface water in fields, the development of a comprehensive solution to remove subsurface water in agricultural settings remains an ongoing research pursuit. In response to this challenge, our current study presents a novel system specifically tailored for eliminating surplus rainwater or stagnant water within agricultural fields. The implementation of this system not only facilitates the extraction of excess water from fields, but also plays a pivotal role in enhancing plant recovery rates and fostering improved plant growth dynamics. Operating discreetly beneath the soil surface, our developed system has been engineered to efficiently collect subsurface water up to a depth of one foot, exhibiting adaptability across diverse soil types. Its operational efficacy lies in its ability to precisely quantify the volume of excess water collected. Furthermore, by using state-of-the-art IoT technology, our system seamlessly integrates with cloud servers, enabling the storage of collected water data for subsequent in-depth analysis and informed decision-making. The integration of this innovative system not only addresses the immediate challenge of excess water in agricultural fields, but also promises a comprehensive approach to optimizing water management practices. This innovative solution has the potential to significantly improve crop yield by effectively using the collected data to inform precise irrigation strategies, thus contributing to sustainable agricultural development and improved resource utilization.

K. Umapathi (✉) · S. Shobana
Department of Biomedical Engineering, KIT-Kalaignarkarunanidhi Institute of Technology, Coimbatore 641402, India
e-mail: umapathi.uit@gmail.com

S. Nishanth
Department of Electrical and Electronics Engineering, M. Kumarasamy College of Engineering, Karur 639113, India

A. K. Yadav et al. (eds.), *Integrated Management of Water Resources in India: A Computational Approach*, Water Science and Technology Library 129,
https://doi.org/10.1007/978-3-031-62079-9_17

Keywords Rainwater harvesting · Excess rainwater collector · Agricultural fields · Iot · Sustainable farming · Environmental sustainability

17.1 Introduction

Agriculture is the backbone of many countries, but there are many factors that affect it, one of the factors is heavy rain. The rain affected in the agriculture field when it rains heavily water gets stored in the farming field if the water stored for a long time, it will affect the growth of the plant and the cause some root disease and root will decay it will almost cause the plant to die. If the water stagnates in the field for more than 2 days or 40 h for some particular crop, the plants or crops in the field will suffer health in the plant and also the root of the plant will start to decay. Rainfall plays a key role in crop growth beyond its impact on diseases. It significantly influences the pace of crop development from seed to harvest. The harmonious synergy between rainfall and efficient irrigation directly influences the speed of plant growth. This balance can significantly reduce the germination period and the overall duration between planting and harvesting.

Skymet reported in the article titled ‘Excess rain during 2019 monsoon damaged 32 lakh acres of agricultural land in 12 states’ 32 lakh hectares of agricultural land in 12 states were damaged due to heavy excess rain during 2019. To protect cotton, paddy, maize and horticultural crops, the government has guided farmers to remove the stagnated excess rainwater from the low-lying agriculture fields. The 4.5 lakh hectares (Guhathakurta et al. 2011) of agriculture field such as standing crops of soybeans, maize, sugarcane, toor and vegetables were damaged due to heavy rain in Maharashtra. In all countries, agriculture fields are suffered from the removal of excess rainwater in surface and subsurface ground.

The drainage of surface and subsurface water from the agriculture field depends up on the soil capacity and its characteristics. Therefore, all agriculture land is not good at draining water in some good cases. Additionally, the land should be a slope to easily drain the water and move towards the lower area of the land to grow the small plants well (Qorani et al. 1990). It would be great if the land was available with slope, but in all cases it is not possible. In practice, plain land is available higher than slope land.

Commonly farmers use water pump to remove the excess water from the agriculture field, but the water pump cannot collect the entire surface water stored in the field some quantity of water are still remain in the field and also the water pump cannot collect the water under the soil. This method also took a lot of energy and manpower. Additionally, various techniques were reported to collect excess rain water stored in the agriculture field during heavy rain and other stagnant conditions of water (Sevenhuijsen 1994).

Yadav et al. (2014) examined uranium concentrations in drinking water in four districts in Uttar Pradesh, India. The determined uranium levels ranged from 0.1 to 19.6 μg/L, well below the US Environmental Protection Agency’s limit of 30 μg/

L. The age-dependent dose of uranium ingestion varied from 0.14 to 48 μSv/year, posing minimal risk. The chemical risk showed that the water is safe for drinking.

The work carried out by Shirin et al. (2021) investigated the characteristics of acid mine drainage (AMD) treated with fly ash. They collected AMD samples from an abandoned mine and treated them with fly ash from thermal power plants in Singrauli Coalfield. The research analyzed the physicochemical parameters and metal content before and after the fly ash treatment. The findings suggest that fly ash has neutralization potential, which makes it effective for AMD treatment and potentially useful for the management of coalfield AMD pits.

Shirin and Yadav (2014) examined the physicochemical parameters of wastewater discharged into the Ganga River at Haridwar. Despite some variations, the quality of the results indicated that the water met permissible limits for domestic use, but vigilance is necessary due to climate changes and pollution. The removal of excess fluoride from water environments due to its impact on human health and the environment. It explores the use of metal–organic framework (MOF) materials for aquatic defluorination, highlighting the need for a global review of artificial water fluoridation practices and improved industrial safety regulations (Kulwant et al. 2023; Kumar et al. 2023). The study investigates (Patel et al. 2023) the use of iron-based materials to remove excess fluoride from water. It covers the preparation, characterization, and practical applications of these materials, emphasizing their potential to address environmental challenges related to fluoride contamination.

Soil moisture sensors are fundamental in optimizing rainwater management. These sensors offer real-time data on soil moisture content, aiding farmers in making informed irrigation and rainwater removal decisions. Advanced soil moisture sensors are now integrating IoT technology for precise data collection. Efficient rainwater management is closely linked to smart irrigation systems. These systems utilize technology to deliver precise amounts of water to crops, reducing waste, and ensuring optimal growth conditions. Weather stations, soil moisture sensors, and IoT connectivity are key components of smart irrigation. Rainwater harvesting complements rainwater management strategies by collecting and storing rainwater for agricultural use. This practice conserves water resources, reduces reliance on external sources, and promotes sustainability.

Efficient rainwater management reduces the strain on local water sources, benefiting both the environment and communities. Water recycling and reuse practices also contribute to sustainability. Proper rainwater management prevents soil erosion, maintaining healthy soil structure vital for crop growth and ecosystem stability. Rainwater management practices that reduce water waste and environmental impact support biodiversity conservation by preserving natural habitats and landscapes.

Rainwater harvesting is a sustainable practice that complements advanced rainwater removal and irrigation technologies. It involves collecting and storing rainwater for later use in agriculture. This practice not only conserves water resources but also reduces reliance on external water sources, such as wells or municipal supplies. Components of Rainwater Harvesting Systems A rainwater harvesting system typically includes the following components: Catchment surface: This surface, often the roof of a building or a specially designed catchment area, collects rainwater and directs it to the storage system. Gutters and Downspouts: Gutters and downspouts guide rainwater from the catchment surface to the storage system. Storage tanks: These tanks store the collected rainwater for later use. They come in various sizes and materials, including concrete, plastic, and metal. Filtration Systems: Filtration systems remove debris and contaminants from collected rainwater, ensuring that it is suitable for agricultural use. Distribution system: Pipes and pumps distribute harvested rainwater to irrigation systems or other areas where it is needed. Rainwater harvesting offers numerous advantages for agriculture: Resource Efficiency: By using rainwater, farmers reduce their dependence on groundwater or surface water sources, conserving these valuable resources. Cost savings: Rainwater harvesting can lead to cost savings by reducing water bills and the need for extensive irrigation systems. Sustainability: Rainwater harvesting is aligned with sustainable agricultural practices and contributes to environmental stewardship. Resilience: During periods of water scarcity or drought, stored rainwater provides a valuable backup supply for irrigation.

Research on climate-resilient rainwater management practices is crucial to help farmers adapt to changing conditions and mitigate the impact of climate-related challenges. Innovations in rainwater management infrastructure, including sustainable materials and eco-friendly construction practices, can improve the environmental sustainability of these systems. Governments play a crucial role in promoting sustainable rainwater management. Incentive programs and supportive policies can encourage the adoption of these practices. Emerging technologies such as artificial intelligence (AI) and blockchain have the potential to improve rainwater management by optimizing resource usage and ensuring transparency. Collaboration among researchers, agricultural experts, and policy makers is vital to address the complex challenges of rainwater management and promote sustainable practices throughout the world.

In the present work, we have developed a novel device to remove excess rain water or stagnant water in agricultural fields. The removal of excess water from agricultural fields improves the rate of recovery of the plant and enhances the growth of the plant. Our system will collect subsurface water up to a one-foot height of one foot under any type of soil. In addition, excess water can be transferred to the required agricultural fields by transporting the container. Our prototype model is built with industry 4.0 technologies for calculating and communicate the removal of excess water. The primary objectives of this research are as follows: To develop an innovative system capable of efficiently removing excess rainwater, both on the surface and subsurface, from agricultural fields. Enhance the growth and health of crops by addressing excess water issues promptly. To use IoT technology to collect data on rainwater removal, enabling precise analysis and informed decision-making for water resource management.

17.2 System Design

The objective of the work is to remove the excess water from both the surface and subsurface up to a foot under the soil. Figure 17.1 shows the proposed model prototype operational block diagram of the proposed model. In this diagram, the top cabin in the device consists of the motor and it can be used to collect water from Core Cylinder. The core cylinder is the main part of the system that collects and stored the surface and subsurface water within the cylinder part. The bottom of the core cylindrical is the spike structure that is firmly fixed inside the underground soil.

In addition, it consists of water intake hole in the bottom surface to collect deep subsurface water. Two water level sensors are attached to the top and bottom of the Core cylinder to identify the water levels and to take outside above the ground surface. For the prototype model, the amount of water removed is updated in cloud server using the IoT technology. Using this innovative system, excess surface and subsurface water in agricultural field can easily remove and the amount of excess water can be easily calculated with the data storage for future analysis.

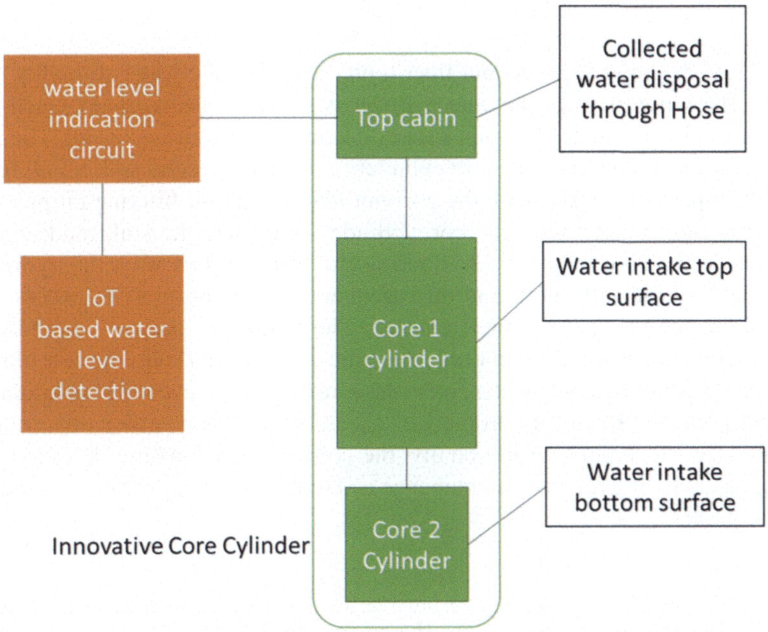

Fig. 17.1 Proposed model to remove subsurface water

17.2.1 Core Cylinder Construction

The core cylinder's top portion features three branch holes connected to small PVC pipes, enabling the collection of subsurface water. These PVC pipes also serve to stabilize the core cylindrical in the underground soil. Within these branch holes, Coir is packed for filtration purposes. This design ensures that only water enters the core cylindrical, preventing contamination from soil particles. Even small droplets of excess water can be efficiently collected using this coir technique. The core cylindrical structure is the heart of excess system to collect the surface and subsurface water from the agricultural field. Figure 17.2a shows the construction of the primary core cylinder. Total height of the system is 2 foot, where 1.5 foot of the cylinder is enclosed in underground of the soil and 0.5 foot top surface of the cylinder is shown in the above surface of the soil which is shown in Fig. 17.2b.

17.2.2 Coconut Fiber (Coir) Arrangements for Collecting the Water

Figure 17.3a, b shows the Coconut fiber (coir) to collect the water smoothly under the soil. The top portion of the core cylinder is also provided with 3 branch holes and connected with small PVC pipes for collecting the subsurface water. The small pipe structure is used to lay the core cylinder in the underground soil. Small branch hole PVC pipes are stacked with the coconut fiber (Coir) for filtering purpose. The coir allows only water inside the core cylinder and filters the soil, mud and dust to get inside the core cylinder. By filtering the water in this way, the cylinder is not affected by the soil, mud, and dust frequently. Even droplets of excess water can be collected using this coir technique. The bottom of the core cylinder is a spike structure that is highly anchored inside the underground soil. The coir filtration mechanism ensures that only water enters the core cylinder, preventing contamination from soil particles. Even tiny droplets of excess water can be efficiently collected using this Coir technique. Additionally, the coir arrangements are designed to be replaceable, allowing for easy maintenance and replacement of the filtration medium when necessary.

17.2.3 Interfacing Sensor Technology Inside the Core Cylinder

The complete setup of the core cylinder interfaced with the water level sensor is shown in Fig. 17.4. There are two water level sensors used in this setup. One is connected to the bottom of the cylinder to enable the indication that excess water is available inside the cylinder, and the second is connected top of the core cylinder

(a)

(b)

Fig. 17.2 **a** Full view of the primary core structure **b** inside the soil

(a)

(b)

Fig. 17.3 **a** and **b** Coconut fiber (Coir) inserted inside of the water collecting parts

to identify the level reached, and the pump motor is should be turned on to remove excess collected water inside the core cylinder.

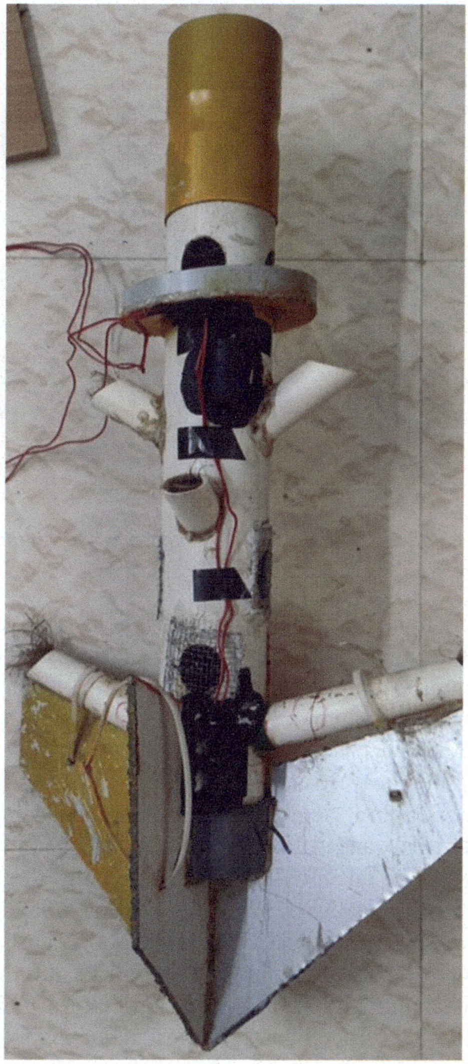

Fig. 17.4 Interfacing the sensors with the core cylinder

17.3 Results and Discussions

To assess the performance of our device, we conducted extensive testing in various agricultural fields. The core cylindrical, embedded in the soil at a depth of 2 feet, efficiently collected excess rainwater. To analyze the device performance, the complete set-up is placed inside the soil in the agriculture field as shown in the figure. In the 2-foot height of the device, the 1.5 foot of the cylinder is surrounded in underground of the soil and the 0.5-foot top surface of the cylinder is shown in the above surface of the soil, which is shown in Fig. 17.5. Testing the device performance and studies

Fig. 17.5 Complete system setup integrated with IoT technology

are done in various agricultural fields practically. Figure 17.6a–d shows the water is surrounded in the device by the surface of the soil, the water removed by the internal pumping motor through the hose, the water can be transfer or stored in some other places through hose, and the dry soil surface after the water is fully removed by the device respectively.

To investigate the performance of the device, the device is placed in various agricultural fields and the results are taken out. We have taken Red, Clay, Black, Laterite, and Alluvial soil fields. The results are shown in figure. Figure 17.7 shows the time taken by the cylinder to collect excess surface and subsurface water from the various fields. We have conducted the three trials in each field. From the figure we can understand that the red soil is absorbing the water very well and easily so the time is taking longer to recollect into the cylinder than any other soil. Red soil is commonly used for most agricultural fields and the crops which is cultivated in the field. Black soil has a less tendency to spread the water. Clay and alluvial soil which can be very hard to observe the water. Thus, the collection of excess water time is less. Our device collects all the excess water on the surface and subsurface of agricultural fields which are shown in figure. The collected water is pumped out using the prototype model pump and it can be transferred where the water requires or may be the highest outer water storage place for future use. The capacity of the prototype model core cylinder is 1 L. The prototype pumping motor will take one to two minutes to remove the water from the cylinder based on the signals received from the water level sensors. This process will be continued until the water is collected from the soils, which is shown in Fig. 17.6d. Since we have used the coir stacks in the water collecting places, even droplets of subsurface water can be collected up to the height of the soils. With the duration of the pump motor turned on and the volume of the cylinder we can calculate the amount of water taken off from the surface and

Fig. 17.6 **a** Water is surrounded in the device by the surface of the soil, **b** water removed by the internal pumping motor through a hose, **c** the water can be transferred or stored in some other places through a hose, and **d** the dry soil surface after the water is fully removed by the device respectively

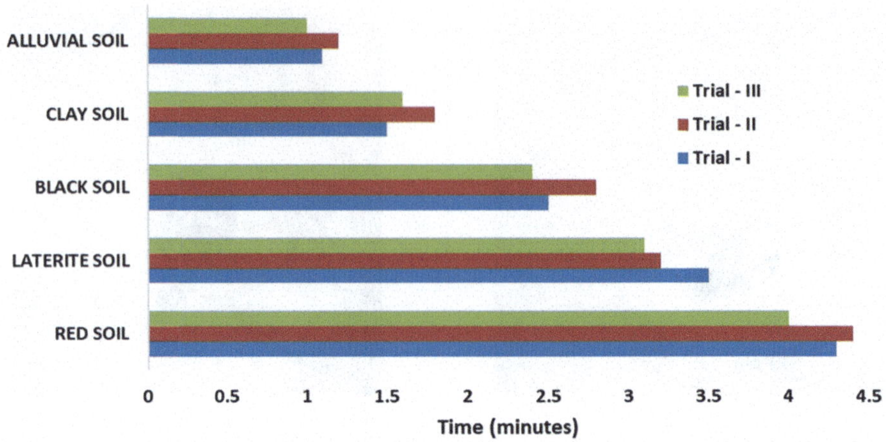

Fig. 17.7 Time taken to remove excess water from the various soils

subsurface of fields. Using the IoT technologies, the details can be stored in the cloud server for the calculation of excess water removed. Future analysis can also be done with references to the storage of the details.

From our observations, it is evident that the type of soil significantly affects the absorption of water and, consequently, the time required for the removal of water. Red soil, commonly used in agriculture, demonstrated higher water absorption rates, resulting in a longer removal time. In contrast, black, clay and Alluvial soils showed faster water removal times. Our core cylindrical system effectively collected excess water from all these soil types.

17.4 Conclusion

The removal of excess rainwater stored on the surface and subsurface of the agriculture field during heavy rain is one of the important requirements for growing the safe growth of plants and resource management of rainwater resources. Our innovative technique will remove excess water from the surface and subsurface of agriculture fields. Our system provides a high chance to protect crops in the agriculture field during the high rainy season and the technique is also applicable for the harvest and management. The removal of excess water from agricultural fields improves the rate of recovery of the plant and enhances the growth of the plant. Additionally, by storing the excess water through pump which can be transported using the water storage vehicles. From the results, it can be understood that the system will collect the subsurface water up to a one-foot height under any type of soil. By extending the number of devices in the equal places, the total agricultural area can cover using our system. The coir allows only water inside the core cylinder and it filters the soil, mud, and dust to get inside the core cylinder. By filtering the water in this way, the cylinder is not affected by the soil, mud, and dust frequently. Even droplets of excess water can be collected using this coir technique. Its hassle-free eco system. Additionally, using the IoT technologies the details can be stored in the cloud server to calculate the amount of excess water removed.

Declaration of Interest The authors declare that they have no known competing financial interests or personal relationships that could have appeared to influence the work reported in this document.

References

Guhathakurta P, Sreejith OP, Menon PA (2011) Impact of climate change on extreme rainfall events and flood risk in India. J Earth Syst Sci 120(3):359–373

Kulwant M, Patel D, Shirin S, Sharma SN, Yadav AK (2023) Future frameworks for fluoride and algorithms for environmental system. In: Advanced treatment technologies for fluoride removal in water. Water Science and Technology Library, Vol 125. Springer, Cham. https://doi.org/10.1007/978-3-031-38845-3_19

Kumar A, Chava R, Gupta S, Shirin S, Jamal A, Yadav AK (2023) Environmental and health effects of fluoride contamination and treatment of wastewater using various technologies. In: Advanced treatment technologies for fluoride removal in water. Water Science and Technology Library, Vol 125. Springer, Cham. https://doi.org/10.1007/978-3-031-38845-3_18

Patel D, Kulwant M, Shirin S, Varshney R, Pandey G, Yadav AK (2023) Fluoride removal from aqueous solution using iron-based materials: Preparation, characterization, and applications. In: Advanced treatment technologies for fluoride removal in water, Water Science and Technology Library, Vol 125. Springer, Cham. https://doi.org/10.1007/978-3-031-38845-3_4

Qorani M, Abdel Dayem MS, Oosterbaan RJ (1990) Evaluation of restricted subsurface drainage in rice fields. Symposium on land drainage for salinity control in arid and semi-arid regions, vol 3. Drainage Research Institute, Cairo, pp 415–423

Sevenhuijsen RJ (1994) Surface drainage systems. In: Ritzema HP (ed) Drainage principles and applications. ILRI Publication 16, pp 799–826. International Institute for Land Reclamation and Improvement, Wageningen, The Netherlands. ISBN:90-70754-33-9

Shirin S, Yadav AK (2014) Physico-chemical analysis of municipal wastewater discharge in Ganga River, Haridwar District of Uttarakhand, India. Current World Environ 9(2):536

Shirin S, Jamal A, Emmanouil C, Yadav AK (2021) Assessment of characteristics of acid mine drainage treated with fly ash. Appl Sci 11(9):3910

Yadav AK, Sahoo SK, Mahapatra S, Kumar AV, Pandey G, Lenka P, Tripathi RM (2014) Concentrations of uranium in drinking water and cumulative, age-dependent radiation doses in four districts of Uttar Pradesh, India. Toxicol Environ Chem 96(2):192–200

Chapter 18
The Role of Community Participation in Sustainable Integrated Water Resources Management: Challenges, Opportunities, and Current Perspectives

Mohd Akhter Ali and M. Kamraju

Abstract Community participation is recognized as a cornerstone of sustainable integrated water resources management (IWRM), yet its effective implementation faces various challenges and offers diverse opportunities. This paper aims to comprehensively assess the multifaceted nature of community participation in the context of IWRM. Through a comprehensive review of the existing literature, encompassing case studies and best practices, the article examines the complexities inherent in community involvement. It analyzes the challenges associated with ensuring meaningful participation, including issues of representation, power dynamics, and capacity building. In addition, it explores the opportunities presented by community engagement, such as leveraging local knowledge, promoting adaptive management strategies, and improving resilience to water-related challenges. In addition, the paper addresses current perspectives that influence community participation in IWRM, including the impacts of climate change, water scarcity, urbanization, and technological advances. By identifying key challenges, opportunities, and perspectives, this document seeks to provide information on enhancing the effectiveness of community participation in sustainable IWRM in evolving environmental and societal contexts, ultimately contributing to the broader goal of achieving sustainable water resources.

Keywords Integrated water resources management · Community participation · Sustainability · Stakeholder engagement · Challenges · Opportunities · Current perspectives

M. A. Ali (✉)
Department of Geography, Osmania University, Hyderabad, India
e-mail: drmohdakhterali@gmail.com

M. Kamraju
Department of Geography, Division of Graduate Studies, Centre for Economic Studies, Nizamiah Observatory Campus, Hyderabad, India

A. K. Yadav et al. (eds.), *Integrated Management of Water Resources in India: A Computational Approach*, Water Science and Technology Library 129,
https://doi.org/10.1007/978-3-031-62079-9_18

18.1 Introduction

18.1.1 Overview of Integrated Water Resources Management

Integrated Water Resources Management (IWRM) is a comprehensive approach to managing water resources in a sustainable and holistic manner (UNESCO 2009). It recognizes the interconnectedness of water resources and aims to balance the social, economic and environmental demands on water (Fig. 18.1) (Global Water Partnership 2000).

IWRM emphasizes the importance of considering the entire water cycle, from the source of water to its various uses and eventual disposal or reuse (UNEP 2002). This approach requires the involvement of multiple stakeholders, including government agencies, communities, industries, and civil society organizations, to ensure that water resources are managed equitably and efficiently (Biswas 2004).

Fig. 18.1 Integrated water resource management (https://www.open.edu/openlearncreate/mod/oucontent/view.php?id=156334&section=4)

At its core, IWRM seeks to address the challenges of water scarcity, pollution, and competing water demands through a participatory and integrated approach (UNESCO 2009). By integrating various aspects of water management, such as water supply, sanitation, irrigation, and conservation, IWRM aims to achieve sustainable development and ensure the availability of water for future generations (Global Water Partnership 2000).

18.1.2 Importance of Community Participation in IWRM

Community participation is a cornerstone of integrated water resources management (IWRM), as it recognizes the significance of involving local communities, stakeholders, and users in decision-making processes related to water resources (UNESCO 2006). This participatory approach recognizes that communities have a deep understanding of their local water resources, including traditional water management practices, ecological knowledge and social dynamics, which are invaluable for developing context-specific and sustainable water management strategies (Biswas 2004; Kumar et al. 2024; Kulwant et al. 2024).

One of the key aspects of community participation in IWRM is the integration of local knowledge and expertise into water management decision making processes (UNESCO 2006). Local communities often have valuable insights into the availability, quality, and use of water resources based on their historical experiences and interactions with the environment. By incorporating this knowledge, IWRM initiatives can benefit from a more comprehensive understanding of local water systems, leading to more effective and sustainable management practices (Ali and Kamraju 2023a, b).

Additionally, community participation in IWRM fosters a sense of ownership and responsibility among local stakeholders for the management and conservation of water resources (Biswas 2004). When communities are actively involved in the decision-making process, they are more likely to take ownership of water management initiatives, leading to increased compliance with regulations and a greater willingness to adopt sustainable water use practices. This sense of ownership can also contribute to the long-term success and sustainability of water management projects by ensuring ongoing community support and participation.

In addition to promoting ownership, community participation in IWRM improves social equity and inclusivity in water management processes (UNESCO 2006). By involving a diverse range of stakeholders, including marginalized groups and indigenous communities, in decision making, IWRM initiatives can ensure that the needs and interests of all members of society are taken into account. This inclusivity can help address historical inequalities in access to water resources and promote more equitable and just water allocation and management practices.

Community participation is vital for the success of IWRM initiatives, as it leverages local knowledge, fosters ownership and responsibility, and promotes social equity and inclusivity in water management decision-making processes. By actively engaging communities in water management, IWRM can achieve more sustainable and effective outcomes that benefit present and future generations.

18.1.3 Objectives of the Paper

This research paper aims to comprehensively analyze the role of community participation in sustainable integrated water resource management. The paper seeks to contribute to a deeper understanding of the importance of community participation in achieving sustainable water management outcomes.

18.2 Theoretical Framework

18.2.1 Theoretical Perspectives on Community Participation

Community participation in the management of water resources is underpinned by various theoretical perspectives that seek to explain its significance and impact on sustainable development. These perspectives provide frameworks for understanding the dynamics of community involvement in water management and offer information on factors that influence its effectiveness.

A prominent theoretical perspective is social capital theory, which emphasizes the role of social networks, trust, and norms in fostering community participation (Pretty and Smith 2004). According to this theory, communities with high levels of social capital are more likely to engage in collective action for common goals, such as sustainable management of water resources. Social capital theory suggests that strong social bonds and networks within communities can facilitate cooperation and collaboration, leading to more effective water management practices.

Another relevant theoretical perspective is the adaptive co-management framework, which emphasizes the need for adaptive and collaborative approaches to natural resource management (Armitage et al. 2007). This framework recognizes the complexity of natural resource systems and the importance of involving local communities in decision-making processes. Adaptive co-management emphasizes the adaptive capacity of communities to respond to changing environmental conditions and the need for flexible governance structures that can accommodate diverse stakeholder interests.

The theory of participation and empowerment focuses on the empowerment of marginalized communities through their active involvement in decision-making processes (Arnstein 1969). According to this theory, meaningful participation goes beyond tokenism and involves genuine power sharing between communities and decision-makers. Participation and empowerment theory highlight the importance of giving voice to marginalized groups in water management processes to ensure that their needs and perspectives are taken into account.

These theoretical perspectives provide valuable frameworks for understanding the complexities of community participation in IWRM. By examining these perspectives, researchers and practitioners can gain insight into the underlying dynamics of community participation in water management and develop strategies that are more responsive to the needs and aspirations of local communities.

18.2.2 Models of Community Engagement in Water Management

Stakeholder Engagement Model: This model emphasizes the participation of stakeholders in water management decisions to consider their interests (Reed et al. 2009). It stresses understanding diverse stakeholder perspectives and fostering collaboration for sustainable outcomes.

Participatory Action Research (PAR): PAR involves active community participation in the identification of water issues and the development of solutions. It empowers communities in decision making, leading to sustainable water practices.

Community-Based Natural Resource Management (CBNRM): CBNRM decentralizes the management of natural resources to local communities (Blaikie 2006). It allows communities to sustainably manage water resources, often through community organizations and local governance structures.

Adaptive Co-Management: This approach emphasizes adaptive and collaborative management of natural resources, including water resources (Armitage et al. 2007). It advocates flexible governance that adapts to environmental changes and stakeholder dynamics.

Integrated Water Resources Management (IWRM): IWRM coordinates the development and management (UNESCO 2009). It involves multi-stakeholder processes integrating various actors' perspectives, including communities.

Community-Based Participatory Research (CBPR): CBPR is a collaborative research approach that involves community, organizational, and researcher participation. In water management, it empowers communities to contribute knowledge, leading to relevant results.

These models offer frameworks for community participation in water management, promoting inclusive, participatory, and sustainable decision-making that addresses local needs and ensures the sustainability of water resources.

18.3 Challenges of Community Participation in IWRM

18.3.1 Socio-Economic Factors Influencing Community Engagement

The participation of communities in water management is influenced by various socio-economic factors that shape the capacity, willingness, and opportunities for communities to participate in decision-making processes. These factors can significantly impact the effectiveness and sustainability of community engagement efforts in water management.

a. **Education and awareness**: Education levels and awareness of water-related issues within communities can significantly influence their involvement in water management (Lubell et al. 2002). Higher levels of education and awareness are often associated with increased participation and informed decision making about water resources.
b. **Income and lifelong living**: The economic status of communities plays a crucial role in their ability to engage in water management activities (Cleaver 2000). Communities with higher incomes and diversified incomes may have more resources and flexibility to participate actively in water management initiatives.
c. **Access to Resources**: The availability and access to water resources and infrastructure can influence community engagement (Meinzen-Dick et al. 2004). Communities with limited access to clean water sources may be more motivated to participate in efforts to improve water management practices.
d. **Social Capital**: Social networks and relationships within communities, often referred to as social capital, can influence their ability to engage in collective action for water management (Pretty and Ward 2001). Strong social networks and bonds can facilitate cooperation and collaboration among community members.
e. **Governance and Institutions**: Governance structures and institutional arrangements in place can impact community involvement in water management (Agrawal and Gibson 1999). Communities with transparent and inclusive governance systems may be more willing to participate in decision-making processes.
f. **Cultural and Social Norms**: Cultural and social norms within communities can also shape their attitudes towards water management (Ferguson and Meinzen-Dick 2003). Traditional practices and beliefs related to water use and management can influence community engagement approaches.
g. **Risk perception**: Perception of risks associated with water scarcity, pollution, or access can influence community participation (Grothmann and Patt 2005). Communities that perceive water-related risks as significant may be more motivated to participate in efforts to address these challenges.

These socioeconomic factors is crucial for designing effective community participation strategies in water management. By considering these factors, policymakers and practitioners can tailor interventions that are responsive to the specific needs and contexts of communities, thereby enhancing the success and sustainability of water management initiatives.

18.3.2 Institutional and Governance Challenges

Community involvement in water management is often influenced by institutional and governance challenges that can hinder effective participation and decision-making processes. These challenges can arise from various factors related to the design and implementation of institutional frameworks for water management.

a. **Policy Frameworks**: The absence of clear and supportive policy frameworks for community participation in water management can be a significant challenge (Agrawal and Gibson 1999). Policies that do not prioritize community involvement or provide adequate mechanisms for participation can hinder effective participation.
b. **Decision-Making Structures**: Centralized decision-making structures that do not incorporate local perspectives and knowledge can limit community participation (Blaikie 2006). Decisions made without meaningful community input may not reflect local needs and priorities.
c. **Resource allocation**: Inadequate allocation of resources for community engagement initiatives can be a barrier (Cleaver 2000). Without sufficient resources, communities may lack the capacity to participate effectively in water management activities.
d. **Capacity Building**: Limited capacity-building efforts to empower communities with the skills and knowledge necessary for meaningful engagement can pose challenges (Ferguson and Meinzen-Dick 2003). Capacity building initiatives are essential to improve community participation in decision-making processes.
e. **Legal frameworks**: Legal frameworks that do not recognize or protect community rights to water resources can undermine engagement efforts (Meinzen-Dick et al. 2004). Clear legal frameworks that secure community rights can support their participation in water management.
f. **Power dynamics**: Power imbalances between different stakeholders, including communities and government agencies, can affect the effectiveness of community engagement (Pretty and Ward 2001). Unequal power dynamics can marginalize community voices in decision-making processes.
g. **Accountability mechanisms**: Weak accountability mechanisms for water management decisions can reduce transparency and trust, hindering community engagement (Reed et al. 2009). Effective accountability mechanisms are essential to ensure that decisions are responsive to community needs.

These institutional and governance challenges are crucial to promoting effective community participation in water management. By recognizing and overcoming these obstacles, policymakers and practitioners can create an enabling environment that empowers communities to participate meaningfully in decision-making processes and ultimately contribute to sustainable water management results.

18.3.3 Knowledge and Awareness Constraints

Community participation in water management can be hindered by knowledge and awareness constraints that limit communities' ability to participate effectively in decision-making processes. A significant constraint is the limited access to accurate and relevant information on water management practices and policies. Without access to critical information, such as water resource data or policy documents, communities can struggle to make informed decisions about water use and conservation (Lubell et al. 2002).

Another constraint is the lack of technical knowledge and expertise among community members. Technical aspects of water management, such as hydrology or engineering, often require specialized knowledge that may not be readily available within communities. This lack of technical expertise can make it challenging for communities to understand complex water management issues and contribute meaningfully to decision-making processes (Reed et al. 2009).

Inadequate educational opportunities, particularly in rural or marginalized communities, can also limit the ability of community members to engage in water management. Education plays a crucial role in providing the foundational knowledge and skills needed to understand water-related issues and participate in decision-making processes (Cleaver 2000).

Insufficient awareness-raising efforts about the importance of water management and conservation can further contribute to low levels of community engagement. Effective awareness campaigns can enhance community understanding and motivation to participate in water management initiatives (Meinzen-Dick et al. 2004).

Language and communication barriers can also impede knowledge sharing and community participation. Clear communication in local languages and through accessible channels is essential to ensure that information about water management reaches all community members (Ferguson and Meinzen-Dick 2003).

Underappreciation of traditional and cultural knowledge related to water management can be another constraint. Traditional knowledge systems often have valuable insights into sustainable water management practices that can complement scientific approaches (Agrawal and Gibson 1999).

Finally, limited capacity-building efforts to improve community knowledge and skills in water management pose significant challenges. Capacity-building initiatives are crucial to empower communities with the tools and knowledge needed to engage effectively in water management processes (Pretty and Ward 2001).

These knowledge and awareness constraints require targeted efforts to improve access to information, provide educational opportunities, increase awareness, and promote the integration of traditional knowledge into water management practices. By addressing these constraints, policy makers and practitioners can improve communities' ability to participate meaningfully in water management processes and contribute to sustainable water resource management.

18.3.4 Conflict Resolution and Decision-Making Processes

Effective conflict resolution and decision-making processes are crucial for fostering community participation in water management. Conflicts over water resources can arise from competing interests, differing priorities, and resource scarcity, which can affect the ability of communities to participate in collaborative decision-making processes. Identifying and addressing conflicts related to water management is essential for promoting community participation (Lubell et al. 2002). Conflicts may arise between different user groups, such as agricultural and urban users, or between upstream and downstream communities. Resolving these conflicts requires mechanisms for dialogue, negotiation, and mediation to find mutually acceptable solutions.

Involving diverse stakeholders in decision-making processes is critical for ensuring that the interests and concerns of all parties are considered (Reed et al. 2009). Stakeholders may include community members, local authorities, government agencies, and nongovernmental organizations. Engaging stakeholders in collaborative decision making can help build consensus and ownership of water management initiatives. Transparent decision-making processes that are accountable to all stakeholders are essential for building trust and confidence (Agrawal and Gibson 1999). Transparent processes ensure that decisions are based on clear criteria and are not influenced by hidden agendas. Accountability mechanisms can hold decision-makers responsible for their actions and ensure that decisions reflect the interests of the community.

Utilizing participatory approaches in decision-making processes can enhance community engagement (Meinzen-Dick et al. 2004). Participatory methods, such as community meetings, workshops, and citizen science initiatives, enable community members to contribute their knowledge and perspectives to decision-making processes. Building the capacity of communities and decision makers in conflict management skills is essential for effectively addressing conflicts (Ferguson and Meinzen-Dick 2003). Training programs focusing on negotiation, communication, and problem solving can help build the capacity of stakeholders to manage conflicts constructively. Clear legal frameworks and effective dispute resolution mechanisms are necessary to address conflicts related to water management (Cleaver 2000). Legal frameworks provide a basis for resolving conflicts, while dispute resolution mechanisms offer formal processes for resolving disputes that cannot be resolved through negotiation.

In terms of these aspects of conflict resolution and decision-making processes, policy makers and practitioners can create an enabling environment for community engagement in water management. This, in turn, can lead to more sustainable and equitable water management practices that benefit both communities and the environment.

18.4 Opportunities to Improve Community Participation

18.4.1 Empowerment and Capacity Building

Empowerment and capacity building are essential to foster meaningful community participation in water management, ensuring that communities are equipped with the knowledge, skills and confidence to engage effectively in decision-making processes and take ownership of sustainable water resource management practices. One key aspect of empowerment is knowledge sharing, where communities are provided with accessible information and education about water management (Ferguson and Meinzen-Dick 2003). This empowers them to understand the complexities of water management decisions, assess their impacts, and actively participate in related discussions and initiatives. In addition, capacity building initiatives play a crucial role in empowering communities. These initiatives focus on enhancing the technical and practical skills of community members in water management, such as training in water conservation techniques or monitoring methods (Meinzen-Dick et al. 2004; Shirin and Yadav 2021; Shirin et al. 2023). By building these capacities, communities are better equipped to respond to water-related challenges effectively.

In addition, promoting leadership and governance skills among community members is essential for their empowerment (Reed et al. 2009). Leadership development initiatives enable individuals to advocate for their community's interests in water management decision-making processes, while governance training improves their understanding of institutional processes and policies related to water management. Furthermore, capacity building efforts should also aim to foster collaboration and partnerships among diverse stakeholders involved in water management (Agrawal and Gibson 1999). This collaborative capacity enables communities to work effectively with various stakeholders, including local authorities and government agencies, towards shared water management goals. (Shirin and Yadav 2014).

Lastly, legal awareness and literacy programs can empower communities by increasing their understanding of their rights and responsibilities with respect to water resources (Cleaver 2000). When communities are aware of their legal rights to access, usage, and protection, they can advocate for fair and sustainable water management practices more effectively. By empowering communities through knowledge sharing, capacity building, leadership development, and legal awareness, stakeholders can create an environment where communities are empowered to actively participate in water management processes, leading to more inclusive and sustainable water resource management.

18.4.2 Technology and Information Sharing

Technology and information sharing play crucial roles in improving community participation in water resource management, providing communities with tools and resources to make informed decisions, improve water management practices, and address challenges effectively. Through the use of modern technologies, such as Geographic Information Systems (GIS), remote sensing, and data analytics, communities can access valuable information about water resources, including water quality, availability, and usage patterns (Lemos and Morehouse 2005). This information enables communities to understand the status of their water resources, identify potential risks or vulnerabilities, and develop targeted strategies for sustainable water management.

Furthermore, technology facilitates real-time data collection and monitoring, allowing communities to track changes in water resources and respond promptly to emerging issues (Reed et al. 2009). For example, sensor networks can provide continuous data on water levels, flow rates, and quality parameters, enabling communities to detect changes in water conditions and take timely actions, such as adjusting water usage or implementing conservation measures. Furthermore, digital platforms and mobile applications can facilitate the sharing of information and knowledge among community members, stakeholders, and experts, fostering collaboration and collective problem solving (Kumar and Singh 2019). These platforms can serve as valuable tools for community engagement, allowing communities to voice their concerns, share local knowledge, and participate in decision-making processes related to water management.

Incorporating technology into water management initiatives can also improve the efficiency and effectiveness of interventions, leading to better outcomes for communities and the environment (Lemos and Morehouse 2005). For example, the use of smart irrigation systems can optimize water usage in agriculture, reducing water wastage, and improving crop yields. Similarly, water treatment technologies can help communities access clean and safe drinking water, improving public health and well-being. By leveraging technology and promoting information sharing, stakeholders can empower communities to actively participate in water resource management, contributing to more sustainable and resilient water systems.

18.4.3 Collaborative Governance Models

Collaborative governance models are increasingly recognized as effective approaches to fostering community participation in water resource management. These models emphasize the importance of collaborative decision-making processes that involve various stakeholders, including government agencies, local communities, nongovernmental organizations, and private sector entities (Emerson et al. 2012). By bringing together diverse perspectives and experience, collaborative governance models seek to address complex water management challenges through collective problem-solving and shared decision-making.

One key feature of collaborative governance is the emphasis on inclusivity and stakeholder engagement (Emerson et al. 2012). These models recognize that effective water management requires the active participation of all relevant stakeholders, including those who may be directly affected by water-related decisions. By including a wide range of perspectives and interests, collaborative governance models aim to develop more holistic and sustainable solutions that consider the needs of all stakeholders.

Furthermore, collaborative governance models often emphasize the importance of building trust and fostering positive relationships between stakeholders (Ansell and Gash 2008). Trust is essential for effective collaboration, as it creates an environment where stakeholders feel comfortable sharing information, engaging in dialogue, and working toward common goals. Building trust requires transparency, open communication, and a commitment to shared values and objectives.

Another key aspect of collaborative governance models is the use of adaptive management approaches (Folke et al. 2005). These approaches recognize the dynamic and uncertain nature of water systems and promote flexibility and learning in decision-making processes. By continuously monitoring and evaluating the results of water management actions, stakeholders can adjust their strategies based on new information and changing conditions, improving the effectiveness and resilience of water management efforts.

Collaborative governance models offer a promising framework to promote community participation in water resources. By fostering inclusivity, trust, and adaptive management, these models can help address complex water challenges more effectively and ensure the long-term sustainability of water resources.

18.4.4 Policy and Legal Frameworks

The policy and legal frameworks play a critical role in shaping the context of community participation in water resource management. These frameworks provide the necessary guidelines, regulations, and incentives to support sustainable water management practices and ensure the participation of communities in decision-making processes.

An important aspect of policy frameworks is the establishment of clear objectives and targets for water management (UNESCO 2009). Policies set the overarching goals for water resource management, such as ensuring water quality, protecting ecosystems, and meeting the water needs of communities. By defining these objectives, policies provide a basis for prioritizing actions and investments in water management.

In addition, policy frameworks can outline the roles and responsibilities of various stakeholders involved in water management, including government agencies, local authorities, and community organizations (UNEP 2002). A clear delineation of the roles helps clarify the responsibilities of each stakeholder in managing water resources, promoting accountability and coordination among different actors.

On the other hand, provide the legal basis for implementing water management policies and regulations (UNESCO 2009). These frameworks include laws, regulations, and legal instruments that govern water rights, allocation, pollution control, and other aspects of water management. Legal frameworks ensure that water management activities comply with legal requirements and provide mechanisms for enforcing regulations and resolving disputes.

Furthermore, policy and legal frameworks can include provisions for public participation in decision-making processes related to water management (UNEP 2002). These provisions may require authorities to consult with stakeholders, including local communities, before making decisions that affect water resources. By involving communities in decision making, policy and legal frameworks can ensure that the perspectives and concerns of local stakeholders are taken into account, leading to more informed and inclusive decision-making processes.

Policy and legal frameworks are essential components of effective water resource management, providing the necessary guidance, regulations, and mechanisms to support sustainable practices and ensure the participation of communities in decision-making processes.

18.5 Current Perspectives on Community Participation

18.5.1 Case Studies and Best Practices

a. In Nepal, community-based water management programs have empowered local communities to improve water access and sanitation in rural areas by building and maintenance of water supply systems and promoting hygiene.
b. India has implemented participatory groundwater management initiatives that involve local communities, government agencies, and NGOs to monitor groundwater levels, promote water conservation, and regulate groundwater use, leading to more sustainable practices.
c. Europe has adopted collaborative river basin management approaches involving multiple stakeholders to develop and implement plans for integrated and sustainable management of transboundary river basins, improving water quality and ecosystem health.

d. Kenya's community-led water management initiatives involve local communities in building, operating and maintaining water supply systems, leading to greater access to water and improved water quality in rural areas.
e. Australia's Indigenous Water Use Plans (IWUPs) recognize indigenous communities' rights to manage water resources according to cultural and environmental values, integrating traditional knowledge for more sustainable water governance.
f. Bangladesh's community-based flood management programs involve communities in building flood protection infrastructure and developing early warning systems, reducing the impact of flooding on local communities.
g. In the United States, community-led water conservation initiatives involve local communities in developing and implementing water conservation plans, promoting sustainable water use, and protecting water resources.
h. Brazil's Water Resources Councils (WRCs) involve stakeholders in decision-making processes related to water management, promoting multistakeholder collaboration for more inclusive and sustainable water governance.
i. Japan's community-based river restoration projects engage local communities in restoring degraded rivers and improving aquatic ecosystems through activities such as river clean-up and habitat restoration, improving ecological health and community stewardship.

18.5.2 Lessons Learned from Successful Community Initiatives

Lessons learned from successful community initiatives in integrated water resources management (IWRM) provide valuable insights into effective strategies and best practices. A key lesson is the importance of local knowledge and participation in decision-making processes, leading to more contextually relevant and sustainable solutions. The capacity building and empowerment of local communities are crucial, as they allow communities to take ownership of water management processes and address their own challenges effectively. Collaborative governance structures that bring together diverse stakeholders promote inclusive decision making and partnerships, ensuring that different perspectives are considered. Flexibility and adaptive management are critical, allowing learning, experimentation, and adjustment based on feedback and new information. Integrating traditional knowledge with modern approaches leads to more sustainable and culturally relevant outcomes. Long-term commitment and sustainability are essential and require long-term partnerships, adequate resources, and a sense of ownership among stakeholders. These lessons offer valuable guidance in promoting community participation in IWRM and in achieving sustainable water management outcomes.

18.5.3 Innovations in Community Engagement Approaches

Innovations in community engagement have transformed integrated water resource management (IWRM), enhancing its effectiveness and inclusivity. Taking advantage of technology, communication, and participatory methods, these innovations promote meaningful collaboration among stakeholders. Key innovations include:

i. **Digital Tools and Platforms**: Mobile applications, web platforms, and GIS allow stakeholders to access and share water data, participate in virtual meetings, and contribute to decision-making from remote locations, improving transparency and efficiency.
ii. **Citizen Science Initiatives**: Engaging locals in data collection and monitoring builds capacity and raises awareness of water issues, empowering communities for sustainable water management.
iii. **Participatory Modeling and Simulation**: Stakeholders co-design and evaluate water management scenarios through simulation models, fostering a deeper understanding and consensus on sustainable strategies.
iv. **Social Media and Online Engagement**: Platforms like social networks and forums facilitate discussions, knowledge sharing, and co-creation of solutions, democratizing the water management dialogue.
v. **Innovative financing mechanisms**: Crowd funding and impact investment funds enable communities to mobilize resources for water projects, empowering them to control funding and implementation.

These approaches improve the effectiveness, inclusivity, and sustainability of community participation in water management, revolutionizing stakeholder collaboration in IWRM.

18.6 Implications for Sustainable IWRM

18.6.1 Role of Community Participation in Water Sustainability

Community participation is a cornerstone of sustainable water management, crucial for ensuring responsible use and conservation of water resources. The participation of local communities in decision-making processes adapts sustainable water practices to meet specific needs and challenges. This approach fosters ownership and responsibility among community members, leading to more effective outcomes (UNESCO 2006).

Community-led initiatives are the key to raising awareness and promoting education on water-related issues. Through campaigns and programs, communities learn about conservation and pollution prevention, empowering them to protect and conserve water resources (UN 2015).

Participation also improves governance related to water resources. Local input is essential to develop effective water policies and strategies. Community-based committees ensure that local voices are heard, enhancing the legitimacy and effectiveness of water governance (UNDP 2006).

Engaging communities is critical to achieving water-related Sustainable Development Goals (SDGs) related to water. Community involvement accelerates progress toward SDG 6, such as improving water quality and ensuring equitable access (UN 2015).

Community participation builds resilience to the impacts of climate change on water resources. Involving communities in developing climate-resilient practices improves sustainability in the face of changing conditions (UNEP 2018).

Addressing climate change and ensuring water security requires coordinated efforts. Strategies include improving water resource management, investing in climate resilient infrastructure, promoting water conservation, and fostering international cooperation (IPCC 2014; UNEP 2016; World Bank 2019).

The achievement of the SDGs through community involvement is crucial. Empowering communities to participate in water management ensures sustainable practices and resilience to climate change impacts (Yadav et al. 2014).

18.7 Conclusion

In conclusion, community participation is essential to achieve sustainable outcomes in Integrated Water Resources Management (IWRM). This study highlights the importance of community involvement in decision making, capacity building, and integrating traditional wisdom. It emphasizes the need for supportive governance structures, technological advancements, and interdisciplinary approaches to enhance community engagement. The article also emphasizes the importance of community empowerment to address water-related challenges, including those from climate change and natural disasters. Implementing the recommendations of this study and addressing research gaps can create an environment conducive to community participation in water management. This, in turn, can lead to more sustainable and equitable outcomes for water resources and communities relying on them.

18.7.1 Summary of Key Findings

The analysis of community participation in sustainable integrated water resources management (IWRM) highlights several key findings. Firstly, community involvement is essential to promote sustainable water practices, as it fosters a sense of responsibility and ownership among locals, leading to more effective management.

Second, community-led initiatives are crucial to raising awareness and educating the public about water-related issues. This empowers communities to take proactive measures to protect and conserve water resources.

Third, community participation enhances governance and decision-making in water management, contributing to more legitimate and effective systems. Inclusive approaches ensure that the voices of local communities are considered in decisions affecting their water resources.

Furthermore, community participation is critical to achieving the water-related Sustainable Development Goals (SDGs), particularly SDG 6. Engaging communities accelerates progress toward ensuring the availability and sustainable management of water and sanitation for all.

Lastly, community participation contributes to the impacts of building resilience to climate change on water resources. By incorporating local knowledge, communities can develop climate-resilient water management practices.

In conclusion, community participation is vital to achieving sustainable water management outcomes. Engaging communities as active stakeholders is essential to ensure long-term sustainability of water resources and address complex water challenges.

18.7.2 Recommendations for Policy and Practice

To enhance community participation in sustainable integrated water resource management (IWRM), policymakers and practitioners should prioritize inclusive decision-making processes involving local communities and stakeholders. Establishing participatory platforms, such as water user associations, can enable stakeholders to contribute to decision making related to water management, ensuring that diverse needs and perspectives are considered.

Efforts should also focus on capacity building and awareness among local communities and stakeholders. Organizing training programs and educational campaigns can empower community members to participate actively in water management initiatives and foster a sense of ownership and responsibility.

Encouraging the integration of traditional knowledge and practices related to water management, particularly from indigenous communities, can complement modern approaches and enhance the effectiveness and sustainability of water management initiatives.

Governments should review and update the legal frameworks related to water management to support community participation and align with sustainability principles. Adequate allocation of resources, including funding and technical supporting, is crucial to support community-led water management initiatives.

Establishing mechanisms to monitor and evaluating the effectiveness of community participation in water management can help identify successful practices, challenges, and areas for improvement, forming future policy and practice. These efforts can create an enabling environment for community participation in water management, leading to more sustainable outcomes for water resources and the communities that depend on them.

18.7.3 Future Research Directions

Future research on community participation in integrated water resources management (IWRM) should focus on impact assessments of community-led initiatives compared to traditional approaches, identifying best practices. It should explore how technology, such as mobile apps and data analytics, can empower communities and improve real-time communication. Research should also investigate community-based adaptation strategies to climate change and their effectiveness in building resilience. Furthermore, governance frameworks supporting community participation, the link between community empowerment and social equity, the integration of cross-sectoral water management, and community resilience in the reduction of disaster risk need further exploration. Lastly, studying successful community participation models, their scalability, and the role of government support in scaling them up nationally or regionally is crucial.

Acknowledgements We are thankful to Department of Geography, Osmania University Hyderabad for the secondary data Assistance. We are also Thankful to Division for Graduate studies of Centre for Economic and Social Studies, Begumpet Hyderabad for their support in literature guidance. We are also thankful to P. Manisha, K. Haritha, O. Pranathi, M. Uma Maheshwari for their support in writing the paper.

References

Agrawal A, Gibson CC (1999) Enchantment and disenchantment: the role of community in natural resource conservation. World Dev 27(4):629–649

Ali MA, Kamraju M (2023) Natural resources and society: understanding the complex relationship between humans and the environment. Springer Nature

Ali MA, Kamraju M (2023b) The role of women in rural development programs. ASEAN J Comm Serv Educ 2(1):67–84

Altman JC, Kerins S (2019) Indigenous water governance in Australia: exploring the potential for Indigenous water rights? Australasian J Environ Manage 26(3):219–233

Ansell C, Gash A (2008) Collaborative governance in theory and practice. J Public Admin Res Theory 18(4):543–571

Armitage DR, Plummer R, Berkes F, Arthur RI, Charles AT, Davidson-Hunt IJ, Wollenberg EK et al (2007) Adaptive co-management for social–ecological complexity. Front Ecol Environ 5(5):232–238

Arnstein SR (1969) A ladder of citizen participation. J Am Inst Plann 35(4):216–224

Biswas AK (2004) Integrated water resources management: a reassessment. Water International 29(2):248–256

Blaikie P (2006) Is small really beautiful? Community-based natural resource management in Malawi and Botswana. World Dev 34(11):1942–1957

Bonney R, Cooper CB, Dickinson J, Kelling S, Phillips T, Rosenberg KV, Shirk J (2009) Citizen science: a developing tool for expanding science knowledge and scientific literacy. Bioscience 59(11):977–984

Cleaver F (2000) Moral ecological rationality, institutions and the management of common property resources. Dev Chang 31(2):361–383

Dabla-Norris E, Kochhar K, Suphaphiphat N, Ricka F, Tsounta E (2015) Causes and consequences of income inequality: a global perspective. International Monetary Fund

Emerson K, Nabatchi T, Balogh S (2012) An integrative framework for collaborative governance. J Pub Admin Res Theory 22(1):1–29

European Commission (n.d.) River basin management in Europe. Retrieved from https://ec.europa.eu/environment/water/participation/index_en.htm

Ferguson H, Meinzen-Dick R (2003) The relevance of social capital for water management. Water Policy 5(6):545–562

Folke C, Hahn T, Olsson P, Norberg J (2005) Adaptive governance of social-ecological systems. Annu Rev Environ Resour 30(1):441–473

Ghosh S (2020) Digital platform for water management: a case study of "Jal Jeevan Mission." Int J Manage, Tech, Soc Sci 5(1):84–90

Gleick PH (2018) Water and climate change. The Ann Rev Environ Res 43:399–417

Global Water Partnership (2000) Integrated water resources management. Retrieved from https://www.gwp.org/en/ToolBox/FAQ/What-is-IWRM/

Grothmann T, Patt A (2005) Adaptive capacity and human cognition: the process of individual adaptation to climate change. Glob Environ Chang 15(3):199–213

Hope R (2018) Evaluating the impact of community-led water management in rural Kenya. Water Alternatives 11(3):681–701

IPCC (2014) Climate change 2014: impacts, adaptation, and vulnerability. contribution of working group II to the Fifth Assessment Report of the Intergovernmental Panel on Climate Change. Cambridge University Press

Kishan A, Ali M, Kamraju M (2022) Sustainable urban development strategies for slum upgrading and environmental improvement using material science technology. ASEAN J Sci Eng Mat 1(2):107–116. Retrieved from https://ejournal.bumipublikasinusantara.id/index.php/ajsem/article/view/346

Kooiman J (2003) Governing as governance. Sage Publications

Koop SHA, Van Leeuwen CJ, Van Bruggen JJA (2013) Water governance in the Netherlands: fit for the future? J Environ Manage 122:1–7

Kulwant M, Patel D, Shirin S, Sharma SN, Yadav AK (2024) Future frameworks for fluoride and algorithms for environmental system. In: Advanced treatment technologies for fluoride removal in water: water purification, pp 343–364. https://doi.org/10.1007/978-3-031-38845-3_19

Kumar A, Singh R (2019) Role of information communication technologies (ICTs) in water management: a review. In: Advances in water resources management. Springer, pp 73–94

Kumar A, Chava R, Gupta S, Shirin S, Jamal A, Yadav AK (2024) Environmental and health effects of fluoride contamination and treatment of wastewater using various technologies. In: Advanced treatment technologies for fluoride removal in water: water purification. Springer Nature Switzerland, Cham, pp 323–341. https://doi.org/10.1007/978-3-031-38845-3_18

Lemos MC, Morehouse BJ (2005) The co-production of science and policy in integrated climate assessments. Glob Environ Chang 15(1):57–68

Lubell M, Zahran S, Vedlitz A (2002) Collective action and citizen responses to global warming. Polit Behav 24(3):269–298

Meinzen-Dick R, Knox A, Place F (2004) Understanding property rights. In: Barrett CB (ed) Food security and environmental quality in the developing world. Springer, pp 147–166

Ministry of the Environment, Japan (2020) Community-based river management. Retrieved from https://www.env.go.jp/en/water/wetland/river_restoration/index.html

OECD (2019) Water governance in Brazil: a multi-level approach. Retrieved from https://www.oecd.org/environment/water-governance-in-brazil-a-multi-level-approach-2019.pdf
Pahl-Wostl C (2009) A conceptual framework for analysing adaptive capacity and multi-level learning processes
Paul SK, Routray JK, Sahoo DK (2019) A community-based flood early warning system for disaster risk reduction: a case study from Bangladesh. Int J Disaster Risk Reduct 36:101096
Pretty J, Ward H (2001) Social capital and the environment. World Dev 29(2):209–227
Pretty J, Smith D (2004) Social capital in biodiversity conservation and management. Conserv Biol 18(3):631–638
Reed MS, Graves A, Dandy N, Posthumus H, Hubacek K, Morris J, Stringer LC et al (2009) Who's in and why? a typology of stakeholder analysis methods for natural resource management. J Environ Manage 90(5):1933–1949
Shirin S, Yadav AK (2014) Physico chemical analysis of municipal wastewater discharge in Ganga River. Haridwar District of Uttarakhand, India. Current World Environ 9(2):536
Shirin S, Yadav AK (2021) Assessment and suitability for the dumping of flyash in an abandoned mine. Environ Challen 4:100111. https://doi.org/10.1016/j.envc.2021.100111
Shirin S, Jamal A, Emmanouil C, Singh VP, Yadav AK (2023) Assessment and characterization of waste material used as backfilling in an abandoned mine. Int J Coal Prep Util 43(8):1402–1410. https://doi.org/10.1080/19392699.2022.2118259
Shah T, Verma S (2008) Co-management of natural resources: organising, negotiating and learning-by-doing. Sage Publications India
Shrestha S, Pattanayak SK, van Houtven G (2016) Participatory water management and governance in Nepal: a review of policy and practice. Water Policy 18(6):1549–1565
U.S. Environmental Protection Agency (n.d.) Community-led water conservation. Retrieved from https://www.epa.gov/sites/default/files/2015-06/documents/conservation-leaflethi-res.pdf
United Nations (2015) Transforming our world: the 2030 agenda for sustainable development. United Nations, New York, NY
United Nations Development Programme (2006) Human development report 2006: beyond scarcity: power, poverty and the global water crisis. UNDP, New York, NY
United Nations Environment Programme (2002) Integrated water resources management. UNEP, Nairobi, Kenya
United Nations Environment Programme (2016) Climate change adaptation in the water sector. UNEP, Nairobi, Kenya
United Nations Environment Programme (2018) The adaptation gap report 2018. UNEP, Nairobi, Kenya
United Nations Educational, Scientific and Cultural Organization (2006) Water for people, water for life: The United Nations World Water Development Report 2. UNESCO, Paris, France
United Nations Educational, Scientific and Cultural Organization (2009) Water in a changing world: The United Nations World Water Development Report 3. UNESCO, Paris, France
United Nations Framework Convention on Climate Change (2015) Paris agreement. Retrieved from https://unfccc.int/process-and-meetings/the-paris-agreement/the-paris-agreement
Voinov A, Bousquet F (2010) Modelling with stakeholders. Environ Model Softw 25(11):1268–1281
World Bank (2019) Water security. Retrieved from https://www.worldbank.org/en/topic/water-security
Yadav AK, Sahoo SK, Mahapatra S, Kumar AV, Pandey G, Lenka P, Tripathi RM (2014) Concentrations of uranium in drinking water and cumulative, age-dependent radiation doses in four districts of Uttar Pradesh, India. Toxicol Environ Chem 96(2):192–200

Chapter 19
Challenges and Opportunities in Integrated Water Resources Management

Krishna Neeti, Reena Singh, Shaz Ahmad, Sakshi, and Abhishek Kumar

Abstract Integrated Water Resources Management (IWRM) is a process that harmonizes the utilization of water, land, and associated resources to enhance economic and social well-being while ensuring ecosystem sustainability. It is crucial to global water resource development and addressing conflicting demands. This chapter analyzes urban challenges, such as water, wastewater, solid waste, and climate change, and assesses the gaps in achieving the SDGs, especially SDG 6 (clean water and sanitation) and SDG 11 (sustainable cities and communities). It provides solution pathways for IWRM, which have many benefits, such as coordinated resource development, environmental protection, economic growth, sustainable agriculture, democratic governance, and human health. It also faces challenges, such as water shortages, quality deterioration, and flood impacts, which are worsened by climate variability and change, urbanization, and consumption and production patterns. It proposes to improve IWRM planning by aligning it with local to national priorities and employing a learning-by-doing approach to innovate and adjust water resources management strategies, tools, and technologies.

Keywords Integrated water resource management · Sustainable development goals · Urban challenges · Water governance · Climate change adaptation

---

K. Neeti · R. Singh (✉) · Sakshi · A. Kumar
National Institute of Technology Patna, Ashok Rajpath, Patna 800005, India
e-mail: reena@nitp.ac.in

S. Ahmad
Government Engineering College, Kaimur, Jaitpur Kalan Bhagwanpur Kaimur, Bhabua 821102, India

A. K. Yadav et al. (eds.), *Integrated Management of Water Resources in India: A Computational Approach*, Water Science and Technology Library 129,
https://doi.org/10.1007/978-3-031-62079-9_19

19.1 Introduction

Water is vital for life, health, food, energy, environment, and development. Water resources face challenges such as demand, pollution, scarcity, variability, and conflicts. These are exacerbated by climate change, population growth, urbanization, and socioeconomic inequalities (Yadav et al. 2014; Kumar et al. 2024). We need a holistic approach and IWRM to address these challenges. It should consider the multiple dimensions and interdependencies of water and balance the diverse needs and interests of different stakeholders (Hameed et al. 2019; Brennan et al. 2021).

IWRM is a strategy that coordinates the planning and use of water, land, and associated resources. It aims to equitably optimize economic and social well-being while safeguarding the sustainability of vital ecosystems. IWRM is grounded in three principles: (1) water is a finite and vulnerable resource that requires integrated management; (2) water management should be participatory and sensitive to gender issues; and (3) water management should be informed by robust knowledge and information. IWRM also comprises three components: (1) a supportive framework of policies, laws, and institutions; (2) the institutional roles of various actors and organizations; and (3) management tools and techniques (Arfan et al. 2020; Kulwant et al. 2024).

Water governance involves the decision making and execution of plans concerning water resources and services. It influences the accessibility, utilization, and administration of water. It is a fundamental aspect of IWRM, determining the allocation, timing and methods of water distribution, as well as the rights and advantages associated with water and its services (Meran et al. 2021). It also influences the effectiveness, efficiency, equity, accountability, and transparency of water management. Improving water governance is a key challenge and opportunity to achieve IWRM.

The Sustainable Development Goals (SDG) are a set of 17 global objectives and 169 targets designed to eradicate poverty, protect the environment, and promote peace and prosperity by 2030 (Halkos and Gkampoura 2021). They are relevant for IWRM, as they recognize the interlinkages and synergies among water, food, energy, health, the environment, and development. SDG 6 and SDG 11 are related to the challenges and opportunities of IWRM in urban settings. Achieving these goals requires implementing IWRM at different scales and sectors, and addressing the trade-offs and cobenefits among water-related objectives and targets. IWRM practices and initiatives are found at various levels and sectors. However, more assessments, solutions, and innovations are needed for IWRM and its links and contributions to the SDGs. They should address the complex and dynamic challenges of water resource management and enhance the resilience and sustainability of water systems (Nagata et al. 2022).

This chapter provides an overview of IWRM and water governance and their linkages with the SDGs, especially SDG 6 and SDG 11. It presents examples and case studies of IWRM implementation in different regions and countries. Explore IWRM and its intersection with the SDGs and the opportunities, challenges, and solutions in water resource management. It prepares the reader for understanding the complex dynamics of sustainable water and urban development.

19.2 IWRM: An Overview

IWRM is a process that emerged in the late twentieth century to address the challenges and complexities of water resource development and management (Arfan et al. 2020). It was introduced by the Dublin Conference in 1992 and endorsed by the Rio and Johannesburg Conferences in 1992 and 2002, respectively (Meran et al. 2021). IWRM is a way to manage water, land, and related resources together to help everyone in a community, without hurting the environment. It is based on the idea that water is really important and that we have to use it carefully because it is limited. It considers the interdependencies and trade-offs between different sectors, stakeholders, and environmental objectives. IWRM integrates the technical, social, economic, and environmental aspects of water issues to achieve efficient, equitable, and sustainable development and management.

19.2.1 The IWRM Paradigm and Its Evolution

The IWRM paradigm emerged in the 1970s to address the interconnected issues of water quantity, quality, and allocation. In the 1980s and 1990s, it gained prominence due to the global water crisis and the environmental and social repercussions of water projects (Godinez-Madrigal et al. 2019; Arfan et al. 2020). The IWRM paradigm was shaped by environmental movements, human rights and gender perspectives, and international conferences and declarations, such as the Dublin Principles and the Rio Declaration. They highlighted the need for a holistic, participatory and sustainable water management approach. The IWRM paradigm was supported and promoted by the international community and various organizations in the 2000s as a key means to achieve the Millennium Development Goals and the World Water Vision (Godinez-Madrigal et al. 2019). The IWRM paradigm has also integrated novel dimensions and challenges, including adaptation to climate change, the water-food-energy nexus, and transboundary water cooperation. The IWRM paradigm is comprehensive, flexible, and reflects widely accepted principles and values. It can provide multiple benefits and solutions to water problems. However, it is ambiguous, complex and lacks clear definition, operationalization, implementation, and evaluation. It also faces practical and institutional barriers and requires balancing interests, values, and conflicts.

19.2.2 A General Framework of IWRM and Its Components

IWRM is a way to achieve sustainable development and management of water resources, endorsed by various international organizations, such as the GWP, the UN and the WWC (Challies and Newig 2022). Figure 19.1 shows the general framework of IWRM, comprising three interrelated components: the Enabling Environment, the Institutional Framework, and the Management Instruments.

i. Enabling environment: This component provides the foundation and direction for water management. It includes policies, laws, and plans at different levels (Jiménez et al. 2019).
ii. Institutional framework: This component defines the organizations and arrangements for water management and service delivery. It includes roles and responsibilities, water rights and permits, and coordination and collaboration among stakeholders. It also includes capacity-building and training (Van Hofwegen and Jaspers 2020).
iii. Management instruments: This component employs various tools and methods to assess and monitor water resources, allocate and distribute water, regulate water use, and resolve conflicts and disputes. It includes data and information systems, water resource assessment and accounting, water demand management,

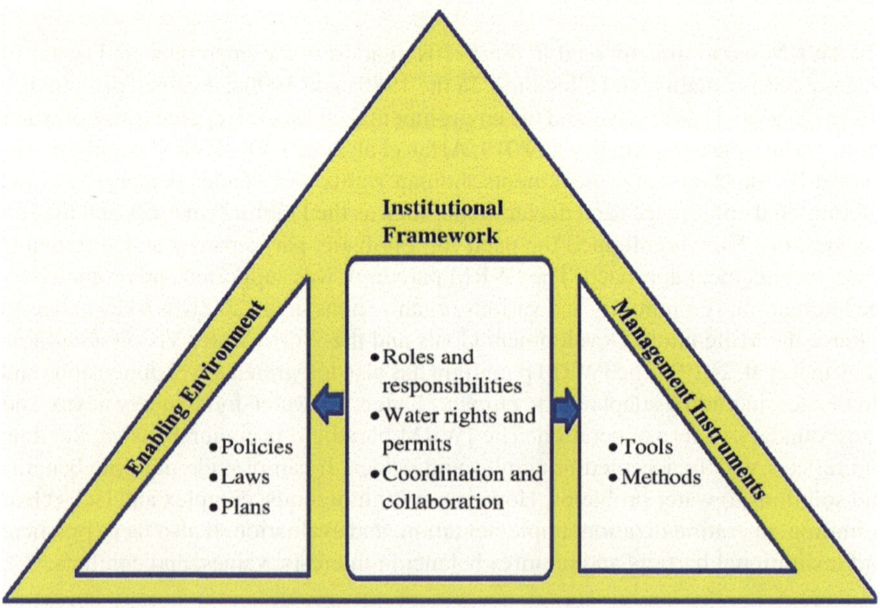

Fig. 19.1 Framework of IWRM and its components

water quality management, water pricing, cost recovery, and conflict resolution mechanisms (Dirwai et al. 2021).

The components of IWRM interact and complement each other and can be applied at different scales and sectors, depending on the context and needs (Badham et al. 2019). The enabling environment gives vision and guidance, the institutional framework implements and enforces policies and laws, the management instruments optimize the use and protection of water resources, and the financing ensures the availability and sustainability of funds. IWRM principles can be effectively implemented on different scales, including local, national, regional, and global levels. These principles span a wide range of water-related domains, encompassing agriculture, industry, energy, health, and environmental contexts.

19.2.3 Approaches and Methods for IWRM Implementation

Various approaches and methods can implement IWRM in different contexts and scales. Some of the main ones are the following:

(a) River basin management: This approach manages water resources within a natural hydrological unit, the river basin. Integrates quantity and quality, surface and groundwater, and upstream and downstream uses and impacts. It can facilitate transboundary cooperation (van den Brandeler et al. 2019).
(b) Adaptive management: This method monitors and evaluates water resource management actions and outcomes, and adjusts plans and strategies based on feedback and learning. It recognizes the uncertainty and complexity of water resources systems and the need to cope with change and challenge. It can improve resilience, flexibility, innovation, and learning (Karthe et al. 2021).
(c) Ecosystem approach: This approach considers the water resource system as part of a larger ecosystem, including the biotic and abiotic components and their interactions. It aims to maintain or restore the structure and function, and the provision of ecosystem services. It can promote conservation and restoration of natural capital and restoration, and environmental and social values (Hülsmann et al. 2019).
(d) Stakeholder participation: This method involves stakeholders in the water resource management process, from planning and decision-making to implementation and evaluation. Stakeholders are individuals or groups that have a stake in the water resource system. Participation can improve legitimacy, transparency, accountability, and ownership and foster social learning and conflict resolution (Sigalla et al. 2021).
(e) Water accounting and auditing: This method measures and reports the status and performance, using physical and monetary indicators. It can provide information on water availability, use, allocation, and productivity, and the costs and benefits of water management interventions. It can support evidence-based planning and

management and efficiency, effectiveness, and equity evaluation (Delavar et al. 2020).

These approaches and methods have advantages and disadvantages and depend on various criteria and conditions, such as the context, data, capacity, and objectives of water resource management.

19.2.4 Application of the Components of IWRM at Different Scales and Sectors

19.2.4.1 IWRM on Different Scale

IWRM can be applied at different scales, such as local, national, regional, and global, with different challenges and opportunities for water management.

(a) Local scale: IWRM can empower communities, protect ecosystems, and improve water efficiency and security. This can be done through integrated urban water management. These involve participation, planning, implementation, and monitoring of water activities and integration of water supply, wastewater, stormwater, and reuse. Challenges include lack of capacity and resources, conflicts among water users and stakeholders, and inadequate institutions and laws (Cacal and Taboada 2022).
(b) National scale: IWRM can provide data, framework, regulation, and pollution control for water resources. This can be done through the evaluation of water resources, the national the policy and strategy, water allocation and licensing, and water quality. These involve the evaluation, allocation, regulation and protection of water resources among different sectors and users, including surface, ground, and rainwater. The challenges include political instability and corruption, weak governance and enforcement, and divergent interests and priorities between sectors and regions (Cacal and Taboada 2022).
(c) Regional scale: IWRM can promote cooperation, harmonization, coordination, and integration of water resources within a basin. This can be done through transboundary water cooperation, regional water agreements and institutions, and basin-wide planning and management. These involve the negotiation, establishment, and implementation of water sharing agreements and joint institutions for water management between different countries and users within a region or river basin. Challenges include cultural, social, economic, political, legal, and environmental barriers to cooperation, power imbalances between countries or actors in a basin, and uncertainty due to impacts of climate change (Cacal and Taboada 2022).
(d) Global scale: IWRM can recognize and protect the global water commons, promote collaboration and knowledge sharing, and monitor and evaluate progress toward sustainable water management. This can be done through international water conventions and protocols, global water partnerships and

networks, and water-related SDGs and indicators. These involve the adoption, implementation, and monitoring of global frameworks and standards for water management, including protection and sustainable use of transboundary water resources. They also involve collaboration of different stakeholders, including governments, civil society, and the private sector, to address water challenges at the global level (Burlacu et al. 2020; Cacal and Taboada 2022).

19.3 Examining IWRM Implementation Across Regions and Projects

19.3.1 Regional Perspectives: IWRM in Different Countries

IWRM has been implemented in different regions and countries, with different success and challenges. Some examples are as follows:

A. South Africa

The 1998 National Water Act created a legal framework for water management, based on IWRM principles. It recognized water as a scarce and public resource and introduced authorizations for water use, water pricing, catchment management agencies, and stakeholder participation (Burlacu et al. 2020). It aimed to ensure an equitable allocation of water for human needs, environment, and development. The challenges included lack of capacity, complex authorization process, resistance from water users, and interbasin and transboundary issues.

B. Brazil

The 1997 National Water Resources Policy and the 2000 National Water Resources Management System established a decentralized and participatory model for water management, engaging various levels of government, water users and civil society (Kratzert et al. 2019). The model introduced water resource planning, information and financing. It embraced the principles of IWRM, including river basin management, integration of quantity and quality aspects, recognition of water as an economic asset, and active participation in social participation and conflict resolution. The results included river basin committees, water resource plans, water charges, and water monitoring and information systems. Challenges included coordination among government levels, land and water integration, enforcement of water rights and regulations, and climate change adaptation.

Table 19.1 summarizes the key aspects of IWRM in South Africa and Brazil. It shows the legal framework, stakeholder participation, challenges, and outcomes of IWRM in these contexts. The analysis compares the variations and commonalities in IWRM approaches and their success and challenges.

Table 19.1 A comparative analysis of the operationalization of IWRM in three diverse regions: South Africa and Brazil

Region	Legal framework	Stakeholder participation	Challenges	Outcomes
South Africa	The National Water Act of 1998	Water use authorizations, water pricing, catchment management agencies	Lack of capacity, complex authorization process, resistance from water users, inter-basin and transboundary issues	Equitable water allocation for human needs, environment, and development
Brazil	The National Water Resources Policy of 1997 and the National Water Resources Management System of 2000	Different levels of government, water users, and civil society	Coordination among government levels, land and water integration, water rights and regulations enforcement, climate change adaptation	River basin committees, water resources plans, water charges, water monitoring and information systems

19.3.2 Project-Level Insights: IWRM Approaches and Methods

IWRM approaches and methods have been integrated in different case studies and projects around the world. Some examples are:

A. Nile Basin Initiative (NBI)

NBI is a regional partnership of 10 countries that share the Nile River, the longest and most complex river system in the world. It was launched in 1999 to foster cooperation and development in the Nile basin, using river basin management and stakeholder participation (Sintayehu 2020). The NBI has implemented various projects and programs that address the basin's challenges and opportunities, such as water security, hydropower, irrigation, navigation, fisheries, wetlands, climate change, and transboundary governance (Bukhari and Brown 2022). The achievements include: strengthening trust and dialogue, knowledge and information generation, improving water management and infrastructure improvement, and creating benefits and opportunities. Challenges include political tensions and disputes over water allocation, lack of a comprehensive and binding legal framework, insufficient funding and capacity, and population growth, urbanization, and climate change impacts.

B. Water Evaluation and Planning (WEAP) system

WEAP is a software tool designed to facilitate the integrated analysis and planning of water resources. It employs adaptive management, water accounting, and auditing techniques to simulate current and future water supply and demand, as well as the potential impacts of various scenarios and policies (Agarwal et al. 2019). It uses indicators and criteria, such as water availability, quality, efficiency, environmental flows, costs, benefits, equity, and adaptation to climate change. It also facilitates

Table 19.2 Comparative summary of IWRM approaches in case studies and projects

Case study	Main focus and objectives	Methodologies and tools	Achievements	Challenges
Nile Basin Initiative (NBI)	Regional cooperation, basin management, stakeholder participation	Shared Vision Program (SVP), Subsidiary Action Program (SAP)	Trust-building, knowledge generation, improved infrastructure	Political tensions, legal framework gaps, funding issues
WEAP system	Integrated analysis and planning of water resources	Adaptive management, water accounting, auditing methods	Scenario simulation, stakeholder involvement, decision support	Trade-offs among water uses, climate change vulnerability
PES scheme	Market-based instrument for rewarding ecosystem services	Ecosystem approach, water accounting, auditing methods	Improved water quality and quantity, livelihood generation	Implementation costs, scalability, stakeholder cooperation

stakeholder participation. It has a user-friendly interface, a transparent and flexible modeling framework, and a scenario-based approach. It has been applied in various regions and countries to support the decision making and planning of water resource management. The outcomes include: trade-offs and synergies identification, alternative options and strategies evaluation, vulnerability and resilience assessment, and dialogue and consensus promotion.

C. Payment for Ecosystem Services (PES) scheme

The PES scheme is a market-based instrument that rewards ecosystem services, such as water quality and quantity, for their conservation and restoration efforts. It uses the ecosystem approach and water accounting and auditing methods. It is based on the principle that the beneficiaries of ecosystem services should pay for them and the providers should be compensated. It can create incentives and mechanisms, promote sustainable and fair management of water resources, and recognize and value the environmental and social benefits of water (Bleeker and Vos 2020).

Table 19.2 compares three IWRM approaches in different case studies and projects. They are NBI, WEAP system, and PES scheme. It summarizes their focus, methods, results, and challenges.

19.4 Role of IWRM in Achieving SDG 6 and SDG 11

The SDGs are 17 global goals for 2030 to end poverty, protect the planet, and ensure peace and prosperity. SDG 6 and SDG 11 are about water resources and management. SDG 6 is for clean water and sanitation, and SDG 11 is for sustainable cities and communities. IWRM can help achieve SDG 6 and SDG 11 (Halkos and Gkampoura 2021).

19.4.1 The Role of IWRM in Achieving SDG 6 (Clean Water and Sanitation)

SDG 6 is a global goal for sustainable development. It aims to ensure universal access to water and sanitation while promoting sustainable management. The scope includes water resources and essential services, which include drinking water, sanitation, hygiene, and wastewater treatment. It also covers water quality, efficiency, IWRM, transboundary cooperation, and ecosystems (Benson et al. 2020). IWRM can help achieve this goal by using an inclusive approach that considers the whole system and involves everyone. It can improve access to water and sanitation, especially for the poor, marginalized, and vulnerable groups. It can respect their human rights and empower them in decision-making.

19.4.2 The Role of IWRM in Achieving SDG 11 (Sustainable Cities and Communities)

SDG 11 is a global goal for sustainable development. It wants to make cities better for everyone. It covers many things that affect urban life. IWRM can improve water supply and sanitation services in cities by building and improving facilities and systems, making them affordable and accessible to all. IWRM can also improve water quality and reduce water pollution in cities by treating, recycling, and reusing wastewater, managing stormwater and pollution, and using green and nature-based solutions. These can clean water and benefit the urban environment and well-being. IWRM can also increase water efficiency and reduce water stress in cities by encouraging rational and fair water use and diversifying water sources. This can meet the water demand of urban people and activities and deal with water scarcity and variability (Benson et al. 2020). IWRM can also improve water governance and management in cities by creating and carrying out policies, plans and strategies that treat water as part of the city system, a natural resource, and an economic and social good.

19.5 Opportunities of IWRM

IWRM has many benefits and opportunities for various sectors and stakeholders. Some of them are as follows:

(a) Coordinated resource development and management: IWRM coordinates water, land, and related resources for equitable and sustainable welfare. It overcomes the fragmented water management approach. It adopts a holistic and cross-sectoral policy. Encourage participation and collaboration. This improves water governance and management.
(b) Environmental protection and ecosystem sustainability: IWRM implements measures to protect and restore water and biodiversity. These improve the resilience and adaptation.
(c) Economic growth and social welfare: IWRM optimizes and maintains water use for various purposes. It improves water supply and services, which are vital for human development and the reduction of poverty. It also supports the creation of employment, income, and livelihoods. Encourages water-related development and innovation.
(d) Economic growth and social welfare: IWRM optimizes and maintains water use for various purposes. It improves water supply and services, which are vital for human development and the reduction of poverty. It also supports the creation of employment, income, and livelihoods. Encourages water-related development and innovation.
(e) Democratic Governance and human rights: IWRM supports democratic governance and human rights to water resources and services. It ensures participation, representation, and empowerment of all water users and stakeholders. Respects and protects the human right to water and sanitation. It resolves and prevents water conflicts and disputes.
(f) Human health and well-being: IWRM improves health and well-being by improving water resources and services, which are vital for survival, hygiene, and sanitation. It prevents and reduces waterborne and water-related diseases by improving safe drinking water and sanitation coverage, and implementing hygiene education and behavior change.

19.6 Challenges and Solutions of IWRM

IWRM faces many challenges in urban settings, where water demand is high, but resources are scarce and polluted. Urban areas have high population density, growth, industrialization, and consumption, which pressure water resources. We will discuss some challenges and solutions for IWRM.

(a) Water scarcity and demand management: Water scarcity occurs when water resources are not sufficient for water demand. A solution is to manage water demand, not increase supply. This involves reducing water losses, improving water efficiency, promoting water conservation, and implementing water pricing and allocation mechanisms.
(b) Water quality and pollution control: The quality is the characteristics of water for its use. Water quality can be affected by natural or human factors. Water pollution is the introduction of harmful substances or microorganisms into water

bodies. A solution is to implement integrated pollution prevention and control measures, or adopt the 'polluter pays' principle.

(c) Water supply and distribution systems: These collect, store, treat, and deliver water to users. They face challenges that can compromise water services, such as aging, deterioration, leakage, vandalism, inadequate maintenance, natural disasters, and human-induced threats. A solution is to invest in and modernize water infrastructure, and enhance its resilience and security.
(d) Wastewater collection and treatment: These collect, convey, and treat wastewater from different activities. They face challenges from insufficient coverage, capacity, or performance, water pollution and health risks, high costs, energy consumption, or greenhouse gas emissions. A solution is to adopt a circular economy approach or implement low-cost and low-energy wastewater treatment systems. These recover, reuse, or recycle wastewater and reduce the environmental footprint.
(e) Municipal solid waste management and resource recovery: These handle urban solid waste. They face challenges from the growing and complex waste stream and the environmental and social impacts of waste disposal. A solution is to apply the waste hierarchy principle or integrate waste management with the water and energy sectors, recovering and using water, nutrients, or energy from waste.
(f) Impacts of climate change and adaptation: The water sector is affected by climate changes. These are the impacts. The water sector also adapts its policies, practices, and infrastructure. These are the adaptations. However, both face challenges, such as climate uncertainty, variability, complexity, and hazards. A solution is to adopt a 'no-regret' or a 'flexible' approach.

Table 19.3 shows a summary of the main challenges and solutions for urban water management in the twenty-first century. The solutions are based on the principles of IWRM.

19.7 Conclusion

This chapter has provided a comprehensive analysis of IWRM and its pivotal role in achieving the UN Sustainable Development Goals (SDGs) 6 and 11. It has traced the evolution of IWRM from its origins to its current status and discussed its main components and implementation methods. It has also revealed the importance of IWRM in fostering sustainable water and urban development, especially in the face of the challenges posed by population growth, urbanization, climate change, and water scarcity and pollution. The chapter has also identified promising opportunities for IWRM to improve resource development, environmental protection, economic growth, and social welfare. It has proposed a variety of technical, institutional and behavioural measures to address the persistent challenges of water management, ensuring the efficiency and sustainability of water use and allocation. In addition, it has highlighted

Table 19.3 Challenges and solutions for urban water management based on IWRM principles

Challenge	Solution
Water scarcity and demand management	Reduce water losses, improve water efficiency, promote conservation, use water pricing reflecting true value
Water quality and pollution control	Enforce environmental regulations, monitor water quality, implement wastewater treatment, "polluter pays" principle
Water supply and distribution systems	Invest in infrastructure rehabilitation, enhance resilience and security with risk assessment
Wastewater collection and treatment	Adopt circular economy for wastewater recovery, use decentralised, low-cost treatment systems
Municipal solid waste management	Apply waste hierarchy principle, integrate waste management with water and energy sectors
Climate change impacts and adaptation	Implement 'no-regret' measures, Adopt a 'flexible' approach with adaptive management

the intersectoral and multiscale linkages and trade-offs of water management, and the need for integrated and adaptive solutions that can accommodate the diversity and complexity of water-related issues.

Acknowledgements The authors express their gratitude to the Department of Civil Engineering at both the National Institute of Technology Patna and the Government Engineering College Kaimur, as well as all individuals who have contributed directly or indirectly to this study.

References

Agarwal S, Patil JP, Goyal VC, Singh A (2019) Assessment of water supply–demand using water evaluation and planning (WEAP) model for Ur River Watershed, Madhya Pradesh, India. J Instit Eng Ser A 100:21–32. https://doi.org/10.1007/S40030-018-0329-0/FIGURES/11

Arfan M, Ansari K, Ullah A et al (2020) Agenda setting in water and IWRM: discourse analysis of water policy debate in Pakistan. Water 12:1656. https://doi.org/10.3390/W12061656

Badham J, Elsawah S, Guillaume JHA et al (2019) Effective modeling for integrated water resource management: a guide to contextual practices by phases and steps and future opportunities. Environ Modell Softw 116:40–56. https://doi.org/10.1016/J.ENVSOFT.2019.02.013

Benson D, Gain AK, Giupponi C (2020) Moving beyond water centricity? Conceptualizing integrated water resources management for implementing sustainable development goals. Sustain Sci 15:671–681. https://doi.org/10.1007/S11625-019-00733-5/TABLES/1

Bleeker S, Vos J (2020) Payment for ecosystem services in Lima's watersheds: power and imaginaries in an urban–rural hydrosocial territory. Rural-Urban Water Struggles 42:144–162. https://doi.org/10.4324/9780429354427-9

Brennan M, Rondón-Sulbarán J, Sabogal-Paz LP et al (2021) Conceptualising global water challenges: a transdisciplinary approach for understanding different discourses in sustainable development. J Environ Manag 298:113361. https://doi.org/10.1016/J.JENVMAN.2021.113361

Bukhari H, Brown CA (2022) A comparative review of decision support tools routinely used by selected transboundary River Basin Organisations. Afr J Aquat Sci 47:318–337. https://doi.org/10.2989/16085914.2021.1976610

Burlacu S, Vasilache C, Velicu ER et al (2020) Management of water resources at global level. Econ Soc Sci 66:998–1009. https://doi.org/10.2478/9788366675162-100

Cacal JC, Taboada EB (2022) Assessment and evaluation of IWRM implementation in Palawan, Philippines. Civil Eng J 8:290–307

Challies E, Newig J (2022) Water, rivers and wetlands: governance paradigms and principles. In: Routledge handbook of global environmental politics, 2nd edn, pp 512–525. https://doi.org/10.4324/9781003008873-43/WATER-RIVERS-WETLANDS-EDWARD-CHALLIES-JENS-NEWIG

Delavar M, Morid S, Morid R et al (2020) Basin-wide water accounting based on modified SWAT model and WA+ framework for better policy making. J Hydrol 585:124762. https://doi.org/10.1016/J.JHYDROL.2020.124762

Dirwai TL, Kanda EK, Senzanje A, Busari TI (2021) Water resource management: IWRM strategies for improved water management. A systematic review of case studies of East, West and Southern Africa. PLoS ONE 16:e0236903

Godinez-Madrigal J, Van Cauwenbergh N, van der Zaag P (2019) Production of competing water knowledge in the face of water crises: revisiting the IWRM success story of the Lerma-Chapala Basin, Mexico. Geoforum 103:3–15. https://doi.org/10.1016/J.GEOFORUM.2019.02.002

Halkos G, Gkampoura EC (2021) Where do we stand on the 17 sustainable development goals? An overview on progress. Econ Anal Policy 70:94–122. https://doi.org/10.1016/J.EAP.2021.02.001

Hameed M, Moradkhani H, Ahmadalipour A et al (2019) A review of the twenty first century challenges in the food-energy-water security in the middle east. Water 11:682. https://doi.org/10.3390/W11040682

Hülsmann S, Sušnik J, Rinke K et al (2019) Integrated modelling and management of water resources: the ecosystem perspective on the nexus approach. Curr Opin Environ Sustain 40:14–20. https://doi.org/10.1016/J.COSUST.2019.07.003

Jiménez A, LeDeunff H, Giné R et al (2019) The enabling environment for participation in water and sanitation: a conceptual framework. Water 11:308. https://doi.org/10.3390/W11020308

Karthe D, Bogardi JJ, Borchardt D (2021) Water resources management: integrated and adaptive decision making. In: Handbook of water resources management: discourses, concepts and examples, pp 365–381. https://doi.org/10.1007/978-3-030-60147-8_12/COVER

Kratzert F, Klotz D, Herrnegger M et al (2019) Toward improved predictions in ungauged basins: exploiting the power of machine learning. Water Resour Res 55:11344–11354. https://doi.org/10.1029/2019WR026065

Kulwant M, Patel D, Shirin S, Sharma SN, Yadav AK (2024) Future frameworks for fluoride and algorithms for environmental system. In: Advanced treatment technologies for fluoride removal in water: water purification, pp 343–364. https://doi.org/10.1007/978-3-031-38845-3_19

Kumar A, Chava R, Gupta S, Shirin S, Jamal A, Yadav AK (2024) Environmental and health effects of fluoride contamination and treatment of wastewater using various technologies. In: Advanced treatment technologies for fluoride removal in water: water purification. Springer, Cham, pp 323–341. https://doi.org/10.1007/978-3-031-38845-3_18

Meran G, Siehlow M, von Hirschhausen C (2021) Integrated water resource management: principles and applications. Springer, New York, pp 23–121. https://doi.org/10.1007/978-3-030-48485-9_3/FIGURES/25

Nagata K, Shoji I, Arima T et al (2022) Practicality of integrated water resources management (IWRM) in different contexts. Int J Water Resour Dev 38:897–919. https://doi.org/10.1080/07900627.2021.1921709

Sigalla OZ, Tumbo M, Joseph J (2021) Multi-stakeholder platform in water resources management: a critical analysis of stakeholders' participation for sustainable water resources. Sustainability 13:9260. https://doi.org/10.3390/SU13169260

Sintayehu F (2020) Prospects for cooperation in the Nile Basin. Adv Afr Econ Soc Polit Develop 7:127–144. https://doi.org/10.1007/978-3-030-42890-7_10/COVER

van den Brandeler F, Gupta J, Hordijk M (2019) Megacities and rivers: scalar mismatches between urban water management and river basin management. J Hydrol 573:1067–1074. https://doi.org/10.1016/J.JHYDROL.2018.01.001

Van Hofwegen PJM, Jaspers FGW (2020) Analytical framework for integrated water resources management : IHE monographs 2. Analytical framework for integrated water resources management. https://doi.org/10.1201/9781003077824

Yadav AK, Sahoo SK, Mahapatra S et al (2014) Concentrations of uranium in drinking water and cumulative, age-dependent radiation doses in four districts of Uttar Pradesh, India. Toxicol Environ Chem 96:192–200. https://doi.org/10.1080/02772248.2014.934247

Chapter 20
Social and Ecological Framework for the Allocation of Water Resources: The Challenge for Water Security in India

Reinhart Philip

Abstract Who gets water, when, how, and under what conditions are fundamental questions before every society, since water is vital for life, good health, and development. India, the most populous country and a fastest growing economy is facing the twin issues of scarcity and allocation of water, as it needs to manage the water requirements of a larger population with just 4% of the water resources. Together with relative and absolute water stress and scarcity, there is the mounting demand for water for domestic uses, agriculture, commerce, and industry, pollution, decline of water bodies, impacts of climate change, sectoral conflicts, and disputes regarding sharing of water resources. So a strategically scarce resource having alternative uses is to be distributed to a population that is diverse, socially, culturally, economically, and politically. When the multiple uses associated with water make its allocation complex, the degree of inclusiveness and exclusiveness in access to water makes it more complicated. This necessitates a policy framework that can address the issues of social equity, efficiency, and ecological sustainability simultaneously, within the broader context of water security.

Keywords Water · Water governance · Water management · Water policy · Water security

20.1 Introduction

India, the most populated country that aspires to be a USD 5 trillion economy in the world, faces the twin issues of scarcity and allocation of a strategic resource that is essential for the proliferation of life, health of the ecosystem, and for sustainable

---

R. Philip (✉)
Department of International Relations, Central University of Kerala, Kasaragod, Kerala 671320, India
e-mail: reinganges@gmail.com

A. K. Yadav et al. (eds.), *Integrated Management of Water Resources in India: A Computational Approach*, Water Science and Technology Library 129,
https://doi.org/10.1007/978-3-031-62079-9_20

development. When the scarcity of water is the product of anthropogenic and natural factors, the issue in its allocation is the product of governance failures and poor policy choices. With an increase in the demand for water due to multiple factors that range from increased population to economic growth, on the other hand, the land faces the issue of depletion and pollution of freshwater sources at a startling rate. The decline in availability and quality happens at a time when the strategic role of water in sustaining life, health, sanitation, economic development, poverty reduction, food security, and development of commerce and industry is very clear, and various international agencies and organizations have endorsed the same through their studies and reports (Yadav et al. 2014). When the demand for water does not match with its flow, the increase in population and the subsequent additional municipal, agricultural, and industrial demands for water exclude a sizable section from access to potable water. Given the potential of water scarcity to sprout out a conflict or can aggravate the already existing conflicts with individuals, communities, and nation-states, no political system can see the issue in isolation. When the multiple uses associated with water make its allocation complex, the degree of inclusiveness and exclusiveness in access to water makes it more complicated, which is also the case for India.

20.2 Water Governance for Water Security

Studies related to water have been the exclusive privilege of hydrologists, structural engineers, environmentalists, and geographers who focused mainly on the properties of water, its availability, scarcity, and multiple uses. The question of governance came to the limelight when modern hydrological studies revealed that the amount of freshwater is not only finite, but also depleting at a faster rate. Since water is critical for development and human welfare, who gets water, when, how, and under what conditions are important questions. This has made issues related to water the focus of policy sciences, democratizing it from the monopoly of technocrats. With the entry of policy sciences, the question of governance got significance as it normally addresses the questions of availability of water, its quality and accessibility. Technically, water governance denotes 'the capacity of the system to mobilize energies, in a coherent manner, for the sustainable development of water resources' (Solanes and Jouravlev 2006). The ability to design socially acceptable and environmentally sustainable water policies and their implementation by involving different actors or stakeholders thus forms the key aspect of water governance. Water governance in this context is an institutional process, as defined by Hurlbert et al. (2009), which defines the organization and management of the interrelationships between society and water resources.

20.2.1 Water Governance: Differential Approaches

Traditionally, the state was tasked with the responsibility of water supply, until it was challenged by the neoliberal economic reforms across the world. With a shift from the state hydraulic model of water supply, the identity of water too changed from that of a commons to that of a commodity. Under the state hydraulic model, the state took the responsibility to provide water to all on grounds of equity and universal access to water, either free or at a subsidized rate (Bakker 2001; Roberts 2008; Swyngedouw 2005; Shirin and Yadav 2014; Shirin et al. 2019; 2021). With the Dublin Conference on Water and Environment, 1992, and the subsequent policy framework adopted by the World Bank, there is an increasing consensus to treat water as an economic good (World Bank Group 1993). This school of market proponents argues for pricing water at its economic value and creating a market where water can be allocated to the best uses (Briscoe 1997; Rogers et al. 2002; Rogers et al. 1998). The supply-driven approach that characterised the state model began to be replaced by demand-driven approach in every day decisions of allocation and consumption.

In the controversies surrounding the 'economic good' approach of the Dublin Conference, another important concept/approach from the same conference moved to oblivion, which is integrated water resources management (IWRM). The conference (second principle) called for "a participatory approach involving uses, planners, and policy makers at all levels, for the development and management" (ICWE 1992). With regard to water, its demand is multisectoral, ranging from agricultural, commercial, and industrial, and the interests are different and at the same time competing in nature. When the quantity of water available remains the same or declines, an additional allocation to one sector can only be at the cost of another sector. The scenario becomes complicated when sectors replace communities and nation states, as it has already been predicted that wars of the twenty-first century will be for water. The positive aspect of the IWRM approach is that it recognizes the competing interests and multiple sectors involved along with the environmental demands of water. That is why IWRM rests upon the three important principles of social equity, economic efficiency, and ecological sustainability (Stockholm International Water Institute 2020). Only when the ecological needs of water are acknowledged and protected (ecological sustainability), a state can ensure equity in access to potable water to sustain human life and well-being (social equity), through available financial and water resources (economic efficiency). When the different uses of a fast-depleting resource is interdependent and mutually conflicting in nature, it demands a coordinated action, and the more holistic approach to water governance is the rationale for the IWRM approach.

20.2.2 Water Security

Water security is a distinguishing challenge, as unclean water and poor sanitation, drought and floods, agricultural and industrial pollution, and degradation of the ecosystem claim the life of millions and threaten the livelihood of many in the world every year. Unclean water and sanitation have been identified as the second leading killer of children around the world, and dirty water kills three times the number of children who die in a conflict zone (UNICEF 2019, p. 3). When water is vital for life and socioeconomic development, the situation is that men, women, and children in large numbers lack access to safe water. Although water is one without alternatives, it has competing uses and users. The diversity in the demand for water needs to be met with the existing resources that are fast depleting. Therefore, the ability of the population in accessing safe drinking water is important and in a simple sense water security denotes the same. A universally acceptable working definition of water security is given by the UN Water (United Nations 2013), which denotes it as 'the capacity of a population to safeguard sustainable access to adequate quantities of acceptable quality water for sustaining livelihoods, human well-being, and socio-economic development for ensuring protection against water-borne pollution and water related disasters, and for preserving ecosystems in a climate of peace and political stability.'

Water security according to this definition requires satisfying a number of conditions. First, there shall be sustainable access to adequate potable water to the population, present and coming; second, water shall be available not only for sustaining life and livelihood, but also for socioeconomic development; third, there shall be protection against waterborne diseases, pollution, and disasters and the water cycle should be kept intact; fourth, the ecological needs of water shall be taken care of, and finally, there shall be effective institutional mechanisms to deal with conflicts and disputes arising out of the competing uses of water. Though it is very easy to mention, at the practical level, it requires long-term multidisciplinary and multi-sectoral efforts, from local to national, regional and international levels. The contributing factors to water security range from ecological and hydrological to socio-political-financial, and even sectors that lie outside the water realm.

With the shift in demographic factors, demand for water also increases, and this is mainly due to the changes in consumption pattern because of increased population, urbanization, and migration. The changes in demographic factors, natural disasters, and climate change affect the hydrological cycle and its availability. Structural competition will arise among different sectors like domestic, irrigation, and industry for water, and this will lead to overexploitation of the water resources including the groundwater. At the same time, the well springs of nature are getting polluted by human intervention. A holistic approach to water related challenges is possible with the term water security as it extends beyond accessibility, availability, and quality to include the different dimensions of water, water related issues, interconnected challenges, and the solutions. Together with this, the term water security can echo the strategic significance of water, a resource inevitable for sustaining life, development and one that has the potential to trigger or aggravate existing conflicts over resource allocation between different users, sectors, or states. Since the concept of water security is multidisciplinary and operates at multi-levels (Bigas 2000, p. 1), it demands a paradigm that is multidisciplinary and collaborative and integrated water resources management suits this.

20.3 Water: Law, Policy, and Governance in India

Water is related to the history and geography of the land so that the state (India) itself got its name on the basis of a river, the Sindhu (Indus). The Aryans, who came to India, first settled in the *Saptha Saindhava* region, the basin of seven rivers: *Indus, Jhelum, Chenab, Ravi, Sutlej, Beas, and the Sarasvati.* Water resource management and practices in India have evolved over these years and it dates back to the Indus Valley Civilization of the Pre Vedic times. The civilization of the Indus Valley was characterised by a network of wells and a drainage system based on scientific principles (Joisea 2006, p. 385). In ancient India, water was considered a nature provided resource for all and indigenous technologies or management practices aim at the efficient use of water. There are numerous records of various kings constructing dams, reservoirs, and lakes for the purpose of irrigation and officers being appointed specifically to take care of river systems. Laws were also made with regard to water, water bodies, and usage associated with water. Since the precolonial times water resource management was indigenous and based on the geographic and climatic conditions of the region, it was capable of supplying adequate fresh water to all areas and sectors.

A sea change has occurred in the approach and attitude of the state toward nature and its resources, with the colonization of the land by the Britishers. The harmony between man and nature in a subsistence economy shifted to one that of exploitation under the commodity-oriented economy advocated by Britain and through various laws the Britishers introduced common law principles in the management of water resources. This shift towards common law principles gave landowners the unrestricted access to water under their holding. With the movement of time, the state began to have control over surface water (riparian rights), weakening the customary rights of people. This was completed with the Madhya Pradesh Irrigation Act of 1931, whereby ‘all rights in the water of any river, natural stream, or natural drainage channel, natural lake, or other natural collection of water got vested in the government’ (Cullet 2007; Shirin et al. 2023). When the colonial framework completely abandoned the region-specific indigenous water management practices in favor of a unified water supply model, largely based on dams, the hydrological cycle began to lose its balance.

Water laws and management practices of the postcolonial times have the legacy of the British rule in India as the center-state division of responsibilities and the powers of the centre to appoint an interstate commission to enquire into the conflicts all date back to the Government of India Act of 1935. Water was made a ‘state subject’ in the Constitution of free India, with states having rights to make laws regarding water supply (Schedule 7, List II, 17) and the Union is entitled to legislate on navigation and shipping and matters specified in Schedule 7, List I (Items 24, 25, 56). Water, being a ‘state subject’, more legislation is from its part than the central government. Even though the central government had no control over ground water, it took a significant step in 1970 by creating the Central Ground Water Board (CGWB), with mandate over groundwater resources and for its protection, augmentation, and prevent the pollution of groundwater. When the lowering of the water table and the increase in groundwater use necessitated the presence of a statutory framework governing groundwater, the government forwarded a model bill in 1970, based on common law principles. This Model Bill for the adoption of states has been revised in 1992, 1996, 2005, 2011 and 2016, but states have shown little interest in adopting it. The Parliament passed the Water (Prevention and Control of Pollution) Act in 1974, to address the issue of water pollution and for maintaining the wholesomeness of water or restoring it.

20.3.1 National Water Policy 2012

Six decades after independence, India realised the need for abandoning the colonial legacy in water resource management and adopting an integrated approach that suits the geography, culture and requirements of the region. The poor water policy and management practices over these years have affected the economic development of the land. Instead of giving water first priority in policy planning, especially freshwater was totally neglected. This could be due to the belief that freshwater is an abundant resource. When modern hydrological studies came to the conclusion that freshwater is a finite resource, enough damage had occurred to freshwater sources in India and millions of people were out of protected drinking water. In fact, the 2012 National Water Policy admits that the water stress and critical situation in many parts of the land is due to mismanagement (Ministry of Jal Sakthi 2022). Together with water stress, there is mounting demand for water, pollution of water bodies, issues related to climate change, sectoral and interregional, interstate, and intra-state disputes in sharing water resources. So, the challenge before the state is really high and it needs a holistic and interdisciplinary approach.

Considering the multidimensional challenges, the Water Policy 2012 gave emphasis to the human and ecological needs of water, and this is a sharp departure from the earlier policy of 2002 that gave emphasis to the economic value of water. The realization is that water is a common resource (*commons*) that is inevitable for food security, livelihood support, and for sustainable development. So instead of fragmented policies, what is required is integrated policies that take into account the multidisciplinary nature of the issue and the multiple stakeholders involved in the same. Together with the principle of equity and social justice, the policy calls for sensitizing the community and securing its partnership for dealing with issues related to water scarcity, water pollution, and adaptation to climate change. To implement the IWRM approach, the policy also calls for community participation and restructuring of existing departments and organizations, at the various levels, on multidisciplinary lines. The policy is comprehensive as it covers areas related to water stress and scarcity, pollution, sanitation, increase in demand for water, climate change, disasters, management of drought and flood, transboundary waters, and finally institutional mechanisms for dealing with water resources.

20.3.2 Governmental Response to Drinking Water Management

Water and water resources, especially for drinking water and domestic uses, were neither a focus area nor a policy priority of the Government of India, immediately after independence. This is quite natural when water resources are abundant to meet the various needs of society. India's per capita water availability in 1951 was estimated to be 5178 m^3 per year (Central Statistical Office 2018, p. 80), which was well enough for an economy under reconstruction. At the same time, constitutionally and legally, 'water and water resources' were not the responsibility of the Union Government. Since drinking water was not on the priority list, a separate Ministry of Water Resources was also not established. For matters pertaining to irrigation the Ministry of Irrigation and Power was established in 1952. But groundwater and irrigation for agricultural activities were under the Ministry of Agriculture and it continued up to 1980. A full-fledged Ministry of Water Resources was established only in 1985 by reconstituting the Ministry of Irrigation (Ministry of Jal Shakti n.d.). Perspective planning will not be possible in the absence of a full-fledged ministry and this is what happened in the country. The investment of the Government of India in rural drinking water was also limited to zero during the first three five-year plans (CAG 2018, p. 1).

The creation of an independent Ministry for Drinking Water and Sanitation, on 13 July 2011, was a significant step in the provision of drinking water in India. This ministry then became the nodal agency for the flagship programs for rural drinking water in India. The increasing water stress and scarcity, pollution of surface and groundwater, increase in burden of waterborne diseases, systematic exclusion of rural population especially scheduled castes and scheduled tribes from accessing safe drinking water, climate change and its impacts, etc. have forced the Government of India to abandon its earlier passive position with regard to water supply. The binding nature of Human Right to Water declaration (2010) and the legally binding responsibility imposed upon the state systems for the realisation of the right might have also influenced the Government to take a lead role in drinking water supply. With more responsibilities the Ministry was reconstituted in 2019 as the Ministry of Jal Shakti and with this reconstitution, all matters related to water, water storage and supply, its development, prevention of pollution, drinking water and sanitation, came under the ambit of a single ministry in an integrated manner, which was previously fragmented under various ministries including agriculture, irrigation, and power.

20.4 Challenges to Water Security in India

Inclusion of each and every individual under protected water is a great challenge for a vast nation like India that is geographically and culturally diverse. Though India is the land of rivers, lakes, forests and mountains, water scarcity arises due to spatio-temporal variations in precipitation as well as poor water management practices. Therefore, with only 4% of water resources, the land needs to manage the demands of 18% of the total world population (The World Bank 2023). Even when access to water has improved considerably in rural and urban spaces, when it comes to safe or protected water, according to the report of NITI Aayog, the country's largest think tank on policy matters, only half of the rural population has access to it (NITI Aayog 2018). It is to bridge this gap the flagship programme of the Jal Jeevan Mission was launched in 2019 and in a short four-year period 10.74 crore new connections were provided in rural areas, bringing 13.98 crore rural households with a piped water connection (Government of India 2024). Given the population of India, only 72% of the total rural population is under a functional household tap connection (FHTC).

The source of water and its flow/stock is the greatest concern for the successful implementation of every water supply scheme. The potential of India in terms of natural runoff is estimated to be 1869 BCM per year and the total usable water resources to 1121; 690 bcm/year as surface water and 431 bcm/year as replenishable groundwater resources (CGWB 2024). But there is a concentration of ground water in 14 states and surface water in six states. The changes in precipitation also affect the availability of ground and surface water as the monsoon is the main source of water in the land, which is estimated to be 3880 BCM (Ministry of Jal Shakti 2021). There is marked variation in rainfall concentration longitudinally, and the distribution of rainfall throughout the year is also not equal. With these structural differences in the availability of water resources, the multiple uses associated with water for an increasing population must be met with. A marked decline in the per capita water availability is visible in these years and from 5178 m^3 in 1951 it declined to 1820 m^3 in 2001 and is expected to be at a lower value of 1367 m^3 by 2031 (Ministry of Jal Shakti 2022), showcasing a water stressed situation. The trend is that when the consumption rate of water doubles every ten years, there is a 15% decline is there in per capita water availability.

Being the largest populated country in the world, the demand for water is also highest in India. An increase in population will naturally increase the demand for food, water, clothing, shelter, and other basic necessities of life, which will have its impact on other sectors like economy, energy, commerce, and industry. When the current pattern of agriculture is water intensive in nature, more water is required to produce additional food grains to meet the increase in population, especially for wheat and rice, the staple foods. It should be noted that the water footprint of rice, wheat, milk and meat, which are consumed mainly in rural areas, is high, as it requires 2800 L of water to produce 1 kg of rice, 1654 L for 1 kg of wheat, 1020 L of water for 1 L of milk, 15,415 L for meat (bovine), 4325 L for chicken and 10,412 L for mutton. The water footprint of processed foods, which have more demand in urban areas, is also very high and the item wise (1 kg) is chocolate (17,196 L), butter (5553 L), bread (1608 L). After food and domestic uses of water, comes energy, where water is used either as a coolant or to run the turbines in the process of producing electricity. Fossil fuels still account for 59.7% of the total energy production in India, followed by renewable energy sources (38.5%) and nuclear (1.7%). Since thermal power plants withdraw a large share of water, the increase in demand for energy needs to be met with either nuclear energy or renewable energy, raising another policy challenge for the government. The food-water-energy relationship/nexus is as complex as all of them are highly water intensive, and managing the rapidly increasing and at the same time competing demands are critical for the ecosystem and for sustainable development.

To keep the pace of economic growth and development intact, more roads, bridges, railways, buildings, airports, seaports, industries, commercial establishments, schools, hospitals, recreation centers, etc. are required, and with this growth the consumptive and non-consumptive uses of water will also increase. According to the ENVIS-Stats-India Environmental Accounts Report, 2018 (Central Statistical Office 2018, p. 61), the projected demand for water in India by 2025 is 611 BCM for irrigation, 62 BCM for drinking water, 67 BCM for industry; 33 BCM for energy and 70 BCM for other activities, making the total demand 843 BCM; an increase in 133 BCM over a period of 25 years. But every usage of water, domestic, agriculture, commerce, industry, and recreation, has a price attached to it, either in the form of withdrawal/scarcity or pollution. Pollution in turn affects not only the recreational uses of water like swimming, boating, surfing, fishing, etc. but also to human health in a fatal manner. When rural water bodies are polluted by agricultural runoffs, urban water bodies are polluted by untreated waste from municipal areas and industries. Since 70% of the wastewater goes untreated and each day 40 million liters of waste water flows directly into water bodies (World Economic Forum 2019), 80% of India's water is severely polluted. 311 river stretches in India are identified as polluted (Ministry of Environment, Forest and Climate Change 2023) and the most polluted rivers include Cooum (Tamil Nadu), Sabarmati (Gujarat), and Bahela (Uttar Pradesh). It is ironic that Indians who believe in the "sacredness" of rivers and water bodies pollute them in multiple ways. The River Ganges, which is emotionally, culturally and religiously attached to the minds of Indians, is one of the most polluted rivers, and ultimately the government has to come up with a flagship program, the Namami Gange Programme with an outlay of Rs. 20,000 crore.

Pollution of water, both surface and groundwater, is a major challenge for the government when it comes to including many families under protected water supply. Since pollution is mainly the product of anthropogenic intervention, without changing the behavior of human beings towards water, pollution will continue in different ways. Being deliberate actions, to deal with pollution and encroachment of water bodies, it requires awareness and strict legal measures. The pressure on land for housing, agriculture, urbanization and other developmental activities leads to encroachment, and as a result streams, ponds, wetlands, watersheds, rivers, and lakes, either shrink in size or disappear. When the traditional sources of water disappear or become polluted, many are forced to rely on water sources which are contaminated by their own untreated human waste, making them victims of water-borne diseases and preventable deaths. With 73 million working days lost every year due to water-borne diseases, its economic cost is estimated to be 600 million USD annually (UNICEF INDIA 2022). Together with the economic cost, pollution bears a social cost, as women and children, especially adolescent girls, have to travel long distances to collect water. Every hour spent collecting water compromises with educational opportunities by taking children away from schools.

Many schemes have been initiated by the Government of India for conservation of water bodies, distribution of water and sustainable availability of potable water to all. National Water Mission, Namami Gange, Atal Bhujal Yojana, Har Khet Ko Pani (HKKP), Jal Jeevan Mission, Jal Shakti Abhiyan, Atal Mission for Rejuvenation and Urban Transformation are the most prominent schemes. The objectives, strategies, and mode of implementation of these programs differ sharply from the traditional water supply schemes, which were highly centralised and bureaucratized. Community participation has been identified as crucial to the successful implementation of each scheme. For example, 'Sahi Fasal' is an innovative campaign under the National Water Mission to convince farmers regarding efficiency in agriculture by using agricultural crops that are less water-intensive in nature. Taking the case of Atal Bhujal Yojana, the objective is sustainable groundwater management by involving communities for budgeting and implementation at Grama Panchayat levels. A large-scale awareness campaign was conducted among the masses regarding water scarcity and the importance of water conservation. Understanding the critical importance of rain water harvesting, the "Catch the Rain" initiative was launched in 2021 with the objective of "Catch the Rain—Where it falls When it Falls" (PIB 2023). This innovative approach of the Ministry of Jal Shakti in the fields of water conservation, management, and drinking water supply is one that abandons colonial legacies in favor of more integrated, participative, and sustainable practices.

20.5 Conclusions

Managing water, a strategic and scarce resource, for multiple uses ranging from domestic, municipal, commercial, industrial, and ecological is a challenge for India due to its population and development priorities. The challenge arises not only because of physical water scarcity but also due to demand–supply mismatch, which is a product of poor policy choices and management. The demand for water is expected to increase in India, but its scarcity can lead to sectoral competition for water resources. Indian agriculture is water intensive, and the failure of every monsoon results in additional extraction of groundwater for irrigation. When the same pattern was followed with regard to commercial and industrial uses, the groundwater level decreased to alarming levels. The pollution of water bodies adds to it, and the discharge of household, municipal, agricultural, and industrial leftovers have converted many of the water bodies into sewage channels. A life-sustaining liquid which is critical for agricultural intensification, economic development, sustainable development and above all functioning of the ecosystem has been converted into toxic waste, despite the holy status accorded to it. The dependence on contaminated sources of water affects life, its quality, health of human beings, and economic productivity due to loss of working hours. When there is unequal distribution of surface and groundwater across the nation, scarcity of water and its pollution has the potential to create new conflicts interregional, intraregional, or can aggravate the already existing ones. Already there exist conflicts between Karnataka and Tamil Nadu over the sharing of Kauvery Waters and between Kerala and Tamil Nadu over Mullaperiyar Dam.

Access to safe drinking water is a basic human right that is closely related to the right to life of an individual. Being a human right and by being a signatory to the United Nations legally binding declarations, the State is under the legal obligation to respect, protect, and fulfill it. Here comes the significance of policy and effective governance in ensuring such a right, in a highly populated country like India that faces the twin issues of scarcity and allocation of water simultaneously. Short-term policies focused on individual aspects cannot address the challenge of bringing all under protected water supply. To address a defining challenge like water security, multiple factors have to be taken care of in the long run, like availability and quality of water, carrying capacity of the water source, conflicting demands for water, pollution of water bodies, changes in precipitation, and hydrometeorological disasters. Together with balancing food-water-energy nexus, the municipal, commercial, recreational, and industrial needs of water have to be ensured to sustain the process of development. Since the population of India is expected to be stable only by the 1960s, to meet the fluctuating demands, a long-term plan with goals fixed for each decade is the viable option. A strategic resource like water demands strategic planning that integrates the multiple uses, multiple sectors, various stakeholders, geography, and climate of the region and finally the ecological dimensions of water. Participation of the local population is critical in designing and implementing projects related to water conservation, restoration, rejuvenation, and distribution of water in a democratic way, to avoid potential conflicts. To avoid the complications that arise due to the degree of exclusion in access to water, there shall be a balance between the human and ecological needs of water, recognising the competing interests, multiple sectors, and stakeholders involved. In short, water security in India demands a prudent water security policy that takes into account the questions of equity, efficiency, and ecological sustainability, which is possible only through Integrated Water Resource Management.

References

Bakker KJ (2001) Paying for water: water pricing and equity in England and Wales. Trans Instit Br Geogr 26(2):143–164

Bigas H (2000) Water security and the global water agenda: a UN-water analytical brief. United Nations University—Institute for Water, Environment and Health

Briscoe J (1997) Managing water as an economic good: rules for reformers. https://citeseerx.ist.psu.edu/document?repid=rep1&type=pdf&doi=7ac157bd9d19812e7999e26a7de9622c31f3e0cd

CAG (2018) Performance audit on national rural drinking water programme in ministry of drinking water and sanitation. Comptroller and Auditor General of India

Central Statistical Office (2018) Envi Stats-India 2018: supplement on environmental accounts. Ministry of Statistics and Programme Implementation

CGWB (2024) FAQ|cgwb. CGWB. https://cgwb.gov.in/faq. Accessed 29 Jan 2024

Cullet P (2007) Water law in India. International Environmental Law Research Centre

Government of India (2024) JJM Brochure. Jal Jeevan Mission. https://jalshakti-ddws.gov.in/sites/default/files/JJM-Brochure-Year-2024.pdf. Accessed 29 Jan 2024

Hurlbert M, Diaz H, Corkal D, Warren J (2009) Climate change and water governance in Saskatchewan, Canada. Int J Clim Change Strat Manag 1:118–132. https://doi.org/10.1108/17568690910955595

ICWE (1992) International conference on water and the environment: development issues for the 21st century. UNEP Document Repository. https://wedocs.unep.org/20.500.11822/30961. Accessed 27 Jan 2024

Joisea KJ (2006) River cult and river management practices in ancient India. In: A history of water: the world of water. I.B. Tauris, pp 385–408

Ministry of Environment, Forest and Climate Change (2023) Press information Bureau. Press Information Bureau. https://pib.gov.in/PressReleseDetailm.aspx?PRID=1941065. Accessed 30 Jan 2024

Ministry of Jal Shakti (2021) Press information Bureau. Press Information Bureau. https://www.pib.gov.in/PressReleasePage.aspx?PRID=1783527. Accessed 29 Jan 2024

Ministry of Jal Shakti (2022) Press information Bureau. Press Information Bureau. https://www.pib.gov.in/PressReleasePage.aspx?PRID=1882796. Accessed 29 Jan 2024

Ministry of Jal Shakti (n.d.) History. Department of Water Resources, River Development and Ganga Rejuvenation. https://jalshakti-dowr.gov.in/history/. Accessed 29 Jan 2024

NITI Aayog (2018) Composite water management index 2018. NITI Aayog

PIB (2023) Jal Shakti Abhiyan: catch the rain campaign. PIB. https://pib.gov.in/PressReleaseIframePage.aspx?PRID=1989132. Accessed 30 Jan 2024

Roberts A (2008) Privatizing social reproduction: the primitive accumulation of water in an era of neoliberalism. Antipode 40(4):535–560

Rogers P, Radhika SD, Ramesh B (2002) Water is an economic good: how to use prices to promote equity, efficiency and sustainability. Water Policy 4:1–17

Rogers P, Ramesh B, Annatte H (1998) Water as a social and economic good: how to put the principle into practice, vol 2. Global Water Partnership

Shirin S, Yadav AK (2014) Physico-chemical analysis of municipal wastewater discharge in Ganga river, Haridwar district of Uttarakhand. India. Curr World Environ 9(2):536–543

Shirin S, Jamal A, Ranjan P, Yadav AK (2019) Study on assessment of slope stability and mixed disposal of overburden in voids of Singrauli Coalfield. Environ Qual Manag 28(3):131–139. https://doi.org/10.1002/tqem.21616

Shirin S, Jamal A, Emmanouil C, Yadav AK (2021) Assessment of characteristics of acid mine drainage treated with fly ash. Appl Sci 11(9):3910. https://doi.org/10.3390/app11093910

Shirin S, Jamal A, Emmanouil C, Singh VP, Yadav AK (2023) Assessment and characterization of waste material used as backfilling in an abandoned mine. Int J Coal Prep Utiliz 43(8):1402–1410. https://doi.org/10.1080/19392699.2022.2118259

Solanes M, Jouravlev A (2006) Water governance for development and sustainability. United Nations

Stockholm International Water Institute (2020) Principles and practices of integrated water resources management. SIWI

Swyngedouw E (2005) Dispossessing H_2O: the contested terrain of water privatisation. Capit Nat Soc 16(1):81–98
The World Bank (2023) How is India addressing its water needs? World Bank. https://www.worldbank.org/en/country/india/brief/world-water-day-2022-how-india-is-addressing-its-water-needs. Accessed 29 Jan 2024
UNICEF INDIA (2022) Clean drinking water. UNICEF. https://www.unicef.org/india/what-we-do/clean-drinking-water. Accessed 30 Jan 2024
UNICEF (2019) Water under fire: emergencies, development and peace in fragile and conflict-affected contexts, vol I. Unicef
United Nations (2013) What is water security? Infographic. UN-Water. https://www.unwater.org/publications/what-water-security-infographic. Accessed 28 Jan 2024
World Bank Group (1993) Water resources management: a world bank policy paper. World Bank
World Economic Forum (2019) Water pollution is killing millions of Indians. Here's how technology and reliable data can change that. The World Economic Forum. https://www.weforum.org/agenda/2019/10/water-pollution-in-india-data-tech-solution/. Accessed 31 Jan 2024
Yadav AK, Sahoo SK, Mahapatra S, Kumar AV, Pandey G, Lenka P, Tripathi RM (2014) Concentrations of uranium in drinking water and cumulative, age-dependent radiation doses in four districts of Uttar Pradesh. India. Toxicol Environ Chem 96(2):192–200. https://doi.org/10.1080/02772248.2014.934247

Chapter 21
Mitigating Urban Floods Through Low Impact Development Techniques: A Comprehensive Study of Biological and Structural Best Management Practices

P. Z. Seenu, Chebrolu Madhu Sudhan, and K. V. Jayakumar

Abstract Due to the combination of frequent high-intensity rainfall and global warming, urban flooding is one of the most damaging problems that can occur anywhere in the world. The region chosen for the study is Amaravati, the legislative capital of Andhra Pradesh, located within the Kondaveeti Vagu's watershed area. This research uses low impact development (LID) strategies based on best management practices (BMP), including biological and structural approaches. The low-impact development methodology is a small-scale method that allows regionalization and is incorporated into the Storm Water Management Model (SWMM). Unlike traditional drainage systems that quickly funnel water away, LID mimics natural processes to manage stormwater at its source. It is inexpensive and simple to build, may be used as a beautiful element in landscaping, and is favorable to the environment. Bio-LID strategies include bioretention, rain gardens, and grasslands. On the other hand, the structural Low Impact Development approaches include the use of permeable pavement, rain barrels, and infiltration trenches. This research also investigates an amalgam of structural and biological LID techniques. It has been noticed, based on the results obtained, that the efficacy and capacity of biological best management practices (BMPs) exceed those of structural BMPs, when it comes to controlling and reducing the amount of runoff. The study's findings will be helpful in the development of flood mitigation measures, which will involve installing drainage systems in metropolitan areas during the development process. Development of flood mitigation measures, which will involve installing drainage systems in metropolitan areas during the development process.

P. Z. Seenu (✉) · C. M. Sudhan
Department of Civil Engineering, Vallurupalli Nageswara Rao Vignana Jyothi Institute of Engineering and Technology, Hyderabad 500090, India
e-mail: seenu_pz@vnrvjiet.in

K. V. Jayakumar
Civil and Infrastructure Engineering, Indian Institute of Technology Dharwad, Dharwad 580007, India

A. K. Yadav et al. (eds.), *Integrated Management of Water Resources in India: A Computational Approach*, Water Science and Technology Library 129,
https://doi.org/10.1007/978-3-031-62079-9_21

Keywords Low impact development · Best management practices: storm water management model · Flood mitigation

21.1 Introduction

Floods are inevitable natural occurrences, but their consequences may be mitigated. Urban flooding worsens due to (i) inadequate urban drainage network design (ii) rising surfaces that are impermeable in cities, and (iii) the destruction or filling of natural water storage facilities like lakes and depressions (Ye et al. 2010). The prevention of flooding in urban areas is of the highest priority since urban regions are the epicenter of economic activity as well as vital infrastructure that has to be safeguarded (Li et al. 2020; La Loggia et al. 2020).

Flood mitigation may be categorized as non-structural and structural methods (Conitz et al. 2021; Abdi-Dehkordi et al. 2021). The objective of flood mitigation in urban areas is to achieve a specified outcome within a set time frame (Ward et al. 2017). Among them are the reduction of flood danger to both individuals and their property, the lessening of the magnitude of flood damage that has already occurred, the diminishment of the pollution of receiving water, the improvement of opportunities for recreation, and the improvement of urban infrastructure (Yadav et al. 2014). BMPs are engineering or structural tools and control systems created to minimize pollutants and regulate stormwater flow (Zhen et al. 2006; Kumar et al. 2023). LID is a land engineering design and planning approach that focuses on controlling storm water runoff using methods including penetration, filtration, storage, evaporation, and detention at the point of origin (NDMA 2012).

LID is the sustainable method of gathering and using precipitation in metropolitan areas (Arjenaki et al. 2021; Tansar et al. 2022). LID addresses consist mainly of vegetative swales, green roofs, bioretention cells, rain gardens, and porous road surfaces. LID approaches are used differently in several nations. LID research in Beijing, China, discovered that porous road surfaces, rain gardens, and can store the water from rain around 60–82% (Randall et al. 2019; Putri et al. 2023). The precision of using LID-BMP strategies in long-duration rainfalls is examined using the SWMM model for relatively rare and long-term rainstorm occurrences (Campisano et al. 2017). In Tehran, Iran, LID-BMP strategies were assessed utilising the SWMM simulation to determine the discharge rate in five different scenarios for various return times. According to research, Green roofs decreased discharge the highest (Poursahebi et al. 2019). Xu et al. (2017) created a novel strategy that uses SWMM to apply block-scale (BMP) for (LID). In Tianjin, the researchers were able to locate and reproduce them for the Haihe River in Tanggu Bay, with the use of SWMM 5.1. The researchers used the "Technical Guide for the Construction of the Sponge City" to establish LID-BMP requirements. Jia et al. (2012) proposed an approach to include an environmentally conscious growth strategy to manage stormwater in the Beijing Olympic Village area. The purpose of this technique is to effectively control urban flooding. The runoff peak rate and decrease in flood volume were evaluated using a coupled SWMM-BMPDSS model, which included the (LID) method and BMP (Yin et al. 2020). An evaluation of low impact development was conducted to assess the effectiveness of (BMPs) including rooftops with green plants and permeable road surfaces (Hager et al. 2019). The recommendation made is to carry out the BMP plan, which might reduce flood volume by 27% while minimizing expenditures. Meena and Anil (2018) outlined a simulation technique that included 4 different kinds of (BMP) at research location. They then calculated how these BMPs reduced flood volume and peak flow. The recommended BMPs have been shown to reduce floods in the urban region using PCSWMM.

The research uses structural and biological LID techniques. Rain garden, bioretention, and grass swale are biological (LID) methods, whilst porous roads, absorption excavation, and rainwater collection are structurally LID methods. This research also investigates a mix of biological and structural approaches (LID). The study's results will assist in creating flood mitigation strategies through the implementation of sewer systems in urban areas.

21.2 Study Area and Methodology

Amaravati, which is the capital of Andhra Pradesh (AP) it is located inside Kondaveeti Vagu's (KV) watershed. The Krishna River is joined by the Amaravati's runoff. The research focuses on the geographical area that covers approximately 420 km^2 of Kondaveeti Vagu. Geographically, the Kondaveeti Vagu catchment region is located in (AP) in India, as shown in Fig. 21.1 of the research area. Its watershed is located between the longitudes of 80°00′00″ and 81°00′00″E, and the latitudes of 16°00′00″ and 17°00′00″N. In the area, May is the warmest month in the region, with an average high of 42 °C and a low of 27 °C. Amaravati gets greater rainfall in July. The average yearly rainfall in the study region is 870 mm. December and January had the lowest mean temperatures of 28 and 15 °C.

The research uses SWMM to model the urban flood. The urban area employs a dynamic rainfall-runoff simulation model (Rossman 2014; Kulwant et al. 2023). Groundwater, transportation, atmosphere, surface water, and transportation are the four main components of the metropolitan system according to the model. The research area is then divided into ground elevation-based subcatchments. For different magnitudes of rainfall, runoff is calculated for every subcatchment. There are three main categories of sub catchments: (i) a watershed area from before, (ii) a waterproof region designed for collecting rainwater without having the capacity to store water in low locations, (iii) a watershed that is impermeable and has reserves in depressions. The SWMM employs a trifecta of flow routing criteria: dynamic wave routing, steady flow routing and kinematic wave routing. This study employs the first routing method.

(LID) refers to a stormwater management strategy that focuses on limiting the adverse impacts of increased flow and pollutants by lowering discharge as much as possible. SWMM uses the concept of water balance to calculate the actual outflow and

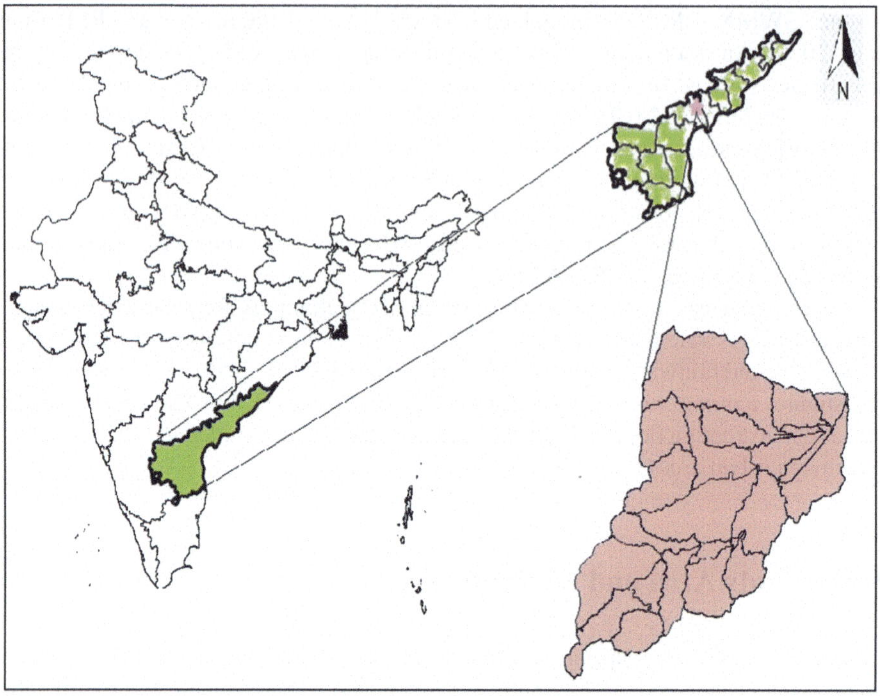

Fig. 21.1 Study area map showing Kondaveeti Vagu catchment

inflow of the subarea in real-time (Rossman 2014; Baumann et al. 2022). LID encompasses a collection of techniques to design sites that reduce the release of water and replicate the natural or original hydrological conditions through retention, harvesting, evapotranspiration, infiltration, and stormwater filtering using small-scale structural practices. These strategies may effectively decrease the intensity and volume of rainwater. Various forms of LID control are available for deployment. Each LID control is associated with certain situations that vary in their degree of implementation. SWMM includes a small-scale localized small-scale measure component of LID. The structure is economical, small, easy to build, sustainable, and looks good in an outdoor environment. Several LID modules are produced in SWMM to be incorporated into the respective sub-area by adjusting the parameters to match the particular circumstances. SWMM offers many LID controllers categorized as structural and biological. Commonly used are bioretention cells, permeable pavements (Xie et al. 2017), green roofs (Peng and Stovin 2017), rain gardens (Aad et al. 2010), and vegetative swales (Lee et al. 2012).

The research investigated three scenarios: Structural BMPs, biological (BMPs), and a mix of both biological and structural BMPs. Bioretention, Rain Garden, and Grass Swale are instances of biological (BMPs). Some examples of structural measures are permeable road surfaces, infiltration trenches, and rain barrels.

Three 5-year scenarios are simulated in five subcatchments encompassing 50% of the watershed's area.

SWMM conducts a water equilibrium assessment for all simulations to calculate the runoff volume that moves between and remains intact throughout the entire system (LID). Rainfall from the uphill area is immediately absorbed by the earth's surface, which also creates surface outflow and stores excess input in its highest layer. The pavement layer refers to continuous porous pavement systems that employ a porous concrete pavement layer. Bioretention cells use a mixture of synthetic soil to facilitate the development of vegetation. The crushed stone or gravel forms the storage layer and serves as a medium for storage in bioretention cells, infiltration trenches, and permeable surface. An under-drain system transports water from bioretention cell gravel storage and infiltration trenches to an output pipe. The goal of installing different LID units is to decrease runoff from the surface and increase the region's capacity to absorb water. Input information from Andhra Pradesh Capital Region Development Authority (APCRDA) and ArcGIS was used for developing the SWMM model. After Kondaveeti Vagu watershed, 25 sub-basins were established. The longest subbasin flow route, sub-catchment centroid, and longest centroidal flow path have been identified.

21.3 Results and Discussion

Water levels and discharge data were not accessible for the ungauged Kondaveeti Vagu watershed. The sole calibrated figure supplied was the highest possible discharge 460 m^3/s and at the spot where the Kondaveeti Vagu and Krishna rivers meet for a return period of one hundred years. The average Manning's n value and the percentage of imperviousness were determined by manual calibration, which included a technique based on trial and error. The goal was to mimic a historical flood inundation event.

The precipitation data was analyzed for frequency. Easyfit Software uses Anderson–Darling, Kolmogorov–Smirnov, and Chi-square tests to identify the optimal fit by reducing the overall value. The data match nicely with the (GEV) distribution. Table 21.1 displays the depth of precipitation for various return times using the Generalised Extreme Value (GEV) method. The research indicates that the rainfall depth is around 240 mm with a 100-year return time.

The highest level of discharge at the outlet of the Kondaveeti Vagu watershed was determined to be 460.4 m^3/s by modeling for a single day with a return duration of 100 years. When exiting and at intersections, hydrographs are produced. For each subbasin, junction, and reach, data on the hundred-year return period's peak discharge, peak time, and volume are required. The Undavalli outlet had a highest discharge of approximately 460 m^3/s.

The net flow (inflow minus outflow) was inconsistent in the simulation. If the continuity error is greater than 10%, the input settings need to be adjusted. Excessive

Table 21.1 Precipitation depth for various time periods using GEV

Return period (years)	P	Xt (mm)
2	0.50	84.47
5	0.80	119.95
10	0.90	145.25
20	0.95	170.96
50	0.98	206.50
100	0.99	234.90
200	0.995	264.81

continuity errors might occur due to lengthy calculation processes or insufficiently short conduits. In this research, a continuity error of under 10% was obtained.

As can be shown in Fig. 21.2, the SWMM model was modified for the particular research region. Gradient and size determine the flow of the channel. Every sub-aqueduct area, computation of runoff hydrographs. The outflow of subcatchment S1 had the highest peak runoff, with peak discharges for each sub-catchment being computed.

Based on SWMM data, low-lying areas such as Undavalli, Nalakapeta, Mangala-giri, Krishnayapalem, and Dolas Nagar were inundated. The Thullur region, along with other areas around the Neerukonda reservoir, was flooded. During intense rains, even high-elevation areas around Vaddamanu and Anathavaram saw little flooding.

This study mainly focused on the results of flood control measures adopted using low impact development techniques in the rapidly urbanizing Amaravati area. To be more precise, a 10-year return period rainfall is used to model the impact of six different kinds of LIDs and combinations. Table 21.2 shows the effect of several approaches on peak runoff. Tables 21.3, 21.4, and 21.5 demonstrate the impacts on maximum discharge for various low impact development (LID) methods.

Subcatchment S14 achieved the greatest decrease in maximum runoff (11.88%) with the implementation of bioretention structures. At 8.73, 8.86, and 3.85% respectively, grass swales manage and reduce runoff volume for S16, S10, and S2. When using an infiltration trench, S2 has the least effect on peak runoff. Research shows that biological BMPs are more effective and have more ability to manage and reduce runoff volume than structural BMPs. Biomass methylpropyl (BMP) products are both ecologically sustainable and economically efficient.

21.4 Summary

- The efficacy and capacity of biological best management practices (LID) were superior in managing and decreasing the amount of runoff compared to structural best management practices.

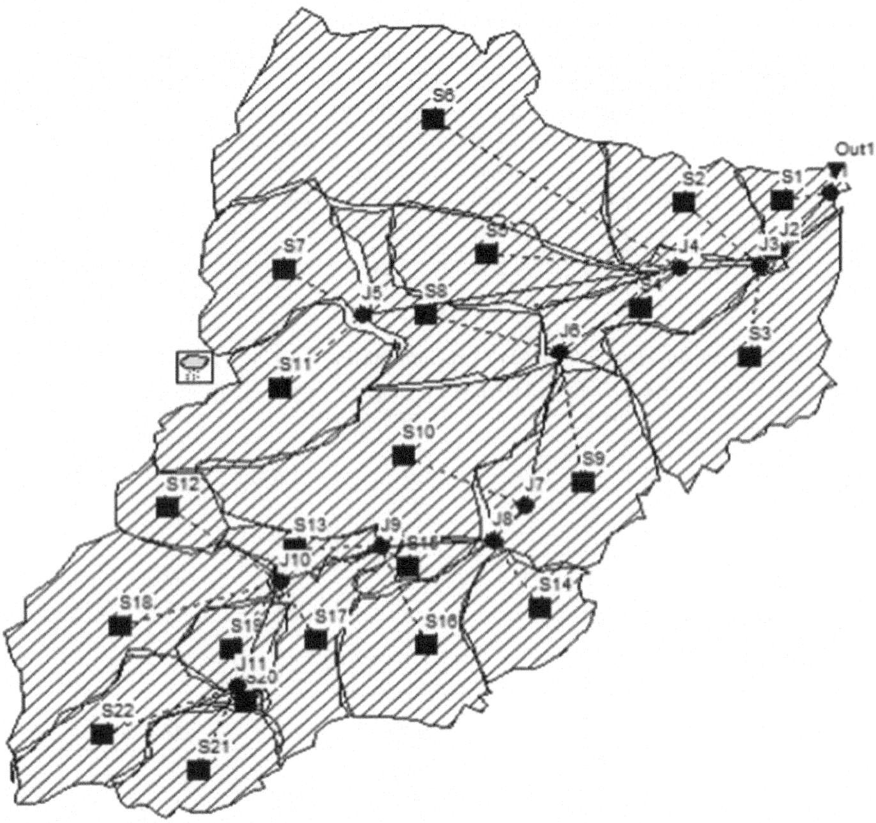

Fig. 21.2 SWMM model for the Kondaveeti Vagu catchment

Table 21.2 Comparison of runoff (m^3/s)

Sub-catchment	No LID	Bioretention	Rain garden	Grass swale	Permeable pavement	Infiltration trench	Rain barrel	Combination
S16	12.6	11.9	12.1	11.5	12.4	12.0	11.7	11.8
S14	10.1	8.9	9.5	9.3	9.6	9.4	9.8	9.1
S10	7.9	7.3	7.5	7.2	7.6	7.5	7.8	7.2
S6	28.7	27.1	26.7	27.2	27.3	27.4	27.8	26.9
S2	23.4	22.8	23.1	22.7	22.8	23.2	22.9	22.5

- Implementing the most effective management approaches identified in this research may help reduce susceptibility to urban floods in the research region.
- This study's technique is applicable to urbanizing cities with comparable characteristics.

Table 21.3 Comparison of runoff (m^3/s) of biological BMPs

Sub-catchment	No LID (m^3/s)	Bioretention (m^3/s)	% reduction	Rain garden (m^3/s)	% reduction	Grass swale (m^3/s)	% reduction
S16	12.6	11.9	5.56	12.1	3.97	11.5	8.73
S14	10.1	8.9	11.88	9.5	5.94	9.3	7.92
S10	7.9	7.3	7.59	7.5	5.06	7.2	8.86
S6	28.7	27.1	5.57	26.7	6.97	27.2	5.23
S2	23.4	22.8	2.56	23.1	1.28	22.7	2.99

Table 21.4 Comparison of runoff (m^3/s) of structural BMPs

Sub-catchment	No LID (m^3/s)	Permeable pavement (m^3/s)	% reduction	Infiltration trench (m^3/s)	% reduction	Rain barrel (m^3/s)	% reduction
S16	12.6	12.4	1.59	12.0	4.76	11.7	7.14
S14	10.1	9.6	4.95	9.4	6.93	9.8	2.97
S10	7.9	7.6	3.80	7.5	5.06	7.8	1.27
S6	28.7	27.3	4.88	27.4	4.53	27.8	3.14
S2	23.4	22.8	2.56	23.2	0.85	22.9	2.14

Table 21.5 Comparison of runoff (m^3/s) of combined BMP

Sub-catchment	No LID (m^3/s)	Combination (m^3/s)	% reduction
S16	12.6	11.8	6.35
S14	10.1	9.1	9.90
S10	7.9	7.2	8.86
S6	28.7	26.9	6.27
S2	23.4	22.5	3.85

Acknowledgements The authors thank APCRDA for the data and all the reviewers and editors during the review process.

References

Aad MPA, Suidan MT, Shuster WD (2010) Modeling techniques of best management practices: rain barrels and rain gardens using EPA SWMM-5. J Hydrol Eng 15:434–443

Abdi-Dehkordi M, Bozorg-Haddad O, Salavitabar A, Mohammad-Azari S, Goharian E (2021) Development of flood mitigation strategies toward sustainable development. Nat Hazards 108(3):2543–2567. https://doi.org/10.1007/s11069-021-04788-5

Arjenaki MO, Sanayei HRZ, Heidarzadeh H, Mahabadi NA (2021) Modeling and investigating the effect of the LID methods on collection network of urban runoff using the SWMM model (case

study: Shahrekord City). Model Earth Syst Environ 7(1):1–16. https://doi.org/10.1007/s40808-020-00870-2
Baumann H, Ravn NH, Schaum A (2022) Efficient hydrodynamic modelling of urban stormwater systems for real-time applications. Modelling 3(4):464–480
Campisano A, Catania FV, Modica C (2017) Evaluating the SWMM LID Editor rain barrel option for the estimation of retention potential of rainwater harvesting systems. Urban Water J 14:876–881
Conitz F, Zingraff-Hamed A, Lupp G, Pauleit S (2021) Non-structural flood management in European rural mountain areas—are scientists supporting implementation? Hydrology 8(4):167
Hager J, Hu G, Hewage K, Sadiq R (2019) Performance of low-impact development best management practices: a critical review. Environ Rev 27(1):17–42. https://doi.org/10.1139/er-2018-0048
Jia H, Lu Y, Yu SL, Chen Y (2012) Planning of LID-BMPs for urban runoff control: the case of Beijing Olympic Village. Separat Purif Technol 84:112–119. https://doi.org/10.1016/j.seppur.2011.04.026
Kulwant M, Patel D, Shirin S, Sharma SN, Yadav AK (2023) Future frameworks for fluoride and algorithms for environmental system. In: Advanced treatment technologies for fluoride removal in water, water science and technology library. Springer, Cham, p 125. https://doi.org/10.1007/978-3-031-38845-3_19
Kumar A, Chava R, Gupta S, Shirin S, Jamal A, Yadav AK (2023) Environmental and health effects of fluoride contamination and treatment of wastewater using various technologies. In: Advanced treatment technologies for fluoride removal in water (water science and technology library), vol 125. Springer, Cham. https://doi.org/10.1007/978-3-031-38845-3_18
La Loggia G, Puleo V, Freni G (2020) Floodability: a new paradigm for designing urban drainage and achieving sustainable urban growth. Water Resour Manag 34:3411–3424. https://doi.org/10.1007/s11269-020-02620-6
Lee JM, Hyun KH, Chio JS (2012) Flood reduction analysis on watershed of LID design demonstration district using SWMM5. Desalin Water Treatment 38:255–261
Li L, Uyttenhove P, Van Eetvelde V (2020) Planning green infrastructure to mitigate urban surface water flooding risk: a methodology to identify priority areas applied in the city of Ghent. Landscape Urban Plann 194:103703
Meena YR, Anil KG (2018) Evaluation of stormwater BMPs performances for flood volume reduction in Bengaluru City, Karnataka. India. SSRG Int J Civil Eng 5(7):5–10
NDMA (2012) National disaster management guidelines. National Disaster Management Authority, Government of India
Peng ZJ, Stovin V (2017) Independent validation of the SWMM green roof module. J Hydrol Eng 22:1–12
Poursahebi A, Niri MZ, Ghoudarzi SM (2019) Simulating of LID-BMP methods on urban runoff (case study: District 22, Tehran). J Water Wastewater 30:3136
Putri FK, Hidayah E, Maruf MF (2023) Enhancing stormwater management with low impact development (LID): a review of the rain barrel, bioretention, and permeable pavement applicability in Indonesia. Water Sci Technol 87:2345–2361
Randall M, Sun F, Zhang Y, Jensen MB (2019) Evaluating Sponge City volume capture ratio at the catchment scale using SWMM. J Environ Manag 246:745–757
Rossman LA (2014) Storm water management model user's manual version 5.1. USEPA, Washington, DC
Tansar H, Duan HF, Mark O (2022) Catchment-scale and local-scale based evaluation of LID effectiveness on urban drainage system performance. Water Resour Manag 12:1–20. https://doi.org/10.1007/s11269-021-03036-6
Ward PJ, Jongman B, Aerts JC, Bates PD, Botzen WJ, Diaz Loaiza A, Winsemius HC (2017) A global framework for future costs and benefits of river-flood protection in urban areas. Nat Clim Change 7(9):642–646
Xie JG, Wu CH, Li H, Chen GT (2017) Study on storm-water management of grassed swales and permeable pavement based on SWMM. Water 9:840

Xu T, Jia H, Wang Z et al (2017) SWMM-based methodology for block-scale LID-BMPs planning based on site-scale multi-objective optimization: a case study in Tianjin. Front Environ Sci Eng 11(1):934. https://doi.org/10.1007/s11783-017-0934-6

Yadav AK, Sahoo SK, Mahapatra S, Kumar AV, Pandey G, Lenka P, Tripathi RM (2014) Concentrations of uranium in drinking water and cumulative, age-dependent radiation doses in four districts of Uttar Pradesh. India. Toxicol Environ Chem 96(2):192–200

Ye B, Sheng DL, Men XY (2010) Causes and countermeasures of urban water logging. J Econ Water Resour 28:62–78

Yin D, Evans B, Wang Q, Chen Z, Jia H, Chen AS, Leng L (2020) Integrated 1D and 2D model for better assessing runoff quantity control of low impact development facilities on community scale. Sci Total Environ 720:137630

Zhen J, Shoemaker L, Riverson J, Alvi K, Cheng M-S (2006) BMP analysis system for watershed-based stormwater management. J Environ Sci Health A 41(7):1391–1403. https://doi.org/10.1080/10934520600657172

Chapter 22
Future of Water Management in India: Emerging Role of IoT, AI, and Data Analytics

Pritam Das, Nirmalya Kumar Nath, Abhinav Kumar, Leena Rani Mishra, Sachin Suryawanshi, and Vinay Kumar Gautam

Abstract India, a nation of diverse landscapes and a burgeoning population, faces immense challenges in managing its water resources. Rising demand, climate change impacts, and regional disparities exacerbate the situation. However, technological advancements are offering promising solutions tailored to India's specific needs. Remote sensing, with its powerhouses like unmanned aerial vehicles (UAVs) and satellite imaging, has become crucial. Authorities leverage these technologies to monitor watersheds, assess drought and flood risks, and gauge vital reservoir levels, enabling proactive responses. Artificial intelligence (AI) algorithms further refine these forecasts, facilitating informed decision-making to prevent water crises. The convergence of AI, data analytics, and sensor technology empowers real-time monitoring and control of water systems. Smart sensor networks, encompassing India's infrastructure, offer continuous water distribution and quality tracking. The Internet of Things (IoT) fosters communication between various water management elements, optimizing resource allocation and minimizing losses. Desalination technology, particularly seawater conversion, has become economically viable due to advancements in membrane technology and energy-efficient processes. This presents a potential solution for India's freshwater-scarce regions. Additionally, blockchain technology is promoting transparency and efficiency in water governance. However, challenges persist, including pricing, scalability, and technological adoption in developing nations like India. Overcoming these requires tailoring technologies like remote sensing, smart sensors, desalination, and blockchain applications to India's unique context. Collaborative efforts between governments, technology developers, and local communities are crucial to ensure equitable access to water resources and navigate this transformative journey.

P. Das (✉) · N. K. Nath · A. Kumar · L. R. Mishra · S. Suryawanshi · V. K. Gautam
Department of Soil and Water Engineering, College of Technology and Engineering, Maharana Pratap University of Agriculture & Technology, Udaipur, Rajasthan 313001, India
e-mail: mepritam1993@gmail.com

A. K. Yadav et al. (eds.), *Integrated Management of Water Resources in India: A Computational Approach*, Water Science and Technology Library 129,
https://doi.org/10.1007/978-3-031-62079-9_22

Keywords Water management · Remote sensing · Artificial intelligence · IOT · Smart meters · Nanotechnology

22.1 Introduction to Water Management

22.1.1 Background

India has a long history of managing water resources, which is ingrained in its agricultural and cultural legacy despite its varied climate and topography. For many centuries, India had an indigenous and community-managed water management system that provided the community's needs for drinking, irrigation, and household water supply (Shankari and Shah 1993; Shirin and Yadav 2014). Historically, to maintain agriculture in the face of erratic monsoons, agricultural cultures in India relied on clever inventions like step wells, tanks, and complex canal systems. A significant change from conventional community-based water management occurred during colonial rule. Large barrages and canals were constructed by the British, but the irrigation systems were managed rather than ruled. Furthermore, they were too big for the communities to have any meaningful influence on how they were run (Chitale 1991).

The shortage and unpredictable nature of water becomes even more problematic due to climate change. The twenty-first century saw a paradigm change towards the integration of cutting-edge technologies into water management as a reaction to these difficulties. This background lays the groundwork for an in-depth investigation of how technology could revolutionize water management methods in India while honoring and learning from ancient wisdom and addressing modern issues. Gaining an appreciation of the dynamic interaction between tradition and technology in the sustainable management of India's valuable water resources requires understanding of this historical background.

22.1.2 Importance of Water Management in India

India's socioeconomic structure is based primarily on water, which is essential for maintaining livelihoods, guaranteeing food security, and promoting economic growth. India, one of the most populous countries on earth, has several challenges in efficiently managing its water resources. Developing nations like India, where more than 60% of the population depends on agriculture, want their economies to be 5 trillion USD overall, with the agricultural sector alone accounting for almost 60% of this growth (Singh et al. 2020). The majority of people in rural India work in agriculture, where crops are increased by irrigation systems and rely on periodic monsoons. In terms of area, India is the world's largest rain-fed nation (71.7 Mha)

(National Rainfed Area Authority [NRAA] 2020) however, it has one of the lowest yields (around 1 t/ha) (Rao et al. 2015). In rainfed and dryland agriculture, the main crops that contribute to overall agricultural production are coarse cereals (85%), pulses (83%), oilseeds (70%) and cotton (65%) (Nagaraj 2013). Yields from irrigation systems are usually at least twice as high as those from rainfed crops (Wani et al. 2009).

Gupta and Deshpande (2004) have also looked at India's water supplies in light of the country's expanding population. Less than 1000 m^3 of available space per person was deemed a shortage situation by international authorities (Gosain et al. 2006; CWC 2009). Based on the theoretical distribution of water resources in the nation, the annual per capita availability of water is 1731 m^3 (CWC 2012). Enough water supplies and effective wastewater treatment are necessary to sustain industrial activities, maintain urban growth without jeopardizing the environment, and safeguard public health (Kumar et al. 2023; Kumari et al. 2023).

Ecosystems, biodiversity, and general ecological balance are all affected by pollution and water scarcity, which pose significant issues. A systematic approach to water management that considers conservation initiatives, regulatory changes, and technology improvements is needed to address these problems. Essentially, the importance of water management in India goes beyond its practical applications to include the conservation of a way of life, the promotion of economic development, and the protection of the environment. A sustainable and resilient future depends on the adoption of innovative water management practices, as the country struggles with the complex interaction of population growth, industrial expansion, and climate change.

22.1.3 Traditional Methods of Water Management

For a long time, traditional water management strategies in India have been vital to the survival of people and agriculture. These practices are based on ancient wisdom and are tailored to the many ecosystems found throughout the subcontinent. These tried-and-true techniques demonstrate a thorough awareness of the complex equilibrium between the demands of nature and humankind.

The construction of step wells, or "baolis," which are common in many parts of India, is one such example. These architectural wonders provide water, but also serve as gathering places for communities. Step wells provide a reliable source of water during dry spells by effectively collecting rainfall throughout the monsoon season. Furthermore, elaborate decorations stop contamination and evaporation. Rural landscapes have always had "talabs" or village ponds, which are traditional water collection devices. These ponds function as catchment areas, collecting rain to replenish groundwater levels and providing community drinking holes for cattle. Another ancient technique for reducing water flow, halting soil erosion, and promoting groundwater recharge is the construction of check dams, or "bandharas," beside rivers and streams.

In addition, historic canal networks, such as the Grand Anicut in Tamil Nadu, demonstrate expertise in hydraulic engineering. These canals demonstrate a comprehensive understanding of the terrain and water flow by effectively distributing water for cultivation.

Recognizing and using these historic traditions into current water management strategies becomes essential in light of today's water issues. Their community-focused and sustainable approach offers insightful guidance on building resilience in India's dynamic water management environment.

22.2 Role of Technology in Water Management

22.2.1 Overview of Technological Advancements

Technological breakthroughs have become transformational agents in the landscape of water management, following India's fast industrialization and the escalation of water-related difficulties. India has experienced a paradigm change in the last few decades from conventional techniques to state-of-the-art technology, bringing about a new era of sustainability, efficiency, and precision in the management of water resources.

Water monitoring and resource evaluation have undergone a radical change thanks in large part to developments in sensor technology, satellite photography, and data analytics. Proactive response to fluctuating water levels, droughts, and floods is made possible by collecting and analysis. Due to its wide coverage, satellite technology has proven to be a useful tool for monitoring large-scale bodies of water and making more accurate predictions about water-related events. Unmanned aerial vehicles (UAVs) have become a flexible platform for the evaluation of water resources. UAVs equipped with high-definition cameras and sensors offer comprehensive views of the terrain, facilitating well-informed choices for effective water resource management.

Every facet of water management has been impacted by artificial intelligence (AI), from data processing to predictive analytics. Large-scale datasets are processed by AI algorithms, which find patterns and anomalies that conventional approaches would miss. Predictive analytics powered by artificial intelligence (AI) improves the capacity to predict patterns of water demand, allocate resources optimally, and raise overall system efficiency. With the introduction of smart sensor networks and the Internet of Things (IoT), technological integration goes beyond monitoring to include physical infrastructure. These networks provide quick solutions to new difficulties by continually monitoring resource allocation, distribution networks, and water quality in real time. The use of blockchain technology in water management is also being investigated. In order to ensure the equitable and effective distribution of water resources, it can offer a transparent and safe platform for water trade. Finally, the treatment of water with nanotechnology has shown potential. It may be used in the development of sophisticated filters and membranes that guarantee a clean and safe water supply by eliminating impurities and pathogens from water.

India's adoption of these technical advancements is paving the way for an all-encompassing and innovative approach to water management that will combine traditional knowledge with state-of-the-art techniques to tackle the intricacies of a rapidly changing water environment.

22.2.2 Impact of Technology on Water Management

An era of increased efficiency, sustainability and resilience has been ushered in by the significant and revolutionary influence of the integration of sophisticated technology into Indian water management techniques. These technical advancements have addressed long-standing issues and offered vital answers for the management of water resources in a setting that is changing rapidly.

The domains of data accuracy and real-time monitoring are among the main effects. The precision of data collection has been greatly improved by the use of sensor technology and satellite imaging, allowing more complex knowledge of water distribution, quality, and levels. Proactive management of water resources now relies heavily on real-time monitoring, allowing prompt reactions to variations, crises, and resource allocation optimization. The ability to anticipate and control water-related catastrophes has greatly benefited from technological improvements. Satellite imaging has proven invaluable for watershed monitoring, flood control, and drought forecasting. India has improved its ability to predict and lessen the effects of natural catastrophes, protecting both rural and urban populations, by using remote sensing technology.

Artificial intelligence (AI) has become a disruptive force in predictive analytics and data processing. Large datasets may be combed through by AI algorithms, which can spot patterns and trends that a human analyst would miss. The water infrastructure has undergone a transformation due to the Internet of Things (IoT) and the deployment of smart sensor networks (Kulwant et al. 2023). Continuous, real-time monitoring of water distribution and quality systems guarantees early identification of anomalies, timely problem solving, and optimal water allocation. This helps to preserve this precious resource, while also increasing the general effectiveness of water delivery systems. Although it is still in its infancy, blockchain technology has the power to completely transform the water trade industry. In order to ensure the equitable and effective distribution of water resources, it can offer a transparent and safe platform for trading water rights.

22.3 Remote Sensing Technologies in Indian Water Management

Indian water management now includes sophisticated instruments such as remote sensing technologies that offer never-before seen capacities for monitoring, evaluation, and predictive analytics. These technologies provide us with a bird's-eye perspective of the surface of the world, allowing us to monitor and manage water resources extensively. This section delves into the many ways satellite images, unmanned aerial vehicles (UAVs), and artificial intelligence contribute to the development of a data-driven strategy for managing water resources.

22.3.1 Satellite Imagery

22.3.1.1 Watershed Management

Any area of land that accumulates and discharges all of the rain and streams into one spot- the entrance of a bay, the reservoir's outflow, or any other point along a stream channel- is referred to as a watershed. In terms of using natural resources within their watershed boundaries, watershed management involves strategic planning and prudent utilization of these land water resources. As it provides a comprehensive perspective of the terrain and water bodies, satellite imaging is essential for the management of watersheds. High-resolution photos provide comprehensive details about the watershed's geography, land use, plant cover, and other physical features. Plans for managing watersheds can be created using this information, with the goal of striking a balance between resource consumption and ecological preservation.

For instance, regions of the watershed that are vulnerable to erosion or sedimentation can be found using satellite photography. This information is essential for developing focused plans to maintain soil moisture, stop runoff, and improve the general health of watersheds. Similarly to this, remote sensing can track changes in plant cover and land use over time, offering important insights into how human activity affects the watershed.

22.3.1.2 Flood Prediction and Management

Satellite imaging is used as a preventive measure for flood control and prediction in areas that are prone to flooding. Satellites monitor variations in soil moisture, precipitation levels, and river dynamics, providing vital information to predict future flood disasters. This makes it possible to conduct timely evacuation plans, allocate resources, and reduce the damage caused by flooding. For example, depending on their geography and proximity to water bodies, places that are at high risk of flooding can be identified using satellite photography. Decisions on land use planning and the creation of maps showing flood danger can be influenced by this. In addition, real-time monitoring of changes in soil moisture content and river levels can be achieved using remote sensing. This can provide early notice of imminent floods, enabling the evacuation of high-risk regions and the deployment of flood prevention measures.

22.3.1.3 Drought Prediction and Management

Early identification and control are greatly aided by satellite-based remote sensing. Satellites assist in detecting areas that are vulnerable to drought conditions by analyzing plant health, soil surface temperatures, and precipitation patterns. Authorities can use this information to launch targeted relief operations and water conservation measures in a timely manner.

22.3.2 Unmanned Aerial Vehicles (UAVs)

22.3.2.1 Applications in Water Resource Assessment

Unmanned aerial vehicles (UAVs) have provided high-resolution aerial pictures of landscapes, revolutionizing the evaluation of water resources. Unmanned aerial vehicles (UAVs) play a crucial role in mapping water sources, evaluating groundwater levels, and pinpointing sites that hold promise for sustainable water extraction in water-stressed countries.

22.3.2.2 Enhancing Decision-Making with UAV Data

The accuracy of water management plans is increased by incorporating UAV data into decision-making procedures. UAVs enable decision makers to create flexible and responsive strategies to maximize water usage and conservation by providing real-time data on land use, vegetation health, and water distribution.

22.4 Smart Sensor Networks for Improved Water Infrastructure

The integration of smart sensor networks with water infrastructure represents a paradigm-shifting move toward proactive, data-driven management. In addition to emphasizing the use of smart sensors and their role in localized reactions and resource optimization through IoT connectivity, this part explores the revolutionary effects of continuous monitoring of water quality and real-time monitoring of water distribution (Vijai and Sivakumar 2016).

22.4.1 Continuous Monitoring of Water Quality

22.4.1.1 Application of Smart Sensors

Smart sensors, which provide real-time data on several factors, including pH levels, dissolved oxygen, and pollutant concentrations, have become essential instruments for ongoing water quality monitoring. Strategically placed across distribution networks, treatment facilities, and bodies of water, these sensors create a network that continuously collects, processes, and disseminates critical data.

22.4.1.2 Localized Responses to Water Quality Anomalies

The capacity of smart sensor networks to initiate localized reactions to abnormalities in water quality is one of its main features. These sensors have the ability to automatically initiate corrective action when they identify violations of predetermined quality standards. This could involve modifying the way water is treated, rerouting its flow, or promptly notifying the appropriate authorities. The agility of water management systems is improved by these localized reactions, which allow quick containment of any problem and reduce the effect on end users.

22.4.2 Real-Time Monitoring of Water Distribution

22.4.2.1 Role of Smart Sensors in Distribution Systems

The optimization of the water distribution system is greatly aided by smart sensors, which provide real-time information on pressure levels, network performance, and any leaks. These sensors, which are strategically positioned throughout the distribution infrastructure, continually measure water flow and pressure to spot system weaknesses and inefficiencies.

Water utilities can quickly identify and resolve problems thanks to the real-time data produced by smart sensors. Smart sensors give water distribution networks more resilience and dependability by providing actionable information in the event of a broken pipe, pressure drop or unanticipated demand changes. Using a proactive approach, water losses are reduced, system efficiency is increased, and end users receive constant service.

22.4.2.2 Optimizing Resource Allocation Through IoT Connectivity

The inclusion of the Internet of Things (IoT) significantly enhances the interconnectedness of smart sensor networks. An ecosystem for responsive and dynamic water management is created when smart sensors and central control systems interact seamlessly thanks to the IoT connection. Due to this connectedness, water utilities are able to make decisions in real time and effectively allocate resources. Smart sensors may send data to centralized control centres via IoT connection, where it is processed in real-time by algorithms and analytics. This improves reaction time, while also making it possible to automate some processes, such as changing valve settings or rerouting water flow to maximize distribution.

Smart sensors have a purpose in water distribution systems that goes beyond solving problems reactively. These sensors help in the creation of adaptive strategies by providing real-time data analysis and ongoing monitoring. For instance, smart sensors can help dynamically rerouting water to prioritize key locations during times of high demand or supply limits, ensuring equal distribution and minimizing service disruptions.

22.5 Data Analytics and Artificial Intelligence

The use of artificial intelligence (AI) and data analytics in water management has created new opportunities to maximize the use of this essential resource. Artificial intelligence (AI)-based systems have demonstrated increased field efficiency and offer scalable solutions to address the agricultural industry's mounting problems, including water management and irrigation (Kim et al. 2008). For instance, AI is leading the way in the digital transformation of water supply to ensure Sustainable Development Goal (SDG)-6 (sustainable management of water and sanitation for all) and SDG-13 (urgent actions to promote climate-related investments to combat climate change impacts) (Pigola et al. 2021). To address important global water management related challenges, such as preventing the aging and degradation of water assets, managing the increase in water tariffs, identifying water resources as a growth-limiting factor in the water-food-energy nexus, both in terms of quality and scarcity under the compounding effects of climate change and increased urbanization, and integrating soil, water, and vegetation at a watershed level to support natural resource management for agriculture, numerical tools based on artificial intelligence applications have been developed (Nourani et al. 2014).

22.5.1 Historical Data and Usage Patterns

Analyzing historical water consumption data while considering factors such as population growth, seasonal fluctuations, and industrial needs is the first step in predictive analysis. Technology can forecast future water usage by identifying trends and correlations by analyzing historical patterns.

22.5.2 Machine Learning Models for Forecasting

Once machine learning models are included in water management systems, they constantly learn from new data to enhance and improve forecasts. These models enable accurate and nuanced projections by taking into account dynamic variables including land use changes, economic advancements, and weather patterns.

22.5.3 Strategic Planning and Resource Allocation

Water management authorities can allocate resources and plan strategically with the use of predictive analytics. For instance, timetables for water treatment, distribution, and infrastructure maintenance can all be adjusted during times of predicted high demand. By ensuring that resources are used effectively, this proactive strategy reduces the possibility of shortages, maximizes energy use, and advances sustainability in general.

22.5.4 Sustainable Management and Conservation

Through the facilitation of conservation efforts and long-term planning, predictive analysis promotes sustainable water management. Authorities can put policies in place to improve water efficiency, raise public awareness, and promote responsible water consumption habits by projecting future needs and identifying possible stress areas in the water supply chain.

22.6 Advancements in Desalination Technologies for India

Desalination technologies have come to light as creative answers to India's growing water problems, particularly in areas where freshwater is in scarce supply. Given its lack of freshwater resources, India could use desalination to address its water needs using its saltwater resources. Its 7517 km coastline is shared by nine states, four union territories, and the ocean on one or more sides. Three major metropolitan areas, Chennai, Mumbai, and Kolkata are coastal and around 25% of the population lives along the coast. These cities can obtain their fresh water needs from desalinated seawater. The population of 41,850,260 people in seven states in India that have brackish groundwater as their primary source of freshwater is dependent on it (Abraham and Luthra 2011).

This section examines the developments in desalination, focusing on energy-efficient procedures, improved membrane technology, the feasibility of converting saltwater economically, and the contribution of desalination to alleviation of water shortages in certain areas.

22.6.1 Energy-Efficient Desalination

22.6.1.1 Improved Membrane Technologies

Developments in membrane technology, which are the foundation of desalination operations, are essential for energy-efficient desalination. Improved membranes have led to an evolution in desalination efficiency, improving processes such as reverse osmosis. For example, thin-film composite and nanostructured membranes have better salt rejection capabilities, which lowers the energy needed for the desalination process. However, the primary issues in developing nations such as India are related to energy supplies and brackish water; therefore, using Reverse Osmosis (RO) or electrodialysis (ED) to desalinate the water makes sense (Gude 2016; Islam et al. 2018).

22.6.1.2 Economic Viability of Seawater Conversion

Desalination's economic feasibility has always been problematic, mostly because of the process's high energy costs. However, the economic viability of seawater conversion is increasing due to the development of energy-efficient desalination systems and the falling cost of renewable energy sources (Manju and Sagar 2017). Reduced energy use through technological advances reduces operating costs, making desalination a financially feasible option for replenishing freshwater supplies.

22.6.2 Desalination as a Solution for Regional Freshwater Shortages

Desalination becomes a vital option in water-short places when traditional freshwater supplies are not sufficient to fulfil the increasing needs. The coastal regions are ideal locations for desalination facilities. These areas may reduce their reliance on finite freshwater supplies by diversifying their water supply sources and using cutting-edge desalination technology.

Strategic placement of desalination plants in areas with limited water resources improves the water security of the region. By offering a consistent and reliable supply of freshwater, these plants can reduce the effects of droughts and guarantee a steady supply of water for communities, businesses, and agriculture. Desalination is now a calculated investment to guarantee water supply in the future as climate change makes the problems with water shortage worse. According to Dhakal et al. (2022), by 2050, other countries may be able to address the problem of water shortage by implementing the desalination water experiences of some countries (e.g., India, China, and South Africa).

22.7 Nanotechnology in Water Treatment

In the field of water treatment, nanotechnology has become a cutting-edge solution because of its extraordinary capacity to eliminate pollutants and pathogens. It is the act of working with atoms at the nanoscale. Within the field of nanotechnology, nanomembranes are employed to remove physical, biological, and chemical impurities, as well as soften water.

22.7.1 Removing Contaminants and Pathogens

22.7.1.1 Nanomaterials for Enhanced Filtration

Water treatment undergoes a paradigm change due to nanotechnology, which uses nanoparticles to improve filtration. Materials at the nanoscale, such as carbon nanotubes and graphene oxide, have special qualities that increase how well they absorb pollutants. The reactivity and large surface area of these materials allow them to effectively remove organic chemicals, herbicides, and heavy metals from water sources.

22.7.1.2 Targeted Removal of Pathogens

The capability of nanotechnology to precisely target and eradicate viruses is one of its outstanding applications in water treatment. The remarkable effectiveness of nanoscale antimicrobial compounds, such as silver nanoparticles, in the disinfection of water, can be attributed to their disruption of bacterial and viral cell membranes. This focused strategy guarantees increased water purity, lowering the possibility of waterborne illnesses, and improving general health.

22.7.2 Ensuring Safe and Clean Water Supply

22.7.2.1 Nano Sensors for Water Quality Monitoring

Monitoring water quality is revolutionized by innovative sensors made possible by nanotechnology. Nanosensors can provide real-time data on water conditions by detecting minute levels of pollutants. These sensors, which are frequently used in smart water systems, allow for ongoing observation and timely response to alterations in the quality of the water. By doing this, a consistently safe water supply is maintained through the adaptation and responsiveness of water treatment procedures.

22.7.2.2 Self-Cleaning Nanomaterials in Water Infrastructure

An innovation in the maintenance of effective and hygienic water distribution systems is the use of self-cleaning nanoparticles in water infrastructure. By using the photocatalytic effect, surfaces coated with nanoparticles, such as titanium dioxide, demonstrate self-cleaning characteristics. In addition to ensuring the lifespan of pipes and conduits in the water supply network, this minimizes the production of biofilm and lowers the demand for chemical cleaning agents.

22.8 Blockchain Technology in Water Trading

The decentralized and transparent nature of blockchain technology presents a significant opportunity to transform water administration in India. This section examines the various ways blockchain technology might improve the accountability, transparency, traceability, and fair resource allocation of the water sector.

22.8.1 *Enhancing Transparency and Traceability*

22.8.1.1 Blockchain Applications in Water Governance

The applications of blockchain technology in water governance offer an unchangeable, secure ledger for logging and validating data and transactions related to water. Blockchain's decentralized structure guarantees that data are dispersed over a network of nodes, removing the possibility of manipulation or a single point of failure. This openness is particularly important for monitoring the distribution, application and administration of water resources.

A tamper-proof record of water-related actions, such as extraction, treatment, and distribution, can be easily created thanks to blockchain technology. This not only increases the data's trustworthiness, but also builds confidence among stakeholders, including the public, government organizations, and water utilities. Blockchain becomes a keystone technology for creating a strong and transparent water governance system by guaranteeing the integrity of data.

22.8.1.2 Improving Accountability in Water Transactions

Ensuring accountability in transactions and resource use is one of the main difficulties facing water governance. Blockchain makes water-related agreements automatic and enforceable by introducing smart contracts, which are self-executing contracts with programmed stipulations. The entire water transaction lifecycle-from extraction licenses to delivery and payments-can be planned and carried out openly with the help of these smart contracts.

Smart contracts provide an auditable and verifiable record of each transaction, which improve accountability. As a result, there is less chance of fraud, illegal water extraction, or inconsistent reporting. By using blockchain technology to automate contractual procedures, parties may ensure that all parties follow predetermined laws and regulations while simultaneously reducing administrative responsibilities.

22.8.2 *Smart Contracts for Water Trading*

22.8.2.1 Secure and Automated Transactions

Smart contracts enabled by blockchain technology provide security and automation to the water trade industry, revolutionizing the conventional resource exchange environment. Smart contracts provide safe and transparent water trade between different stakeholders, such as farmers, industry, and water utilities, by using a decentralized and consensus-driven method.

Without the need for middlemen, smart contracts enable smooth, automated transactions that guarantee that water trades meet predefined standards. By using blockchain technology, the water trade process is free from the possibility of disagreements, hold-ups, or inconsistencies. This not only makes trading easier, but also makes water transfers more reliable and efficient overall.

22.8.2.2 Promoting Equitable Water Resource Distribution

Beyond efficiency, blockchain technology plays a role in water trade by facilitating fair allocation of water resources. Water allocations may be dynamically modified based on real-time data via transparent and automatic smart contracts, guaranteeing that resources are dispersed equitably across rival consumers.

The decentralized ledger of blockchain technology guarantees precise recording of water allocations and its accessibility to all relevant parties. The possibility of water hoarding or unjust distribution methods is decreased by this transparency. Blockchain makes water governance more transparent and accountable by offering an auditable record of water transactions and allocations.

22.9 Conclusion

The complicated and multidimensional problem of water management is growing more and more significant as a result of population expansion, economic expansion, and climate change. The examination of technological developments in water management in the Indian setting ends with the conclusion that creative solutions are changing the face of water resource governance. The amalgamation of technologies, including nanotechnology, blockchain, and remote sensing, presents unparalleled prospects to address water-related predicaments and lay the groundwork for a robust and sustainable future.

22.9.1 Summary of Key Findings

The key findings emphasize how technical advances in water management have a revolutionary effect. UAVs and satellite imaging are examples of remote sensing technology that offer real-time data for effective monitoring and predictive analytics. By facilitating constant monitoring of water distribution and quality, smart sensor networks optimize water infrastructure. Artificial intelligence and data analytics provide decision makers with the ability to anticipate water use patterns and manage in real time. Freshwater scarcity can be effectively addressed by desalination technologies, which are driven by improvements in energy efficiency and membrane quality. To ensure the removal of impurities and pathogens for a safe water supply, nanotechnology is essential to the water treatment process. With smart contracts, blockchain technology promotes fair resource allocation and improves transparency, traceability, and accountability in water administration.

22.9.2 Challenges and Future Directions

22.9.2.1 Scalability and Affordability Issues

Although technical improvements seem promising, there are obstacles to overcome before these solutions can be made affordable and scalable. Implementing a large-scale project might need a lot of resources, and finding a balance between cost and efficiency is still a difficult task. It is critical to address price and scalability concerns in order to guarantee that technological advancements benefit a variety of groups and geographical areas.

22.9.2.2 Technology Adoption in Developing Regions

The deployment of these technologies in developing countries presents another challenge. Although modern technologies may be easily adopted in metropolitan areas, infrastructure, awareness, and financial resources present challenges in rural locations. Maintaining inclusive water management practices and avoiding access inequities require closing the technology adoption gap between urban and rural communities.

22.9.2.3 Future Research Directions

The subject of technological exploration in water management is always developing and future research approaches are essential to achieve ongoing progress. The development of context-specific solutions, multidisciplinary method exploration, and technology refinement should be the main priorities for researchers and practitioners. Concerns about new pollutants, desalination energy efficiency, and the resilience of blockchain applications for water governance should be prioritized.

In the Indian context, integrating technology into water management offers a promising path to resolve water-related issues. With cutting-edge monitoring technologies and creative treatment options, fair and sustainable water resource management is possible. It is crucial to address issues with scalability, price, and technology acceptance in various geographic locations in order to fully realize this promise. Subsequent research initiatives will be crucial in improving and broadening the reach of technical advancements, guaranteeing a water-secure future for India and other regions.

Acknowledgements The authors extend their gratitude to the scientists, researchers, and reviewers who helped to shape the pertinent subjects covered in this chapter.

References

Abraham T, Luthra A (2011) Socio-economic and technical assessment of photovoltaic powered membrane desalination processes for India. Desalination 268(1–3):238–248

Chitale MA (1991). Peoples' participation in large irrigation projects. Farmers in the management of irrigation systems. Sterling Publishers, Private Limited, New Delhi

CWC (2009) Integrated hydrological data book (Non-classified river basins). Hydrological Data Directorate, Information System Organisation, Water Planning & Projects Wing, Central Water Commission, Govt. of India, New Delhi

CWC (2012) Integrated hydrological data book (Non-classified river basins). Hydrological Data Directorate, Information System Organisation, Water Planning & Projects Wing, Central Water Commission, Govt. of India, New Delhi

Dhakal N, Salinas-Rodriguez SG, Hamdani J, Abushaban A, Sawalha H, Schippers JC, Kennedy MD (2022) Is desalination a solution to freshwater scarcity in developing countries? Membranes 12(4):381

Gosain AK, Rao S, Basuray D (2006) Climate change impact assessment on hydrology of Indian river basins. Curr Sci 346–353

Gupta SK, Deshpande RD (2004) Water for India in 2050: first-order assessment of available options. Curr Sci 1216–1224

Gude VG (2016) Desalination and sustainability—an appraisal and current perspective. Water Res 89:87–106

Islam MS, Sultana A, Saadat AHM, Shammi M, Uddin MK (2018) Desalination technologies for developing countries: a review. J Sci Res 10(1):77–97

Kim Y, Evans RG, Iversen WM (2008) Remote sensing and control of an irrigation system using a distributed wireless sensor network. IEEE Trans Instrum Meas 57(7):1379–1387

Kulwant M, Patel D, Shirin S, Sharma SN, Yadav AK (2023) future frameworks for fluoride and algorithms for environmental system. In: Yadav AK, Shirin S, Singh VP (eds) Advanced treatment technologies for fluoride removal in water. Water Science and Technology Library, vol 125. Springer, Cham. https://doi.org/10.1007/978-3-031-38845-3_19

Kumar A, Chava R, Gupta S, Shirin S, Jamal A, Yadav AK (2023) Environmental and health effects of fluoride contamination and treatment of wastewater using various technologies. In: Yadav AK, Shirin S, Singh VP (eds) Advanced treatment technologies for fluoride removal in water. Water Science and Technology Library, vol 125. Springer, Cham. https://doi.org/10.1007/978-3-031-38845-3_18

Kumari S, Dhankhar H, Abrol V, Yadav AK (2023) Effect of fluoride-contaminated water on the living being and their surroundings. In: Yadav AK, Shirin S, Singh VP (eds) Advanced treatment technologies for fluoride removal in water. Water Science and Technology Library, vol 125. Springer, Cham. https://doi.org/10.1007/978-3-031-38845-3_12

Manju S, Sagar N (2017) Renewable energy integrated desalination: a sustainable solution to overcome future fresh-water scarcity in India. Renew Sustain Energy Rev 73:594–609

Nagaraj N (2013) Rapporteurs' report: development of dryland agriculture: technological, institutional, infrastructural and policy imperatives. Indian J Agric Econ 68(3):1–16

National Rainfed Area Authority (NRAA) (2020) Prioritization of districts for development planning in India, a composite index approach. Task force on revisiting prioritization of rainfed areas, New Delhi, India

Nourani V, Baghanam AH, Adamowski J, Kisi O (2014) Applications of hybrid wavelet—artificial intelligence models in hydrology: a review. J Hydrol 514:358–377

Pigola A, da Costa PR, Carvalho LC, Silva LFD, Kniess CT, Maccari EA (2021) Artificial intelligence-driven digital technologies to the implementation of the sustainable development goals: a perspective from Brazil and Portugal. Sustainability 13(24):13669
Rao CS, Lal R, Prasad JV, Gopinath KA, Singh R, Jakkula VS et al (2015) Potential and challenges of rainfed farming in India. Adv Agron 133:113–181
Shankari U, Shah E (1993) Water management traditions in India. PPSt Foundation, Madras
Shirin S, Yadav AK (2014) Physico chemical analysis of municipal wastewater discharge in Ganga River. Haridwar District of Uttarakhand, India. Curr World Environ 9(2). https://doi.org/10.12944/CWE.9.2.39
Singh AK, Upadhyaya A, Kumari S, Sundaram PK, Jeet P (2020) Role of agriculture in making India $5 trillion economy under corona pandemic circumstance: role of agriculture in Indian economy. J AgriSearch 7(2):54–58
Vijai P, Sivakumar PB (2016) Design of IoT systems and analytics in the context of smart city initiatives in India. Proc Computer Sci 92:583–588
Wani SP, Rockström J, Oweis T (eds) (2009) Rainfed agriculture: unlocking the potential. CABI

Part IV
Advanced Optimization and Efficiency in Water Management

Chapter 23
Dynamic Search Space Reduction Methodology to Improve the Computational Efficiency of Evolutionary Algorithms for Optimal Pipe Sizing of Large Water Distribution Networks

Rajesh Gupta, Laxmi Gangwani, Nikita Palod, and Shilpa Dongre

Abstract A general optimal pipe sizing problem of a water distribution network (WDN) is a combinatorial optimization problem in which the sizes of different pipes are selected from a set of available pipe sizes. The number of combinations in a network with X number of pipes and Y number of available pipe sizes are Y^X. Complete enumeration to check the cost and feasibility of each combination using a network solver to search for the global optimal solution is practically not possible even for a moderate-size WDN, as the number of network analysis [called the number of functional evaluations (NFE)] will be Y^X. Several evolutionary algorithms (EAs) have been suggested in the last three decades, in which the search for the best in the search space (SS) is carried out by mimicking the process involved in a natural phenomenon or adopting some other strategy. Although the NFE required in EAs is substantially less as compared to that involved in complete enumeration, the computational time and efforts required to obtain the global optimal solution for large WDNs are high. Therefore, attempts have been made to reduce SS without losing the global optimal solution and to improve the efficiency and effectiveness of EAs. Various

R. Gupta (✉) · N. Palod · S. Dongre
Department of Civil Engineering, Visvesvaraya National Institute of Technology, Nagpur 440010, India
e-mail: rajeshgupta@civ.vnit.ac.in

N. Palod
e-mail: npalod@gmail.com

S. Dongre
e-mail: shilpadongre@civ.vnit.ac.in

L. Gangwani
Department of Civil Engineering, Shri Ramdeobaba College of Engineering and Management, Nagpur 440013, India
e-mail: gangwanilr@rknec.edu

A. K. Yadav et al. (eds.), *Integrated Management of Water Resources in India: A Computational Approach*, Water Science and Technology Library 129,
https://doi.org/10.1007/978-3-031-62079-9_23

search space reduction (SSR) methodologies are described and the application of a simple dynamic search space reduction (DSSR) is shown in this chapter with two different EAs—a parameter-based Genetic Algorithm (GA) and a parameter-less Rao-II algorithm. The DSSR methodology is generic and can be applied to other EAs as well.

Keywords Evolutionary algorithms · Genetic algorithm · Rao-II algorithm · Search space reduction · Water distribution network

23.1 Introduction

Optimization of the design of the water distribution network (WDN) is one of the most heavily researched areas due to its cost, which constitutes more than 60% of the total outlay of water supply project. The optimal design problem of looped WDN is considered as a NP-hard (Nondeterministic Polynomial-time hard), as it is a multimodal nonlinear-constrained problem (Yates et al. 1984). The available methodologies for the optimal design of WDNs can be classified as: (1) Deterministic search techniques; and (2) Evolutionary algorithms (EAs). In deterministic search techniques (linear, non-linear, and dynamic programming), the search for optimum begins by identifying a feasible solution and then moving to a better solution using some mathematical functions. The search terminates when no further improvement in cost is observed. As the WDN optimization problem is multi modal, the search depends on initial starting point and often terminates at a local optimal solution. Several EAs have been suggested in the last three decades in which the search starts with multiple points in the search space, moves iteratively towards better solutions, and terminates when the convergence criteria are achieved. The EAs have more probability of identifying the global optimal solution, however, cannot guarantee to obtain global optimal solution. The main drawback in EAs is the computational effort often represented by the number of functional evaluations (NFE), i.e., the number of hydraulic analyses carried out to identify the optimal solution. In recent past, several suggestions have been made to improve the efficiency of EAs. However, still their applicability to design large size networks is limited. One of the reasons for poor efficiency is the large search space that expands exponentially with the increase in the number of pipes in the network. A lot of time is consumed in evaluating the potentially unfit solutions, and search efficiency can be improved by eliminating potentially unfit solutions, i.e. by reducing the search space. However, reduction in search space should not eliminate the possibility of reaching global optimal solutions.

The pipes near the source will rarely have minimum size. Similarly, pipes at the ends will rarely have the largest size. Such sizes can be eliminated, and most appropriate sizes from the list of candidate pipe sizes can be selected for each link. Several methods have been suggested for reducing search space by restricting the number of candidate pipe sizes for each link in the network. These methods are described in this chapter. Here, a simple dynamic search space reduction method suggested by Gangwani et al. (2024) is applied to two different types of evolutionary algorithms. The genetic algorithm (GA) is one of the first types that involves several problem-specific parameters like cross-over and mutation probabilities. Such parameters require fine-tuning each time a new problem is solved. The parameterless Rao-II algorithm is of the second type in which the parameters are decided on its own during execution.

23.2 Formulation of the WDN Optimization Model

Consider a looped WDN with ND demand nodes, NP pipes, and NL loops. The minimum cost optimization problem for the above network can be formulated as:

$$\text{Minimize}\, f(D_1, \ldots, D_x) = \sum_{x=1}^{NP} c(D_x) \times L_x \tag{23.1}$$

where, $c(D_x) =$ cost per meter length of pipe x of diameter D_x; $L =$ pipe length. The objective function is subject to the constraints consisting of the flow continuity equations, loop headloss equations and that satisfying minimum pressure requirements at the demand nodes. The continuity equation states that algebraic summation of flows at demand nodes should be zero, and can be written as

$$\sum_{x \in j} Q_x + q_j^{req} = 0; j = 1, ..., ND \tag{23.2}$$

where j denotes the demand node; $Q =$ pipe discharge; ${q_j}^{req} =$ nodal demand at node j. The energy equation states that algebraic summation of head losses in closed loop should be zero, and can be written as

$$\sum_{x \in y} h_x = 0; y = 1, ..., NL \tag{23.3}$$

where, y represents the loop; $h =$ head loss in the pipe. The available HGL at all the demand nodes should be more than the minimum desired HGL. This can be written as

$$H_j^{\text{avl}} \geq H_j^{\text{des}}; j = 1, \ldots, ND \tag{23.4}$$

where ${H_j}^{avl} =$ available HGL at node j, ${H_j}^{des} =$ minimum desirable HGL at node j. Finally, the diameters should be selected from the available pipe sizes. This can be written as

$$D_x \in \{D_{\min} \ldots .D_{\max}\}; x = 1, \ldots, NP \tag{23.5}$$

where, {} denotes a set of available pipe sizes; $D_{\min}$ and $D_{\max}$ = minimum and maximum diameters of available pipes, respectively. The head loss in pipe can be obtained by the Hazen-Williams equations as:

$$h_l = \frac{\omega L Q^{\alpha}}{C_{HW}^{\alpha} D^{\beta}} \tag{23.6}$$

where, C_{HW} = Hazen-Williams pipe roughness coefficient; and ω, α, and β are constants.

The EPANET hydraulic solver (Rossman 2000) can handle pipe flow governing Eqs. (23.2 and 23.3). Furthermore, the selection of pipe sizes can be made from the available pipe sizes to satisfy Eq. 23.5. The penalty-based method is used to satisfy Eq. 23.4 thus converting constrained optimization problem into an unconstrained one. Penalty cost reduces the fitness of an infeasible solution, and in the process of optimization infeasible solutions are improved to feasible solutions. The objective function can be modified as

$$Minimize\, f(D_1, \ldots, D_x) = \sum_{x=1}^{NP} c(D_x)L_x + \sum_{j=1}^{ND} p \times (q_j^{req}) \times \{\max(H_j^{des} - H_j^{avl}, 0)\} \tag{23.7}$$

where, p is penalty multiplier. There are several types of penalty and methods to evaluate them (Abebe and Solomatine 1998, Montesinos et al. 1999, Wu and Walski 2005, Mohan and Jineshbabu 2009; Kadu et al. 2008; Abdy Sayyed et al. 2021; Palod et al. 2021 and Sangroula et al. 2022). The selection of a proper penalty coefficient value for the penalty function is a key to improve the efficiency of EAs. A larger coefficient imposes a larger penalty than a smaller coefficient for the same magnitude of constraint violation. A large penalty may quickly eliminate infeasible solutions from the population, and thus infeasible solutions may not be given enough opportunity to improve. On the contrary, a smaller penalty may allow the infeasible solutions to remain in the population for longer, thereby delaying the convergence, or the process may even converge at an infeasible solution if the penalty cost is too low. The self-adaptive or self-organizing penalty is recommended to avoid their fine-tuning. Kadu et al. (2008) suggested obtaining the penalty multiplier p by calculating the capitalized energy cost per unit of flow and per unit of head.

The other crucial point of concern in EA for WDN optimization is the search space defined by the number of possible combinations. A two-source network considered by Kadu (2008) has 34 pipes and 14 available pipe sizes. The search space is 14^{34}. As the number of pipes in the network cannot be reduced, attempts are made to limit the number of candidate sizes for each pipe (Vairavamoorthy and Ali 2005; Kadu et al. 2008; Haghighi et al. 2011; Zheng et al. 2011; Barlow and Tanyimboh 2014; Reca et al. 2017; Tanyimboh and Czajkowska 2018). By limiting the candidate pipe sizes to 5 for each pipe, the SS will be substantially reduced to 5^{34} (= 6.26×10^{-14}% of the original search space). The EA search in this reduced space would certainly be more effective. However, this reduction in SS should not eliminate the region of the global optimal solution. This aspect is further considered with two different types of EAs.

23.3 Parameter-Based Genetic Algorithm

Genetic algorithm work on the principle of 'survival of fittest". Initially, the selected population is reproduced by giving more probability of selection to more fit members. Population is modified by crossing over and mutation. While cross-over explores the newer region of search space, mutation is considered to exploit the nearby search space by making small changes. Besides the population size and number of generations, the GA search is governed by several other parameters such as type of selection, cross-over, and mutation operators, probability of cross-over, and probability of mutation. A high probability of cross-over and low probability of mutation are used. The stepwise methodology can be written as

(1) Generation of an initial random population.
(2) Calculate the cost of the network for each member of the population.
(3) Check the feasibility of the solution through hydraulic analysis and compute the penalty cost; compute total cost for infeasible solutions by adding penalty cost.
(4) Use reproduction, crossover, mutation, elitism and their derivatives to generate new solutions.
(5) Repeat steps 2–4 for the specified number of generations.

23.4 Parameter Less Rao-II Algorithm

Rao (2020) developed few optimization algorithms (Jaya, Rao-I, Rao-II, Rao-III, TLBO) in which no problem-specific parameters are required. One of the algorithms considered in this chapter is Rao-II. Parameters such as the number of iterations and the population size are the only requirements of this algorithm. Rao-II begins just like any other EA by generating a random initial population. It works by finding the best and worst solution from the entire population. The solutions for the next iteration are updated using Eq. 23.8. The solutions with minimum and maximum total cost are selected as the best and worst solutions, respectively. The total cost of any solution is the summation of the network cost and penalty (due to the violation of the constraint).

$$X'_{j,k,i} = X_{j,k,i} + r_{1,j,i}\left(X_{j,best,i} - X_{j,worst,i}\right) + r_{2,j,i}\left(\left|X_{j,k,i} or X_{j,l,i}\right| - \left|X_{j,l,i} or X_{j,k,i}\right|\right) \tag{23.8}$$

The best and worst values for any iteration i, for a variable j, are represented by $X_{j,best,i}$ and $X_{j,worst,i}$ respectively. $X'_{j,k,i}$ and $X_{j,k,i}$ represents the updated and old value for any variable j. $r_{1,j,i}$ and $r_{2,j,i}$ are random numbers distributed uniformly. The range of $r_{1,j,i}$ and $r_{2,j,i}$ is (0,1). l is any random solution from population set, and $X_{j,l,i}$ represents its value for variable j at i^{th} iteration. Solutions k and l are compared based on the total cost. If solution l is better than solution k (total cost of solution l < total cost of solution k), then the information must be exchanged from l to k; *therefore,* the term $X_{j,k,i}$ or $X_{j,l,i}$ is simplified to $X_{j,l,i}$ and the term $X_{j,l,i}$ or $X_{j,k,i}$ is simplified to $X_{j,k,i}$. This delineates the transfer of information from a superior solution to an inferior solution. In contrast to this, if the total cost of solution k < solution l, the term $X_{j,k,i}$ or $X_{j,l,i}$ is simplified to $X_{j,k,i}$ and the term $X_{j,l,i}$ or $X_{j,k,i}$ is simplified to $X_{j,l,i}$. Once the updated values of all the solutions in the population are determined, a completely new population set is created. The updated values are compared with the previous values of the solution, and the better performing solution (solution with minimum total cost) is directed to the next iteration. This process of updating and comparing the solutions is continued until termination criteria, which is generally a fixed number of iterations or when over a larger number of iterations, no improvement is observed in the fitness value of the best solutions. For the present case, the termination criteria is considered as fixed number of iterations to avoid premature convergence, as fitness value of best solutions at later stage is observed to remain constant over a large number of iterations before change.

23.5 Search Space Reduction (SSR) Methodologies

Several methods have been developed to reduce the search space by restricting candidate pipe diameters for different pipes. One type of method uses a pre-algorithm to lower and upper bounds on candidate pipe diameters for each pipe (Vairavamoorthy and Ali 2005; Kadu et al 2008; Haghighi et al. 2011; Zheng et al. 2011; Reca et al. 2017). Vairavamoorthy and Ali (2005)

used a pipe index vector based on the relative importance of all pipes in terms of their impact on the hydraulic performance of the network and used it for SSR. Kadu et al. (2008) used critical path method (Bhave 1978, 2003), Zheng et al. (2011) used nonlinear programming (NLP) approach coupled with graph theory, and Reca et al. (2017) used quadratic programming for SSR.

In other type of methods, initially FSS is used to identify the optimal solution/ near optimal solutions; and then using the obtained solutions, lower and upper bounds are fixed to execute the algorithm with reduced search space (Cisty and Bajtek 2009; Barlow and Tanyimboh 2014). Both, Cisty and Bajtek (2009) and Barlow and Tanyimboh (2014) obtained the optimal solution considering FSS and GA. Cisty and Bajtek (2009) used velocity criteria to decide lower and upper bounds on candidate pipe diameters of each link, Barlow and Tanyimboh (2014) reduced the number of decision variables by fixing the pipe sizes of those pipes that did not change across the solutions obtained with FSS.

The GA was coupled with ILP by Haghighi et al. (2011) in which primary pipes that carry major flows and secondary pipes, just necessary to form loops were identified first. GA was used to obtain the diameters of the secondary pipes only. The diameters of the primary pipes were obtained using ILP. This approach required iterative use of two optimization algorithms in sequence.

These methodologies in which the lower and upper bounds remain constant in the subsequent optimization runs were termed static search space reduction (SSSR) methodologies (Gangwani et al. 2024). These may have an over-restriction of candidate sizes of some of the pipes as observed by Abdy Sayyed et al. (2021), and the possibility of reaching a global optimal solution reduces. Therefore, the dynamic search space reduction (DSSR) methodologies in which SS keeps updating are better than SSSR.

Tanyimboh and Czajkowska (2018) used the maximum entropy formalism (Jaynes 1957) to dynamically update the active search space in the optimization process. The active solution space was updated by Gangwani et al. (2024) after completion of a predefined number of generations using the frequency of selection of a particular diameter for a particular pipe in the solutions of the last generation before updating. The DSSR methodology of Gangwani et al. (2024) is generic, simple, and obviates the need of any additional considerations as required in other approaches. This DSSR methodology is applied to design two networks with two different types of EAs.

23.6 Application of Dynamic Search Space Reduction Methodology

23.6.1 A Two-Reservoir Network: *Kadu et al. (2008)*

The network shown in Fig. 23.1 has 2 reservoirs, 24 demand nodes, 34 pipes, and 9 loops. Nodes 1 and 2 are the source nodes that have water levels of 100 m and 95 m, respectively. Pipe data, node data are given in Tables 23.1 and 23.2, respectively. The available pipe sizes and their respective unit costs are given in Table 23.3. Head loss is calculated using H-W formula with the HW pipe roughness coefficient as 130 for all pipes.

Some of the solutions obtained by different researches using different EAs and considering FSS/SSR are given in Table 23.4. Only those solutions which are better than the best available at that time are listed. The following points can be observed in Table 23.4.

1. It is only five occasions when an improved solution to Kadu's network is obtained.
2. Barlow and Tanyimboh (2014) obtained the least cost solution of Rs. 124.69 million using SSSR in 142,000 NFE. However, their solution was found to be infeasible with current EPANET parameters (Gangwani et al. 2024). The solution obtained from Gangwani et al. (2024) with GA and DSSR in 1,125,300 NFE has the least cost of Rs. 125.019 million. This solution was obtained with a self-adaptive penalty approach based on combined flow and head deficit-based penalty (Abdy Sayyed et al. 2021, Gangwani et al. 2024).
3. Gangwani et al. (2023b) obtained the same solution using the Rao-II algorithm with FSS in only 33,525 NFE using the self-adaptive penalty approach. As the

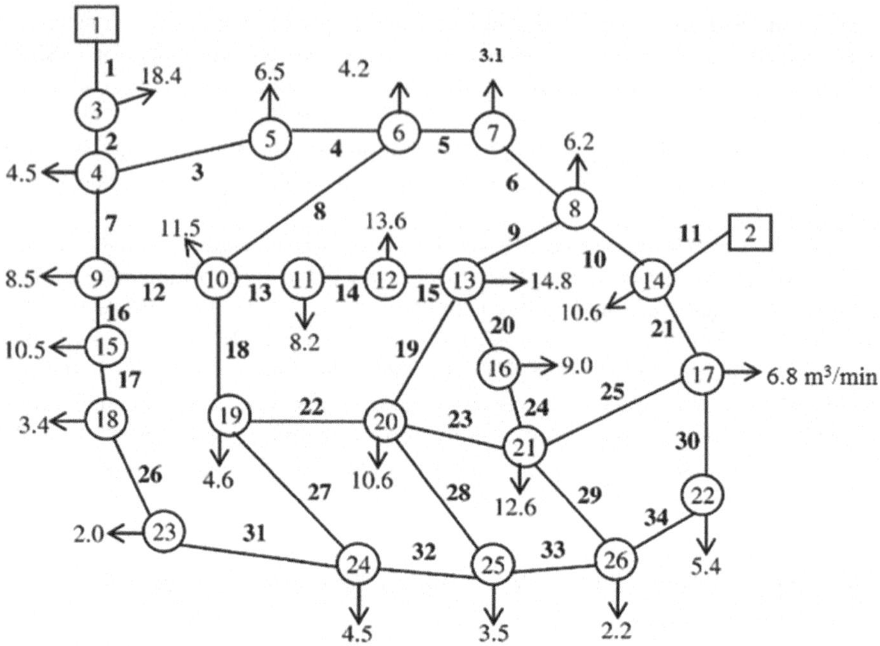

Fig. 23.1 Layout of Kadu water distribution network

same minimum cost solution as obtained with DSSR and GA is obtained with FSS and Rao II, the DSSR methodology is not attempted on this problem.

The diameters obtained for the best cost solution by GA with DSSR and Rao-II with FSS is given in Table 23.1 and corresponding available HGLs at nodes are given in Table 23.2.

23.6.2 Ramnagar Network: *Gangwani et al. (2024)*

Next, we take a real-world Ramnagar WDN as shown in Fig. 23.2 for the optimal design with DSSR. The gravity network fed by a Ground Service Reservoir (supply level = 327.205 m) has 375 pipes, 292 junctions. The network is designed to meet the minimum pressure requirement of 8 m at all demand nodes. The costs of the available diameters are given in Table 23.3.

Gangwani et al. (2024) obtained two solutions using GA with FSS and DSSR. Ten GA runs were performed and the best solution in 10 runs was reported. The results obtained for the network are given in Table 23.5.

The Rao II algorithm was used by Gangwani et al. (2023b) for the optimal design of the Ramnagar network with FSS. Fifteen trials were carried out with a population size of 200 and number of generations of 1000. The best solution obtained from 15 trials was reported. The NFE in 15 trials before a decision made are 15 $\times$ 200

Table 23.1 Link details for the Kadu network

Link index	Connecting nodes		Length	Diameter (mm)
	From	To	(m)	Gangwani et al. (2024)
1	1	3	300	900
2	3	4	820	900
3	4	5	940	350
4	5	6	730	300
5	6	7	1620	150
6	7	8	600	250
7	4	9	800	800
8	6	10	1400	150
9	8	13	1175	450
10	8	14	750	500
11	14	2	210	750
12	9	10	700	700
13	10	11	310	500
14	11	12	500	500
15	12	13	1960	150
16	9	15	900	500
17	15	18	850	350
18	10	19	650	400
19	13	20	760	150
20	13	16	1100	150
21	14	17	660	700
22	19	20	1170	150
23	20	21	980	450
24	16	21	670	350
25	21	17	1080	700
26	18	23	750	250
27	19	24	900	250
28	20	25	650	300
29	21	26	1540	200
30	17	22	730	300
31	23	24	1170	150
32	24	25	1650	150
33	25	26	1320	150
34	26	22	3250	150

Table 23.2 Node details for the Kadu network

Node no	Demand (m^3/min)	Minimum HGL (m)	Available HGL as per Gangwani et al. (2024)
1	–	100	100
2	–	95	95
3	18.4	85	98.28
4	4.5	85	95.04
5	6.5	85	87.33
6	4.2	85	85.35
7	3.1	82	85.21
8	6.2	82	87.81
9	8.5	85	91.13
10	11.5	85	88.31
11	8.2	85	86.44
12	13.6	85	85.15
13	14.8	82	82.38
14	10.6	82	92.95
15	10.5	85	88.01
16	9	82	82.17
17	6.8	82	89.56
18	3.4	85	85.45
19	4.6	82	86.11
20	10.6	82	82.33
21	12.6	82	86.35
22	5.4	80	85.09
23	2	82	83.01
24	4.5	80	80.17
25	3.5	80	80.52
26	2.2	80	80.22

× 1000 (=30,00,000). When applying DSSR, ten trial runs with population size of 500 is considered, and the active search space is updated after every 500 generations with a total number of generations of 5000. Thus, the NFE considered with DSSR are 25,00,000. The lower and upper bounds are changed and a random population is generated after each update. This brings an entirely new population to search the best in updated search space. Because NFEs are counted from the beginning of the process of DSSR, it shows that more NFEs are required with DSSR compared with FSS. However, several trials are also necessary in the case of FSS. Herein, the NFE of all trials is counted in making comparison of FSS with DSSR. The design solutions of the Ramnagar network are also shown in Table 23.5 for comparison.

Table 23.3 Available pipe diameters and their unit costs

Diameter in mm	Cost/m in Rs		Diameter in mm	Cost/m in Rs	
	Kadu network	Ramnagar network		Kadu network	Ramnagar network
80	–	830	450	5172	5922
100	–	1102	500	6092	6890
150	1115	1454	600	8189	9012
200	1600	1975	700	10,670	11,619
250	2154	2660	750	11,874	13,071
300	2780	3295	800	13,261	14,225
350	3475	4060	900	16,151	17,386
400	4255	4866	1000	19,395	21,023

Table 23.4 Solutions for Kadu network

Reference papers	Method used	Full SS/static/dynamic SSR (FSS/SSSR/DSSR)	Optimal cost (Rs.)	Functional evaluation (NFE)
Kadu et al. (2008)	Genetic algorithm	SSSR	131,678,935	120,000
Siew et al. (2014)	Multi objective evolutionary algorithm	FSS	125,460,980	436,000
Barlow and Tanyimboh(2014)	GA based memetic algorithm	SSSR	124,690,000[a]	142,000
Palod et al. (2021)	Rao II algorithm	FSS	125,434,170	19,700
Gangwani et al. (2023a)	Genetic algorithm	FSS	125,209,860	157,760
Gangwani et al. (2024)	Genetic algorithm	DSSR	125,019,790	1,125,300
Gangwani et al. (2023b)	Rao II algorithm	FSS	125,019,790	33,525

[a] infeasible solution with current EPANET parameters $(\alpha, \beta, \omega) = (1.852, 4.871, 10.667)$

The merit of DSSR over FSS methodology can be clearly observed from the design results of the Ramnagar network. The best price was Rs. 37.84 million for GA with FSS obtained in 881,600 NFE in 106.3 min of run time in the best run. However, the total NFE in 10 trials amounted to 96,00,000. A 9.38% cheaper solution is obtained with GA and DSSR (Rs. 34.29 million) in 382,500 NFE in 18.67 min of run time in the best run. The total NFEs for 10 trials were 25,00,000.

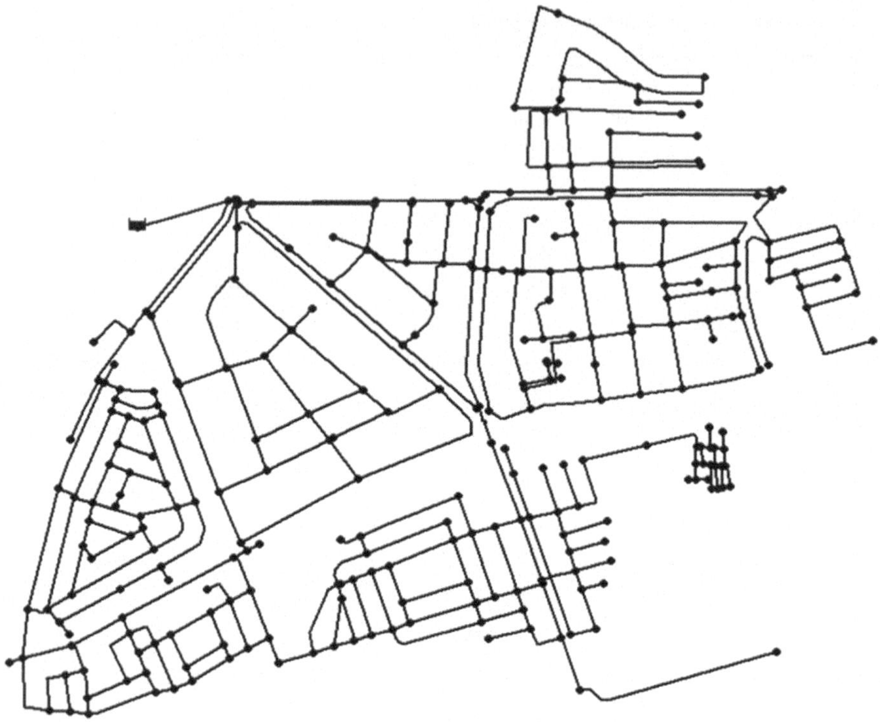

Fig. 23.2 Layout of Ramnagar water distribution network

Table 23.5 Solutions for Ramnagar network

Methodology	Optimal cost (Rs.)	Number of trial	Number of function evaluations		Residual Head (m)	
			Total	In best run	Max	Min
1	2	3	4	5	6	7
GA with FSS (Gangwani et al. 2024)	37,837,223	10	96,00,000	881,600	18.72	8.03
GA with DSSR (Gangwani et al. 2024)	34,289,227	10	25,00,000	382,500	15.92	8.07
Rao II with FSS (Gangwani et al. 2023b)	37,189,448	10	30,00,000	200,000	16.41	8.13
Rao II with DSSR	34,913,280	-	25,00,000	995,000	14.59	8.03

The results of Rao II algorithm show that the best cost of Rs. 37.19 million obtained in 30,00,000 NFEs with FSS in 15 trials. The DSSR provided a 6.13% cheaper solution, which is Rs. 34.91 million in the 25,00,000 NFE. This best solution appeared after 995,000 NFE. The better-cost solution shows the efficiency improvement of Rao II with DSSR.

23.7 Summary and Conclusions

The EAs have the advantage of global search capacity and identifying an optimal global solution for WDN design. However, computational efforts increase manifold as the size of network increases and the efficiency of EAs to identify the global optimal solution decreases. The SSR is suggested by many researchers by having upper and lower bounds on each pipe of the network. Most of the SSR methodologies requires a different algorithm to get a nearby feasible solution to fix the upper and lower bounds. Furthermore, these bounds remain static throughout the EA run and therefore there is a possibility of potential combination leading to a global optimal solution. This requires DSSR in which the active SS is periodically changed.

The advantages of DSSR over SSSR using GA are seen with the solutions of both, the Kadu and Ramnagar networks. Gangwani et al. (2024) observed that DSSR provided nine new best-known feasible solutions to the Kadu network that were not previously reported in the literature. The advantage of DSSR over FSS using Rao-II can be seen very clearly with the Ramnagar network in terms of a reduction in network cost of 6.13%.

Acknowledgements The authors are thankful to their prestigious institutes and other people who are involved in this study for their direct and indirect contributions.

References

Abdy Sayyed MAH, Gupta R, Tanyimboh TT (2021) Combined flow and pressure deficit-based penalty in GA for optimal design of water distribution network. ISH J Hydraul Engg. https://doi.org/10.1080/09715010.2019.1604180

Abebe AJ, Solomatine DP (1998) Application of global optimization to the design of pipe networks. In: Proceedings of 3rd international conference of hydroinformatics, pp 989–996

Barlow E, Tanyimboh TT (2014) Multi-objective memetic algorithm applied to the optimisation of water distribution systems. Water Resour Manag 28(8):2229–2242. https://doi.org/10.1007/s11269-014-0608-0

Bhave PR (1978) Noncomputer optimization of single-source networks. J Environ Eng Div 104(4):799–814

Bhave PR (2003) Optimal design of water distribution networks. Narosa Publishing House Pvt. Ltd. New Delhi, India; and Alpha Science International Ltd., Pangbourne UK

Cisty M, Zbynek B (2009) Optimization of the water distribution Networks with search space reduction. Slovak University of Technology, Bratislava

Gangwani L, Dongre S, Gupta R et al (2023a) Improved design solutions for benchmark networks using genetic algorithm involving penalty based on combined flow and pressure deficit. In: Timbadiya PV, Patel PL., Singh VP, Mirajkar AB (eds) Geospatial and soft computing techniques. HYDRO 2021. Lecture notes in civil engineering, vol 339. Springer, Singapore (2023a). https://doi.org/10.1007/978-981-99-1901-7_24

Gangwani L, Palod N, Dongre S, Gupta R (2023b) Optimal pipe-sizing design of water distribution networks using Modified Rao-II Algorithm. In: 4th international conference on futuristic and sustainable aspects in engineering and technology (FSAET-2023), 28–30 Nov 2023. GLA University, Dept. of Civil Engineering, Mathura

Gangwani L, Dongre S, Gupta R et al (2024) Design optimization of water distribution networks with dynamic search space reduction GA. Water Resour Manage 38:63–79. https://doi.org/10.1007/s11269-023-03648-0

Haghighi A, Samani HM, Samani ZM (2011) GA-ILP method for optimization of water distribution networks. Water Resour Manag 25(7):1791–1808

Jaynes ET (1957) Information theory and statistical mechanics. Phys Rev 106:620–630 and 108:171. https://doi.org/10.1103/PhysRev.106.620

Kadu MS, Gupta R, Bhave PR (2008) Optimal design of water networks using a modified genetic algorithm with reduction in search space. J Water Res Plann Manag ASCE 134(2):147–160

Montesinos P, Gracia-Guzman A, Ayuso JL (1999) Water distribution network optimization using a modified genetic algorithm. Water Res Research 35(11):3467–3473. https://doi.org/10.1029/1999WR900167

Mohan S, Jineshbabu KS (2009) Water distribution network design using heuristic-based algorithms. J Comp in Civ Eng 23(5):249–257

Palod N, Prasad V, Khare R (2021) Redefining the application of an evolutionary algorithm for the optimal pipe sizing problem. J Water Clim Change 12(6):2299–2313

Rao R (2020) Rao algorithms: three metaphor-less simple algorithms for solving optimization problems. Int J Ind Eng Comput 11:107–130. https://doi.org/10.5267/j.ijiec.2019.6.002

Reca J, Martinez J, Lopez R (2017) A Hybrid water distribution networks design optimization method based on a search space reduction approach and a genetic algorithm. Water 9(11):845. https://doi.org/10.3390/w9110845

Rossman LA (2000) EPANET 2 user's manual. Water Supply and Water Resources Division, National Risk Management Research Laboratory, Cincinnati. OH

Sangroula U, Hon RK, Koo KM et al (2022) Optimization of water distribution networks using genetic algorithm-based SOP-WDN Program. Water 14. https://doi.org/10.3390/w14060851

Siew C, Tanyimboh TT, Seyoum AG (2014) Assessment of penalty-free multi-objective evolutionary optimization approach for the design and rehabilitation of water distribution systems. Water Resour Manag 28(2):373–389. https://doi.org/10.1007/s11269-013-04888

Tanyimboh TT, Czajkowska AM (2018) Self-adaptive solution-space reduction algorithm for multi-objective evolutionary design optimization of water distribution networks. Water Resour Manage 32:3337–3352. https://doi.org/10.1007/s11269-018-1994-5

Vairavamoorthy K, Ali M (2005) Pipe index vector: a method to improve genetic-algorithm-based pipe optimization. J Hydraul Eng 131:1117–1125. https://doi.org/10.1061/(ASCE)0733-9429(2005)131:12(1117)

Wu ZY, Walski T (2005) Self-adaptive penalty approach compared with other constraint-handling techniques for pipeline optimization. J Water Res Plann Manag ASCE 131. https://doi.org/10.1061/(ASCE)0733-9496(2005)131:3(181)

Yates D, Templeman A, Boffey T (1984) The computational complexity of the problem of determining least capital cost designs for water supply networks. Eng Optim 7:143–155

Zheng F, Simpson AR, Zecchin AC (2011) A combined NLP-differential evolution algorithm approach for the optimization of looped water distribution systems. Water Resour Res 47:W08531. https://doi.org/10.1029/2011WR010394

Chapter 24
Comparison of Six Parameter-Less Evolutionary Algorithms for the Optimal Design of Water Distribution Networks

Nikita Palod and Rajesh Gupta

Abstract Water distribution networks (WDNs) play a crucial role in the development of any society. These are expensive infrastructures, and therefore cost optimization is desirable. In the previous decade, the use of Evolutionary Algorithms (EAs) as an optimization tool has grown substantially and several new algorithms have been proposed. EAs work on metaheuristics based on some natural phenomenon, which may be physical, chemical, or biological inspired, or may be based on some other strategy. These algorithms require the use of some parameters to mimic the behavior of adopted metaheuristic. These parameters are problem-specific and require fine-tuning either by the user each time a new problem is solved or done internally on its own during execution of the algorithm to avoid the necessity of fine-tuning by the user. Obviously, internal fine-tuning reduces computational efforts. Thus, EAs can be broadly classified as parameter-based and parameter-less algorithms depending on the need for fine-tuning of parameters by users. In this chapter six parameter-less EAs namely Jaya, Rao-I, Rao-II, Rao-III, Black hole optimization (BHO), and Teaching learning-based optimization (TLBO) are described and compared for the optimal pipe sizing problem. Among the six EAs, BHO and TLBO were not applied earlier for the optimization of WDN. All six EAs are applied on a well-known benchmark network taken from the literature to compare the optimal cost and the required number of functional evaluations (hydraulic analysis) to reach the best cost solution by fixing the population size and termination criteria (a fixed set of number of iterations). Among the six EAs, it was found that Rao-II outperformed the other five EAs. BHO is observed to fail to converge to the optimal results for the example WDN.

Keywords Evolutionary algorithms · Optimal design · Parameter-less EAs · Rao-algorithms · Water distribution networks

---

N. Palod · R. Gupta (✉)
Department of Civil Engineering, Visvesvaraya National Institute of Technology, Nagpur 440010, India
e-mail: rajeshgupta@civ.vnit.ac.in

A. K. Yadav et al. (eds.), *Integrated Management of Water Resources in India: A Computational Approach*, Water Science and Technology Library 129,
https://doi.org/10.1007/978-3-031-62079-9_24

24.1 Introduction

Water distribution networks (WDNs) are among the basic requirements for a developed society, as they make water available on the doorstep of a consumer. However, designing a WDN is a complex task because of the involvement of various components such as pipes, pumps, valves, reservoirs, etc. The non-linear relation between the head and discharge, large search space, and discrete diameter options further complex the problem (Savic and Walters 1997). Moreover, WDNs constitute roughly about 70–80% (Cunha and Sousa 1999) of any water supply schemes, which indicates that these are costly infrastructures and demand optimization.

With the advancement in technology, there is always a hunt for a better solution; hence many meta-heuristic techniques such as evolutionary algorithms (EAs) have been developed and applied for the design of WDNs. A few of them are Genetic Algorithm (GA) (Dandy et al. 1996), Particle Swarm Optimization (PSO) (Suribabu and Neelkantan 2006), Water Cycle Algorithm (WCA) (Sadollah et al. 2014), Soccer League Competition (SLC) (Moosavian and Roodsari 2014), Improved Mine Blast Algorithm (MBA) (Sadollah et al. 2015), Shuffled Frog Leaping Algorithm (SFLA) (Eusuff and Lansey 2003), Immune Algorithm (IA), Harmony Search (HS) (Geem 2006), Memetic Algorithm (MA) (Banos et al. 2010), Jaya Algorithm (JA) (Rao 2016) and many more. These techniques are observed to perform better when compared with many deterministic and heuristic methods (Lin et al. 2007; Shirin et al. 2019, 2023) in obtaining the optimal cost.

In this chapter, six parameter-less techniques, namely Jaya, Rao-I, Rao-II, Rao-III, Teaching Learning Based Optimization (TLBO), and Black Hole Optimization (BHO) are applied and tested for the optimization of a well-known benchmark network. BHO and TLBO are applied for the first time for the optimal design of WDN. These algorithms are tested for the same number of function evaluations (NFEs), and the optimal costs for the network are obtained and compared.

24.2 Pertaining Literature

EAs are broadly classified on the parameters that are utilized in the evolution process. Two types of parameters are generally observed in the EAs. These are common controlling parameters and algorithm-specific parameters. Common controlling parameters such as population size and termination criteria are required for most of the algorithms. EAs that require the tuning of algorithm-specific parameters are called parameter-based EAs. In contrast, EAs are called parameter-less or self-adaptive when these algorithm-specific parameters are self-adaptive or generated randomly in the evolution process.

24.2.1 Parameter-Based EAs

As stated above, the efficiency of parameter-based EAs is largely dependent on their parameters. The first step in implementing all these techniques is the determination of optimal controlling parameters. This is a tedious and cumbersome process that requires multiple trial and hit tests. Not only this, but many techniques have more than one controlling parameter, due to which multiple permutations and combinations of these parameters are possible and need to be evaluated. Further, these parameters are problem-specific, hence optimal parameters once found may not be suitable for other problems. The computational efforts reported by these algorithms to reach the optimal solution are independent of the efforts vested in obtaining the optimal parameters. The table below lists various parameter-based EAs applied for the optimal design of WDN. Suribabu (2010) states that the efficiency of Differential Evolution (DE) largely depends on its parameter, and if not tuned properly, it may not yield optimal results.

24.2.2 Parameter-Less/Self-Adaptive EAs

The limitations of parameter-based EAs are overcome by parameter-less techniques. These techniques are free from parameter tuning, as the parameters are either self-adaptive, i.e., they improve themselves, or they are generated randomly with the progress of the algorithm. The efficiency of these algorithms does not depend on the parameters and can be easily applied to various problems without any prior tuning.

24.3 Problem Formulation and General Steps of EA for the Optimal Design of WDN

The goal of WDN design is to minimize the cost of the network, as given in Eq. 24.1, while ensuring that all the constraints are satisfied. A general problem can be formulated as follows:

$$Mincost = \sum_{i=1}^{np} C_i(D_i) \times L_i \tag{24.1}$$

$$\sum Q_{in} - \sum Q_{out} = q_j, \forall j \in nn \tag{24.2}$$

$$\sum_{i \in loopl} hf_i = 0, \forall l \in nl \tag{24.3}$$

$$H_j \geq H_j^{min}, \forall j \in nn \tag{24.4}$$

$$x = [x(1), x(2), \ldots\ldots x(S)], \forall i \in np \tag{24.5}$$

The hydraulic network solver, EPANET 2.0 (Rossman 2000), satisfies Eqs. 24.2 and 24.3, i.e., continuity and loop head loss equations. Eq. 24.4 (constraint of minimum pressure requirement) and Eq. 24.5 (discrete pipe diameters) are addressed externally.

The general steps followed for the optimization of a WDN using an EA are as follows.

1. The population for any EA is randomly generated within the lower (d_i^L) and upper bound (d_i^U) of the decision variables (Eq. 24.6). The generated population is continuous in nature, nevertheless, to determine the fitness value (Eq. 24.1), discrete diameters are required (Eq. 24.5). Thus, the obtained continuous diameters are either rounded to the nearest commercially available diameter, or any method can be adopted to convert the continuous diameters to discrete diameters. In this paper, the bound constraint of the average point as described in Eq. (24.7) is used to obtain the discrete diameters.

$$x_{ij} = d_i^L + r \times \left(d_i^U - d_i^L\right) \tag{24.6}$$

$$x_{ij} = \begin{cases} x_i(k) & if\ x_{ij} \leq \frac{x_i(k)+x_i(k+1)}{2} \\ x_i(k+1) & otherwise \end{cases} \tag{24.7}$$

r is any random number in the range of [0, 1] and x_i is the diameters of commercially available pipes.

2. Once discrete diameters are obtained, the network cost is determined using Eq. (24.1). The diameters obtained are passed to the hydraulic solver (EPANET in the present case) to check for the minimum pressure constraint (Eq. 24.4). Since solutions are generated randomly, specific solutions may not be able to satisfy this constraint. These solutions are penalized and the total cost of these solutions is equal to the sum of the network cost and the penalty cost. It is observed that penalty largely affects the convergence of an EA (Wu and Walski 2005; Abdy Sayyed et al. 2021). The penalty cost in the present work is determined using Eq. (24.8).

$$C_p = \sum_{j=1}^{NN} p\, q_j^{req} \left\{ \max\left(H_j^{des} - H_j^{avl}, 0\right) \right\} \tag{24.8}$$

where p is the capitalized energy cost for a unit flow and for a unit of head (p). q_j^{req} is the demand at node j, H_j^{des} and H_j^{avl} are the desirable and available pressure at node j.

3. Solutions are now sorted according to the total cost of the network. If required, a solution with minimum and maximum cost is identified. The solutions are then passed to the optimization model of EAs which differ from EA to EA. The obtained solutions are again passed to Step 1 (only for discretizing) and Step 2.
4. Newly generated solutions are compared with the previous solutions, and the one performing better is moved to the next iteration.
5. The above steps are recursively called until the termination criteria are met.

24.4 Parameter-Less EAs

All parameter-less algorithms follow the general steps outlined earlier. However, they differ in their respective evolutionary processes, which are described below. Each algorithm follows a unique approach to convergence that distinguishes them from each other.

24.4.1 Jaya

The Jaya algorithm was developed by Rao (2016). Jaya is based on selecting the best, $X_{j,best,i}$ and worst, $X_{j,worst,i}$ solutions from a set of population. Solutions with the maximum and minimum objective function values are referred to as the worst and the best solution, respectively, since optimal design of WDN is a minimization problem. New solutions in the Jaya algorithm are generated using Eq. 24.9, as follows:

$$X_{j,k,i'} = X_{j,k,i} + r_{1,j,i}\left(X_{j,best,i} - \left|X_{j,k,i}\right|\right) - r_{2,j,i}\left(X_{j,worst,i} - \left|X_{j,k,i}\right|\right) \quad (24.9)$$

$X_{j,k,i}$ and $X_{j,k,i'}$ here denotes the old and updated values of *the* variable of kth solution respectively. r_1 and r_2 are uniformly distributed random numbers generated between 0 and 1. The algorithm involves adding the difference between the best and current values to the former value. In contrast, the difference between the worst and the current solution is subtracted from the current value. This means that the algorithm strives to approach the better solution and distance itself from inferior solutions. Consequently, the algorithm becomes victorious as it proceeds; hence, it is named Jaya which is derived from a Sanskrit word for victory. The updated solutions are compared with the previous solutions, and the solution with better fitness is moved to next iteration. As shown in Eq. 24.9, Jaya operates without the involvement of any constants; instead, it relies solely on random numbers. This characteristic of Jaya hops over the cumbersome process of parameter tuning and simplifies the algorithm to a greater extent. This simplification has helped to improve the optimization process to a larger degree.

24.4.2 Rao-I

Rao (2020) developed three metaphor-less and parameter-less algorithms named Rao algorithms. This incorporates Rao-I, II, and III. Like Jaya, the Rao algorithms relies on the best and worst solution from the entire population. Once the best and worst solutions are identified, for any variable in Rao-I, the evolved value is determined by adding to its previous value the difference between the best and worst variable value as shown in Eq. (24.10). This contrasts with Jaya where for any variable the difference is determined from its best and worst variable value, respectively. $r_{1,j,i}$ is any random number generated between 0 and 1. The evolved solutions are compared with the previous solutions, and better performing solutions, i.e., solutions with lesser total network cost are sent to the next iteration.

$$X_{j,k,i\prime} = X_{j,k,i} + r_{1,j,i}\left(X_{j,best,i} - X_{j,worst,i}\right) \quad (24.10)$$

24.4.3 Rao-II

The representation of the variables in the Rao-II algorithm remains the same as in the Jaya algorithm. The updated value $X'_{j,k,i}$ for any variable j is determined from the previous value of the solution for variable j, i.e. $X_{j,k,i}$ similar to both Jaya and Rao-I. However, Rao-II does not entirely depend on the value of best and worst solutions, instead to determine the new value of any variable, a solution is randomly selected from the entire population set. The randomly picked solution is compared with the current solution, and the transfer of information takes place from a superior solution to an inferior solution. Thus, Rao-II fetches more randomness than Rao-I. The detailed explanation of the procedure is as follows; the term $X_{j,l,i}$ in the Eq. 24.11 is any randomly chosen solution, l, from the population. This solution is compared with the solution k based on the fitness value which is the total cost of the network. If solution l is of higher quality than solution k (the cost of solution l is less than that of solution k), indicating that the information must be exchanged from l to k, the term $X_{j,k,i}$ or $X_{j,l,i}$ becomes $X_{j,l,i}$ and the term $X_{j,l,i}$ or $X_{j,k,i}$ becomes $X_{j,k,i.}$ However if solution k is of higher quality, information must be exchanged from k to l and thus the term $X_{j,k,i}$ or $X_{j,l,i}$ becomes $X_{j,k,i}$ and the term $X_{j,l,i}$ or $X_{j,k,i}$ becomes $X_{j,l,i..}$ Rao-II enhances the search space horizon by introducing more randomness to the solutions than previously discussed algorithms. Again, $r_{1,j,i}$ and $r_{2,j,i}$ are random numbers distributed uniformly between (0, 1).

$$X_{j,k,i\prime} = X_{j,k,i} + r_{1,j,i}\left(X_{j,best,i} - X_{j,worst,i}\right) + r_{2,j,i}\left(\left|X_{j,k,i} or X_{j,l,i}\right| - \left|X_{j,l,i} or X_{j,k,i}\right|\right) \quad (24.11)$$

24.4.4 Rao-III

The fundamental principles of the Rao-II and Rao-III algorithms closely resemble one another. The only distinction between the two is the placement of the modulus in the evolution equation, as shown in Eq. 24.12. Hence the solutions are evolved using Eq. 24.12 and compared with the previous solutions. The better performing solutions are moved to the next iteration. This process of evolution and selection continues until the termination criteria are reached.

$$X_{j,k,i'} = X_{j,k,i} + r_{1,j,i}\left(X_{j,best,i} - |X_{j,worst,i}|\right) + r_{2,j,i}\left(\left|X_{j,k,i} or X_{j,l,i}\right| - (X_{j,l,i} or X_{j,k,i}\right) \tag{24.12}$$

24.4.5 Teaching Learning-Based Optimization (TLBO)

TLBO was developed by Rao et al. (2012). The population in TLBO is called a class, where each solution represents a student and each variable represents a subject. The student with the best fitness value (which in the case of WDN is the minimum total cost) is assigned as the teacher. In a class of students, it is assumed that students can improve their knowledge by learning from the teacher or by mutual discussion with one another. Hence, the algorithm is divided into two parts. The first part is called the teacher phase, and the second is called the student phase. When the discussion with the teacher is over, the students tend to discuss with each other to clear their doubts and enhance their knowledge. Thus, the student phase starts when the teacher phase ends, and this is a general observation in a class. The fitness value of each student is calculated using Eq. 24.1. The student with the best fitness is assigned as the teacher. The teacher will try to improve the mean of the class, i.e., the teacher will try to bring the student to his or her capacity. This is achieved by determining the difference mean (DM) of each subject (variable) as shown in Eq. 24.13. This DM is then added to the old population to generate a new population (Eq. 24.15). Depending upon the fitness value of the two same students in the old and new population, the student with better fitness will move to the student phase and this completes the teacher phase.

$$DM_j = r\left(x_{j,T} - T_F x_{j,m}\right) \forall j \tag{24.13}$$

where DM_J represents the difference mean of jth variable, r is a random number between 0 and 1, $x_{j,T}$ is the marks of teacher for j^{th} subject, $x_{j,m}$ is mean marks of class for subject j, T_F can take value 1 or 2, and is decided randomly using Eq. 24.14.

$$T_F = round[1 + r\{2 - 1\}] \tag{24.14}$$

$$X_{new,j} = X_{old,j} + DM_j \tag{24.15}$$

if$X_{new,j} < X_{old,j}$; $X_{new,j}$ is selected and move forward to the student phase and vice-versa.

In the second phase, students increase their subject knowledge by interacting with each other, which can be through doubt solving, seminars, presentation, etc. The sharing of knowledge will be from good student to weak student, and this is decided by their fitness value. Thus, from the entire population two students are randomly selected, let us say *m* and *n,* and new value of *n* is to be determined.

If$Fitness(m) < Fitness(n)$ then knowledge transfer will be from *m* to *n*

$$X_{new,n} = X_{old,n} + r(X_m - X_n) \tag{24.16}$$

If$Fitness(m) > Fitness(n)$ then knowledge transfer will be from *n* to *m*

$$X_{new,n} = X_{old,n} + r(X_n - X_m) \tag{24.17}$$

The new population is determined and compared with the teacher phase. The solution with a better fitness value is selected and moved for the next iteration. The process of evaluating and updating solutions is continued until the termination criteria is met.

24.4.6 Black Hole Optimization (BHO)

BHO (Hatamlou 2013) is a nature inspired population-based metaheuristic algorithm. Each solution in BHO is called a star and the best solution is known as a black hole (BH). The movement of stars around the BH is governed by the principle of gravitational pull. The position of stars is updated using Eq. 24.18.

$$X_{j,k,i'} = X_{j,k,i} + r_{1,j,i}\left(X_{j,BH,i} - X_{j,k,i}\right) \tag{24.18}$$

If the fitness of the updated solution is better than the fitness of the BH, the updated solution becomes the BH, and the remaining solutions are updated according to the new BH. Thus, in BHO, the position of BH i.e. the best solution keeps on changing and is not constant like in Jaya and Rao algorithms. However, if the fitness of BH is better than the updated solution, the position of BH remains intact and the horizon distance (i.e., the Euclidean) between the BH and the star is determined. This step is carried out to determine whether the star will remain in its position or it will get absorbed into the BH. The horizon distance calculated is compared with the radius of horizon (R) (Eq. 24.19) and if the horizon distance is less than the R for that star, the star will get absorbed into the BH and a randomly generated star will occupy its position; this will facilitate the exploration in BHO. However, if the horizon distance is greater than R of BH, the star will remain at its position and will be exploited further. In this manner, the exploration and exploitation for any solution exhibits in the BHO. It should be remembered that the radius of horizon is determined with the updated or evolved fitness value of the star. The BHO algorithm is parameterless, as no algorithm specific constants are required for determining the new position of stars and horizon distances.

$$R = \frac{f_{BH}}{\sum_{a=1}^{N} f_a} \tag{24.19}$$

where, f_{BH} is fitness of BH, N is population size, f_a is the fitness of solution a.

24.5 Application to a Benchmark Network

All the six EAs described in the present work are applied and tested on a multi-source network shown in Fig. 24.1 taken from Kadu et al. (2008), hereinafter referred as Kadu network. The network consists of 34 pipes, 24 demand nodes, and two reservoirs. The hydraulic data of the network is given in Table 24.1 which includes junction elevation, pipe lengths, and its connectivity. The Kadu network has been used by many researchers in the past for determining the efficiency of various EAs. The commercially available diameters and their corresponding unit costs are shown in Table 24.2.

The Kadu network is designed for optimal cost by six parameterless EAs. A population size of 75 and the number of generations are considered as 1000 to have uniformity during comparison. The penalty cost is taken as Rs. 121,016.66. These common controlling parameters are provided in Table 24.3. The EAs were repeated 15 times and the best solution in each trial is provided in Table 24.4, along with minimum cost, maximum cost, mean and median costs, and standard deviation. The details of the best cost solution by different EAs are provided in Tables 24.5 and 24.6.

The convergence curve in the best runs of each of six EAs is shown in Fig. 24.2. The residual heads available at different nodes are also plotted and shown in Fig. 24.3.

24.6 Discussion on Results

The Kadu network is a typical two-source network, for which only in 5 occasions better solutions than the available at that time was obtained in the last 15 years. Gangwani et al. (2024) were the first to obtain a minimum cost of 125.019 million Rs using the Genetic Algorithm with self-adaptive penalty and dynamic search reduction method. They obtained nine better solutions than the previous best solution available at that time. However, the required function evaluations were 11,25,300. The following can be observed from the results presented in Tables 24.4, 24.5 and 24.6 and shown in Figs. 24.2 and 24.3.

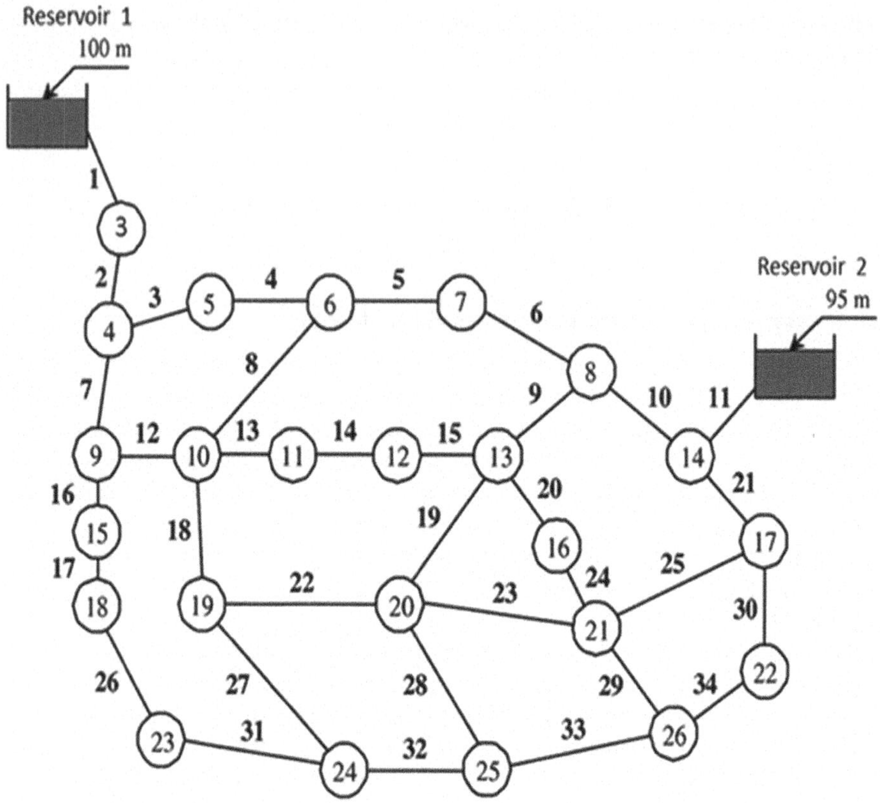

Fig. 24.1 Kadu network

1. Rao-II was the only parameter less EA to obtain the same solution as Gangwani et al. (2024). This solution was repeated twice in 15 trials, showing the superiority of Rao-II algorithm over another five-parameter less EAs.
2. The BHO showed poor convergence. In most of the cost trials, the obtained cost was more than the best solutions obtained by other researchers using different algorithms. The first-best solution reported by Kadu et al. (2008) had a cost of Rs. 131.68 million Rs.
3. The solutions obtained using Jaya, Rao-I, Rao-II and TLBO are comparable as they differ about 1% in cost.
4. The convergence in TLBO was very fast, as can be seen in Fig. 24.2. It provided the solution in 102 iterations (= 7650 NFE) versus 462 iterations (= 33,525) required by Rao-II. Furthermore, it was only 0.035% more expensive than the best cost solution obtained by Rao-II. Thus, it can be considered that TLBO has the same potential to produce the best cost solution as Rao-II.
5. Very marginally high residual pressure was observed at the most critical nodes, showing that all algorithms worked well and ended up providing feasible solution.

Table 24.1 Network details for Kadu network

Junction No	Demand (m^3/min)	Minimum HGL (m)	Link index	Connecting nodes	Length (m)
1	–	100	1	1,3	300
2	–	95	2	3,4	820
3	18.4	85	3	4,5	940
4	4.5	85	4	5,6	730
5	6.5	85	5	6,7	1620
6	4.2	85	6	7,8	600
7	3.1	82	7	4,9	800
8	6.2	82	8	6,10	1400
9	8.5	85	9	8,13	1175
10	11.5	85	10	8,14	750
11	8.2	85	11	14,2	210
12	13.6	85	12	9,10	700
13	14.8	82	13	10,11	310
14	10.6	82	14	11,12	500
15	10.5	85	15	12,13	1960
16	9.0	82	16	9,15	900
17	6.8	82	17	15,18	850
18	3.4	85	18	10,19	650
19	4.6	82	19	13,20	760
20	10.6	82	20	13,16	1100
21	12.6	82	21	14,17	660
22	5.4	80	22	19,20	1170
23	2.0	82	23	20,21	980
24	4.5	80	24	16,21	670
25	3.5	80	25	21,17	1080
26	2.2	80	26	18,23	750
			27	19,24	900
			28	20,25	650
			29	21,26	1540
			30	17,22	730
			31	23,24	1170
			32	24,25	1650
			33	25,26	1320
			34	26,22	3250

Table 24.2 Commercially available diameters along with unit cost for Kadu network

Diameter (mm)	Cost (Rs./m)
150	1,115
200	1,600
250	2,154
300	2,780
350	3,475
400	4,255
450	5,172
500	6,092
600	8,189
700	10,670
750	11,874
800	13,261
900	16,151
1000	19,395

Table 24.3 Common algorithm parameters considered for the optimal design of six EAs

Population size	75
Number of iterations	1000
Penalty cost	Rs. 121016.66
Number of variables	34
Number of trial runs	15

6. All the parameter-less algorithms worked well to identify the optimal solution from the full search space with comparatively smaller NFEs as compared to that required by GA with dynamic search space reduction methodology (Gangwani et al. 2024).

24.7 Summary and Conclusions

The computational efforts required in EAs are extensive, especially when parameter-based algorithms are used. They require fine-tuning of algorithm specific parameters, each time a new problem is solved. Parameter-less algorithms scores well over parameter-based algorithms, as no parameter is to be fine-tunned during the run. This simplifies the optimization process to a greater extent. In this chapter, six parameter-less techniques Jaya, Rao-I, Rao-II, Rao-III, TLBO and BHO—for the optimal design of WDN are described and applied to a typical benchmark Kadu network. All algorithms except BHO showed promising results. However, Rao-II in terms of cost and

Table 24.4 Best cost solutions obtained in different trials for the six EAs

Trial No	Cost in rupees for 15 trials run for six EAs					
	Jaya	Rao-I	Rao-II	Rao-III	TLBO	BHO
1	127,963,310	129,757,165	126,318,200	135,502,665	130,601,650	130,610,075
2	133,480,587	133,834,685	126,670,035	125,767,850	134,969,910	140,421,120
3	128,501,990	129,874,935	127,789,790	126,100,350	125,541,270	144,955,025
4	127,828,025	128,235,595	**125,019,790**	125,737,240	125,601,835	139,902,485
5	126,845,230	128,149,705	126,423,695	125,737,635	127,974,740	139,587,025
6	126,489,765	130,747,695	127,025,010	128,550,600	132,834,720	142,630,855
7	126,529,525	129,136,780	125,436,610	128,129,520	127,410,720	138,397,850
8	137,506,915	127,437,490	125,549,380	125,758,440	128,145,230	137,512,740
9	130,632,455	127,554,100	126,480,480	127,684,100	129,495,680	136,850,180
10	128,714,450	127,539,795	129,165,399	125,901,255	125,461,200	139,882,020
11	126,484,405	129,289,310	127,104,770	125,795,360	125,985,020	137,179,195
12	127,553,455	130,751,720	125,436,610	146,747,560	127,378,890	135,644,720
13	127,670,415	131,123,515	127,974,740	125,624,550	127,437,605	138,313,745
14	131,634,890	126,448,340	126,384,755	126,462,005	126,398,210	133,670,575
15	127,770,525	127,490,945	**125,019,790**	130,023,175	125,486,043	137,906,040
Minimum	126,484,405	126,448,340	**125,019,790**	125,624,550	125,461,200	130,610,075
Maximum	137,506,915	133,834,685	129,165,399	146,747,560	134,969,910	144,955,025
Mean	129,006,720	129,144,842	126,514,895	128,526,321	128,019,216	138,190,684
Median	127,828,025	129,136,780	126,423,695	126,100,350	127,410,720	138,313,745
Standard deviation	2,983,865.7	1,858,896.6	1,131,575.7	5,461,114	2,746,924.8	3,326,262.7
Percentage costlier with best cost	1.16%	1.14%	–	0.05%	0.035%	4.4%

TLBO in terms of convergence scored over other five algorithms for Kadu network. Their application to other large-size network is to be seen.

Table 24.5 Optimal diameters obtained for Kadu network using six EAs

Pipe No	Diameters of pipes (mm)					
	Jaya	Rao-I	Rao-II	Rao-III	TLBO	BHO
1	1000	1000	900	900	900	900
2	900	900	900	900	900	900
3	350	350	350	350	400	300
4	250	300	300	300	250	250
5	150	150	150	150	150	200
6	250	200	250	250	200	300
7	750	800	800	800	800	800
8	150	150	150	150	150	250
9	600	450	450	450	450	400
10	700	500	500	500	500	600
11	1000	750	750	1000	900	800
12	700	700	700	700	700	700
13	500	600	500	500	500	600
14	500	450	500	500	500	500
15	150	150	150	150	150	200
16	500	450	500	500	500	500
17	350	400	350	350	350	350
18	400	400	400	400	350	400
19	450	150	150	150	150	150
20	150	150	150	150	150	200
21	600	700	700	800	700	700
22	150	150	150	150	150	250
23	150	450	450	450	450	500
24	400	350	350	350	350	400
25	500	700	700	600	700	600
26	250	200	250	250	250	250
27	250	350	250	250	300	300
28	350	250	300	350	300	300
29	200	250	200	200	200	200
30	300	250	300	250	250	300
31	150	150	150	150	150	150
32	150	150	150	150	150	150
33	150	150	150	150	150	200
34	150	150	150	150	150	150
Cost (Rs.)	126,484,405	126,448,340	125,019,790	125,624,550	125,461,200	130,610,075
Iterations	909	966	447	681	102	568

Table 24.6 Residual pressure obtained above the minimum HGL by six EAs

Junction number	Jaya	Rao-I	Rao-II	Rao-III	TLBO	BHO
3	14	13.97	13.28	13.29	13.28	13.26
4	10.88	10.72	10.04	10.07	10.03	9.95
5	3.93	3.8	2.33	2.62	5.78	0.53
6	0.35	0.26	0.35	0.81	0.2	0.01
7	6.27	5.49	3.21	4.05	0.88	7.14
8	9.89	6.93	5.81	7.02	7.08	9.27
9	5.69	6.75	6.13	6.18	6.15	5.67
10	3	3.82	3.31	3.37	3.36	2.4
11	1.25	1.96	1.44	1.51	1.5	1.6
12	0.09	0.67	0.15	0.24	0.23	0.22
13	4.79	1.13	0.38	1.16	1.51	0.17
14	12.48	12.5	10.95	12.49	12.16	11.53
15	2.6	1.7	3.01	3.07	3.08	2.6
16	0.18	0.56	0.17	0.08	1.3	0.02
17	8.67	10.8	7.56	10.75	8.75	8.35
18	0.09	0.47	0.45	0.52	0.61	0.15
19	3.94	4.23	4.11	4.19	2.23	2.34
20	1.47	0.86	0.33	0.13	1.31	0.16
21	2.08	4.41	4.35	4.08	5.48	2.15
22	6.13	8.13	5.09	2.4	0.92	5.82
23	0.76	1.49	1.01	1.11	1.36	0.94
24	0.47	2.62	0.17	0.42	1.45	1.23
25	2.39	0.01	0.52	1.18	1.49	0.46
26	0.21	0.01	0.22	0.06	0.34	0.01
Minimum pressure (m)	0.09	0.01	0.15	0.06	0.2	0.01
Maximum pressure (m)	14	13.97	13.28	13.29	13.28	13.26
Most critical node(s)	12, 18	25, 26	12	26	6	6, 26

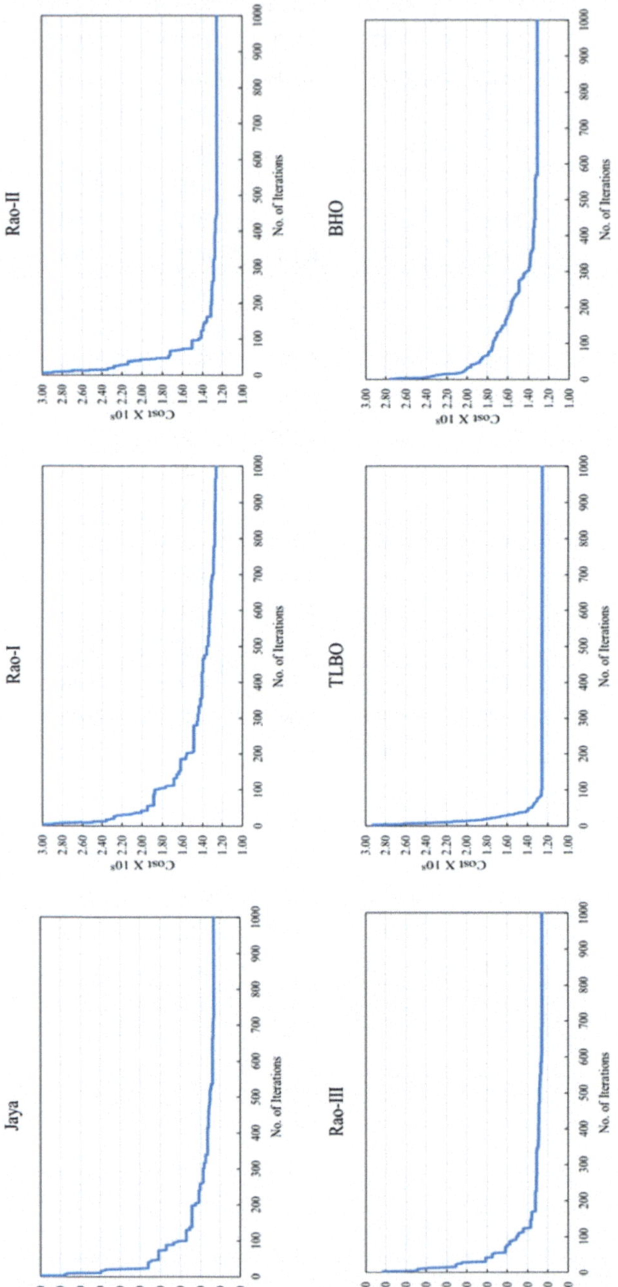

Fig. 24.2 Convergence curve for minimum solution obtained by each EA

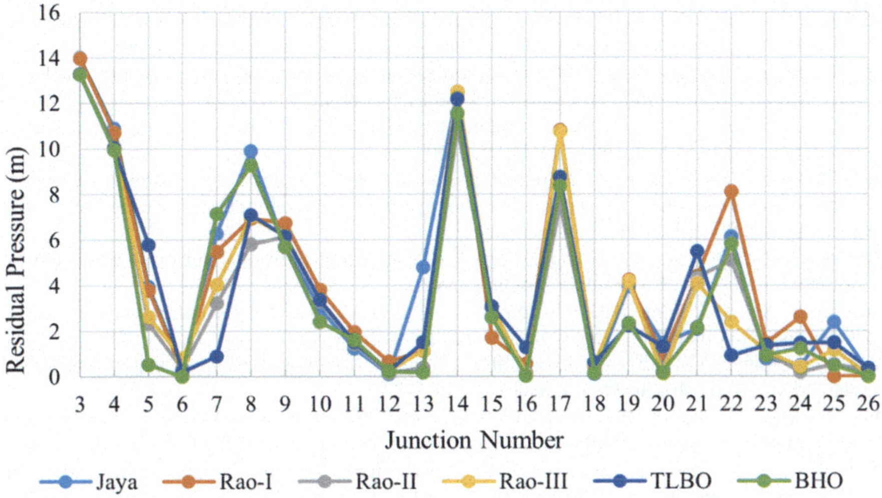

Fig. 24.3 Residual pressure obtained above the minimum HGL using six EAs

Acknowledgements The authors extend gratitude towards Director, VNIT for providing all the facilities and encouragement to complete this research work.

References

Abdy Sayyed MAH, Gupta R, Tanyimboh T (2021) Combined flow and pressure deficit-based penalty in GA for optimal design of water distribution network. ISH J Hydraulic Eng 27(sup1):146–156

Banos R, Gil C, Reca J, Montoya FG (2010) A memetic algorithm applied to the design of water distribution networks. Appl Soft Comput 10(1):261–266

Cunha MDC, Sousa J (1999) Water distribution network design optimization: simulated annealing approach. J Water Resour Plan Manag 125(4):215–221

Dandy GC, Simpson AR, Murphy LJ (1996) An improved genetic algorithm for pipe network optimization. Water Resour Res 32(2):449–458

Eusuff MM, Lansey KE (2003) Optimization of water distribution network design using the shuffled frog leaping algorithm. J Water Resour Plan Manag 129(3):210–225

Gangwani L, Dongre S, Gupta R, Abdy Sayyed MAH, Tanyimboh T (2024) Design optimization of water distribution networks with dynamic search space reduction GA. Water Resour Manage 38(1):63–79

Geem ZW (2006) Optimal cost design of water distribution networks using harmony search. Eng Optim 38(03):259–277

Hatamlou A (2013) Black hole: a new heuristic optimization approach for data clustering. Inf Sci 222:175–184

Kadu MS, Gupta R, Bhave PR (2008) Optimal design of water networks using a modified genetic algorithm with reduction in search space. J Water Resour Plan Manag 134(2):147–160

Lin MD, Liu YH, Liu GF, Chu CW (2007) Scatter search heuristic for least-cost design of water distribution networks. Eng Optim 39(7):857–876

Moosavian N, Roodsari BK (2014) Soccer league competition algorithm: a novel meta-heuristic algorithm for optimal design of water distribution networks. Swarm Evol Comput 17:14–24

Rao R (2016) Jaya: a simple and new optimization algorithm for solving constrained and unconstrained optimization problems. Int J Ind Eng Comput 7(1):19–34

Rao R (2020) Rao algorithms: three metaphor-less simple algorithms for solving optimization problems. Int J Ind Eng Comput 11(1):107–130

Rao RV, Savsani VJ, Balic J (2012) Teaching–learning-based optimization algorithm for unconstrained and constrained real-parameter optimization problems. Eng Optim 44(12):1447–1462

Rossman LA (2000) EPANET 2: users manual

Sadollah A, Yoo DG, Yazdi J, Kim JH, Choi Y (2014) Application of water cycle algorithm for optimal cost design of water distribution systems

Sadollah A, Yoo DG, Kim JH (2015) Improved mine blast algorithm for optimal cost design of water distribution systems. Eng Optim 47(12):1602–1618

Savic DA, Walters GA (1997) Genetic algorithms for least-cost design of water distribution networks. J Water Resour Plan Manag 123(2):67–77

Shirin S, Jamal A, Emmanouil C, Singh VP, Yadav AK (2023) Assessment and characterization of waste material used as backfilling in an abandoned mine. Int J Coal Prep Util 43(8):1402–1410

Shirin S, Jamal A, Ranjan P, Yadav AK (2019) Study on assessment of slope stability and mixed disposal of overburden in voids of Singrauli Coalfield. Environ Qual Manage 28(3):131–139

Suribabu CR (2010) Differential evolution algorithm for optimal design of water distribution networks. J Hydroinf 12(1):66–82

Suribabu CR, Neelakantan TR (2006) Design of water distribution networks using particle swarm optimization. Urban Water J 3(2):111–120

Wu ZY, Walski T (2005) Self-adaptive penalty approach compared with other constraint-handling techniques for pipeline optimization. J Water Resour Plan Manag 131(3):181–192

Chapter 25
Comparative Analysis of SPI and SPEI as Meteorological Drought Indices in the Highland Region of Kerala

Sruthi D. Sivan and S. K. Pramada

Abstract Drought is a disastrous natural phenomenon that can devastate agricultural production and cripple the regional socioeconomic well-being. It will occur due to insufficient rainfall for a prolonged period of time, resulting in water shortages for various activities. Therefore, the analysis and future projection of drought events are necessary to adopt the appropriate mitigation activities. This study evaluates the performance of the Standardized Precipitation Index (SPI) and the Standardized Precipitation Evapotranspiration Index (SPEI) for assessing drought in the highland region of Kerala at different timescales. The study area is Wayanad district which is located in the northeastern part of Kerala, linking the eastern ghats and western ghats. For the analysis of drought indices, GLDAS gridded precipitation and temperature data were used over a span of 67 years. Trend analysis of SPI and SPEI values was detected using the Mann-Kendal trend test and the Sen's slope method. The study findings revealed that the study area is subjected to frequent droughts of varying severity on the timescales considered. The years 2016 and 1983 witnessed notable drought events, effectively demonstrated by the two indices. Additionally, a decreasing trend in SPI and SPEI values was detected at all the gridded points in the study area which indicates an increasing severity of drought events. The results obtained from this study will help identify the regions that require more attention and help the authorities to take appropriate measures.

Keyword Meteorological drought · SPI · SPEI · Mann–Kendall test

---

S. D. Sivan (✉) · S. K. Pramada
Department of Civil Engineering, National Institute of Technology Calicut, Kozhikode 673601, India
e-mail: sruthimangalathu1525@gmail.com

A. K. Yadav et al. (eds.), *Integrated Management of Water Resources in India: A Computational Approach*, Water Science and Technology Library 129,
https://doi.org/10.1007/978-3-031-62079-9_25

25.1 Introduction

Drought, as a creeping natural calamity, has the potential to result in significant loss of life and property. Due to increased greenhouse gas emissions, temperatures are expected to increase every decade, leading to extreme climatic events around the world (Thomas et al. 2015). This will cause remarkable challenges in the socio-economic aspects of a large population. Among hydroclimatic extremes, droughts exhibit a higher economic impact compared to other disasters (Sharafati et al. 2020). India is an agrarian country, where a large population depends on agriculture, experiencing drought every three years for the past few decades (Mishra and Singh 2010). Furthermore, the pollution of available water bodies exacerbates water scarcity problems, significantly impacting the livelihood of many people (Shirin and Yadav 2014; Yadav et al. 2014). Droughts are mainly classified as meteorological, agricultural, hydrological and socio-economic droughts (Wilhite and Glantz 1985). Meteorological drought is the basic form of drought and all other types of droughts occur in succession to meteorological drought (Muthuvel and Mahesha 2021). So, an in-depth examination of meteorological drought is crucial for adopting proper measures to mitigate water stress conditions. It can be quantified using different indices such as the standardized precipitation index (SPI), Palmer Drought Severity Index (PDSI), rainfall anomaly index (RAI), the standardized precipitation evapotranspiration Index (SPEI), etc. Among these, the SPI stands out as the most frequently employed drought index (Thomas and Prasannakumar, 2016), relying solely on precipitation data. However, to address this limitation, SPEI can be employed, which involves both precipitation and temperature data. A drought study based solely on a single drought index may not yield accurate results, potentially impacting mitigation and preparedness measures. Therefore, comparing different drought indices helps identify the best-suited index for a particular climatic condition (Pathak and Dodamani 2020). Therefore, the purpose of this study is to evaluate and compare two frequently employed drought indices, SPI and SPEI, to determine the most appropriate index for the study area.

The Wayanad district is located in the highland region of Kerala which connects the eastern ghats and western ghats. The average yearly precipitation of Wayanad is 2300 mm and the district has a purely agricultural dependent economy. However, recent drought incidents resulted in a huge loss of agricultural production. The Catastrophe of Farmers committing suicides shows an increase in the number attributed to the crop loss suffered from drought events. So, it is crucial to identify regions that are severely impacted by drought to adopt proper mitigation measures. This study focuses primarily on meteorological drought in the Wayanad district, comparing two drought indices. Additionally, the Mann–Kendall trend test and Sen's slope estimator were conducted on drought indices values to detect the significance and magnitude of the trend.

25.2 Study Area and Data

Wayanad district, situated in the northeastern region of Kerala, which extends between latitudes 11°27′ and 11°8′ north and longitudes 75°48′ and 76°26′ east, covering a total area of 2132 km^2. It consists of four administrative blocks, namely Kalpetta, Mananthavadi, Panamaram, and Sulthan Bathery. The district has a purely agriculture-dependent economy and is mainly dominated by cash crops such as coffee, tea, pepper, and cardamom. Wayanad receives an average yearly precipitation of 2300 mm and the average maximum and minimum temperatures are 29 °C and 18 °C, respectively. The main river flowing through the district is the Kabini River, a tributary of the Kaveri River. The Wayanad district map is depicted in Fig. 25.1.

In this study, the grid-based monthly precipitation and temperature with spatial resolution of 0.25° × 0.25° is used from 1954 to 2020 for 7 gridded points. The data was obtained from the Global Land Data Assimilation System, which is an integration of ground and satellite observations. Seven gridded points are considered for this study which is named GP1 (11.375°N, 76.125°E), GP2 (11.625°N,75.875°E), GP3 (11.625°N,76.125°E), GP4 (11.625°N,76.375°E), GP5 (11.875°N,75.875°E), GP6 (11.875°N,76.125°E) and GP7 (11.875°N,76.375°E).

25.3 Methodology

25.3.1 Standardized Precipitation Index

The Standardized Precipitation Index (SPI) is the most frequently used drought index recommended by the World Meteorological Organization (WMO) to analyze meteorological drought. Mckee et al. (1993) developed this index and is characterized as the extent to which the observed precipitation value deviates, in terms of standard deviations, from the long-term mean precipitation at a specific time scale. The only input required for the calculation are long-term precipitation data, which are fitted to a probability distribution. Subsequently, it is transformed into a normal distribution, ensuring a zero-mean value for a specified period and location. The primary strength of this index lies in its simplicity, making it applicable on a range of timescales, including 1-, 3-, 6-, and 12-month periods. The shorter timescale indicates the deficiency in soil moisture, and the longer timescales reflect the condition of stream flow and groundwater levels. A negative value of the SPI represents a dry event and a positive SPI represents a wet event. A drought event is identified by a persistent negative SPI value, reaching a threshold of − 1 or lower. The classification of drought events into mild, moderate, severe, and extreme conditions is determined based on the SPI values.

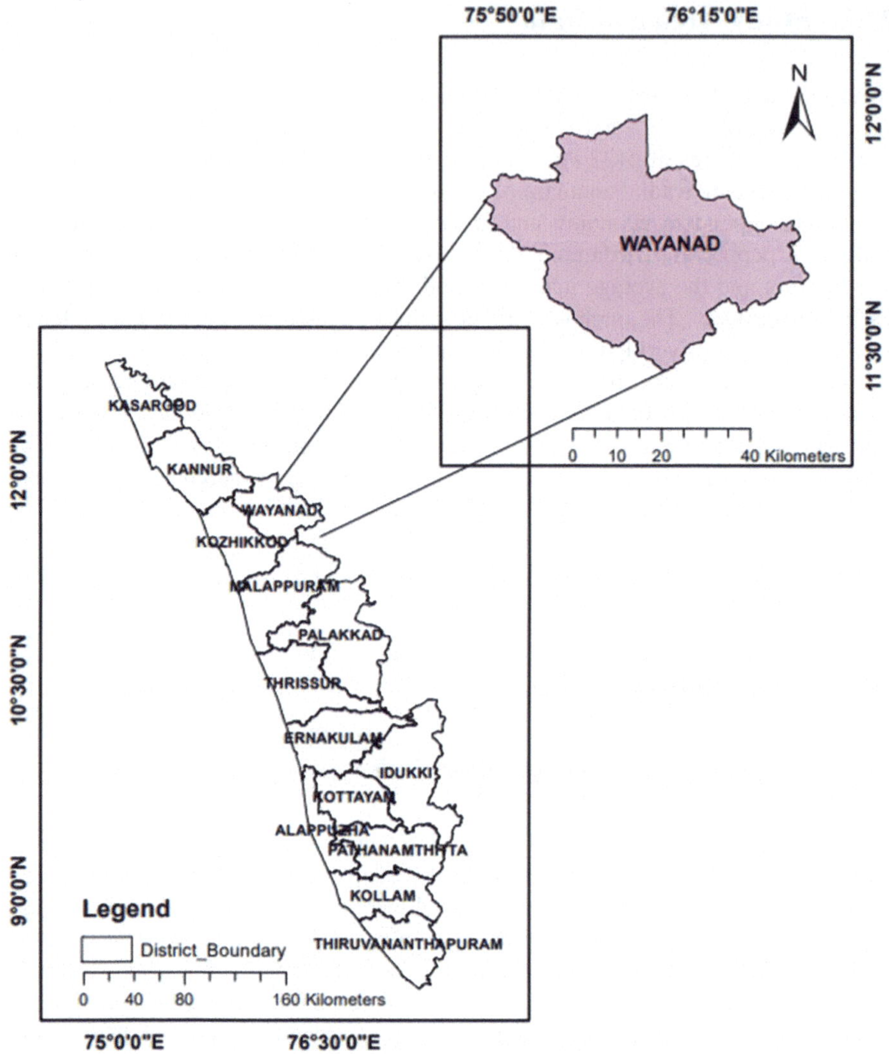

Fig. 25.1 Wayanad district map

25.3.2 Standardized Precipitation Evapotranspiration Index

SPEI was formulated by Vicento-Serrano et al. (2010), which involves both precipitation and temperature data for the analysis. The major limitation of SPI is that it uses only precipitation data for analysis. The primary reason for climate change in different regions worldwide is the increase in temperature resulting from the emission of greenhouse gases. Thus, the analysis of drought in the context of climate change scenario is significantly influenced by temperature. Therefore, the drawback of SPI can be overcome by using an index SPEI, which calculates the monthly discrepancy between rainfall and potential evapotranspiration (PET), which was fitted to a three-parameter log-logistic distribution for the reconstruction of drought. For the calculation of PET, the Thornthwaite method was used, which is simple and requires only mean temperature data for analysis. In this research, 3-month and 12-month time steps are adopted as they effectively represent short- and long-term drought events.

25.3.3 Trend Analysis

The SPI and SPEI time series trends were evaluated using the Mann–Kendall (MK) trend test, and the magnitude of the trend was calculated using Sen's slope method. The Mann–Kendall trend test was extensively used to detect trends in hydrological and climatic data series and is non-parametric (Abraham and Kundapura 2022). Assesses the null hypothesis suggesting no trend against the alternate hypothesis proposing the presence of either an increasing or decreasing trend. Algorithms used for Mann–Kendall test statistics S, are:

$$S = \sum_{i=1}^{n-1} \sum_{j=i+1}^{n} sign(z_j - z_i) \tag{25.1}$$

$$sign(z_j - z_i) = \begin{cases} 1 if\,(z_j - z_i) > 0 \\ 0 if\,(z_j - z_i) = 0 \\ -1 if\,(z_j - z_i) < 0 \end{cases} \tag{25.2}$$

where n represents the length of data set and z_j and z_i indicate sequential data at time j and i respectively. The test statistic Z_{MK} can be determined using S and the variance of S.

$$Z_{MK} = \begin{cases} \frac{S-1}{\sqrt{V(S)}} when S > 0 \\ 0 when S = 0 \\ \frac{S+1}{\sqrt{V(S)}} when S < 0 \end{cases} \tag{25.2}$$

If the calculated value of Z_{MK} is greater than $Z_{\alpha/2}$, the null hypothesis can be rejected at a significance level of α. The Sen's slope estimator can determine the magnitude of the trend by computing the slope in a data series by calculating the median of slopes between every pair of data points.

25.4 Results and Discussion

25.4.1 Drought Analysis Using SPI and SPEI

In the present study, two indices were used to conduct meteorological drought analysis, SPI and SPEI for all seven gridded points within the study area, considering 3-month and 12-month timescales. The results indicate that each gridded point within the study area experienced a drought of varying severity and the frequency of drought events has risen over the last two decades for both time steps. Figures 25.2 and 25.3 illustrate the fluctuations of SPI and SPEI values for a representative grid point over the period from 1954 to 2020, for the 3-month time scale. From the analysis, it is evident that the extreme drought that occurred in the Wayanad district in the recent past was in 2016. Also, in the years 1983 and 2002, the Wayanad region experienced extreme drought conditions.

The drought severity maps for 2016 were generated using Inverse Distance Weighted Interpolation (IDW) method in ArcGIS for both timescales (Figs. 25.4 and 25.5). The results obtained from the SPI analysis indicate, that all the regions in the Wayanad district suffer from extreme drought (SPI < -2) condition for both time scales during the year 2016. Additionally, a decrease in severity was observed when the time scale extended from 3 months to 12-months. The findings from the SPEI analysis indicate, that all the regions in the Wayanad district experience extreme drought conditions in a 3-month time scale. But on a 12-month time scale, none of the regions experienced extreme drought conditions. On this time scale, most of the regions face severe drought conditions (− 1.5 > SPEI > − 1.99) and moderate (− 1 > SPEI > − 1.49).

From the analysis of SPI and SPEI, it is clear that the eastern part of Wayanad district is more affected by the 2016 drought event compared to other regions. Taking

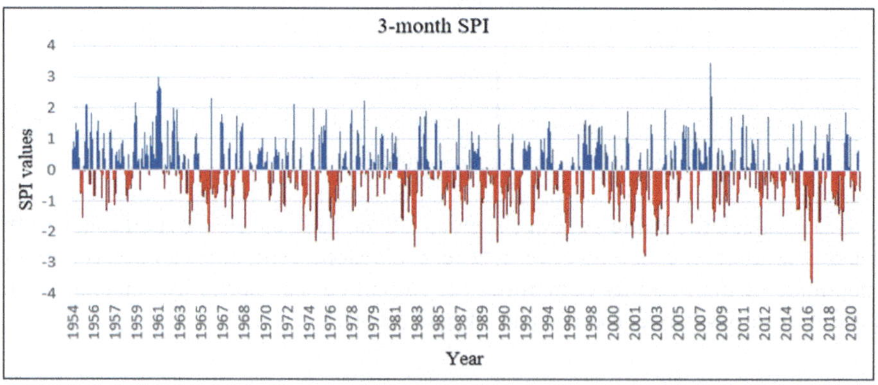

Fig. 25.2 Temporal variation of the 3-month SPI at the GP2 grid point

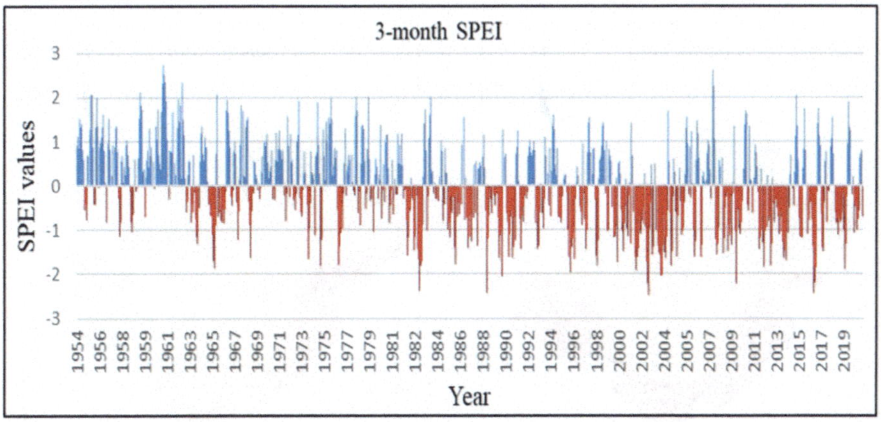

Fig. 25.3 Temporal variation of the 3-month SPEI at the GP2 grid point

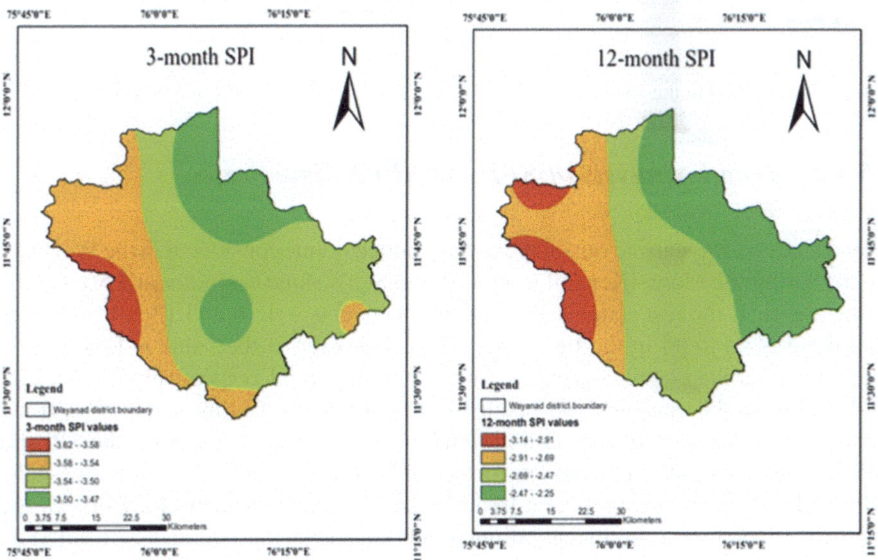

Fig. 25.4 Variation in the spatial pattern of 3-month and 12-month SPI for 2016 drought

into account the administrative block boundaries, the Kalpetta block is in this region. So, this region requires more attention for better adaptation and mitigation measures.

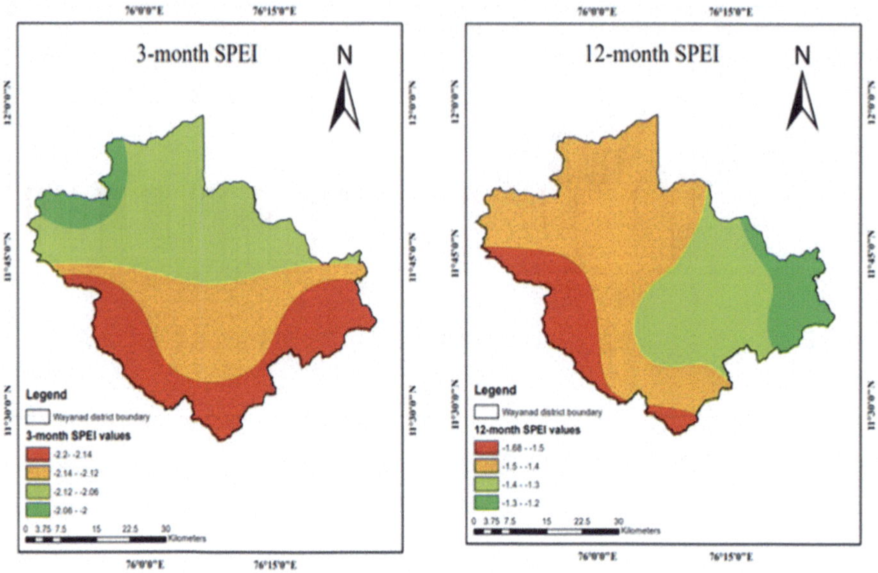

Fig. 25.5 Variation in the spatial pattern of 3 month and 12-month SPEI for 2016 drought

25.4.2 Trend Analysis of SPI and SPEI Time Series

Trend analysis was carried out on the seven gridded points considered in the Wayanad district using the Mann–Kendall trend test and Sen's slope estimator at a 5% significance level. If the test statistic Z value falls between − 1.96 and 1.96, it indicates that there is no trend in the time series data. However, if the value is less than − 1.96 or greater than 1.96 it suggests a falling or rising trend, respectively. Tables 25.1 and 25.2 show the trend analysis results for the SPI and SPEI time series. The Trend analysis revealed a significant falling trend in 12-month time scales for the SPI and SPEI values across all gridded points in the study area. But for the 3-month timescale, GP4 and GP7 show no significant trend in the SPI time series, but for SPEI all the gridded points show a significant falling trend. This downward trend in SPI and SPEI indicates that the severity of drought is increasing in the Wayanad district.

25.4.3 Comparison of SPI and SPEI

The correlation coefficient obtained between SPI and SPEI for both timescales is greater than 0.9 which represents a strong positive correlation between the two indices. Figures 25.6 and 25.7 represent the comparison graphs of SPI and SPEI for the three month and 12-month timescales. If we analyze the data from 1954 to 2020 the extreme drought that occurred in the SPI time series occurred in the year

Table 25.1 Results of the trend analysis for SPI values

	3-month SPI			12-month SPI		
Grid points	Sen's slope (mm/month)	Test statistic, Z	Trend	Sen's slope (mm/month)	Test statistic, Z	Trend
GP1	− 0.0008	− 4.88	Falling	− 0.0021	− 14.48	Falling
GP2	− 0.0007	− 4.43	Falling	− 0.0020	− 14.59	Falling
GP3	− 0.0006	− 3.92	Falling	− 0.0020	− 13.74	Falling
GP4	− 0.0003	− 1.82	No trend	− 0.0013	− 8.06	Falling
GP5	− 0.0006	− 4.17	Falling	− 0.0021	− 15.59	Falling
GP6	− 0.0005	− 3.48	Falling	− 0.0020	− 13.78	Falling
GP7	− 0.0002	− 0.98	No trend	− 0.001	− 6.61	Falling

Table 25.2 Results of the trend analysis for SPEI values

	3-month SPEI			12-month SPEI		
Grid Points	Sen's slope (mm/month)	Test statistic, Z	Trend	Sen's slope (mm/month)	Test statistic, Z	Trend
GP1	− 0.0017	− 10.4	Falling	− 0.0027	− 17.8	Falling
GP2	− 0.0015	− 9.22	Falling	− 0.0025	− 16.9	Falling
GP3	− 0.0013	− 7.03	Falling	− 0.0024	− 14.47	Falling
GP4	− 0.001	− 5.64	Falling	− 0.0017	− 10.19	Falling
GP5	− 0.0013	− 7.91	Falling	− 0.0024	− 16.3	Falling
GP6	− 0.0015	− 7.34	Falling	− 0.0023	− 14.92	Falling
GP7	− 0.001	− 5.15	Falling	− 0.0015	− 8.97	Falling

2016 and for SPEI it was in the year 1983. Also, the severity of the drought obtained from SPI is higher as compared to SPEI. This discrepancy in the SPI and SPEI is obtained for all the gridded points. From the analysis, it is concluded that the results obtained from the SPI analysis align well with the actual drought scenario of the Wayanad district compared to SPEI.

25.5 Conclusions

This study compares the meteorological drought in the Wayanad district of Kerala using two indices, SPI and SPEI for 3-month and 12-month timescales. The results indicate that the recent extreme drought condition occurred in the Wayanad district during 2016 and is more affected in the Kalpetta block. A good correlation coefficient was obtained between SPI and SPEI for both timescales at all the gridded points. However, a significant difference in severity was observed between the two indices. In addition, trend analysis of SPI and SPEI was performed for both timescales. A

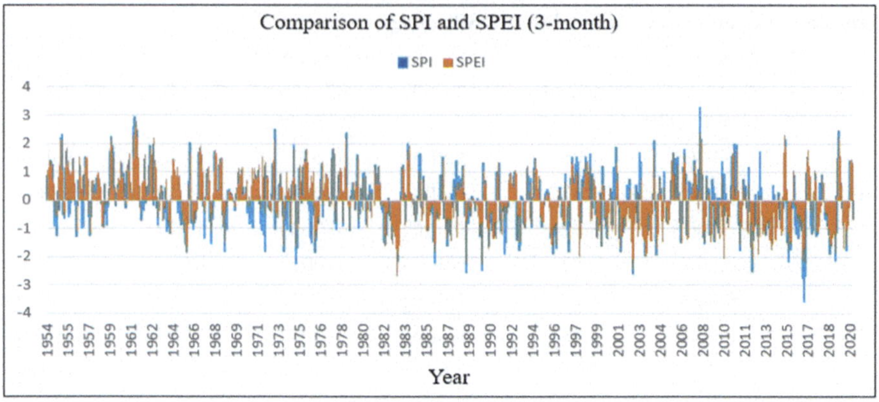

Fig. 25.6 Comparison of SPI and SPEI at GP1 for a 3-month time scale

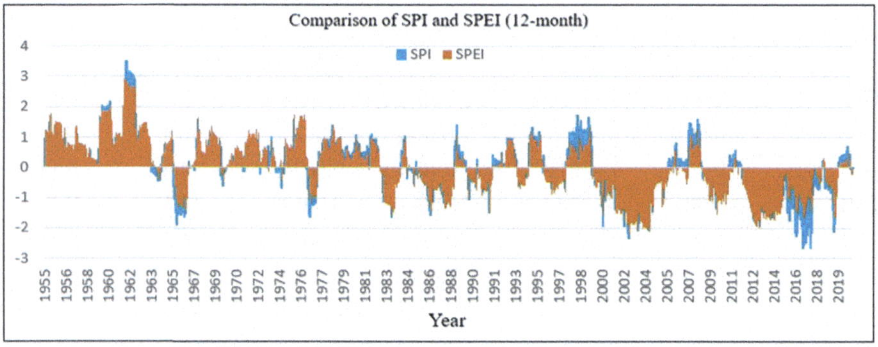

Fig. 25.7 Comparison of SPI and SPEI in GP1 for 12-month timescale

downward trend in SPI and SPEI values was observed for 3-month and 12-month timescales, indicating an increasing severity of drought in the future. Therefore, it is necessary to adopt proper mitigation measures to minimize the impact caused by drought.

Acknowledgements The authors express their gratitude to the reviewers and editor for dedicating their valuable time to enhancing this work.

References

Abraham A, Kundapura S (2022) Evaluating the long-term trends of the climatic variables over three humid tropical basins in Kerala, India. Arabian J Geosci 15:811. https://doi.org/10.1007/s12517-022-10056-y

McKee TB, Doesken NJ, Kleist J (1993) The relationship of drought frequency and duration to time scales. In: Proceeding of 8th Conference applied climatology. American Meteorological, Society, Boston, pp 179–183

Mishra AK, Singh VP (2010) A review of drought concepts. J Hydrol 391(1–2):202–216. https://doi.org/10.1016/j.jhydrol.2010.07.012

Muthuvel D, Mahesha A (2021) Spatiotemporal analysis of compound agrometeorological drought and hot events in India using a standardized index. J Hydrol Eng 26(7):1–15. https://doi.org/10.1061/(asce)he.1943-5584.0002101

Pathak AA, Dodamani BM (2020) Comparison of meteorological drought indices for different climatic regions of an Indian river basin. Asia-Pacific J Atmospheric Sci 56(4):563–576. https://doi.org/10.1007/s13143-019-00162-5

Sharafati A, Nabaei S, Shahid S (2020) Spatial assessment of meteorological drought features over different climate regions in Iran. Int J Clim 40(3):1864–1884. https://doi.org/10.1002/joc.6307

Shirin S, Yadav AK (2014) Physico chemical analysis of municipal wastewater discharge in Ganga River, Haridwar District of Uttarakhand, India. Curr World Environ 9(2):536

Thomas J, Prasannakumar V (2016) Temporal analysis of rainfall (1871–2012) and drought characteristics over a tropical monsoon-dominated State (Kerala) of India. J Hydrol 534:266–280. https://doi.org/10.1016/j.jhydrol.2016.01.013

Thomas T, Nayak PC, Ghosh NC (2015) Spatiotemporal analysis of drought characteristics in the Bundelkhand region of central India using the standardized precipitation index. J Hydrol Eng 20(11):1–12. https://doi.org/10.1061/(asce)he.1943-5584.0001189

Vicente-Serrano SM, Beguería S, Lopez-Moreno JI (2010) A multiscalar drought index sensitive to global warming: the standardized precipitation evapotranspiration index. J Clim 23(7):1696–1718. https://doi.org/10.1175/2009JCLI2909.1

Wilhite DA, Glantz MH (1985) Understanding: the drought phenomenon: the role of definitions. Water Int 10(3):111–120. https://doi.org/10.1080/02508068508686328

Yadav AK, Sahoo SK, Mahapatra S, Kumar AV, Pandey G, Lenka P, Tripathi RM (2014) Concentrations of uranium in drinking water and cumulative, age-dependent radiation doses in four districts of Uttar Pradesh, India. Toxicol Environ Chem 96(2):192–200. https://doi.org/10.1080/02772248.2014.934247

Chapter 26
Exploring Water's Role in Sustainable Electricity Generation for Power in Future

Prathamesh S. Utturkar, Vaishnavi R. Prajapat, Sachin V. Jadhav, and Haresh G. Manyar

Abstract Water plays a crucial role in energy generation as a part of the energy-water-food nexus. In fulfilling the global demand for sustainable energy solutions, the role of water cannot be neglected. India is taking steps towards dependence on renewable energy sources and has committed to meet 50% of the requirements from renewable sources by 2030. More than 80% of the non-agricultural freshwater withdrawal is reported only for energy generation, specifically electric energy. Hydropower, solar PV, Marine wave energy, biomass energy, nuclear energy, etc., can be significant energy resources in the future. In each of them, water is involved directly or indirectly. Hydropower generation has a direct impact on water resources. Compared to countries like Brazil, the hydropower in India is only 13%. Cleaning and maintenance of PV panels, biofuel production, cooling systems in thermal power plants, CCS technology to control emissions, and electrolysis for hydrogen generation are processes in power generation with significant water footprint. Beyond hydropower, solar photovoltaic and conventional energy processes, emerging technologies such as advanced bioenergy and geothermal energy present unique challenges and opportunities in terms of water use. Bioenergy production, particularly advanced biofuel production, involves intricate water-intensive processes for biomass cultivation and conversion. Although considered a low carbon energy source, geothermal energy requires careful water management for improved efficiency and sustainable utilization. Understanding the intricate relationship between water and power generation is crucial, especially considering India's current water resource challenges. Given the increasing water footprint in power generation, coupled with existing water scarcity concerns, proactive measures are imperative to address and mitigate potential future water shortages. This chapter explores the intricate dynamics of water's

P. S. Utturkar · V. R. Prajapat · S. V. Jadhav (✉)
Department of Chemical Engineering, Institute of Chemical Technology, Matunga, Mumbai 400019, India
e-mail: sv.jadhav@ictmumbai.edu.in

H. G. Manyar
School of Chemistry and Chemical Engineering, Queen's University Belfast, Stranmillis Road, Belfast BT9 5AG, UK

A. K. Yadav et al. (eds.), *Integrated Management of Water Resources in India: A Computational Approach*, Water Science and Technology Library 129,
https://doi.org/10.1007/978-3-031-62079-9_26

role in various facets of power generation, aiming to inform sustainable practices and policies for a water-conscious energy future.

Keywords Water-footprint · Future-power-generation · Sustainable-future · Water-scarcity · Renewable-energy

26.1 Introduction

26.1.1 Overview of Water Consumption for Power Generation in India

Water is essential for the existence of human life. Energy is also an integral aspect of human existence, manifesting itself in various forms such as electricity, liquid and gas fuels, and other diverse sources essential for sustaining life and technological advancements. Water plays a crucial role in energy generation as a part of the energy-water-food nexus. For any power generation, water plays a vital role, directly or indirectly.

The global energy requirement is expected to reach 10 EJ/year by 2050, with a 5–10% increase in annual demand. To meet the global energy demand, every nation is working to achieve maximum energy production with minimal environmental effects. This led to sustainable development for the welfare of future generations. In this progress, India has committed to fulfilling 50% of its energy requirements through renewable energy sources. In India, the substantial economic growth witnessed in the last two decades has been closely associated with a rapid increase in energy demand. Over a seven-year period (2014–2021), annual national electricity consumption increased by over 23%, reaching 1230 TWh. Notably, industrial consumption accounted for 42%, followed by domestic use at 27%, and agriculture at 18% (Srinivasan et al. 2018).

In 1973, the global oil crisis prompted countries worldwide to shift their focus towards renewable energy sources. The transition to renewable sources gained momentum in the early 2000s, driven by a sudden increase in the global population. Along with this, India was also showing significant progress in the energy generation sector. Future estimates indicate that India's electricity demands will continuously increase in the coming decades due to economic growth and increased access to electricity for the population. Policymakers in India are increasingly recognizing the challenge of growing water demand for electricity production, as evidenced by government regulations aimed at minimizing water consumption in power plants. There is a significant water footprint in the electricity generation sector, which can be further divided into water withdrawal and consumption. Withdrawal is the use of freshwater from water sources, permanently or temporarily, and the return to the source after use. At the same time, consumption is the withdrawal of water from sources without returning it to the source (Srinivasan et al. 2018).

Excessive utilization of natural water sources is causing resource depletion, with 54% of groundwater wells in India experiencing elevated levels of water stress, ranging from high to extremely high (Srinivasan et al. 2018). The issue arises due to unpredictable rainfall patterns, the use of subsidized electricity for irrigation pumps, and unsustainable practices in managing groundwater resources (World Resources Institute 2015). Also, the transformation of India into an industrial hub, marked by high total dissolved solids and COD levels in the water, shows the critical need for sustainable water management strategies in future energy generation for the region (Shirin and Yadav 2014).

In this backdrop, the energy sector is expected to use an increasingly significant portion of available water resources. The sector's proportionate water consumption is expected to rise from 1.4% to 9% between 2025 and 2050, going from 15 billion m^3 to 130 billion m^3 annually. All sectors, including irrigation, energy, and drinking water, are experiencing a growing demand for water. By 2050, the projected total water demand is estimated to reach 1400 billion m^3. In comparison, the annual usable surface water capacity is approximately 690 billion m^3, and the potentially replenishable annual groundwater capacity amounts to 433 billion m^3 (Central Water Commission: Government of India 2015).

The disparity between demand and supply has the potential to impact water availability in various uses, heightening the risk of competition, especially in areas with limited water resources. Furthermore, with rising demand, the necessity to extract water from more distant or deeper sources may arise, consequently amplifying energy requirements (World Resources Institute and International Renewable Energy Agency. 2018). Therefore, it is essential to estimate the possible footprint of water in future energy generation technologies to ensure the sustainable development of the nation.

The amount of cooling water that the plant requires depends on its thermal efficiency, regardless of the type of fuel used. The higher the efficiency, the smaller the water footprint of the power plant. It is independent of the type of fuel used. Regarding sustainability in future generations, it is necessary to improve the efficiency of power generation units to increase the production rate and reduce the footprint of water in the environment.

26.1.2 Overview of Energy Consumption in India

India is rich in natural resources and energy options, but they are not evenly spread across the country. To ensure a steady supply of electricity for every household, a strategic use of these resources is necessary. India accounts for 17% of the global population, and the economic growth of India has been the highest in the last two decades. Electricity plays a crucial role in India's goal of achieving more than 10% of economic growth. In this country, coal has been playing an important role in electricity generation. It is one of the energy sources that has been developed at a commercial level. Currently, the per capita consumption of around 85% of electricity generation in this country occurs through thermal power plants using fossil fuels like coal, natural gas, and oil and nuclear power plants. In 2015, a pivotal moment occurred as leaders from 200 nations met under the Paris Agreement. This historic gathering marked a commitment to achieving net zero emissions between 2050 and 2100. Moreover, During the CoP 26 summit in Glasgow in 2021, India declared its commitment to attaining a non-fossil energy capacity of 500 GW by 2030. The goal is to meet 50% of the country's energy needs through renewable sources by the same year. Furthermore, India aims to decrease total anticipated carbon emissions by one billion tonnes from today until 2030, achieve a reduction in the intensity of carbon of its economy by less than 45% by 2030, and reach the Net Zero target by 2070 (Government of India 2024). The Ministry of Renewable Energy is providing several schemes to increase solar-based energy generation at both domestic and commercial scales. One of the largest solar parks in the world, the Bhadla Solar Park is located in the Jodhpur District of Rajasthan and has a capacity of 2.25 GW. This is an example of development in recent decades (Government of India: Ministry of Power 2024).

Renewable energy, derived from sources such as sunlight, wind, rain, tides, and geothermal heat, is replenished naturally. However, the output of wind and solar power can be unpredictable because of changing weather conditions. This variability means that operators cannot control the power generation consistently. As a result, additional energy is needed to balance the supply and demand on the grid in real-time, along with ancillary services like frequency regulation and voltage support. This is why a complete shift to renewable energy sources is challenging due to their fluctuating behavior. But to attain sustainability, it is crucial to look for the factors affecting the existing energy production ways and alternate possible ways with the least impact on the environment.

26.1.3 Electricity Generation Within Countries

Brazil has 2.36 MWh of electricity consumption per inhabitant. Surprisingly, 55% of electricity generation in the country Brazil is from Hydropower only. Due to large rivers, high precipitation volumes and the development of new technologies, it was possible to build dams and use gravity to control the flow of water. In 2021, around 362.8 TWh of electricity was produced by dams and hydropower technologies. Furthermore, the country shows a significant contribution of renewable energy sources in this sector. In the same year, 144 TWh of energy was produced by renewable sources such as solar, wind and biofuel (BP p.l.c. 2022) (Figs. 26.1, 26.2 and 26.3).

In 2022, the total electricity generation in the United States was approximately 4.24 trillion kilowatt hours (kWh). The US is also using fossil fuels as a major energy source, and the total contribution to electricity generation is about 60%, which is more than the contribution in the case of India. Although the country uses fossil fuels the major fuel is natural gas, which has a much lower environmental impact as compared to coal-based systems. This is the only country with nuclear

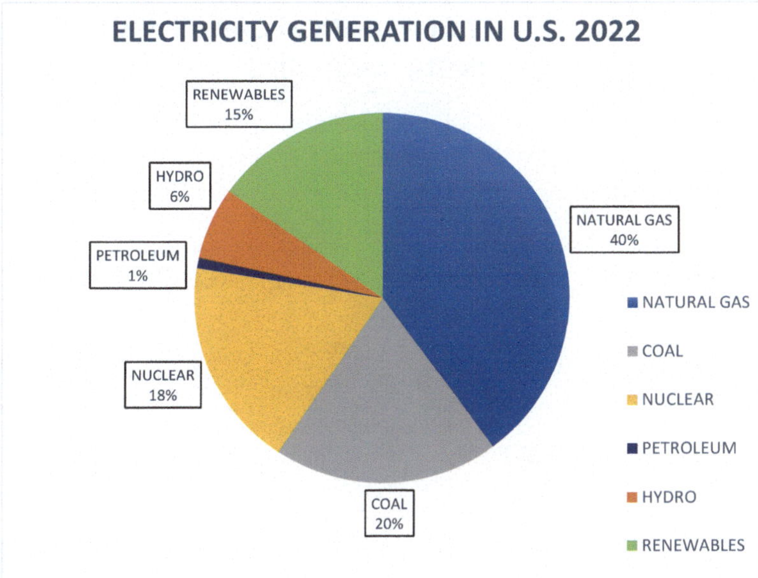

Fig. 26.1 Electricity generation from various resources in United States in year 2022

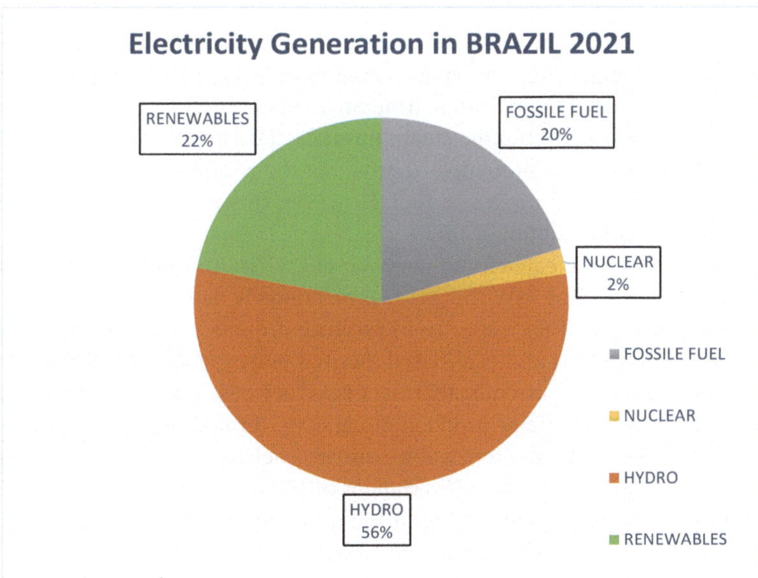

Fig. 26.2 Electricity generation from various resources in Brazil in year 2021

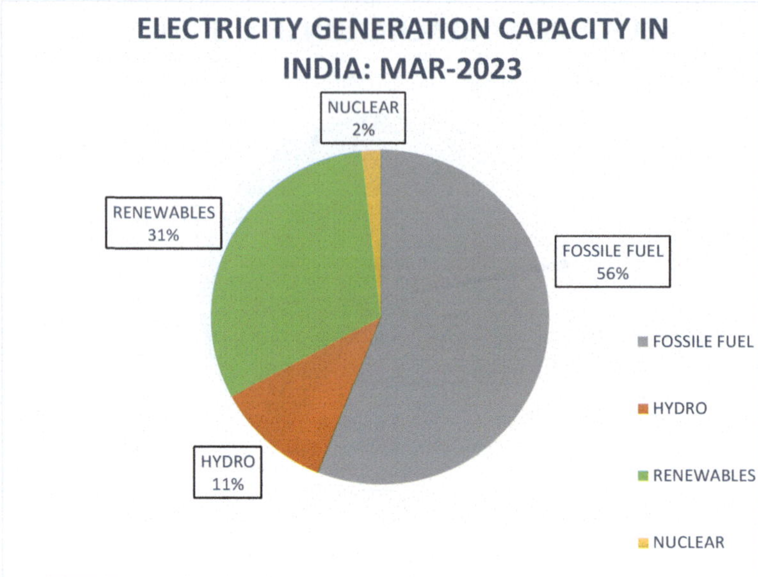

Fig. 26.3 Electricity generation from various resources in India in year 2023

energy sources that are as high as 18%. Neither Brazil nor India has developed from a nuclear energy perspective, with a contribution of only 2% in total. The U.S. has around 1% of total generation from petroleum sources. In this, residual fuel oil and petroleum coke are used to run the steam turbines. These also involve diesel-based generators. This can be an alternative source, but the residual content in petroleum refineries is not efficient for use as a fuel. However, this is not a sustainable option because of the polluting nature of fuel.

India has installed the capacity to generate electricity through coal-based power plants, reaching up to 211 GW. In 2022, the country's natural gas thermal power plants in the country had the capacity to produce around 297.6 GW of electricity. However, the contribution of diesel-based thermal power plants was limited to only 6.6 GW in the same year. Currently, thermal fossil fuel-based power plants contribute 56% of India's total electricity production capacity. In addition to thermal energy sources, hydropower and other renewable sources, such as solar and wind, contribute more than 40% to total capacity (Government of India: Ministry of Power 2024).

When comparing both countries with India, it can be concluded that India is a more carbon emitter than the other two countries. Developing cleaner and more efficient technologies can contribute to reducing carbon emissions and improving the overall environmental impact of the energy industry (U.S. Energy Information Administration 2023).

26.1.4 Section Remarks

With global energy demand expected to reach 10 EJ/year by 2050, India aims to meet 50% of its energy needs through renewables. However, the water consumption is projected to grow significantly, straining resources and exacerbating water stress issues. The mismatch between demand and supply could lead to competition and increased energy requirements for water extraction. India's electricity consumption has increased, with 42% in industry, 27% in domestic production, and 18% in agriculture. The water consumption is estimated to increase from 1.4% to 9% between 2025 and 2050. Sustainable development requires estimating future water footprints in energy generation technologies. Improving the efficiency of power plants is vital to reduce the water footprint. Despite India's commitment to renewable energy, coal still dominates, but efforts include the development of large solar parks. The variability poses challenges, necessitating additional energy for grid balancing. Balancing sustainability and energy demands requires strategic use and efficient technologies.

26.2 Role of Water in Conventional Energy Generation

26.2.1 Thermal Energy

India has significant coal reserves, ranking second globally in fossil fuel production. The primary coal-rich regions include Jharkhand, West Bengal, and Chhattisgarh. Coal extracted in India is primarily immature. Approximately 65% of the country's coal production is dedicated to electricity generation, and the power sector absorbing 85% of lignite consumption (Ministry of Power Government of India 2023).

Given the immature nature of Indian coal, extensive pretreatments are necessary to transform it into a cleaner fuel source. Its application in thermal power plants results in a coal consumption rate of approximately 0.7 kg per kWh of electricity generated. In comparison, the United States achieves a higher electricity output of 1 kWh from only 0.45 kg of coal because of its coal's higher calorific value. The variation in coal consumption can be attributed to the Gross Calorific Value (GSV), which stands at around 4500 kcal/kg for Indian coal. It is worth noting that many other countries have coal of better quality than Indian coal.

India has the potential to generate a significant amount of electricity through coal-based thermal power plants, reaching up to 211 GW. In 2022, the country's natural gas thermal power plants in the country had the capacity to produce around 297.6 GW of electricity. However, the contribution of diesel-based thermal power plants was limited to only 6.6 GW in the same year. Currently, thermal fossil fuel-based power plants contribute 57% of India's total electricity production capacity (Ministry of Coal Government of India 2021).

In the context of thermal power plants, water plays a pivotal role, being extensively withdrawn for steam generation to propel steam turbines. Additionally, water is employed to manage ash in these processes. This emphasizes the critical reliance on water resources. In a typical thermal power plant featuring a 500 MW coal-fired unit, the cooling water necessary for condenser and auxiliary cooling amounts to around 60,000 cubic meters per hour, accompanied by a temperature increase of approximately 9.5 °C across the condenser. Earlier generations of thermal power plants used cooling technologies, resulting in significant water withdrawals. India imposed a ban on the construction of once-through cooling systems using fresh water in 1999, although the implementation took place after 2015 (Ministry of Coal Government of India 2024).

The Ministry of Environment, Forests, and Climate Change of India changed the limit for water withdrawals for the thermal power sector, mandating the installation of cooling towers. After 2017, the limit of water withdrawal for any new thermal power plant was fixed to 2.5 cubic meters/MWh, with conditions of net zero water discharge from the plant.

In the case of cooling tower systems, cooling systems had low consumption but a very high amount of water withdrawal. Recirculating systems, on the other hand, exhibited low water withdrawal but slightly higher consumption. Both water consumption and withdrawal are negligible in the case of dry cooling systems. Therefore, this cooling system is a promising option for the future of power generation.

For dry cooling technology, the water footprint is minimal, but finding a waterless way for ash handling is equally necessary. Wetting injectors are used in dry cooling systems to make the ash slightly moist and prevent the carry-over of dust ash along with air. Traditionally, thermal power plants remove fly ash and bottom ash in ash ponds as wet slurry. The disposal of fly ash from thermal power plants is a significant environmental concern. Several studies suggested use of abandoned mines by filling them with a mix of 70% overburden and 30% fly ash (Shirin et al. 2021, 2023). This not only effectively tackles the fly ash waste issue but also improves acid mine drainage. The case study on Singrauli Coalfield emphasizes the potential of fly ash in neutralizing acidic water, offering a sustainable solution to environmental challenges and aiding in water treatment (Shirin et al. 2019). Use of such byproducts can reduce environmental challenges and useful in water treatment. However, because of environmental concerns, there has been a shift towards implementing measures such as reducing water usage for wet ash disposal. Modern plants now employ the disposal of wet ash with a slurry concentration of 30% for fly ash and 25% for bottom ash. Recirculation of ash water in these plants allows for the recovery and reuse of approximately 70% of ash pond water in the ash handling plant, significantly reducing the net water required for ash disposal. Limited statistical studies concluded an increase in electricity consumption beyond 50,000 TWh in the next 35 years. Around 1200 GW of electricity generation capacity is required to meet energy needs in 2050. Without additional water regulation, average water withdrawals and consumption are expected to grow five times. Freshwater withdrawal may reach up to 250 billion cubic meters (Srinivasan et al. 2018).

Taking into account future power sources, the operational water withdrawal intensity for solar photovoltaic (PV) systems is approximately 0.08 cubic meters per megawatt-hour (m3/MWh), mainly attributed to panel cleaning. This value represents only 0.5% of the average water withdrawal associated with thermal power generation. In the case of wind power, there is no water withdrawal during operations. Biofuels have a significant water footprint because of the involvement of biological processes in them. Hydropower is directly dependent on the availability of water sources, but it has very low water footprint. Such emerging renewable resources and their advanced technologies may surpass the contribution of thermal power plants to energy generation in the future. Hence, reliance on the energy generated by fossil fuels may decrease the environmental impact in the future with future advancements and technologies, but it may not result in sustainable energy generation.

26.2.2 Hydropower

As of June 20, 2023, India is ranked 137th among 218 nations in terms of energy consumption. The per capita electricity usage in India is 1208 kWh/year, significantly less than that of Canada (14,612 kWh/year), the United States (12,154 kWh/year), China (5885 kWh/year), and Brazil (2830 kWh/year). This disparity in electricity consumption is mainly attributed to India's limited generation capacity and supply infrastructure. According to projection models, the electricity demand–supply gap in India, which was approximately 20.5% in 2021–22, is expected to decrease to 17.4% by 2026–27 and further decrease to 14% by 2036–37 (Pandit et al. 2023).

Hydropower is a clean and reliable renewable energy source, having a high degree of flexibility and reliability in operation, but in India it is not fully utilising its potential. Currently, the country has 36 GW of hydroelectric capacity and another 13 GW is under construction. However, approximately 67% of the potential capacity remains untapped. In 1966, hydropower contributed 45% of electricity generation, but due to the development of fossil fuel-based plants, it has dropped to less than 20%. This shows the negligence in the growth of hydropower although the potential (Pricewaterhouse Coopers Private Limited 2014).

The government has tried to promote hydropower, but challenges forced them to stop investing in this sector. Clearances, investment decisions, legal issues, and geological surprises caused delays in the 10th Hydropower Plan projects for the duration of 2002–2007. Additional concerns include interstate disputes related to water sharing, observed by the Sutlej-Beas conflict between Punjab and Haryana, as well as the Mullaperiyar dam disagreement between Kerala and Tamil Nadu. Another significant issue is the rehabilitation of people affected by projects, particularly prominent in storage-based hydropower projects, as witnessed during the development of the Tehri dam. The construction and operation of dams have adverse effects on natural river ecosystems and wildlife, contributing to ecological and socioeconomic crises. The focus of hydropower development in India is mainly on Himalayan rivers, with dams and associated activities rumored to be responsible for various disasters in the region, including floods, landslides, and other natural hazards (Hydro Engineering & Technology Development and Renovation & Modernization Division 2023).

Furthermore, hydroelectric projects require increased initial investments due to the need to address greater intricacies in design, engineering, and the mitigation of environmental and social impacts (Pricewaterhouse Coopers Private Limited 2014). The construction period for most hydro projects typically spans five to six years. Due to these challenges, India is now more on renovating and modernizing existing hydroelectric plants. With the new R&M plan, the government is focusing on enhancing the performance of the hydroelectric plant by adopting new technologies like plasma thermal spray coating on underwater parts, Cupro-nickel tubes in coolers, cyclone separators in cooling water systems, etc. (Fig. 26.4).

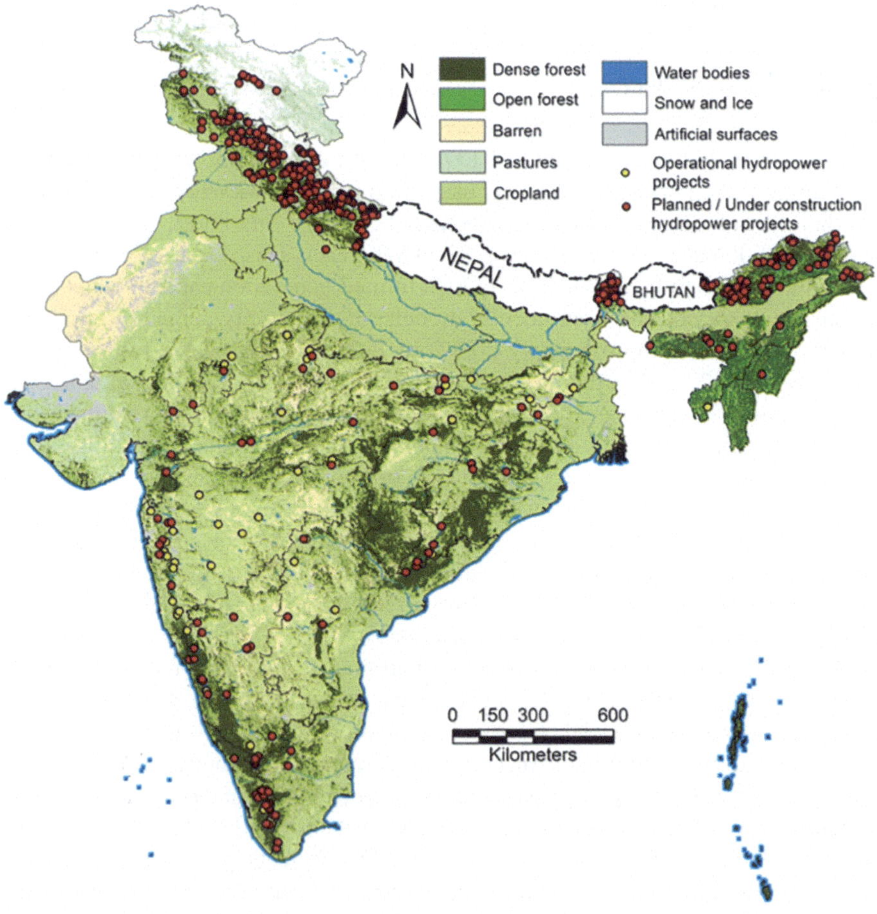

Fig. 26.4 Graph of planned or under construction and operational hydropower projects in India from Hydropower: A low-hanging sour–sweet energy option for India (2023) (Pandit et al. 2023)

26.2.3 Nuclear Energy

Nuclear energy, derived through controlled nuclear reactions, emerges as a pivotal force in shaping India's future energy landscape. The symbiotic relationship with water within nuclear power plants is fundamental, serving as both a coolant and a neutron moderator. Water's critical role in maintaining reactor stability and enhancing the efficiency of nuclear fission positions it as a linchpin in sustainable energy endeavors, not only in India but globally.

In the United States, where nuclear reactors are classified into pressurized water reactors and boiling water reactors, the cooling process is indispensable. Both systems rely on water to generate steam, a crucial step in electricity production. However, it should be noted that nuclear plants consume more water per unit of electricity compared to coal plants, primarily due to their lower operating temperature and turbine efficiency.

The statistics underscore the significance of water in nuclear energy, revealing the gallons required per megawatt hour of electricity produced in different cooling scenarios (Table 26.1):

The advantages of nuclear power extend beyond cooling considerations. Streamlining fuel logistics offer unparalleled location flexibility, a unique advantage over coal-fired counterparts. The pivotal role persists in transferring heat within the reactor core to steam turbines and dissipating excess heat from the steam circuit. Despite inherent energy loss in shared steam/Rankine cycle technology, addressing this limitation is essential to optimize nuclear plant efficiency.

The strategic integration of water in nuclear energy underscores its role in balancing safety, environmental responsibility, and operational efficiency. Site selection becomes a meticulous process, encompassing geological stability, water proximity, population density, environmental impact, emergency preparedness, security, and regulatory compliance. This strategic approach positions nuclear energy as a key player in India's future energy generation landscape.

The nation faces a pivotal decision, a reliance on renewables acknowledging their limitations, or a strategic integration of nuclear energy. This choice, which balances the demand for clean power with practical constraints, defines the trajectory of India's energy future, marking a critical juncture in sustainable energy endeavors. India's nuclear power plants, with a capacity of 6780 MW, make up less than 3% of the nation's electricity generation, and this percentage is expected to decrease as other sources contribute more (Bhattacharyya 2020).

Table 26.1 Water required per megawatt hour of electricity produced in different cooling scenarios. *Source* (Union of Concerned Scientists 2010)

Cooling method	Water withdrawal (in litres)	Water consumption (in litres)
Once-through	95,000–2,27,000	400–1,500
Recirculating	3,000–9,800	2,300–3,000

India's nuclear energy aspirations, outlined in Dr. Homi Bhabha's three-stage program, face formidable challenges, with progress stagnating in the initial phase for seven decades. The envisioned transition to fast breeder reactors (FBRs) and the Thorium cycle remains elusive, primarily hindered by delays in the prototype fast breeder reactor (PFBR) construction, driven by safety concerns related to liquid sodium coolant. While India plans additional nuclear projects, both domestically and through international collaboration, hurdles remain. Foreign partnerships are constrained by the 2010 nuclear liability legislation. As a result, India's nuclear capacity over the upcoming two decades is unlikely to exceed 20,000 MW.

Although clean and depends on the disposal of spent fuel, the cost-effectiveness of nuclear energy is in doubt. Nuclear plants offer a steady base load, but their economic viability is questionable compared to more cost-effective and environmentally friendly alternatives such as wind and solar power.

India is at a crossroads, where the promise of nuclear energy clashes with economic and practical challenges. The envisioned Thorium cycle, a beacon of sustainable energy, remains distant. As the nation charts its energy course, a critical decision looms: persist with the three-stage nuclear program or pivot toward immediate, sustainable alternatives. The choice will profoundly shape India's energy landscape in the coming decades (Ramesh 2019) (Fig. 26.5).

26.2.4 Section Remarks

India is the second largest producer of fossil fuels. Despite requiring extensive pretreatment, Indian coal, with low carbon and minimal sulfur, contributes 65% to electricity generation, consuming approximately 0.7 kg per kWh. The water-intensive nature of thermal power plants in India required a shift to cooling towers, with a limit of 2.5 m^3/MWh and a mandate for net zero water discharge after 2017. Hydropower, a cleaner energy source, faces challenges in India, with only 36 GW utilized of a potential 67 GW. The focus on renovating existing stations rather than new projects is due to obstacles such as clearances, legal issues, and ecological impacts. Nuclear energy, comprising less than 3% of India's electricity generation, faces challenges in progressing the tertium cycle. Economic viability concerns, coupled with environmental considerations, question the role of nuclear energy as a sustainable alternative. The decision between nuclear energy and immediate, cost-effective alternatives such as wind and solar energy will significantly shape India's energy future.

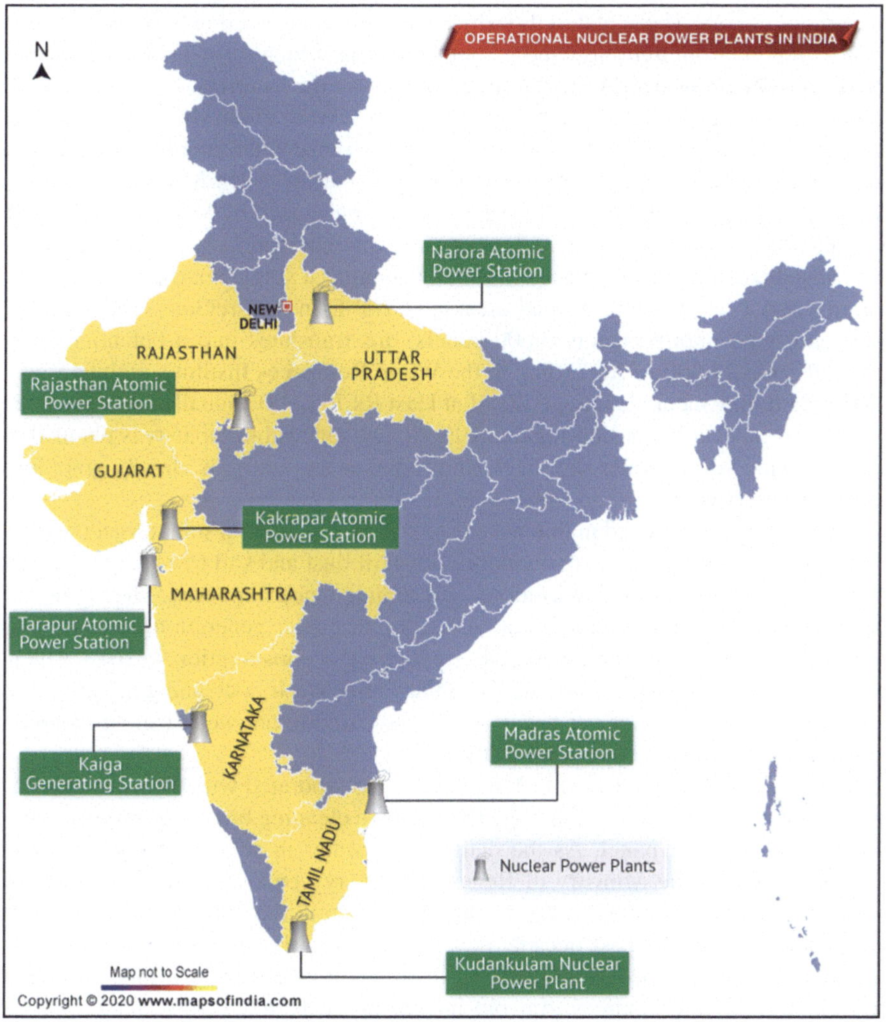

Fig. 26.5 Graph showing operational nuclear power plants in India (Bhattacharyya 2020)

26.3 New and Emerging Technologies for Power Generation

26.3.1 Solar Energy

Solar Photovoltaic Technology, also referred to as SPV technology, is used to convert solar energy from the sun's radiations directly into electrical power. Specific materials called semiconductors (like germanium, silicon, etc.) are used in the conversion of the form of energy. These materials (which can be termed a cell in SPV technology) absorb photons from light rays and release electrons, which form an electric current (direct current). This energy can be stored in the storage batteries.

India's solar capacity has increased from 2.6 GW to about 46 GW in the last 7.5 years. MNRE provides several schemes to increase solar energy generation at both domestic and commercial scales. Schemes like the Ultra Mega Solar Power Project and the Development of Solar Parks were introduced by MNRE (Ministry of New and Renewable Energy 2024). In contrast to this, solar parks have been estimated to be installed in the driest places in India. Cleaning a single solar panel typically requires 3–5 L of water in normal areas and 7–8 L in arid regions. For a 1 MW solar park with approximately 3,000 panels, this translates into a potential water usage of up to 24,000 L. According to the World Resources Institute, maintaining a 1 MW plant may incur cleaning costs of at least Rs 738,000 annually. This becomes significant, especially considering that 79% of new renewable capacity is planned in regions facing severe water shortages, and India, as a country, is already grappling with a water crisis.

Effective operations and maintenance are crucial for optimal solar project performance, with a key aspect being regular removal of dust and soil from panels, which makes up 25–35% of total operations and maintenance expenses. The failure to clean the panels results in a substantial decline in power generation. Recognizing this, innovative solutions have emerged to reduce water consumption. In 2018, American Polywater Corporation introduced Solar Panel Wash, a cleaning solution that uses only 230 ml of water per panel. In contrast, traditional wet cleaning methods consume at least 2 L of water per solar panel.

Inspire Clean Energy, based in Mumbai, has collaborated with the Central Institute of Plastic Energy and Technology to develop a cleaning brush that requires only 300 ml of water per panel. As the demand for water in solar projects increases, it is essential that the Ministry of Renewable Energy encourages the adoption of waterless cleaning systems in future projects. Despite the substantial increase in solar capacity over the past decade, developers currently do not receive information on water procurement. Future tender documents should address this, providing information on water sources to avoid increasing costs and access challenges for developers. Additionally, the government should actively support research and development in waterless cleaning technologies to make them economically competitive for all projects (Macknick and Cohen 2015; Renewable Watch 2020).

26.3.2 Ocean Energy (Marine and Tidal)

Approximately 70% of the Earth's surface is covered by oceans, presenting a substantial opportunity for the provision of sustainable and eco-friendly energy worldwide. Various forms of ocean energy, such as wave, tidal, marine current, thermal gradient, and salinity gradient, hold significant promise. Notably, ocean energy exhibits greater predictability compared to solar and wind sources. In addition to expanding energy generation, its utilization can contribute to a decrease in carbon emissions and the creation of employment opportunities. The mechanical energy of ocean waves can be harnessed by wave energy converters, which convert it into electricity through hydraulic or pneumatic turbines. Key types of wave energy converters include attenuators, point absorbers, overtopping devices, and oscillating water columns. A Ministry of New and Renewable Energy assessment suggests a tidal energy potential of around 8000 MW in India. This includes 7000 MW in the Gulf of Khambhat, 1200 MW in the Gulf of Kutch in Gujarat and 100 MW in the Sunderban delta. The report on Study on Tidal and Wave Energy in India: Survey on the Potential and Proposition of a Roadmap, by IREDA, AFD and IIT, Madras submitted in December 2014 estimated tidal power capacity in India is approximately 12,500 megawatts, with a focus on the use of tidal stream technology in the Kutch and Khambhat. Additionally, there is potential to implement tidal barrage technology in specific locations throughout India, particularly in areas characterized by substantial backwaters (Goswami and Deshmukh 2017).

Tidal and wave energy installations can alter tidal regimes and sedimentation patterns and pose a potential threat to aquatic life. These installations may alter the temperature and flow balance, and both noise and electromagnetic effects can disturb marine ecosystems (Boehlert and Gill 2010; Frid et al. 2012). The extraction of energy through tidal turbines has the potential to modify water movement patterns, subsequently affecting sedimentation scenarios and potentially influencing benthic habitats (Goswami and Deshmukh 2017).

26.3.3 Geothermal Energy

Geothermal energy originates from hot water beneath the Earth's surface, reaching temperatures of 100 to 300 degrees Celsius, which can be harnessed to generate electricity. On the surface, hot water with temperatures ranging from 50 to 100 °C is obtained. Similar to hydropower, geothermal energy relies on water for power generation. Identifying areas abundant in this energy source is essential for its utilization. Leading countries in the field of geothermal energy, including the USA, Indonesia, Philippines, Turkey, and New Zealand, have well-established power generation systems (Jayaraju 2018).

India's Ministry of Renewable Energy has estimated a theoretical geothermal potential of 10 GW, as per GSI (Ministry of New and Renewable Energy 2024). The country reported around 340 hot springs, primarily found in five regions: the Himalayan arc system (extending to Andaman Nicobar Island), the Son-Narmada-Tapti lineament, the continental margin and surroundings, Gondwana grabens, and Delhi fold regions (Jayaraju 2018).

The geothermal potential in India is evident with a 25 MW geothermal power project in the coastal hot spring belt and another in Himachal Pradesh. Beas Valley, Tirthan Valley, and Spiti Valley, Puga in Leh, J & K, has promising geothermal resources with natural springs exhibiting reservoir temperatures ranging from 180 to

260 °C. Despite these opportunities, the exploration and development of geothermal energy in India is in their early stages, showing signs of future potential (Yadav and Sircar 2021).

This source is considered the cleanest, as it does not emit carbon by burning of fuel. Although clean energy, electricity generation from its thermal form again involves water withdrawal and its consumption. Therefore, considering the water footprint of the system, this energy source contributes equally to the footprint same as other thermal power plants. Also, most of the hot springs are found with high mineral content (e.g., chloride content), which may give other issues like fouling of equipment and corrosion. This may be a reason for the least development of such established natural sources (Yadav and Sircar 2021).

As discussed above, the efficiency of a thermal power plant depends directly on the footprint of the water. It is necessary to find efficient electricity generation technology from geothermal energy sources that have the least water consumption (Fig. 26.6).

26.3.4 Bio Energy

Biomass energy is a promising solution for converting biodegradable waste into usable energy, offering a viable alternative to fossil fuels. One well-known method is anaerobic digestion, where microorganisms facilitate oxygen-free decomposition of organic matter, producing methane. Other methods include incineration, gasification, and hydrothermal reforming, among others.

Currently, bioenergy accounts for approximately 9% of the world's primary energy supply, while its contribution in the India was less than 2% in year 2022 (Government of India 2024). However, concerns have emerged regarding biomass-related air pollution, energy costs and, notably, its impact on water resources. Both the cultivation and conversion of biomass require substantial water, posing challenges in the face of global freshwater scarcity. With the global freshwater footprint already exceeding its limits, expanding biomass production could exacerbate water stress. Therefore, the evaluation of water use is crucial for the sustainable development of bioenergy, particularly given its competition with other sectors for limited water resources (Yang et al. 2018).

Numerous studies have examined water use at various levels, highlighting the water intensity of bioenergy compared to non-renewable sources. Traditional first generation (1-G) bioenergy, derived conventionally, exhibits significant water demand, especially during feedstock cultivation. On average, to produce 1 L of ethanol production from sugar cane takes 2500 L of water, which has a very large water footprint. In response to these limitations, the focus has shifted towards more sustainable second generation bioenergy (2-G biomass), which focuses on feedstocks such as agricultural and forestry residues with lower water intensity (Gerbens-Leenes et al. 2009).

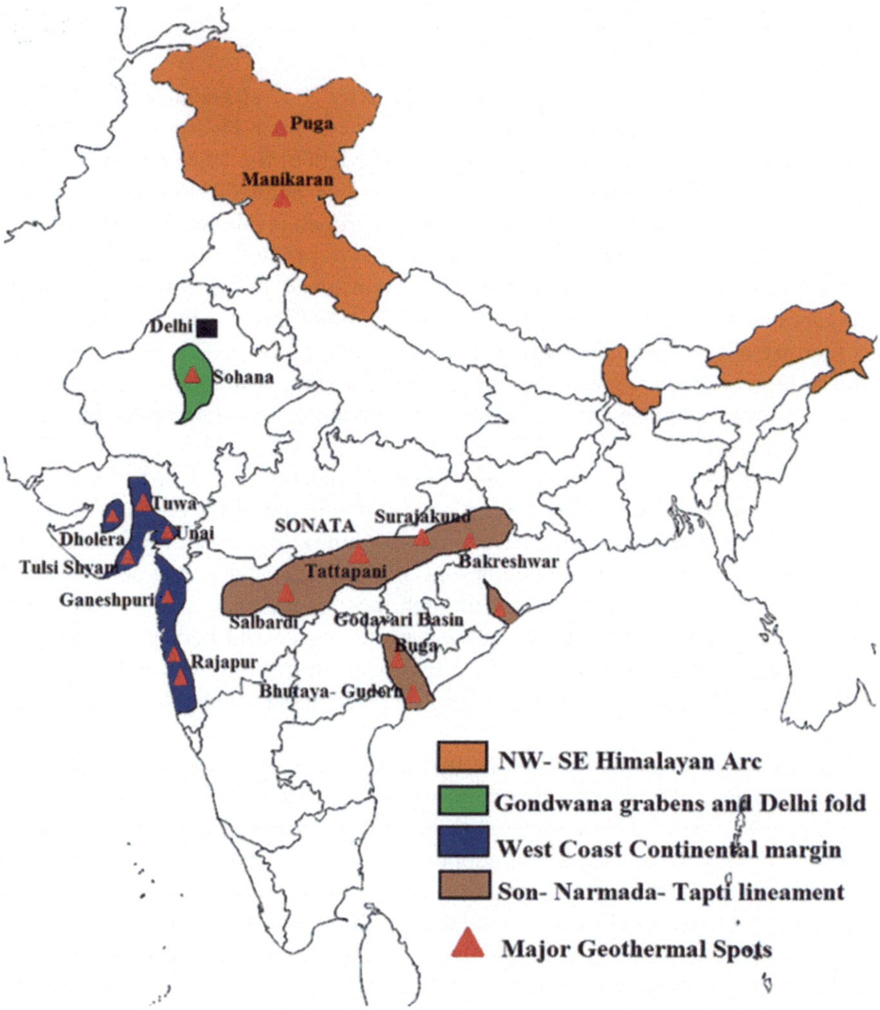

Fig. 26.6 Geothermal provinces in India: taken from Geothermal energy provinces in India: A renewable heritage(2021) (Yadav and Sircar 2021)

To address these challenges, the government of India is actively promoting agricultural food processing industries, such as the sugar cane industry. The emphasis is on not burning agroresidues like bagasse but instead treating them to produce ethanol or other chemicals, employing methods with lower water footprints. MNRE has been implementing the National Bioenergy Programme since 2021, anticipating an increase in this energy source in the next generation. Considerable research and technological advancements are being applied to make bioenergy commercially scalable with the least impact on natural resources, especially water.

26.3.5 Hydrogen Fuel

India has set a goal to attain energy independence by 2047 and achieve Net Zero by 2070, with a notable emphasis on the pivotal role of Green Hydrogen. The Indian National Green Hydrogen Mission, backed by a budget of Rs. 19,744 crores, aspires to generate 5 million metric tons of green hydrogen annually. Being a fuel with high calorific value (33.6 kWh per kg). Green hydrogen is produced by electrolysis of water, splitting atoms of Hydrogen and oxygen to produce molecular gases. For hydrogen production, a large amount of water is consumed. In electrolysis, pure water is required and direct electrolysis of salt water is not efficient. After combustion of this fuel, we obtain thermal energy and water as a by-product. This by-product can again be used to produce green hydrogen. Even if water consumption in this process is not significant, a large amount of freshwater is required to produce green hydrogen (Beswick et al. 2021).

Electrolysis emerges as a compelling option for generating carbon-free hydrogen using natural sources. According to the reaction stoichiometry, the production of one kilogram of hydrogen requires the consumption of nine kilograms of water. In various applications, hydrogen is either combusted or passed through a fuel cell, where it is converted into energy and water. Although most of the water can be recovered, it is practically not returned to its original water source and hence is considered as consumption. The only sector where hydrogen utilization does not return the complete water feedstock through combustion is chemical synthesis (Oliveira et al. 2021).

The projected annual demand in the future is 2.3 gigatons of hydrogen. To meet this requirement, 20.5 billion cubic meters of water withdrawal is essential, representing a reduction of 33% compared to current water usage in the generation of fossil fuel energy. It is evident that adopting hydrogen as a means to transition to renewable energy will result in significant water savings compared to conventional energy derived from fossil fuels (Greenlee et al. 2009; MILLER 2003).

As the available freshwater on the planet is only 1%, it would be better to avoid having an additional burden on its use. In this case, it necessitates the technology to produce hydrogen from sea water. However, water quality is crucial in electrolysis to ensure the production of high-purity hydrogen gas and impurities in saline water can affect the production efficiency. One way can be to convert seawater to pure water by the removing salts by reverse osmosis method. Again, this desalination method requires electricity but can be recovered with the energy from hydrogen production (Cherif and Belhadj 2018).

Currently, electrolysis is not a widely adopted method as it is limited due to high costs. By 2050, green hydrogen is expected to meet 18% of global energy demand, driven by its high energy density and cost-effective manufacturing. Its carbon-free energy production makes it a crucial source in various industries.

More than 95% of hydrogen production takes place by steam reforming. Though it is a cost-effective way for present, but releases around 10 kg carbon dioxide per kg of hydrogen generated. Coal gasification is another way to use coal for energy production. In this method, coal is subjected to high temperature to give hydrogen and other products such as methane, carbon mono, and dioxide. Methane from this gasification further undergoes steam reforming. Water also participates in gasification reactions to produce the synthesis gas. Therefore, the water footprint of this process is significant compared to others. In addition, the drawback of this technology is its higher emission output.

Due to the scalability limits of methods like photobiological and photoelectrochemical water splitting are still at the research level, but may develop in future with various solutions to those problems. Solar reforming of methane is another method that uses concentrated sunlight to unlock hydrogen from natural gas. Scientists are optimizing the process for maximum hydrogen yield and cleaner results to have a scalable choice of hydrogen production. As all other methods involve either high carbon emissions or use of electricity. Advanced reactor designs and improved catalyst materials can improve efficiency and reduce carbon emissions by 35–40% compared to traditional reforming methods that rely on fossil fuels.

All of the methods mentioned above have direct use of water in their hydrogen generation process. Even if this type of fuel has a high-water footprint, it is the cleanest fuel that gives back the water molecule after combustion. As a result of this important advantage, India looks forward to large investments in hydrogen production. With the Indian National Green Hydrogen Mission by the Government of India, hydrogen will be an important fuel for future generations.

Figure 26.5 shows the hydrogen map of the world, indicating maximum hydrogen producing plants in Europe compared to other regions (Fig. 26.7).

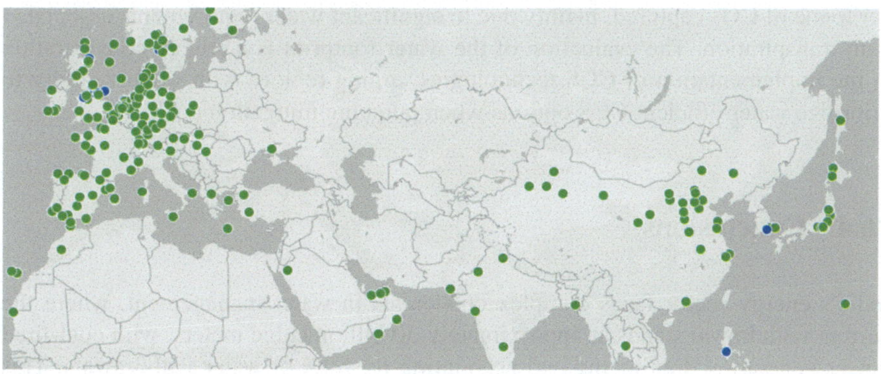

Fig. 26.7 Global hydrogen map graph by Pillsbury Winthrop Shaw Pittman LLP. https://www.thehydrogenmap.com/?operational=1&country=339

26.3.6 Carbon Capture and Storage

The significant role of carbon capture and storage (CCS) is evident in its contribution to reducing CO_2 emissions from both the electricity and the industrial sectors. Additionally, it plays an active role in the removal of atmospheric CO_2. Various technologies for CCS exist based on the source of CO_2. These systems also involve the use of captured carbon to generate new fuels. This technology specifically generates a circular economy of carbon (Rosa et al. 2021).

Several case studies have highlighted the harm caused by coal mining and its use in urban areas of India, emphasizing the urgent need to investigate sustainable water-focused energy solutions. Prioritizing eco-friendly alternatives is crucial to alleviate environmental damage and ensure responsible power generation for the future (Shirin et al. 2021). With the increasing importance of low-carbon electricity generation, water is becoming a crucial element that affects humanity's capacity to meet social requirements and tackle climate change. CCS systems, while crucial for CO_2 reduction, are characterised by their substantial energy and water requirements. If implemented, these technologies would increase overall water use, which would require increased attention to water scarcity concerns (Rosa et al. 2020). The water-intensive nature of CCS is evident both in the cooling process at power plants and in the integral role of additional water in carbon capture processes (Rosa et al. 2020).

The predominant factor contributing to the water footprint in CCUS technology is the cooling of exit gas streams. The water footprint of CCS can vary by up to 575 cubic meters of water per tonne of CO_2, depending on the specific technology utilized. In particular, CCS bioenergy stands out due to its highest water footprint per tonne of CO_2 captured, mainly due to significant water requirements associated with transpiration. The evaluation of the water footprint is a crucial consideration in the implementation of CCS technologies, urging regions with water scarcity to prioritize water-efficient CCS options when pursuing mitigation goals.

26.4 Conclusions

India's energy sector faces complex challenges in water management, where the current withdrawal stands at approximately 30 billion cubic meters, with coal-fired power generation dominating and accounting for 80% of water withdrawals. This presents a complex interplay of water dynamics in the context of diverse energy sources, shaping the nation's envisioned future energy landscape.

In the Sustainable Development Scenario, a comprehensive energy outlook anticipates significant shifts. The projection foresees a substantial decrease in water withdrawals, declining to less than 3 billion cubic meters in 2040. This decline is attributed to a reduction in coal-fired power generation, with solar photovoltaic and wind taking precedence due to their low water requirements. By 2040, renewables are projected to constitute 60% of the energy generation mix, a noteworthy leap from their current share of less than 10%. This shift not only reduces water withdrawals but also curtails water consumption by around 40%.

On the contrary, the Stated Policies scenario envisions a moderate decrease in water withdrawals, settling at less than 7 billion cubic meters in 2040. The prolonged existence of coal infrastructure, coupled with an increase in nuclear capacity after 2030, prevents a more substantial decline in water withdrawals. Water consumption, however, increases by 45% by 2040, driven by India's policy mandating thermal power plants to adopt tower cooling.

To mitigate additional strains on the water system arising from water stress, transformative measures are imperative. Incorporating dry or non-freshwater cooling in new thermal plants, retrofitting older plants, and adopting water-recycling practices emerge as crucial strategies. India has already initiated steps in this direction, with some coal power plants using desalinated seawater and treated wastewater for cooling, notably the Mundra UMPP. However, only a limited percentage of coal plants currently have access to treated wastewater, which underscores the importance of broadening these practices.

The exploration of dry cooling technologies, particularly in water-stressed regions, introduces a water-saving avenue. While dry cooling eliminates the need for water, its adoption faces trade-offs, including higher capital costs and a reduction in power production efficiency. Currently, only 2% of capacity utilizes dry-cooling technology, signaling the need for supportive policies to facilitate upscaling.

As India leans toward a preference for clean energy technologies, careful alignment with water stress mitigation strategies becomes imperative. In particular, not all low-carbon technologies and fuels exhibit low water requirements. The careful consideration of water demands, both present and future, is paramount in energy planning. Nuclear, bioenergy, and concentrated solar power, integral to India's sustainability goals, may encounter limitations based on the cooling technology deployed. Therefore, a nuanced approach that accounts for competing water demands is vital in designing and location of energy plants. Prioritizing water-stressed areas with abundant solar irradiation and wind becomes a strategic move for the development of new renewable capacity.

In conclusion, India is at a pivotal juncture in shaping its future energy, intricately interwoven with considerations of water. The path toward sustainable energy hinges on the judicious management of water resources, the embrace of innovative technologies, and the alignment of energy planning with environmental imperatives. As the nation charts its course, balancing the quest for clean energy with the preservation of water resources will be instrumental in achieving a harmonious and resilient energy landscape for future generations.

Acknowledgements The authors declare no conflict of interest. This work did not receive any specific grant from funding agencies in the public, commercial, or not-for-profit sectors. The authors acknowledge the Institute of Chemical Technology, Mumbai, and Queen's University Belfast, UK, for the resources provided in writing this chapter.

References

Beswick RR, Oliveira AM, Yan Y (2021) Does the green hydrogen economy have a water problem? ACS Energy Lett 6(9):3167–3169. https://doi.org/10.1021/acsenergylett.1c01375

Boehlert G, Gill A (2010) Environmental and ecological effects of ocean renewable energy development—a current synthesis. Oceanography 23(2):68–81. https://doi.org/10.5670/oceanog.2010.46

BP p.l.c. (2022) 72nd edition of the statistical review of world energy. Brazil

Central Water Commission: Government of India (2015) Central Water Commission: Water Resources Information System Directorate: Water Planning & Project Wing

Cherif H, Belhadj J (2018) Environmental life cycle analysis of water desalination processes. In: Sustainable desalination handbook, pp 527–559. Elsevier

Bhattacharyya D (2020) Operational nuclear power plants in India. My India

Frid C, Andonegi E, Depestele J, Judd A, Rihan D, Rogers SI, Kenchington E (2012) The environmental interactions of tidal and wave energy generation devices. Environ Impact Assess Rev 32(1):133–139. https://doi.org/10.1016/j.eiar.2011.06.002

Gerbens-Leenes W, Hoekstra AY, van der Meer TH (2009) The water footprint of bioenergy. Proc Natl Acad Sci 106(25):10219–10223. https://doi.org/10.1073/pnas.0812619106

Goswami P, Deshmukh SP (2017) Assessment of wave and solar energy potential along western coast of India. Res J Eng Technol 8(3):197. https://doi.org/10.5958/2321-581X.2017.00032.0

Government of India (2024) Ministry of new and renewable energy. Retrieved 9 Mar 2024 (https://mnre.gov.in/)

Government of India: Ministry of Power (2024) Central electricity authority. Government of India. Retrieved 9 Mar 2024 (https://cea.nic.in/?lang=en)

Greenlee LF, Lawler DF, Freeman BD, Marrot B, Moulin P (2009) Reverse osmosis desalination: water sources, technology, and today's challenges. Water Res 43(9):2317–2348. https://doi.org/10.1016/j.watres.2009.03.010

Hydro Engineering & Technology Development and Renovation & Modernization Division (2023) Renovation and modernisation of hydro power stations status/programme for the period 2022–27 and 2027–32. Quarterly progress report July–Sept

Jayaraju N (2018) Geothermal energy in India: past, present and future scenarios

Ramesh M (2019) Why nuclear when India has an 'Ocean' of energy. The Hindu, June 30

Macknick J, Cohen S (2015) Water impacts of high solar PV electricity penetration. Golden, CO (United States). https://doi.org/10.2172/1225894

Miller J (2003) Review of water resources and desalination technologies. Albuquerque, NM, and Livermore, CA (United States). https://doi.org/10.2172/809106

Ministry of Coal Government of India (2021) Coal: India energy choice. Ministry of coal: Government of India. Retrieved 9 Mar 2024 (https://coal.nic.in/en/major-statistics/coal-indian-energy-choice)

Ministry of Coal Government of India (2024) Coal: production and supplies. Ministry of Coal: Government of India. Retrieved 9 Mar 2024 (https://coal.gov.in/en/major-statistics/production-and-supplies)

Ministry of New and Renewable Energy (2024) New technologies: geothermal energy. Ministry of New and Renewable Energy (MNRE). Retrieved 9 Mar 2024 (https://mnre.gov.in/new-technologies/)

Ministry of Power Government of India (2023) Power sector at a glance: All India. Ministry of Power: Government of India. Retrieved 9 Mar 2024 (https://powermin.gov.in/en/content/power-sector-glance-all-india)

Oliveira AM, Beswick RR, Yan Y (2021) A green hydrogen economy for a renewable energy society. Curr Opin Chem Eng 33:100701. https://doi.org/10.1016/j.coche.2021.100701

Pandit MK, Manish K, Singh G, Chowdhury A (2023) Hydropower: a low-hanging sour-sweet energy option for India. Heliyon 9(6):e17151. https://doi.org/10.1016/j.heliyon.2023.e17151

Pricewaterhouse Coopers Private Limited (2014) Hydropower in India: key enables for a better tomorrow
Renewable Watch (2020) Waterless way: dry cleaning solutions for solar panels. Renewable Watch. Retrieved 9 Mar 2024 (https://renewablewatch.in/2020/12/03/waterless-way/)
Rosa L, Reimer JA, Went MS, D'Odorico P (2020) Hydrological limits to carbon capture and storage. Nat Sustain 3(8):658–666. https://doi.org/10.1038/s41893-020-0532-7
Rosa L, Sanchez DL, Realmonte G, Baldocchi D, D'Odorico P (2021) The water footprint of carbon capture and storage technologies. Renew Sustain Energy Rev 138:110511. https://doi.org/10.1016/j.rser.2020.110511
Shirin S, Jamal A, Emmanouil C, Singh VP, Yadav AK (2023) Assessment and characterization of waste material used as backfilling in an abandoned mine. Int J Coal Prep Util 43(8):1402–1410. https://doi.org/10.1080/19392699.2022.2118259
Shirin S, Jamal A, Emmanouil C, Yadav AK (2021) Assessment of characteristics of acid mine drainage treated with fly ash. Appl Sci 11(9):3910. https://doi.org/10.3390/app11093910
Shirin S, Jamal A, Ranjan P, Yadav AK (2019) Study on assessment of slope stability and mixed disposal of overburden in voids of Singrauli coalfield. Environ Qual Manage 28(3):131–139. https://doi.org/10.1002/tqem.21616
Shirin S, Yadav A (2014) Physico chemical analysis of municipal wastewater discharge in Ganga river, Haridwar District of Uttarakhand, India. Curr World Environ 9(2):536–543. https://doi.org/10.12944/CWE.9.2.39
Srinivasan S, Kanudia A, Nazar R, Dharmala N (2018) Impact of power sector growth on water resources
Union of Concerned Scientists (2010) How it works: water for nuclear. Union of Concerned Scientists. Retrieved 9 Mar 2024 (https://www.ucsusa.org/resources/water-nuclea)
U.S. Energy Information Administration (2023) Electricity in the United States. Retrieved 9 Mar 2024 (https://www.eia.gov/energyexplained/electricity/electricity-in-the-us.php)
World Resources Institute (2015) India water tool
World Resources Institute and International Renewable Energy Agency (2018) Water use in India's power generation: impact of renewables and improved cooling technologies to 2030
Yadav K, Sircar A (2021) Geothermal energy provinces in India: a renewable heritage. Int J Geoheritage Parks 9(1):93–107. https://doi.org/10.1016/j.ijgeop.2020.12.002
Yang Q, Liang J, Li J, Yang H, Chen H (2018) Life cycle water use of a biomass-based pyrolysis polygeneration system in China. Appl Energy 224:469–480. https://doi.org/10.1016/j.apenergy.2018.05.009

Chapter 27
Future Directions and Innovations in Computational Water Management

Sumit Bansal, Vandana Sindhi, and Bhim Sain Singla

Abstract The field of computational water management stands at the crossroads of unprecedented challenges and opportunities. In this era of rapidly changing climate patterns, burgeoning urban populations, and evolving environmental concerns, the need for innovative approaches to water resource management is paramount. This chapter explores future directions and innovations in computational water management. We begin by discussing technological innovations in water management, the role of emerging technologies such as artificial intelligence (AI), machine learning (ML), and the Internet of Things (IoT) in enhancing the efficiency and accuracy of water resource forecasting, distribution, and quality assessment. These technologies offer the potential to transform traditional water management systems into smarter data-driven solutions capable of optimizing resource allocation and reducing wastage. Furthermore, the chapter also explores advanced modeling techniques combined with high-performance computing for enhanced flood and drought forecasting and ecological impact assessment in complex hydrological systems. In addition to technological advancements, we examine the increasing importance of sustainable water management practices. The role of decentralized and nature-based solutions, such as green infrastructure and rainwater harvesting, is discussed, highlighting their potential to mitigate water scarcity and reduce the impact of urbanization on water resources. The chapter highlights the importance of cross-disciplinary collaboration among researchers, policymakers, and stakeholders to create innovative and holistic water management solutions. It stresses the importance of public engagement, education, and policy reform to ensure the long-term sustainability of water resources. In this new era of computational water management, the chapter

S. Bansal (✉)
Department of Computer Science and Engineering, Gurukul Kangri (DU), Haridwar 249404, Uttarakhand, India
e-mail: pup.sumit@gmail.com

V. Sindhi
Department of Computer Science, Punjabi University, Patiala 147002, Punjab, India

B. S. Singla
Department of Computer Science, Punjabi University, Baba Jogipeer Neighbourhood Campus, Ralla 151510, Punjab, India

A. K. Yadav et al. (eds.), *Integrated Management of Water Resources in India: A Computational Approach*, Water Science and Technology Library 129,
https://doi.org/10.1007/978-3-031-62079-9_27

provides valuable information on potential directions for research, policy development, and technological advancements, all critical for the sustainable and equitable management of this precious resource, water.

Keywords Internet of things · Artificial intelligence · Energy efficiency · Sustainable practices · Energy monitoring · Predictive analytics

27.1 Introduction to Computational Water Management

The chapter begins with an exploration of the contemporary landscape of challenges and opportunities, positioning the field of computational water management at the nexus of these dynamic forces. Against the backdrop of escalating concerns stemming from climate change, burgeoning urban populations, and evolving environmental issues, the imperative for innovative approaches to water resource management is underscored (Abdelfattah and El-Shamy 2024). The narrative unfolds by delving into the pivotal role that emerging technologies—specifically artificial intelligence (AI), machine learning (ML), and the Internet of Things (IoT) in revolutionizing water management paradigms (Khalil et al. 2021)(Patel et al., 2023). The discussion outlines the transformative potential of these technologies to improve the efficiency and accuracy of crucial aspects of water resource management, including forecasting, distribution, and quality assessment. The chapter contends that the integration of AI, ML, and IoT has the power to elevate traditional water management systems into smarter data-driven solutions capable of optimizing resource allocation and reducing waste (Sun and Scanlon 2019; Rane et al. 2023). Furthermore, the exploration extends to advanced modeling techniques synergized with high-performance computing, focusing on their application in improving flood and drought forecasting as well as ecological impact assessment within intricate hydrological systems (Shirin and Yadav 2014). The chapter underscores the critical importance of staying at the forefront of technological advancements to address the complex challenges posed by water resource management. As the discourse progresses, attention is focused towards sustainable water management practices. The chapter examines the burgeoning importance of adopting decentralized and nature-based solutions, such as green infrastructure and rainwater harvesting (Langergraber et al. 2021; Kulwant et al. 2023). These alternative approaches are highlighted for their potential to mitigate challenges from water scarcity and ameliorate the adverse impacts of urbanization on water resources.

Furthermore, the chapter emphasizes the necessity of cross-disciplinary collaboration among researchers, policymakers, and stakeholders. It elucidates the role of public participation, education, and policy reform in ensuring the long-term sustainability of water resources (Pahl-Wostl et al. 2007). The collaborative philosophy is positioned as critical to crafting innovative and holistic water management solutions that transcend individual disciplines and perspectives (Shirin et al. 2019). Conclusively, the chapter sets the stage for an exploration of potential future directions in computational water management. It anticipates offering valuable information on future areas of research, policy development, and technological advancements that are deemed critical to the sustainable and equitable management of the precious resource, water.

27.2 Technological Innovations in Water Management

This section provides an in-depth exploration of how these technologies contribute to enhancing the efficiency and accuracy of various aspects of water resource management, offering a glimpse into the future of the field.

27.2.1 Role of Emerging Technologies (AI, ML, IoT)

This section provides a detailed discussion of the role emerging technologies.

27.2.1.1 AI in Water Resource Management

The chapter examines how artificial intelligence is revolutionizing water resource forecasting, distribution, and quality assessment. AI's capacity to analyze vast datasets enables the development of predictive models for water availability and demand (Nova 2023)(Shirin and Yadav 2021). This section highlights specific applications, such as predictive analytics to optimize resource allocation and reducing wastage.

27.2.1.2 Machine Learning for Water Systems Optimization

Machine learning, with its ability to learn from data patterns, is explored for its role in optimizing water management systems (Drogkoula et al. 2023). The chapter discusses ML applications to improve the accuracy of forecasting models, understanding consumption patterns, and refining decision-making processes related to water distribution. Figure 27.1 illustrates how technologies are taking on new and evolving roles in various fields.

27.2.1.3 IoT-Enabled Water Monitoring and Assessment

The Internet of Things is investigated for its role in real-time water monitoring and quality assessment. IoT devices and sensors play a crucial role in collecting data from various points within water systems, providing continuous insights into water

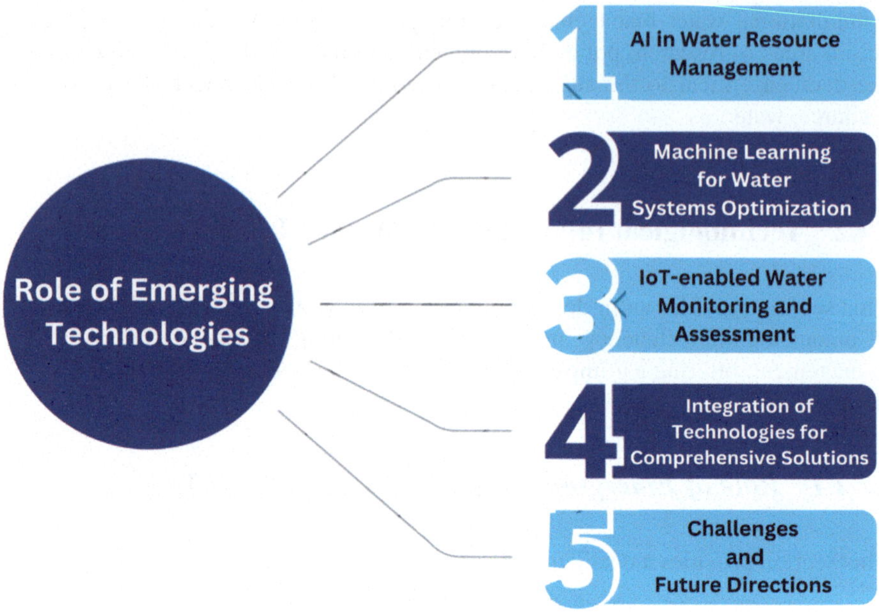

Fig. 27.1 C: Diverse Functions of Cutting-edge Technologies

quality, usage, and potential issues (Glasgow et al. 2004; Kumar et al. 2023). This section highlights the integration of IoT to create smart water management systems.

27.2.1.4 Integration of Technologies for Comprehensive Solutions

An important aspect of the chapter is the examination of how these technologies synergize to create comprehensive solutions. The integration of AI, ML, and IoT is explored for its potential to provide holistic insights into water systems, allowing for better decision making and resource optimization.

27.2.1.5 Challenges and Future Directions

Recognizing the significant contributions of these technologies, the chapter also discusses challenges such as data privacy, security, and the need for robust infrastructure. In conclusion, it offers information on potential future directions, considering ongoing technological advances and their evolving role in shaping the landscape of water resource management (Yadav et al. 2014).

27.2.2 Transformation of Traditional Systems into Data-Driven Solutions

This section provides an in-depth exploration of the evolution of conventional water management systems into advanced data-driven solutions. This transformative journey is marked by the integration of modern technologies, emphasizing the use of data to optimize decision-making processes and improve overall efficiency in water resource management.

27.2.2.1 Overview of Traditional Water Management Systems

It begins by establishing a baseline understanding of conventional water management practices. It delves into the historical context of traditional systems, highlighting their limitations and challenges in addressing the complexities of contemporary water resource management.

27.2.2.2 Emergence of Data-Driven Paradigm

Explore the reasons behind this transition, emphasizing the increasing need for more accurate, timely, and actionable information in managing water resources.

27.2.2.3 Integration of Data Technologies

The heart of the chapter lies in detailing the integration of data technologies that drive this transformation. It explores how advanced data analytics, cloud computing, and sensor technologies are incorporated into existing systems, paving the way for a more dynamic and responsive approach to water management (Bibri 2018).

27.2.2.4 Case Studies and Real-World Applications

To provide tangible insight, the chapter incorporates case studies and real-world applications. These examples showcase successful implementations of data-driven solutions in diverse water management scenarios, illustrating the tangible benefits and positive outcomes achieved.

27.2.2.5 Challenges and Considerations

Recognizing the transformative potential, the chapter also addresses the challenges and considerations associated with this shift. It discusses issues such as data privacy, cybersecurity, and the need for skilled personnel to manage and interpret the flow of data.

27.2.2.6 Future Implications and Sustainability

It concludes by discussing the future implications of this transformation and its sustainability. Explore how ongoing advancements in technology and evolving data-driven strategies will continue to shape the landscape of water resource management. It offers a comprehensive examination of the process through which traditional water management evolves into a more responsive, adaptive, and efficient system through the integration of data-centric technologies.

27.2.3 Optimization of Resource Allocation and Waste Reduction

In the context of water resource management, this concept involves the deployment of advanced technologies and methodologies to ensure that water resources are allocated in the most effective and sustainable manner while simultaneously minimizing unnecessary or inefficient usage (Shang et al. 2016). Figure 27.2 demonstrates the rationalization of resource allocation and the reduction of waste through optimization.

27.2.3.1 Efficient Resource Allocation

This involves using tools such as Artificial Intelligence (AI) and Machine Learning (ML) to analyze historical data, consumption patterns, and environmental variables (Cioffi et al. 2020). By understanding these factors, the system can intelligently allocate water resources, ensuring that each area receives an appropriate amount according to its needs. This optimization process helps prevent overuse in some regions while addressing demand in others.

27.2.3.2 Predictive Analytics for Water Demand

The optimization process is further refined through the use of predictive analytics, a component of AI. Predictive models can anticipate fluctuations in water demand based on various factors, such as climate conditions, population growth, and historical usage patterns. This foresight allows for proactive resource allocation, avoiding shortages, and ensuring a more balanced distribution.

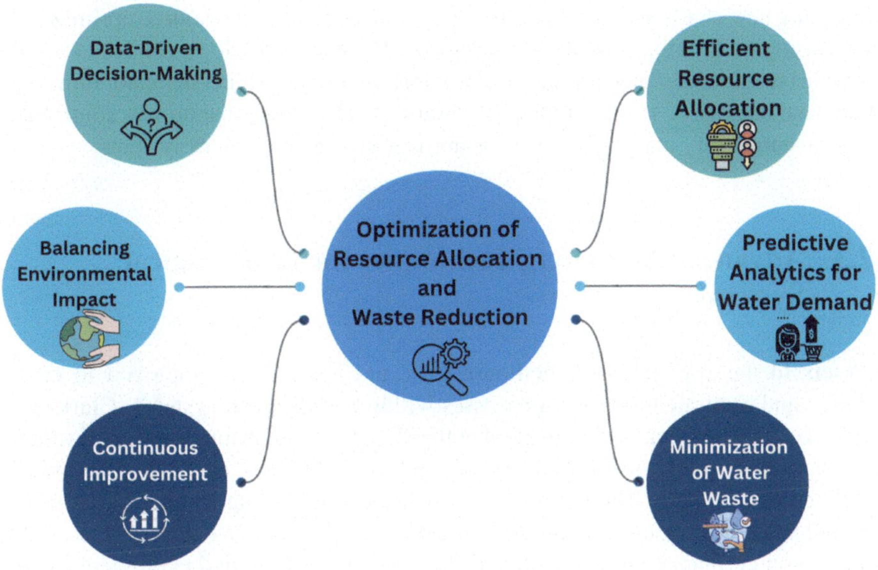

Fig. 27.2 C: Optimization of resource allocation and waste reduction

27.2.3.3 Minimization of Water Waste

Waste reduction is a crucial aspect of resource optimization. Advanced technologies, including IoT sensors and smart infrastructure, can detect and address leaks, inefficiencies, or unauthorized water use in real time. By quickly identifying and resolving issues, the system minimizes waste of water resources, contributing to overall conservation efforts.

27.2.3.4 Data-Driven Decision-Making

Optimization is based on data-driven decision-making processes. Through continuous monitoring and analysis of relevant data, decision makers can identify trends, inefficiencies, and opportunities for improvement. This information enables them to make informed decisions about resource allocation strategies, ensuring a dynamic and responsive water management system.

27.2.3.5 Continuous Improvement

Optimization is an ongoing process that involves constant refinement based on feedback, changing conditions, and evolving technologies. Continuous improvement ensures that the water management system remains adaptive and resilient, capable of addressing new challenges and incorporating advancements in resource optimization methodologies. Finally, it involves leveraging advanced technologies and data-driven strategies to allocate water resources efficiently, predict and meet demand, minimize waste, and balance environmental considerations. This holistic approach contributes to the sustainability and resilience of water resource management systems.

27.3 Advanced Modelling Techniques for Hydrological Systems

It refers to the sophisticated methodologies and approaches employed to create detailed and accurate models of processes within hydrological systems (Clark et al. 2015a, b). Hydrological systems involve the circulation, distribution and movement of water within the atmosphere of the Earth, on its surface, and below the ground (Clark et al. 2015a, b). The application of advanced modeling techniques enhances our understanding of these complex systems, aiding in better decision-making for water resource management, flood and drought forecasting, and ecological impact assessments. Here's a breakdown of key components within this concept:

27.3.1 High-Performance Computing (HPC)

Utilizing high-performance computing resources allows for the simulation of intricate hydrological processes with increased precision and efficiency. Complex calculations and simulations, which would be computationally intensive for conventional systems, can be performed rapidly, enabling more detailed and realistic hydrological models.

27.3.2 Flood Forecasting

Advanced modeling techniques facilitate the development of sophisticated flood forecasting models. These models integrate data from various sources, including weather patterns, topography, land use, and historical flood events. The use of high-performance computing enables real-time or near-real-time simulations, providing timely and accurate predictions of potential flooding events.

27.3.3 Drought Forecasting

Similar to flood forecasting, advanced models are employed for drought prediction. These models consider factors such as precipitation patterns, soil moisture levels, and vegetation health to assess the likelihood and severity of drought conditions (Belal et al. 2014). By incorporating various parameters, these models contribute to more effective water resource planning and management during dry periods.

27.3.4 Sensitivity Analysis and Uncertainty Assessment

Advanced models often incorporate sensitivity analysis to identify the parameters that most significantly influence the outcome of the model. Additionally, uncertainty assessments help quantify the reliability of model predictions, acknowledging the inherent complexities and variability within hydrological systems. These techniques contribute to more accurate predictions, better management of water resources, and a comprehensive assessment of the ecological impacts associated with hydrological changes.

27.4 Sustainable Water Management Practices

The section refers to the adoption of strategies and methods that ensure the responsible use, conservation, and protection of water resources while considering ecological, social, and economic factors (Kalogiannidis et al. 2023; Shirin et al. 2021). The goal is to meet current and future needs for water without compromising the well-being of the environment or depleting water sources. The essential components are outlined below.

- **Sustainable Approach**: The term 'sustainable' implies practices that can be maintained over the long term without causing harm to the environment or depleting resources. In the context of water management, this involves finding a balance between meeting human water needs and preserving the health of ecosystems.
- **Environmental Impact**: Practices take into account the environmental impact of water use. This involves considering the health of aquatic ecosystems, biodiversity, and the overall ecological balance. The purpose of this project is to minimize the negative effects on aquatic habitats, flora, and fauna.
- **Public Engagement and Education**: Sustainable water management practices often involve educating the public about the importance of water conservation, promoting awareness of water-related issues, and encouraging community participation in sustainable water practices. In essence, it is a holistic and forward-thinking approach to using and preserving water resources, taking into account

environmental, social, and economic factors to ensure the long-term health and availability of water for current and future generations.

27.4.1 *Importance of Sustainable Approaches*

It states the significance of adopting methods and strategies that are environmentally, socially and economically sustainable. In various contexts, this can apply to areas such as agriculture, energy, development and, as mentioned in our previous question, water management. Here is a general breakdown of the importance of sustainable approaches:

27.4.1.1 Environmental Conservation

Sustainable approaches prioritize the well-being of the environment. They aim to minimize negative impacts on ecosystems, reduce pollution, and preserve biodiversity (Kumar et al. 2021). By adopting practices that are ecologically sustainable, we help protect natural resources and maintain a healthy planet.

27.4.1.2 Resource Efficiency

Sustainability emphasizes efficient use of resources. Whether it is water, energy, raw materials, or other resources, sustainable approaches aim to optimize usage, reduce waste, and ensure that resources are available for future generations (Shirin et al. 2023).

27.4.1.3 Social Equity

Sustainable approaches often address social considerations, aiming for fairness and social equity. This includes ensuring that benefits and burdens are distributed equitably among different communities, considering social justice in decision-making processes.

27.4.1.4 Innovation and Resilience

Adopting sustainable approaches encourages innovation in technology, business models, and policies. Sustainability challenges often drive the development of new and improved solutions, fostering resilience in the face of evolving environmental and social conditions.

27.4.1.5 Regulatory Compliance

As awareness of environmental and social issues increases, regulatory frameworks are evolving to incorporate sustainability principles. Adopting sustainable approaches helps organizations and individuals comply with regulations and standards. Finally, the importance of sustainable approaches lies in their potential to create a harmonious balance between environmental health, social equity, and economic stability, ensuring that current and future generations can thrive in a resilient and resource-efficient world.

27.4.2 Role of Decentralized and Nature-Based Solutions

It refers to the participation and impact of approaches that distribute responsibility and utilize natural processes to address various challenges (Barredo Arrieta et al. 2020). In the context of water management or environmental solutions, this concept emphasizes the move away from centralized systems and the integration of methods that harness natural elements. Below is an analysis of the essential elements:

27.4.2.1 Decentralized Solutions

Instead of relying on a single, centralized system to manage resources, decentralized solutions distribute responsibilities across various local or smaller-scale entities. For example, in water management, this might involve local communities having more control over their water resources, rather than relying solely on a large, centralized water treatment plant.

27.4.2.2 Reducing Environmental Impact

Decentralized and nature-based solutions often aim to minimize the environmental impact of human activities. For water management, this could involve approaches that mimic or work in harmony with natural water cycles, reducing the need for extensive artificial interventions.

27.4.2.3 Ecosystem Services

By incorporating nature-based solutions, there is an acknowledgment of the multiple benefits that ecosystems provide, such as water purification, habitat for biodiversity, and overall ecological balance. These ecosystem services are often crucial to human well-being and can be harnessed in sustainable management practices. In a nutshell, the role of decentralized and nature-based solutions involves a shift towards more localized, adaptive, and environmentally friendly approaches in addressing challenges, with a focus on sustainability and harmony with natural processes.

27.4.3 Green Infrastructure and Rainwater Harvesting

Green Infrastructure and Rainwater Harvesting are sustainable water management practices that involve utilizing natural systems and technologies to manage water resources efficiently. Here's a clarification for each:

27.4.3.1 Green Infrastructure

Green infrastructure involves the utilization of natural or nature-inspired systems to effectively manage water, mitigate environmental impacts, and offer diverse ecosystem services. Examples of green infrastructure include features like green roofs, permeable pavements, urban forests, wetlands, and vegetated swales. These characteristics emulate natural processes, facilitating the absorption of rainwater, minimizing runoff, and fostering overall environmental sustainability. The benefits of green infrastructure extend to the reduction for flooding, enhancement of water quality through pollutant filtration, the promotion of biodiversity, and the overall improvement of well-being in both urban and natural environments.

27.4.3.2 Rainwater Harvesting

Rainwater harvesting involves the collection and storage of rainwater for various nonpotable purposes, such as landscape irrigation, toilet flushing, and industrial processes. This system comprises key components such as collection surfaces (roofs), gutters, downspouts, and storage tanks or cisterns. Rainwater collected can undergo filtration and treated before utilization. The practice of rainwater harvesting is instrumental in alleviating the demand for conventional water sources, particularly in regions grappling with water scarcity. Additionally, it addresses stormwater runoff issues, contributes to groundwater recharge, and establishes a decentralized and sustainable water supply.

27.5 Cross-Disciplinary Collaboration for Holistic Solutions

It refers to the integration of expertise and perspectives from multiple fields of study or disciplines to develop comprehensive and well-rounded solutions to complex challenges. The breakdown of this concept is as follows:

27.5.1 Collaborative Efforts Among Researchers, Policymakers, and Stakeholders

It entails the coordinated and cooperative actions taken by individuals and groups involved in research, policy development, and those with a vested interest or stake in a particular issue or field. The finer intricacies of this concept are provided below.

27.5.1.1 Collaborative Efforts

Collaboration involves collective efforts aimed at a shared goal or objective. In the realms of research, policymaking, and stakeholder engagement, it signifies joint endeavors to tackle intricate issues.

27.5.1.2 Key Participants

Researchers play a pivotal role in advancing knowledge through systematic investigation and study. They bring scientific or scholarly expertise to the table and conduct studies that inform evidence-based decision making. Policymakers, on the other hand, are individuals tasked with formulating, implementing, and evaluating policies. This group includes government officials, legislators, and decision makers who hold the authority to shape and enact policies. Additionally, stakeholders, including individuals, communities, businesses, advocacy groups, and organizations, have a vested interest or stake in specific issues. This diverse group directly influences or is directly affected by policies and research outcomes, making their perspectives crucial in shaping informed decisions.

27.5.1.3 Importance of Collaboration

Informed Decision Making is a pivotal outcome of collaboration, as it grants policymakers access to cutting-edge research findings, enabling them to make decisions grounded in evidence and expert insights (Lampoltshammer et al. 2023). This synergy extends to real-world relevance, where researchers gain valuable insights into actual societal challenges by engaging with policymakers and stakeholders. This interaction ensures that research remains pertinent and applicable to the pressing needs of society. Additionally, collaboration provides crucial Implementation Support, as stakeholders contribute valuable perspectives on the practical implications of policies and research findings, thereby enhancing the prospects of effective implementation.

27.5.2 Public Engagement, Education, and Policy Reform

It encompasses a holistic strategy designed to engage the public in addressing social issues, raise awareness through education, and advocate for policy changes to achieve positive outcomes. The details of each element are discussed below.

27.5.2.1 Public Engagement

The participation of the public plays a pivotal role in decision-making processes, discussions, and actions related to diverse issues. This involvement underscores the importance of active participation from the general public in the shaping of policies and initiatives. Furthermore, community participation is promoted, urging communities to contribute their unique perspectives, experiences, and ideas. This inclusive approach not only enhances the richness of decision making, but also fosters a sense of ownership and shared responsibility within communities, creating a more collaborative and holistic framework for addressing various challenges.

27.5.2.2 Policy Reform

Changes in policies involve the examination, adjustment, or establishment of institutional, local, or governmental policies to effectively address identified issues. This dynamic process is essential to adapt regulations to the evolving needs and values of the community. Gleichzeitig, advocacy for change emerges through public participation and education, fostering a collective voice that advocates for policy adjustments. This advocacy aims to ensure that policies not only responsive to current challenges, but also align with the broader interests and values of the community, promoting a more inclusive and effective governance approach.

27.5.2.3 Key Aspects

Public engagement goes beyond the mere dissemination of information; it involves a two-way communication process, actively involving both the sharing of information and listening to public concerns and feedback. Education plays a pivotal role in empowerment, offering individuals the knowledge and tools necessary for active participation in decision-making processes. Furthermore, advocating for systemic changes is integral to policy reform, with the aim of broader impacts that address societal issues at their core. Together, these elements contribute to a more inclusive and effective approach to addressing complex challenges in society.

27.5.2.4 Benefits

Informed decision making is pivotal as an informed and engaged public plays a crucial role in enhancing decision-making processes, ensuring that policies align effectively with community needs. This, in turn, contributes to long-term impact, where education and policy reform become catalysts for enduring societal changes. This transformative process nurtures a more informed, proactive, and responsive society over time. Additionally, fostering community ownership is integral, as an engaged and educated public develops a sense of ownership and responsibility for societal issues. This sense of ownership leads to the emergence of sustainable, community-driven solutions, further solidifying the connection between informed civic participation and positive societal outcomes.

27.5.3 Ensuring Long-Term Sustainability of Water Resources

It involves implementing strategies and practices that guarantee the continued availability and quality of water for current and future generations while considering ecological, social, and economic aspects. Here's an elaboration:

27.5.3.1 Conservation and Efficient Use

Efforts to address water conservation are being actively pursued through the implementation of measures aimed at reducing unnecessary water consumption. This involves the promotion and adoption of efficient water utilization practices in various sectors, including agriculture, industry, and households. Emphasizing the dual approach of using advanced technologies and raising awareness through programs, initiatives are underway to optimize water use and minimize wastage. By encouraging responsible water practices at both individual and systemic levels, the goal is to foster a sustainable approach to water management and ensure the judicious use of this essential resource.

27.5.3.2 Protection of Ecosystems

Promoting the conservation and rehabilitation of natural ecosystems, including wetlands and forests, is crucial to sustaining water quality and effectively managing water flow. This involves efforts to preserve these ecosystems and restore any areas that have been degraded. Equally important is the commitment to abstain from activities that pose a threat to aquatic ecosystems, biodiversity, and habitats associated with water. By prioritizing these principles, we contribute to the preservation of a balanced and resilient environment, fostering the well-being of both terrestrial and aquatic ecosystems.

27.5.3.3 Incentives for Sustainable Practices

Encourage the adoption of sustainable water practices involves implementing incentives for both businesses and individuals. This can include the promotion and support of water-efficient technologies and practices. By offering incentives, such as tax benefits or subsidies, for the implementation of water saving measures, organizations and individuals are motivated to invest in and adopt environmentally friendly approaches to water use. This approach aims to foster a collective commitment to sustainable water management, ensuring the responsible and efficient use of this vital resource for the benefit of both present and future generations. By implementing these measures, the goal is to create a comprehensive and adaptive approach to managing water resources that ensures their long-term sustainability, considering the needs of both human populations and the ecosystems that depend on these vital resources.

27.6 Future Directions in Computational Water Management

Standing at the confluence of technological innovation, environmental challenges, and the imperative for sustainable resource management, the landscape of computational water management is poised for transformative advancements. The section explores the future directions that will shape the trajectory of how we model, analyze and manage water resources.

27.6.1 Integration of Artificial Intelligence and Machine Learning

Leveraging the capabilities of artificial intelligence (AI) and machine learning (ML) algorithms is paramount to advance precise and adaptive water resource forecasting.

By incorporating AI-driven decision support systems, real-time responses to dynamic hydrological conditions become feasible. This integration of cutting-edge technologies not only enhances the accuracy of predictions but also enables proactive decision-making, contributing to improved water resource management in response to ever-changing environmental factors.

27.6.2 Internet of Things (IoT) for Real-Time Data Streams

The initiative involves the widespread deployment of Internet of Things (IoT) devices to establish a comprehensive network of sensors dedicated to real-time monitoring of critical water parameters, including quality, flow rates, and environmental conditions (Paepae et al. 2021). At the same time, efforts are directed towards developing IoT-driven systems capable of facilitating rapid responses to evolving water challenges. These systems take advantage of continuous data acquisition, allowing for timely insights and adaptive measures to address emerging issues in the realm of water management and environmental sustainability.

27.6.3 Blockchain Technology for Water Governance

This initiative involves a comprehensive exploration of how to use blockchain technology to increase transparency, security, and efficiency in the realm of water governance and allocation. The focus is on developing decentralized systems that are inherently tamper-proof, specifically designed to track and record water transactions and usage (Bhushan et al. 2020). By integrating blockchain, a distributed ledger technology, into water management practices, the aim is to establish a robust framework that ensures trust, accountability, and streamlined processes for monitoring and allocating water resources.

27.6.4 Education and Capacity Building

The importance of education and skill development in the use of computational tools for effective water management is paramount. This involves prioritizing training initiatives aimed at professionals, researchers, and policy makers. By establishing comprehensive training programs, we can equip the next generation with the necessary skills to navigate the dynamic and evolving landscape of computational water management. This strategic focus on education and capacity building ensures a competent and forward-thinking approach to address the complex challenges in water resource management. As we venture into these future directions, the convergence of technological innovation, interdisciplinary collaboration, and a commitment to sustainability will define computational water management. This journey holds the promise of not only addressing current water challenges, but also shaping a resilient and adaptive future of water for our planet.

27.7 Conclusion

In conclusion, the integration of emerging technologies, such as the Internet of Things (IoT), Artificial Intelligence (AI), and advanced sensing capabilities, is a pivotal force in revolutionizing water management practices. The chapter presented emphasize the transformative impact of these technologies on various aspects of water resource management. Exploring IoT sensors and networks highlights real-time monitoring capabilities that enable decision makers to respond promptly to changes in water quality and distribution. Artificial intelligence and machine learning bring predictive analytics to the forefront, enhancing our ability to forecast water usage patterns, identify leaks, and optimize distribution networks. Data analytics plays a crucial role in understanding behavioral patterns related to water consumption, driving targeted conservation efforts. Cloud-based water management platforms offer integrated solutions for collaborative decision-making among stakeholders. Advanced water recycling and reuse technologies ensure efficient utilization of water resources. Innovative applications of blockchain technology bring transparency to water transactions, while gamification engages and educates the public on water conservation practices. Collectively, these technological advancements form a dynamic toolkit that not only addresses current water challenges but also lays the foundation for sustainable, resilient, and equitable water management practices.

As we navigate the complexities of water scarcity, climate change, and growing urbanization, the chapters presented underscore the importance of harnessing technological innovations. The holistic and interdisciplinary nature of these advances places us at the forefront of a new era in water management, one characterized by efficiency, adaptability, and a deep commitment to safeguarding this precious resource for current and future generations.

Acknowledgements We extend our heartfelt appreciation to all those who contributed to the realization of this book chapter. Our sincere gratitude goes to Dr. Bhim Sain Singla for his guidance, support, and invaluable insights throughout the writing process. We are also grateful to our institutes for providing the necessary resources and environment for our research endeavors. Finally, we would like to express our deepest appreciation to our families for their unwavering love and understanding during this journey.

References

Abdelfattah I, El-Shamy AM (2024) Review on the escalating imperative of zero liquid discharge (ZLD) technology for sustainable water management and environmental resilience. J Environ Manag 351(Nov 2023):119614. https://doi.org/10.1016/j.jenvman.2023.119614

Barredo Arrieta A, Díaz-Rodríguez N, Del Ser J, Bennetot A, Tabik S, Barbado A, Garcia S, Gil-Lopez S, Molina D, Benjamins R, Chatila R, Herrera F (2020) Explainable artificial intelligence (XAI): concepts, taxonomies, opportunities and challenges toward responsible AI. Inf Fusion 58(Oct 2019):82–115. https://doi.org/10.1016/j.inffus.2019.12.012

Belal AA, El-Ramady HR, Mohamed ES, Saleh AM (2014) Drought risk assessment using remote sensing and GIS techniques. Arab J Geosci 7(1):35–53. https://doi.org/10.1007/s12517-012-0707-2

Bhushan B, Khamparia A, Sagayam KM, Sharma SK, Ahad MA, Debnath NC (2020) Blockchain for smart cities: a review of architectures, integration trends and future research directions. Sustain Cities Soc 61(June):102360. https://doi.org/10.1016/j.scs.2020.102360

Bibri, S. E. (2018). The IoT for smart sustainable cities of the future: an analytical framework for sensor-based big data applications for environmental sustainability. Sustain Cities Soc 38(Oct 2017):230–253. https://doi.org/10.1016/j.scs.2017.12.034

Cioffi R, Travaglioni M, Piscitelli G, Petrillo A, De Felice F (2020) Artificial intelligence and machine learning applications in smart production: progress, trends, and directions. Sustainability (Switzerland) 12(2). https://doi.org/10.3390/su12020492

Clark MP, Fan Y, Lawrence DM, Adam JC, Bolster D, Gochis DJ, Hooper RP, Kumar M, Leung LR, Mackay DS, Maxwell RM, Shen C, Swenson SC, Zeng X (2015a) Improving the representation of hydrologic processes in earth system models. Water Resour Res 51(8):5929–5956. https://doi.org/10.1002/2015WR017096

Clark MP, Nijssen B, Lundquist JD, Kavetski D, Rupp DE, Woods RA, Freer JE, Gutmann ED, Wood AW, Brekke LD, Arnold JR, Gochis DJ, Rasmussen RM (2015b) A unified approach for process-based hydrologic modeling: 1. Modeling Concept. Water Resour Res 51(4):2498–2514. https://doi.org/10.1002/2015WR017198

Drogkoula M, Kokkinos K, Samaras N (2023) A comprehensive survey of machine learning methodologies with emphasis in water resources management. Appl Sci 13(22):12147. https://doi.org/10.3390/app132212147

Glasgow HB, Burkholder JAM, Reed RE, Lewitus AJ, Kleinman JE (2004) Real-time remote monitoring of water quality: a review of current applications, and advancements in sensor, telemetry, and computing technologies. J Exp Mar Biol Ecol 300(1–2):409–448. https://doi.org/10.1016/j.jembe.2004.02.022

Kalogiannidis S, Kalfas D, Giannarakis G, Paschalidou M (2023) Integration of water resources management strategies in land use planning towards environmental conservation. Sustainability 15(21):15242. https://doi.org/10.3390/su152115242

Khalil RA, Saeed N, Masood M, Fard YM, Alouini MS, Al-Naffouri TY (2021) Deep learning in the industrial internet of things: potentials, challenges, and emerging applications. IEEE Internet Things J 8(14):11016–11040. https://doi.org/10.1109/JIOT.2021.3051414

Kulwant M, Patel D, Shirin S, Sharma SN, Yadav AK (2023) Future frameworks for fluoride and algorithms for environmental system, pp 343–364. https://doi.org/10.1007/978-3-031-38845-3_19

Kumar A, Chava R, Gupta S, Shirin S, Jamal A, Yadav AK (2023) Environmental and health effects of fluoride contamination and treatment of wastewater using various technologies, pp 323–341. https://doi.org/10.1007/978-3-031-38845-3_18

Kumar R, Verma A, Shome A, Sinha R, Sinha S, Jha PK, Kumar R, Kumar P, Shubham Das S, Sharma P, Prasad PVV (2021) Impacts of plastic pollution on ecosystem services, sustainable development goals, and need to focus on circular economy and policy interventions. In: Sustainability (Switzerland), vol 13, issue 17, pp 1–40. https://doi.org/10.3390/su13179963

Lampoltshammer TJ, Maurer H, Pulda N, Klimek P, Hurt J, Rosenbichler U (2023) Challenges toward evidence-based policymaking using agent-based modeling for federal sports grants: a self-reflection from a transdisciplinary project. Sustainability (Switzerland) 15(4). https://doi.org/10.3390/su15042853

Langergraber G, Castellar JAC, Pucher B, Baganz GFM, Milosevic D, Andreucci MB, Kearney K, Pineda-Martos R, Atanasova N (2021) A framework for addressing circularity challenges in cities with nature-based solutions. Water (Switzerland) 13(17):1–31. https://doi.org/10.3390/w13172355

Nova K (2023) AI-enabled water management systems: an analysis of system components and interdependencies for water conservation. Eigenpub Rev Sci Technol 7(1):105–124. https://studies.eigenpub.com/index.php/erst/article/view/12

Paepae T, Bokoro PN, Kyamakya K (2021) From fully physical to virtual sensing for water quality assessment: a comprehensive review of the relevant state-of-the-art. Sensors 21(21). https://doi.org/10.3390/s21216971

Pahl-Wostl C, Craps M, Dewulf A, Mostert E, Tabara D, Taillieu T (2007) Social learning and water resources management. Ecol Soc 12(2). https://doi.org/10.5751/ES-02037-120205

Patel D, Kulwant M, Shirin S, Varshney R, Pandey G, Yadav AK (2023) Fluoride removal from aqueous solution using iron-based materials: preparation, characterization, and applications, pp 71–92. https://doi.org/10.1007/978-3-031-38845-3_4

Rane N, Choudhary S, Rane J (2023) Leading-edge artificial intelligence (AI), machine learning (ML), blockchain, and internet of things (IoT) technologies for enhanced wastewater treatment systems. SSRN Electron J. https://doi.org/10.2139/ssrn.4641557

Shang Y, Wang J, Liu J, Jiang D, Zhai J, Jiang S (2016) Suitability analysis of China's energy development strategy in the context of water resource management. Energy 96:286–293. https://doi.org/10.1016/j.energy.2015.12.079

Shirin S, Jamal A, Emmanouil C, Singh VP, Yadav AK (2023) Assessment and characterization of waste material used as backfilling in an abandoned mine. Int J Coal Prep Util 43(8):1402–1410. https://doi.org/10.1080/19392699.2022.2118259

Shirin S, Jamal A, Emmanouil C, Yadav AK (2021) Assessment of characteristics of acid mine drainage treated with fly ash. Appl Sci (Switzerland) 11(9). https://doi.org/10.3390/app11093910

Shirin S, Jamal A, Ranjan P, Yadav AK (2019) Study on assessment of slope stability and mixed disposal of overburden in voids of Singrauli Coalfield. Environ Qual Manage 28(3):131–139. https://doi.org/10.1002/tqem.21616

Shirin S, Yadav A (2014) Physico chemical analysis of municipal wastewater discharge in Ganga River, Haridwar District of Uttarakhand, India. Curr World Environ 9(2):536–543. https://doi.org/10.12944/cwe.9.2.39

Shirin S, Yadav AK (2021) Assessment and suitability for the dumping of flyash in an abandoned mine. Environ Challenges 4(March):100111. https://doi.org/10.1016/j.envc.2021.100111

Sun AY, Scanlon BR (2019) How can big data and machine learning benefit environment and water management: a survey of methods, applications, and future directions. Environ Res Lett 14(7). https://doi.org/10.1088/1748-9326/ab1b7d

Yadav AK, Sahoo SK, Mahapatra S, Kumar AV, Pandey G, Lenka P, Tripathi RM (2014) Concentrations of uranium in drinking water and cumulative, age-dependent radiation doses in four districts of Uttar Pradesh, India. Toxicol Environ Chem 96(2):192–200. https://doi.org/10.1080/02772248.2014.934247

Chapter 28
Enhancing Water Conservation and Management in India: A Holistic Approach

Ankit Kumar and Akhilesh Kumar Yadav

Abstract India, as a subcontinent, exhibits diverse climate patterns, physiography, and agro-ecological zones. This natural variability significantly influences the distribution of water resources and rainfall across the country. However, the growing demands of agriculture, industry, and services have led to a concerning decline in both surface water and groundwater availability. Unfortunately, water management and conservation efforts in India have often been fragmented and irrational, driven by individual decisions or local democratic bodies. To address this challenge, the country is revisiting its National Water Policy and exploring innovative solutions, including artificial recharge of aquifers using natural methods. In this context, it becomes imperative to adopt a comprehensive and coordinated approach to decentralize water conservation and management initiatives. In this chapter, we have tried to investigate. Drawing inspiration from historical conservation practices, we can reimagine sustainable strategies for India's natural resources. Additionally, understanding the impact of behavioral changes related to water use is crucial. By synergizing surface water, groundwater, and energy, our objective is to improve long-term water efficiency and accessibility.

A. Kumar (✉)
Lohum Cleantech Pvt. Ltd., Kasna Industrial Area, D-184, Site V, Greater Noida 201306, India
e-mail: ankit.rs.che14@iitbhu.ac.in

Department of Chemical Engineering & Technology, Indian Institute of Technology (Banaras Hindu University), Varanasi 221005, India

A. K. Yadav
Department of Environmental Engineering and Management, Chaoyang University of Technology, Taichung 413310, Taiwan

A. K. Yadav (✉)
Department of Mining Engineering, Indian Institute of Technology (Banaras Hindu University), Varanasi 221005, India
e-mail: yadavbasti@gmail.com

Environmental Science and Engineering Department, Indian Institute of Technology Bombay, Mumbai 400076, India

A. K. Yadav et al. (eds.), *Integrated Management of Water Resources in India: A Computational Approach*, Water Science and Technology Library 129,
https://doi.org/10.1007/978-3-031-62079-9_28

Keywords Groundwater resource · Climate change · Groundwater fluctuation · Water conservation · Management · Policy

28.1 Introduction

Water, being a fundamental element of our surroundings, holds immense importance. The existence of life on Earth is intricately connected to the accessibility of water. It is an essential requirement for human life, playing a crucial role in reducing poverty, hunger, and disease. Moreover, the significance of its economic utilization is emphasized, highlighting its overall value (Chinnaiah 2015).

India's water management system and practices have a long and complex history, closely intertwined with its agricultural and cultural traditions. Throughout the course of many centuries, indigenous and community-based systems have played a crucial role in meeting fundamental requirements such as providing drinking water, irrigation, and household resources. Specifically, these region-specific methods flourished despite the contrasting weather conditions and geographical differences. Step wells, tanks, and complex canal networks were innovative solutions that helped maintain agricultural productivity, especially in the face of unpredictable monsoons (Shankari and Shah 1993). Nevertheless, a crucial transformation took place throughout the period of colonial governance. The British government implemented extensive barrages and canals, which resulted in significant changes to the water management environment. Despite their scalability, these systems were deficient in terms of community engagement. Local communities had minimal control on the operation and maintenance of decisions, since they were centralized (Chitale 1991). The various physiography, climate, and agro-ecology of India result in variances in rainfall and water availability within the nation. Nevertheless, the growth of agriculture, industry, and services in the twenty-first century has resulted in a fast depletion of both surface water and groundwater supplies. In order to guarantee sustainable water efficiency and availability, it is essential to distribute conservation efforts and use time-tested strategies. As a result, there has been a significant change in the approach to incorporating state-of-the-art technologies into water management methods. This transition aims to integrate timeless knowledge with contemporary solutions, effectively bridging the divide between traditional practices and innovative approaches (Pani et al. 2021). In India, the irrigation systems for rapid growth that rely on groundwater has resulted in a significant increase in the extraction of groundwater, ultimately leading to its excessive utilization. Therefore, the consequences of climate change provide additional significant dangers to this valuable asset (Wang et al. 2013). Hence, it is essential to comprehend the dynamics and extent of the groundwater system in the face of potential climate changes and other modifications in order to develop a successful water management plan (Aizebeokhai et al. 2017).

This chapter objective is to investigate for effects of changes in behavior about water usage on the total availability of water. Our goal is to accomplish sustainable water management by analyzing the interactions and interdependencies among surface water, groundwater, and energy resources. This aim will be addressed through the discussion of precise policies, planning, and interventions at the program level. In addition, we will examine how climatic variability influences water management practices, taking into account the current disagreement between technology and policy (Pani et al. 2021; Kumar et al. 2022).

28.1.1 Climate Change Situation in India

Although there are significant variations in wind patterns, temperature changes, and rainfall throughout different regions, the climate of India is often categorized as a 'tropical monsoon type'. The primary reason for this categorization is the seasonal change in wind direction, which leads to alternating periods of high humidity and low humidity. The categorization is also influenced by the cyclical patterns of the seasons and the level of humidity or aridity. The Indian climate is characterized by the monsoon, although it is very complicated, as noted by Khullar (2005) and Rajaram (2006).

The monsoon season in India, which lasts from June to mid-September, accounts for nearly three-fourths of the country's total annual rainfall. This phenomenon occurs as a result of the southwest monsoon winds that start from the Indian Ocean and migrate towards the continent. During seasons other than this one, the nation has very little rainfall, resulting in mostly dry conditions in most areas (Lal 2001; Khullar 2005). The monsoon winds are an intricate system that is affected by several causes at the local, regional, and global scales. As a result, these winds have a substantial influence on the climate of India, leading to a high level of diversity and variability. Climate change adds intricacy to this system, resulting in increasingly erratic and diverse weather patterns. The changes in the geographical and temporal patterns of India's climatic variables might potentially create difficulties for the sustainability of the country's enormous population.

Within the framework of global climate change, several studies have closely examined the changing trends of climatic indices in India. Researchers have analysed historical meteorological data to detect patterns in climatic factors at the local, regional, and national scales. In contrast, a few studies have forecasted future climate scenarios by analyzing existing climatic patterns. Hence, India's climate change situation can be examined through two key distinct: what's happening now and what's projected for the future. This technique offers a thorough comprehension of both historical and prospective climatic scenarios.

28.1.2 Climate Change Observation Trend

Rathore et al. (2013) investigated climate change trends in India from 1951 to 2010, focusing on temperature and rainfall patterns. This analysis was part of the Meteorological Monograph Series of Indian Meteorological Department. The researchers used data from 282 temperature recording stations and 1451 rainfall recording stations.

To identify patterns and assess their statistical significance in rainfall and temperature data, the researchers used two nonparametric statistical techniques: Sen's slope estimator and the Mann–Kendall trend test. The analysis of these trends was conducted on a yearly, seasonal, and monthly basis. The study's main discoveries are outlined in Fig. 28.1.

1. The average highest temperature each year exhibited a consistent upward trend, varying from 0.01 °C to 0.06 °C annually, throughout the majority of Indian states. Nevertheless, several states in the northern region exhibited a declining pattern of 0.01 °C–0.02 °C annually.
2. The yearly average lowest temperature had a notable rise (0.01–0.07 °C/year) in many states located along the eastern coastline, northeastern provinces, and semi-arid desert areas. A substantial decline (0.01–0.07 °C/year) was detected in northern India, although no discernible pattern was recorded in other states.
3. The average temperature in most Indian states, especially in the western, central, southern, and eastern coastal regions, has been steadily increasing at a rate of 0.01–0.07 °C/year. Most of the northeast provinces exhibited a noticeable increase in trends. A notable decline trend (0.01 °C/year) was detected in northern India, whilst six other states had no discernible trend.
4. The annual average rainfall exhibited a rising pattern (0.04–14.68 mm/year) mainly in the southern, eastern, northwestern areas, and the riverine plain region of the northeastern states. An overall decline in precipitation was seen in the central area and along the western coast, with a range of 0.05–7.77 mm/year. A substantial reduction was also noted 14.68 mm/year in Himalayan foothill area of northeast India and 7.77 mm/year in Andaman and Nicobar Islands.

28.1.3 Climate Change Projected Trend

The 2013 assessment by the Intergovernmental Panel on climatic Change (IPCC 2013) states that future climatic conditions are already being affected by past warming from human activities and natural variations (Christensen et al. 2013). Continued greenhouse gas release is expected to further increase temperatures, leading to significant changes in the interconnected components of the climate system. The IPCC provides forecasts for changes in average surface temperature and other climate factors by the twenty-first century, compared to 1986–2005. These forecasts are

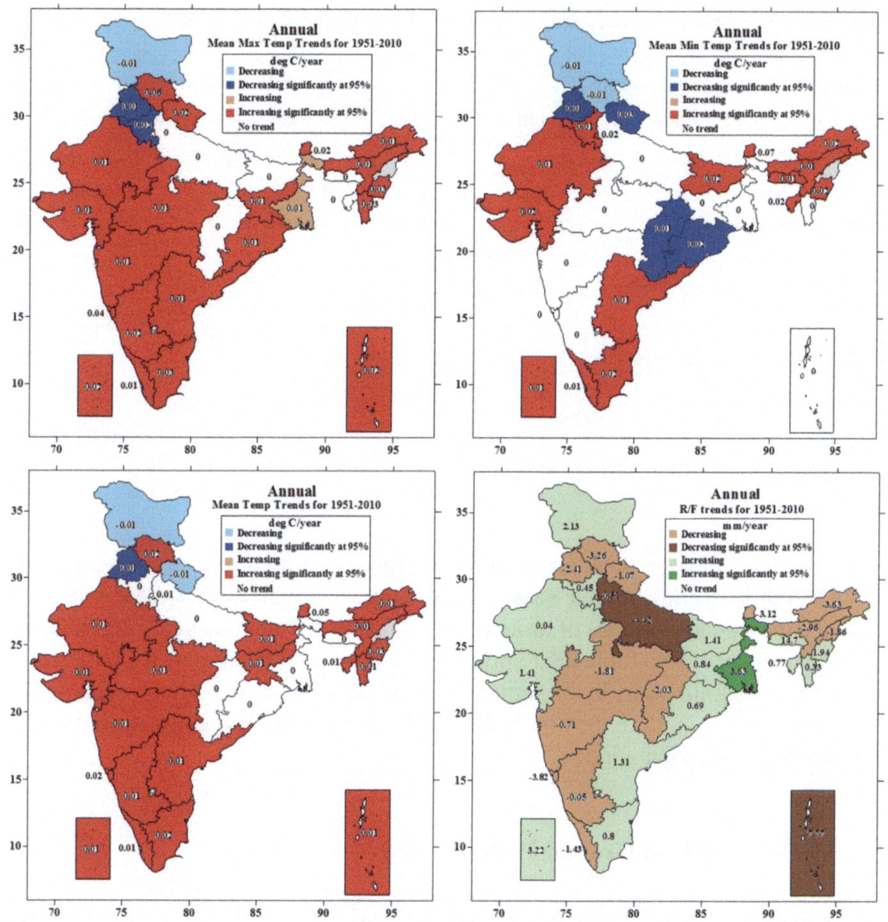

Fig. 28.1 State level trend of annual mean maximum temperature, annual mean minimum temper nature, annual mean temperature, and annual rainfall (1951–2010) (Rathore et al. 2013)

based on various emissions scenarios (RCPs) for greenhouse gases, aerosols, and reactive gases.

The analysis predicts that by end of the twenty-first century (2081–2100), the average global surface temperature is expected to rise between 0.3 and 1.7 °C under RCP 2.6, 1.1–2.6 °C under RCP 4.5, 1.4–3.1 °C under RCP 6.0, and 2.6–4.8 °C under RCP 8.5. The IPCC predicts a significant rise in temperature in India, namely during the winter season.

According to the CMIP5 model, under the RCP 4.5 scenario, the average surface temperature of South Asia is expected to rise between 0.2 and 1.3 °C by 2035, compared to the period between 1986 and 2005. By 2065, the annual mean surface temperature of South Asia is predicted to climb by 0.8–2.5 °C. By the end of the twenty-first century, it is forecast to rise by 1.3–3.5 °C.

Table 28.1 Projected temperature and precipitation by CMIP5 global models in South Asia relative to 1986–2005

Month	Year	Temperature (°C)			Precipitation (%)		
		Minimum	Maximum	Median	Minimum	Maximum	Median
DJF	2035	0.1	1.4	1	− 18	8	− 1
DJF	2065	0.6	2.6	1.8	− 17	4	13
DJF	2100	1.4	3.7	2.3	− 14	28	8
JJA	2035	0.3	1.3	0.7	− 3	9	3
JJA	2065	0.9	2.6	1.3	− 3	33	7
JJA	2100	0.7	3.3	1.7	− 7	37	10
Annual	2035	0.2	1.3	0.8	− 2	7	3
Annual	2065	0.8	2.5	1.6	− 2	26	7
Annual	2100	1.3	3.5	2.1	− 3	27	10

IPCC (2013)

The analysis also predicts a rise in the frequency of high temperature days and nights throughout the summer season. The forecast indicates that there will be a gradual rise in yearly rainfall of around 3%, 7%, and 10% by the years 2035, 2065, and 2100, respectively. Table 28.1 shows that the amount of rainfall is predicted to increase more significantly in the summer compared to the winter. This is because the higher moisture content in the atmosphere will lead to greater precipitation through the southwest monsoon wind.

28.1.3.1 The State of Ground Water Resources in India

India's hydrogeological landscape is highly diverse due to its varied topographic, geological, and climatic conditions (CGWB 2017). This diversity, along with the complex tectonic structure and various hydro chemical conditions, results in a complex groundwater system (Jha 2007).

Different geological formations, each possessing unique hydrogeological characteristics, serve as varied aquifer systems (CGWB 2017). The Central Groundwater Board of India (CGWB 2011) has classified for country into 14 primary aquifer systems and 42 major aquifers, based on these different attributes. The Alluvial aquifer system is the most extensive and prevalent, encompassing approximately 30% of the nation, predominantly in the regions of the Indus-Ganga–Brahmaputra basin. The Banded Gneissic Complex (BGC) and gneiss aquifers cover around 20% of the national territory and are found in the majority of the peninsular and Himalayan states. Other significant aquifers include basalt (16%), sandstone (8%), shale (7%), and limestone (2%). Additional aquifers, such as schist, granite, quartzite, charnockite, khondalite, laterites, and intrusive, constitute the remaining 17% of the land (Fig. 28.2).

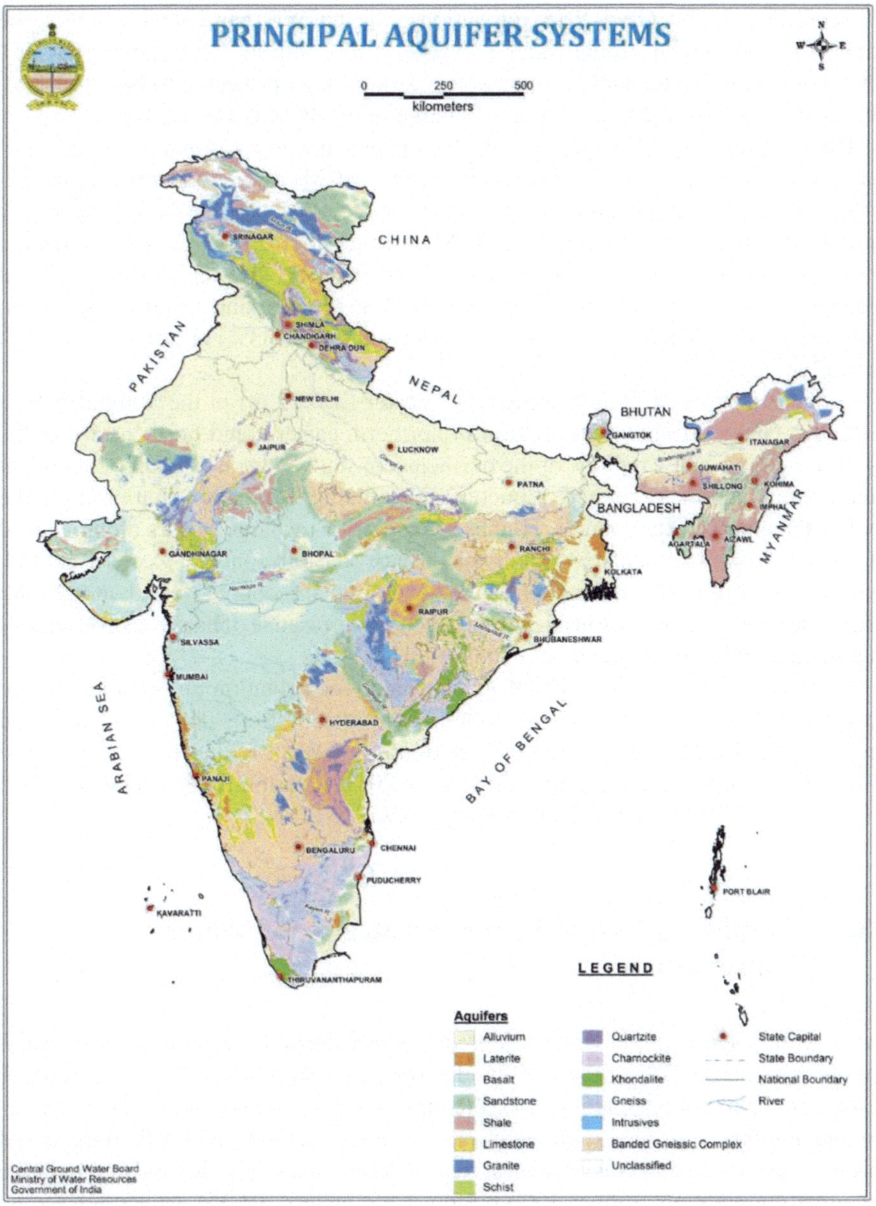

Fig. 28.2 Distribution pattern of principal aquifer system of India 2019 (*Source* CGWB)

According to the GEC-2015 regulations, the CGWB has calculated that the country's total annual groundwater recharge is 432 billion cubic meters (BCM). The entire annual extractable groundwater resources are projected to be 393 BCM, after accounting for the total natural discharge of 39 BCM (CGWB 2019).

The monsoon rainfall is the primary factor responsible for replenishing groundwater, accounting for approximately 58% (251.9 BCM) of the total annual groundwater recharge. The annual groundwater recharge is mostly derived from yearly rainfall, accounting for 67% (288.2 BCM) of the overall recharge, encompassing both monsoon and nonmonsoon seasons. Groundwater is also replenished by canal seepage, return flow from irrigation, and by tanks, ponds, and water conservation structures, which together contribute an estimated 143.6 (33%) billion cubic meters annually.

Groundwater recharge is significantly greater (> 0.25 m) in the alluvial belt of the Indus-Ganga–Brahmaputra region in northern, eastern, and northeastern India, as well as in the valley regions of the northeastern states. This is a result of abundant precipitation and the existence of unconsolidated alluvial deposits that promote the replenishment of water sources. Higher groundwater recharge occurs along coastal plains with loose sediments, especially on the east coast. In contrast, western, central, and much of peninsular India experience lower annual recharge (less than 0.25 m). This is due to two main reasons: less rainfall and the presence of hard rock formations that make it difficult for water to seep through.

Presently, the over extraction of groundwater for irrigation and residential use has emerged as a significant peril to the nation, leading to its depletion. Based on the estimate of the Central Ground Water Board (CGWB), India extracted a total of 248.69 billion cubic meters of groundwater in 2017. Out of this, around 89% (221.5 BCM) was utilized for agriculture purposes (CGWB 2019).

28.2 Examining Past to Future Strategies for Water Conservation

India, while having ample water resources, mainly depends on rainfall for its water supply, catering to 17% of the worldwide population with just 4% of world's water resources (Kumar and Bharat 2014). Precipitation is collected and stored in underground aquifers, waterways, and man-made storage facilities, and is then dispersed through various channels within or across different areas. The nation's cumulative precipitation amounts to over 4000 billion cubic meters (BCM). Nevertheless, the allocation of water in India exhibits temporal and regional variations. The majority of the yearly precipitation, around 80%, takes place during the monsoon period spanning from June to September. Because of the limited storage capacity, the government faces difficulties in storing sufficient water during the dry season. The mean annual natural flow, accounting for evaporation, soil absorption, and seepage, is around 1869 BCM (Sivanappan 2006).

The utilization of water is restricted to 433 BCM from groundwater and 690 BCM from surface water due to many factors including geography, population increase, insufficient monitoring of water resource development projects, and excessive extraction (Fig. 28.3). Out of the total amount of surface water that may be used, almost 40% is found in the Ganga–Brahmaputra–Meghna system, according to the Central Water Commission of the Government of India in 2017. The per capita water availability has shown a consistent decline over the years. Specifically, the figures for the years 2001, 2011, 2025, and 2050 are 1818, 1567, 1340, and 1140 m^3 respectively, as shown in Fig. 28.4 (Garg and Hassan 2007). The Central Water Commission categorizes the 22 river basins into surplus and deficit groups for the purpose of artificial recharge. The groundwater supply of a basin is determined by factors such as the amount of water that flows over the surface, the amount that seeps into the ground, the properties of the water, and changes in the level of the water table. Climate change is anticipated to exacerbate issue by inducing erratic weather patterns that have the potential to substantially diminish both rainfall and glaciers (Amarasinghe and Smakhtin 2014).

The Prime Minister of India recently launched the Jal Jeevan Mission to promote water conservation techniques across the country. As a result, the Ministry of Water Resources, River Development, and Ganga Rejuvenation was established. Water conservation, which refers to the meticulous administration of the existing water resources, is not a novel notion. For centuries, Indians have constructed various infrastructures to collect, confine, and save monsoon rainfall for utilization during arid periods. Although these ancient approaches are less prevalent in modern times, they nonetheless maintain their effectiveness.

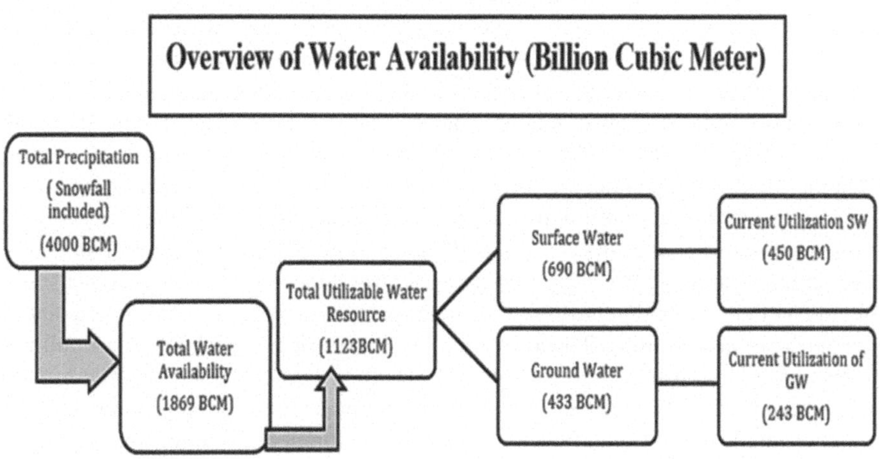

Fig. 28.3 Water availability schematic representation in India (based on MoWR)

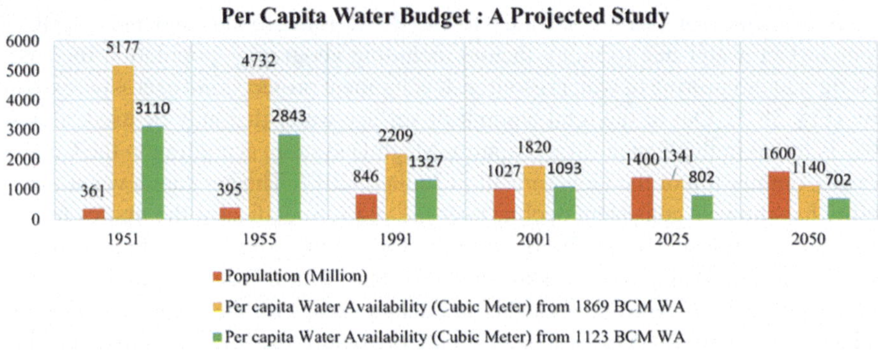

Fig. 28.4 Water availability (per capita) in India [Projected by CWC (2017)]

28.3 Coordination Between Groundwater and Surface Water Management

Agriculture is a significant consumer of water, representing 70% of the total world outflow of freshwater. In India, nearly all (90%) of people in rural areas and half (50%) of people in urban areas rely on groundwater as their main source of drinking water. With the reduction of surface irrigation areas, there is a growing dependence on groundwater, which aids in the reduction of environmental dangers. Due to its substantial population and significant agricultural significance, South Asia is extremely susceptible to the impacts of climate change. Climate change disrupts historical precipitation patterns, putting stress on irrigation systems. This necessitates a shift towards more water-efficient irrigation practices. Intensified agriculture may lead to local problems such as erosion, depletion of soil fertility, and reduction in biodiversity, as well as regional concerns like poisoning of groundwater. India and Mexico are the leading consumers of groundwater globally and are confronted with substantial issues related to overdraft. The depletion of global groundwater reserves is leading to a significant decline in water levels, resulting in escalated expenses for extracting water for agricultural irrigation and drinking purposes. The energy needed to lift water from greater depths is significantly more than that required from shallower depths. Implementing appropriate agricultural techniques, practicing sustainable intensification, and adhering to seasonal patterns can enhance the efficiency of water utilization in farming. The utilization of groundwater has enhanced agricultural production and the variety of crops grown, resulting in increased economic growth and higher salaries for both farm laborers and farmers who rely on groundwater.

According to the National Water Resources Development Commission, in order to attain food self-sufficiency by 2025, the irrigated area has to be expanded by approximately 61%, which is equivalent to 35 million hectares (GOI 2002). Nevertheless, the potential for increasing the overall cultivable land in India is restricted. Consequently, farmers in regions with less rainfall rely on groundwater to make up for the substantial water requirements necessary for achieving high crop yields. Groundwater, while it has a smaller impact on agriculture compared to surface water, has some unique benefits like dependability, easy access, availability whenever needed, minimal initial investment, and high efficiency. These attributes aid in minimizing the utilization and extraction of surface water. Insufficient water management may lead to excessive use of groundwater resources. At present, over 70% of groundwater is being used, exerting significant strain on the existing resources. India, a crucial part of South Asia, faces growing worries about food security and long-term agricultural health due to the overuse of groundwater. This overuse stems partly from government subsidies or free electricity, which have encouraged excessive pumping, particularly in western India and other agricultural regions. The result is a concerning decline in groundwater levels. The advantages of inexpensive or complimentary power tend to disproportionately favor affluent farmers who have the financial means to purchase pumps and tube wells. It is observed that numerous states, namely Punjab, Jharkhand, West Bengal, Gujarat, Uttar Pradesh, Maharashtra, Odisha, Chhattisgarh, and all southern states, have reservoirs with storage capacity below the average level.

The depletion of groundwater levels in some states presents a serious challenge to India's hydro institutional framework, namely in the management of catchment regions (Bhaduri et al. 2006; Pani 2020). Improving water storage is a crucial method for adapting to changing conditions. The use of irrigation tanks has been shown to have both advantages and challenges in implementing this approach (Palanisami et al. 2010).

Surface irrigation enhances the replenishment of groundwater in locations where canals are used for irrigation and in regions that rely on rainfall. Nevertheless, this study will evaluate the magnitude of this recharge and its significance in facilitating the growth of sustainable irrigation, as seen in Fig. 28.5.

Given the effectiveness of groundwater cultivation for irrigation, future surface water irrigation projects are expected to improve groundwater usage and intensify land use (Roy and Shah 2002). To balance drainage and recharge, it is necessary to formulate appropriate policies and procedures (Agarwal 2008). This requires a comprehensive database that details user profile, usage statistics, groundwater extraction rates, water conditions, water table depth, and groundwater quality. These data are crucial for identifying regions with underdeveloped groundwater resources and the necessary infrastructure for groundwater harvesting.

Future planning and development should consider the socioeconomic dynamics of these locations (Hanjra and Qureshi 2010; Kumar et al. 2024). Hence, it is imperative for India to adopt both physical and institutional measures, such as the implementation of micro irrigation techniques, in order to develop effective water markets. This measure would effectively control the distribution of electricity and prevent excessive extraction of water in many river valleys and regions with abundant water resources (Hope et al. 2008).

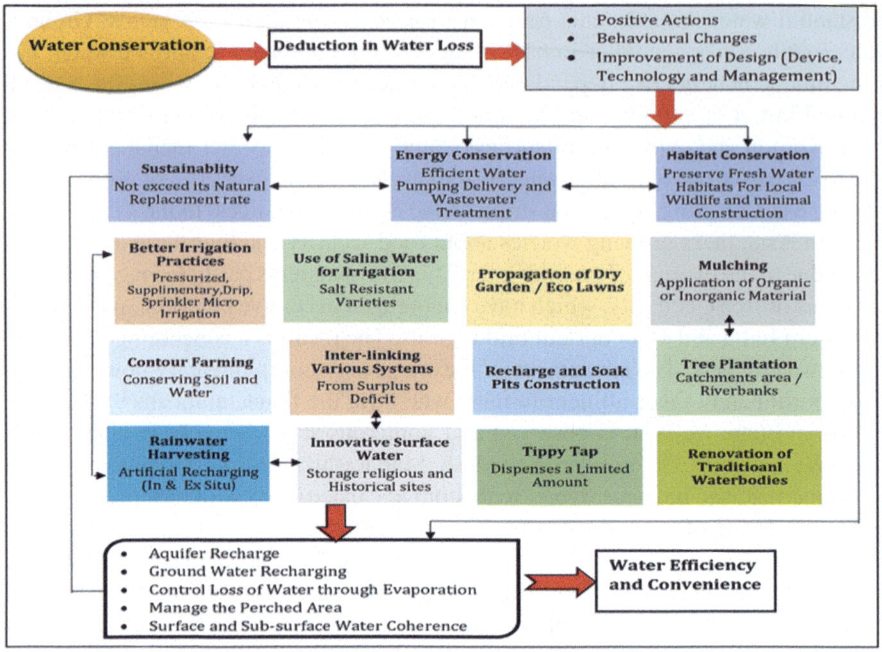

Fig. 28.5 Water conservation framework in India (Pani et al. 2021)

28.4 Water Development Policy and Management

The 2018 Composite Water Management Index (CWMI) of NITI Aayog predicts that by 2020 (NITI Aayog 2020), groundwater will be depleted, affecting over 100 million people and causing a 6% decrease in India's GDP. Climate change is anticipated to worsen water supply issues in South Asia, leading to more frequent occurrences of intense droughts and floods (Sivakumar and Stefanski 2010). Improved water storage is an essential technique for adapting to changing conditions, as demonstrated by the achievements and difficulties encountered with irrigation tanks (Pani 2020; Kulwant et al. 2024). Nevertheless, in order to effectively plan and develop in India, it is crucial to take into account several elements such as climate adaptability, crop type, evapotranspiration, topography, gradient, depth, and soil type, owing to the country's different physical and economic features (Alam and Murray 2005). Sustainable water management in conflict zones requires careful coordination. This includes factors like water source behavior, resettlement needs of displaced populations, and future land-use planning. Effective natural resource management (NRM) strategies and strong governance are especially critical in these areas due to the complex web of tensions that need untangling.

The combination of rapid and uncontrolled urbanization, together with antiquated strategies for managing urban water, presents risks to the quantity and quality of our current water resources (Kumari and Singh 2017). Hence, the preservation of regionally accessible resources, such as underground water and water on the Earth's surface, is vital for long-term viability. It is ensure long-term production and safeguard soil and environmental health, we need targeted soil and water conservation (SWC) strategies that address specific natural resource challenges (Gachene et al. 2019; Yadav et al. 2014). The effectiveness of grass strips and stone rows in minimizing runoff and managing soil erosion has been demonstrated (Zougmoré et al. 2004). Furthermore, the enhancement of the long-term execution and administration of SWC acts provides economic, environmental, and social advantages. The UNSDGs offer a chance to incorporate ideas of soil into the wider SDGs agenda post-2015 (Gachene et al., 2019).

The practical utilization of natural resources results in intricate associations that encompass many geographical and temporal aspects. The dynamics of these connections are shaped by several elements, such as livestock breeders, farmers, investors, legislators, and practitioners (Arimoro and Musa 2020). To maintain sustainable production, it is vital to acknowledge and protect the inherent connection between trees and water, as well as the longstanding relationship between agriculture and water. Technology, governmental policies, and institutional arrangements play a crucial role in closing the gap between knowledge and application in the management of natural resources, namely water utilization.

The development of water law has occurred in a fragmented and arbitrary fashion, including a range of statutory and informal rules, conventions, and principles. Over the last two decades, there has been a concentrated effort to reform water legislation, driven by a set of principles aimed at guiding the overall development of water policy (National Water Policy 2013). In India, the responsibility for water management is mostly handled by the state governments, with the federal government having a limited involvement that is mainly focused on resolving issues related to water sharing between different states. The federal government has implemented environmental legislation that specifically addresses pollution prevention, which has direct implications for water usage and the exploitation of groundwater (India federal Water Commission 2013).

The twenty-first century national water policy places importance on acknowledging water as a valuable national asset in order to accomplish national development objectives and strategizing (Pani et al. 2021). The approach promotes decentralized water management, working in partnership with local communities and state governments. This method provides comprehensive principles that are applicable to many settings, including distinct agroclimatic zones, pollution levels, industrial activities, population density, urban areas in each river basin and watershed (Kathpalia and Kapoor 2002). Water management policy at the national level should take into account the balance between the amount of water that is available and the preservation of the overall health and functioning of ecosystems. Countries with limited water resources can achieve more sustainable water use by strategically importing goods that are water-intensive to produce and exporting goods that require less water (Asthana 2009). There is a necessity for substantial alterations in the policy framework concerning water resources. Criticisms encompass the classification of water as a commodity with economic value and the necessity to integrate cutting-edge technical advancements for accurate on-site verification. India's gender-inclusive policies enable women to lawfully engage in official political procedures that govern resource management.

Water policy should include appropriate price mechanisms that consider the needs of the poor and socially disadvantaged, while ensuring access to water is not denied. Power is often delegated to appropriate communities to facilitate effective co-management of resources (Robinson 2016). Several factors influence the effectiveness of water resource management in Indian communities, including diverse values, cultural heterogeneity, socioeconomic disparities (encompassing class, gender, and social stratification), and fragmented, inconsistent water legislation. Case studies illustrate these challenges. India's precarious water situation, with high demand and limited resources, necessitates optimal water use efficiency and public awareness campaigns promoting conservation. This emphasizes the need for national policies that integrate sustainable planning, development, and management of water resources (Table 28.2). Studying and analyzing the impact of subsidies, such as those for water, energy, and other associated resources, on water usage, particularly in agriculture, is essential. The traditional water management techniques we utilize are both sustainable and responsible. These systems operate based on the notion of reusing and recycling water resources (Hanjra and Qureshi, 2010). It is essential to improve the potential for harmonious interactions among governments, communities, regulatory bodies, scientists and charities.

28.5 Implementation of Selected Water Conservation Methods at a Nationwide Level in India

Traditional water management systems, characterized by sustainability and accountability, are crucial for maintaining the hydrological cycle. Reviving these systems requires the identification and development of supportive policies. Water harvesting projects can significantly enhance the living conditions of disadvantaged communities, especially rural women and small-scale farmers. The establishment of legal, regulatory, and institutional frameworks for water resource management is a critical responsibility of governments. This contributes to the efficient functioning of rural economies and communities (Sreedevi et al. 2019). Encouraging the participation of community and local self-governing institutions in local resource management can lead to the creation and maintenance of local water conservation infrastructures (Sharawat et al. 2019).

The Ministry of Water Resources (MOWR) helps state governments find funding from international organizations like the World Bank and Asian Development Bank (ADB) to fill resource shortfalls and adopt modern water management technologies. These funds recently aided the National Bank for Agriculture and Rural Development

Table 28.2 Various national policies and plans related to water management in India from the year 1987 to 2018 (Pani et al. 2021)

Name of the Provision	Year	Thrust area
National Water Policy I	1987	Country wise zonation of water resources, drinking water allocation, Maximize water efficiency
National Water Policy II	2002	Established a robust relationship between water and ecosystem, Safe drinking water supply and access
National Environment Policy	2006	Asylums surface water, groundwater and wetlands, Importance on groundwater conservation, Advancement of efficient water use techniques by sprinklers/drip irrigation, Understanding the Water and Energy nexus (electricity and diesel pricing)
National Water Mission	2008	Water conservation, augmentation and preservation by state machinery, Water recycling
National Water Policy III	2012	Treat water as an economic good, Privatization of water-delivery services, Widespread of water conservation in urban, industrial and agricultural uses, Watershed Management through an integrated way of development
12th Five Year Plan	2012–2017	Approval for new institutions, groundwater laws, Repair, renovation and restoration bodies, Cheap accessible technology intervention for water conservation practices
National Mission on Sustainable Habitat	2018	Waste-water reuse and storm-water management, Eco-service valuation of water in different strata, Sustainable development of energy &water matrix

(NABARD) and the Ministry of Rural Development (MORD) in developing strategies, training personnel, and pursuing ongoing research in this field. The implementation of water conservation initiatives has resulted in the development of significant infrastructure and irrigation capabilities, with the objective of enhancing water availability and ensuring food security within the nation (Das 2015). In recent times, there has been a growing trend of integrating bioengineering approaches with conventional mechanical structures in order to effectively manage soil erosion and preserve water resources (Hadda et al. 2000). The choice of structural measures is contingent upon several criteria, such as the extent of erosion, the soil composition, the terrain, and the climatic conditions specific to a particular agroclimatic zone. Hence, the objective of the agroecosystem redesign is to optimize the harmonious interaction of climatic, ecological, geomorphological, economic, and social factors, while simultaneously avoiding conflicts in the sustainable preservation and control of water resources over an extended period (Pani and Mishra 2019).

In water-scarce parts of India, where there is limited water supply due to unfavorable geoclimatic conditions and water usage patterns, certain conservation strategies have demonstrated their effectiveness. These techniques encompass contour bunding, antisilt fences, dugout ponds, and rainfed ponds, each possessing distinct attributes and advantages. These approaches have the potential to greatly improve water conservation efforts in a comprehensive way.

Bhungroo, a water management method, is specifically intended to store excess rainwater underground for subsequent utilization in irrigation during seasons of drought. With a capacity of up to two crore liters, this device can store significant amounts of rainwater. By diluting saline groundwater with freshwater rainwater, it reduces overall salinity, making the water suitable for agriculture in coastal Gujarat and Andhra Pradesh. Masonry control dams, though expensive to build, are widely used in India's watershed management programs. These enduring edifices are specifically engineered to manage the erosion of gullies, collect water, and replenish underground water sources. In the world of alternative irrigation systems, a new design called the "hapa" or "5% water collection structure" is showing promise. This innovative approach could revolutionize sustainable irrigation in the deserts of West Bengal and Jharkhand in eastern India. Retention ditches are most efficient in cultivating high-water-demand crops in semi-arid areas, particularly in soils that are light, well-drained, deep, and stable (Pani and Mishra 2019). Rooftop harvesting is a simple method that enables the collection and storage of significant quantities of rainwater for later use in seasons of low rainfall. While often employed for household needs, stored water may also be employed for the production of high-value horticultural crops on a limited scale. These crops are frequently vulnerable to drought conditions. Utilizing this method in combination with drip irrigation yields significant results (Manasi and Umamani 2013).

Artificially recharging aquifers is important in dry and semi-arid regions to replenish groundwater. Typically, the rainfall intensity is not enough to provide an excess of moisture when it comes to infiltration conditions (Bevir and Rhodes 2001). The effectiveness of water conservation programs, regardless of their scale, relies on factors such as technical expertise, workforce, efficient administration, and the ability and desire to cover the associated expenses. Although the government has made significant efforts to initiate various initiatives for rainwater collecting and artificial aquifer recharge, there is a widespread lack of enthusiasm for water conservation among the general population (Bhattacharyya et al. 2016). In order to restore these water resources, diligent endeavors are necessary to ensure their appropriate upkeep and exploitation. Society has a responsibility to proactively engage in water conservation by utilizing a blend of conventional and contemporary methods (Sharma and Kanwar 2009).

28.6 Conclusion

Water is fundamental for human survival and national development. However, the availability of this resource is threatened by climate change, insufficient groundwater replenishment, saltwater intrusion, and natural disasters. Despite recent advancements, policy and institutional shortcomings have hindered effective water management. Therefore, it is essential to continuously restore groundwater through methods such as recharging, rainwater harvesting, micro-irrigation, efficient water markets, and limiting unregulated water extraction. Employing strategies such as agro-aquaculture, supplemental irrigation, growing drought-resistant crops, soil conservation, tank rejuvenation, and zero-till farming can boost local-level water use efficiency. A holistic approach is vital for conservation and management. This includes creating a dedicated "institutional space" where small farmers and local institutions can collaborate to explore more efficient resource management methods. The establishment of effective local institutions, promotion of public awareness, and enhancement of capacity building are key to managing and controlling water resources. Furthermore, it is necessary to improve the database and information system by collecting, organizing, and analyzing hydrological data to ensure effective monitoring. While an agricultural system that evaluates water use, maximizes its economic value, and prioritizes efficiency by minimizing waste is crucial, successful conservation of natural resources ultimately hinges on the active participation of local communities throughout the development, implementation, and monitoring processes, alongside effective governance.

Acknowledgements The authors express their heartfelt gratitude to our esteemed institute and all individuals who have played a role, directly or indirectly, in contributing to this study.

References

Agarwal A (2008) Role of NGOs in the protection of environment. J Environ Res Dev 2(4):933–938

Aizebeokhai AP, Oyeyemi KD, Adeniran A (2017) An overview of the potential impacts of climate change on groundwater resources. J Inform Math Sci 9(2):437–453

Alam M, Murray LA (2005) Facing up to climate change in South Asia. IEED

Amarasinghe UA, Smakhtin V (2014) Global water demand projections: past, present and future, vol 156. IWMIs

Arimoro AE, Musa H (2020) Towards sustainable water resource management in rural Nigeria: the role of communities. J Sustain Dev Law Policy (The) 11(1):1–17

Asthana V (2009) Water policy processes in India: discourses of power and resistance. Routledge

Bevir M, Rhodes RA (2001) A decentered theory of governance: rational choice, institutionalism, and interpretation

Bhaduri A, Amarasinghe U, Shah T (2006) Groundwater irrigation expansion in India: an analysis and prognosis. Draft prepared for the IWMI-CPWF project on "Strategic Analysis of National River Linking Project of India"

Bhattacharyya R, Ghosh BN, Dogra P, Mishra PK, Santra P, Kumar S, Sarkar D (2016) Soil conservation issues in India. Sustainability 8(6):565

CGWB (2011) Dynamic ground water resources of India. Central Ground Water Board, Ministry of Water Resources, Government of India, New Delhi

CGWB (2017) Report of the Ground Water Resource Estimation Committee (GEC-2015). Central Ground Water Board, Ministry of Water Resources, RD & GR, Government of India, New Delhi

CGWB (2019a) National compilation on dynamic ground water resources of India, 2017. Central Ground Water Board, Ministry of Jal Shakti, Department of Water Resources, RD & GR, Government of India, Faridabad

Chinnaiah Y (2015) Analysis of ground water levels and water quality in Musi River Basin, Telangana State. doctoral thesis. Sri Krishnadevaraya University, Anantapuram, A.P., India

Chitale MA (1991) Peoples' participation in large irrigation projects. Farmers in the Management of Irrigation Systems. Sterling Publishers, Private Limited, New Delhi

Christensen JH, Krishna Kumar K, Aldrian E, An S-I, Cavalcanti IFA, de Castro M, Dong W, Goswami P, Hall A, Kanyanga JK, Kitoh A, Kossin J, Lau N-C, Renwick J, Stephenson DB, Xie S-P, Zhou T (2013) Climate phenomena and their relevance for future regional climate change. In: Stocker TF, Qin D, Plattner G-K, Tignor M, Allen SK, Boschung J, Nauels A, Xia Y, BexV, Midgley PM(eds) Climate change 2013: the physical science basis, Contribution of Working Group I to the fifth assessment report of the Intergovernmental Panel on Climate Change. Cambridge University Press, Cambridge, UK and New York

CWC (2017) Reassessment of water availability in India using space inputs (volume II). Government of India (GoI), Central Water Commission, New Delhi, India. http://cwc.gov.in/sites/default/files/volume-ii-compressed.pdf

Das S (2015) Dr. Rajendra Singh: His tryst with water and peace. J Geol Soc India 85(5):521

Gachene CK, Nyawade SO, Karanja NN (2019) Soil and water conservation: an overview. Zero hunger. Encyclopedia of the UN sustainable development goals. Springer, Cham

Garg NK, Hassan Q (2007) Alarming scarcity of water in India. Curr Sci 93:932–994

Government of India (2002) National water policy, Ministry of Water Resource, GoI, New Delhi, India

Hadda MS, Khera KL, Kukal SS (2000) Soil and water conservation practices and soil productivity in north-western sub-montaneous tract of India: a review. Indian J Soil Conserv 28(3):187–192

Hanjra MA, Qureshi ME (2010) Global water crisis and future food security in an era of climate change. Food Policy 35(5):365–377

Hope R, Borgoyary M, Agarwal C (2008) Smallholder Preferences for Agri-environmental Change at the Bhoj Wetland. India. Dev Policy Rev 26(5):585–602

India CWC, & Central Water Commission (2013). Water and related statistics

IPCC (2013) In: Stocker TF, Qin D, Plattner G-K, Tignor M, Allen SK, Boschung J, Nauels A, Xia Y, BexV,Midgley PM(eds) Climate change 2013: the physical science basis, Contribution of Working Group I to the fifth assessment report of the Intergovernmental Panel on Climate Change. Cambridge University Press, Cambridge, UK and New York, 1535 pp

Jha BM (2007) Groundwater development and management strategies in India. Bhu-Jal News 22:1–15

Kathpalia GN, Kapoor R (2002) Water policy and action plan for India 2020: an alternative. Alternative Future, Delhi, pp 1–33

Khullar DR (2005) India: a comprehensive geography. Kalyani, New Delhi

Kumar SV, Bharat GK (2014) Perspectives on a water resource policy for India. Energy Resour Institute, India

Kulwant M, Patel D, Shirin S, Sharma SN, Yadav AK (2024) Future frameworks for fluoride and algorithms for environmental system. Adv Treat Technol Fluoride Removal in Water Water Purif 343–364. https://doi.org/10.1007/978-3-031-38845-3_19

Kumar A, Shirin S, Ansari MI, Pandey G, Sharma SN, Yadav AK (2022) Fuzzy and neural network model-based environmental quality monitoring system: past, present, and future. In: Modeling and simulation of environmental systems, pp 153–176). CRC Press

Kumar A, Chava R, Gupta S, Shirin S, Jamal A, Yadav AK (2024) Environmental and health effects of fluoride contamination and treatment of wastewater using various technologies. In: Advanced

treatment technologies for fluoride removal in water: water purification, pp 323–341. Springer Nature Switzerland, Cham. https://doi.org/10.1007/978-3-031-38845-3_18
Kumari M, Singh J (2017) Water conservation: strategies and solutions. Int J Adv Res Rev 1(4):75–79
Manasi S, Umamani KS (2013) Water conservation in urban areas: a case study of rainwater harvesting initiative in Bangalore city. In: Knowledge systems of societies for adaptation and mitigation of impacts of climate change. Springer, Berlin, Heidelberg, pp 303–328
Lal M (2001) Climatic change–implications for India's water resources. J Soc Econ Dev III 1:57–87
Palanisami K, Meinzen-Dick R, Giordano M (2010) Climate change and water supplies: options for sustaining tank irrigation potential in India. Econ Polit Wkly 183–190
Pani A (2020) Linking agro—ecological infrastructure to cropping practices for sustainable agricultural growth: do policies and local level institutional participation matter? J Glob Res 6(2):72–80. https://doi.org/10.46587/JGR.2020.v06i02.011
Pani A, Mishra P (2019) Hapa irrigation for promoting sustainable agricultural intensification: experience from Bankura district of India. GeoJournal 86(1):1–24
Pani A, Ghatak I, Mishra P (2021) Understanding the water conservation and management in India: an integrated study. Sustain Water Resour Manag 7(5):77
Rajaram K (2006) Geography of India. Spectrum Books, New Delhi
Rathore LS, Attri SD, Jaswal AK (2013) State level climate change trends in India. Meteorological monograph no. ESSO/IMD/EMRC/02/2013, India Meteorological Department
Robinson EJ (2016) Resource-dependent livelihoods and the natural resource base. Annu Rev Resour Econ 8:281–301
Roy AD, Shah T (2002) Socio-ecology of groundwater irrigation in India. Intensive use of groundwater challenges and opportunities, pp 307–335
Shankari U, Shah E (1993) Water management traditions in India. PPSt Foundation
Sharawat I, Dahiya R, Dahiya RP, Sreekrishnan TR, Kumari S (2019) Policy options for managing the water resources in rapidly expanding cities: a system dynamics approach. Sustain Water Resour Manage 5(3):1201–1215
Sharma N, Kanwar P (2009) Indigenous water conservation systems—a rich tradition of rural Himachal Pradesh. Indian J Tradition Knowl 8(4):510–513
Sivakumar MV, Stefanski R (2010) Climate change in South Asia. Climate change and food security in South Asia. Springer, Dordrecht, pp 13–30
Sivanappan RK (2006) Rainwater harvesting, conservation and management strategies for urban and rural sectors. In: National seminar on rainwater harvesting and water management, vol 11, issue 2, p 1
Sreedevi PD, Sreekanth PD, Ahmed S, Reddy DV (2019) Evaluation of groundwater quality for irrigation in a semi-arid region of South India. Sustain Water Resour Manage 5(3):1043–1056
Wang X, Farooq SH, Keramat M, Lubis RF, Akhter MG (2013) Climate change vulnerability and adaptation in ground water dependent irrigation system in Asia-Pacific Region. Asia-Pacific Network for Global Change Research, Project Reference Number: ARCP2013-25NSY-Shahid
Yadav AK, Sahoo SK, Mahapatra S, Kumar AV, Pandey G, Lenka P, Tripathi RM (2014) Concentrations of uranium in drinking water and cumulative, age-dependent radiation doses in four districts of Uttar Pradesh, India. Toxicol Environ Chem 96(2):192–200. https://doi.org/10.1080/02772248.2014.934247
Zougmoré R, Ouattara K, Mando A, Ouattara B (2004) Interaction between nutrients and water and soil conservation techniques in runoff control and soil fertility management in Burkina Faso. Sci Change Planét Sécher 15(1):41–48

SPRINGER NATURE

GPSR Compliance

The European Union's (EU) General Product Safety Regulation (GPSR) is a set of rules that requires consumer products to be safe and our obligations to ensure this.

If you have any concerns about our products, you can contact us on ProductSafety@springernature.com

In case Publisher is established outside the EU, the EU authorized representative is:

Springer Nature Customer Service Center GmbH
Europaplatz 3
69115 Heidelberg, Germany

The manufacturer's authorised representative in the EU is Springer Nature Customer Service Centre GmbH, Europaplatz 3, 69115 Heidelberg, Germany. If you have any concerns regarding our products, please contact ProductSafety@springernature.com

Printed and bound by CPI Group (UK) Ltd, Croydon, CR0 4YY
30/09/2026
02225598-0014